CLASSICAL OPTICS AND ITS APPLICATIONS

This book covers a broad range of the major topics of classical optics, in the form of 37 self-contained chapters. The chapters in the first half of the book deal primarily with the basic concepts of optics, while those in the second half describe how these concepts can be used in a variety of technological applications.

In each chapter, Professor Mansuripur introduces and develops a specialized topic in a comprehensive, clear and pedagogical style. The mathematical content is kept to a minimum as the book aims to provide the reader with insightful discussions of optical phenomena, at a level which is both accessible and interesting. This is aided by the numerous illustrations throughout in the form of diagrams, graphs and powerful computer simulation images. Topics covered include classical diffraction theory, optics of crystals, peculiarities of polarized light, thin-film multilayer stacks and coatings, geometrical optics and ray-tracing, various forms of optical microscopy, interferometry, coherence, holography, and nonlinear optics.

As such, this book will constitute the ideal companion text for graduate-level courses in optics, providing supplementary reading material for teachers and students alike. Industrial scientists and engineers developing modern optical systems will also find it an invaluable resource.

MASUD MANSURIPUR received a Bachelor of Science degree in Electrical Engineering from Arya-Mehr University of Technology in Tehran, Iran (1977), a Master of Science in Electrical Engineering from Stanford University (1978), a Master of Science in Mathematics from Stanford University (1980), and a Ph.D. in Electrical Engineering from Stanford University (1981). He has been Professor of Optical Sciences at the University of Arizona since 1988. His areas of research include: optical data storage, optical signal processing, magneto-optical properties of thin magnetic films, and the optical and thermal characterization of thin films and stacks. A Fellow of the Optical Society of America, he has published more than 200 papers in various technical journals, holds four patents, has given numerous invited talks at international scientific conferences, and is a contributing editor of *Optics & Photonics News*, the magazine of the Optical Society of America. Professor Mansuripur's published books include *Introduction to Information Theory* (1987), and *The Physical Principles of Magneto-optical Recording* (1995).

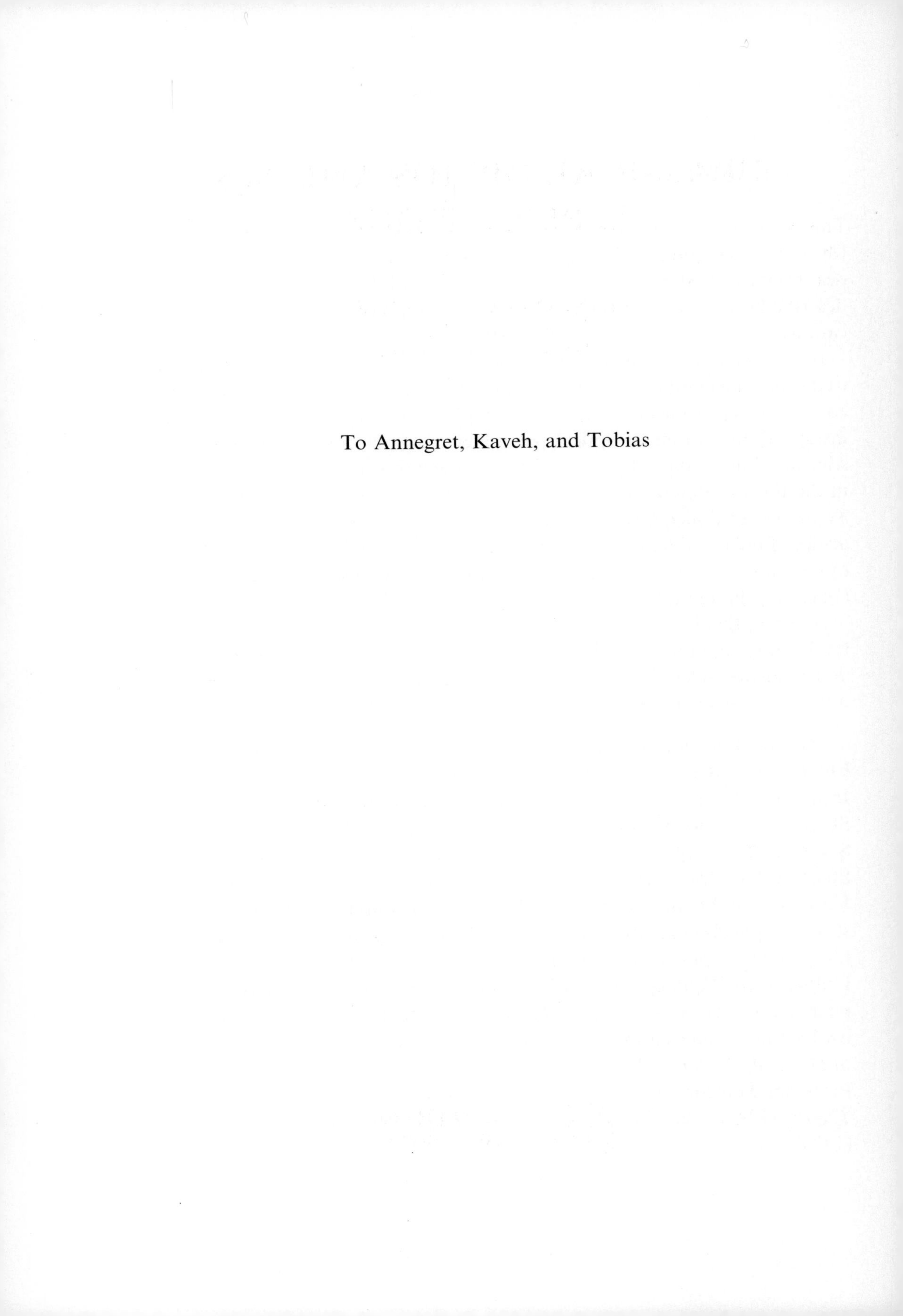

To Annegret, Kaveh, and Tobias

CLASSICAL OPTICS AND ITS APPLICATIONS

MASUD MANSURIPUR

Optical Sciences Center
University of Arizona, Tucson

PUBLISHED BY THE PRESS SYNDICATE OF THE UNIVERSITY OF CAMBRIDGE
The Pitt Building, Trumpington Street, Cambridge, United Kingdom

CAMBRIDGE UNIVERSITY PRESS
The Edinburgh Building, Cambridge CB2 2RU, UK
40 West 20th Street, New York, NY 10011-4211, USA
477 Williamstown Road, Port Melbourne, VIC 3207, Australia
Ruiz de Alarcón 13, 28014 Madrid, Spain
Dock House, The Waterfront, Cape Town 8001, South Africa

http://www.cambridge.org

First published 2002

Printed in the United Kingdom at the University Press, Cambridge

Typeface Times 11pt on 14pt *System* 3B2 [KW]

A catalogue record for this book is available from the British Library

Library of Congress Cataloguing in Publication data

Mansuripur, Masud, 1955–
Classical optics and its applications / Masud Mansuripur.
p. cm.
Includes bibliographical references and index.
ISBN 0 521 80093 5 (hb.) – ISBN 0 521 80499 X (pb.)
1. Optics. I. Title.
TA1520 M37 2001
621.36–dc21 2001025766 CIP

ISBN 0 521 80093 5 hardback
ISBN 0 521 80499 X paperback

Contents

Preface

I started writing the Engineering column of *Optics & Photonics News* (*OPN*) in early 1997. Since then nearly forty articles have appeared, covering a broad range of topics in classical optical physics and engineering. My original goal was to introduce students and practising engineers to some of the most fascinating topics in classical optics. This I planned to achieve with minimal usage of the mathematical language that pervades the literature of the field. I had met many bright students and practitioners who either did not know or did not fully appreciate some of the major concepts of classical optics such as the Talbot effect, Abbe's sine condition, the Goos–Hänchen effect, Hamilton's internal and external conical refraction, Zernike's method of phase contrast, Michelson's stellar interferometer, and so on. My columns were going to have little mathematics but an abundance of pictures and pedagogical arguments, to bring forth the essence of the physics involved in each phenomenon. In the process, I hoped, the readers would appreciate the beauty of the subject and, if they found it interesting, would dig deeper by searching the cited literature.

A unique tool available to me for this purpose was the computer programs DIFFRACT™, MULTILAYER™, and TEMPROFILE™, which I have developed in the course of my research over the past 20 years. The first of these programs simulates the propagation of light through optical systems consisting of discrete elements such as lasers, lenses, mirrors, prisms, phase/amplitude masks, gratings, polarizers, wave-plates, multilayer stacks, birefringent crystals, diffraction gratings, and optically active materials. The other two programs simulate the optical and thermal behavior of multilayer stacks. I have used these programs to generate graphs and pictures to explain the various phenomena in ways that would promote a better understanding.

The articles have been successful beyond my wildest dreams. While I had hoped that a few readers would find something useful in this series, I have received notes, e-mails, and verbal comments from distinguished scholars around the world who have found the columns stimulating and helpful. Some teachers informed me that they use the articles for their classroom teaching, and I have heard of several readers who collect the articles for future reference. All in all, I have been pleasantly surprised by the positive reaction of the *OPN* readers to these columns.

Optics & Photonics News is not an archival journal and, therefore, will not be widely available to future students. Thus I believe that collecting the articles here in one book, which provides for ease of cross-referencing, will be useful. Moreover, the book contains additional explanations of topics that were originally curtailed for lack of space in *OPN*; it includes corrections to errors discovered afterwards and incorporates some comments and criticisms made by *OPN* readers as well as my answers to these criticisms.

This book covers a broad range of topics: classical diffraction theory, the optics of crystals, the peculiarities of polarized light, thin-film multilayer stacks and coatings, geometrical optics and ray-tracing, various forms of optical microscopy, interferometry, coherence, holography, nonlinear optics, etc. It could serve as a companion to the principal text used in a number of academic courses in physics, engineering, and optics; it should be useful for university teachers as a guide to selecting topics for a graduate-level course; it should be useful also for self-study by graduate students. It could be used fruitfully by engineers who develop optical systems such as laser printers, scanners, cameras, displays, image-processing equipment, lasers and laser-based systems, telescopes, optical storage and communication systems, spectrometers, etc. I believe anyone working in the field of optics could benefit from this book, by being exposed to some of the major concepts and ideas (developed over the last three centuries) that shape our modern understanding of optics.

Some of the original *OPN* columns were written jointly with colleagues and students; these are identified in the footnotes and the corresponding co-authors acknowledged. I thank Ewan Wright and Rongguang Liang of the Optical Sciences Center, Lifeng Li of Tsinghua University, Mahmoud Fallahi of Nortel Co., and Wei-Hung Yeh of Maxoptix Co. for their collaboration as well as for giving permission to publish our joint papers in this collection. I also would like to acknowledge the late Peter Franken, Pierre Meystre, Yung-Chieh Hsieh, Dennis Howe, Glenn Sincerbox, Harrison Barrett, Roland Shack, José Sasian, Michael Descour, Arvind Marathay, Ray Chiao, James Wyant, Marc Levenson, Ronald Gerber, James Burge, Ferry Zijp and Dror Sarid, who shared their valuable insights with me and/or criticized the drafts of several articles prior to publication. Needless to say,

I am solely responsible for any remaining errors and inaccuracies. For their help with graphics and word processing, I am grateful to our administrative assistants Patricia Gransie, Nonie Veccia, Marylou Myers, and Amanda Palma.

Last but not least, I am grateful to my wife, Annegret, who has tolerated me with love and patience over the past four years while this book was being written. It is to her and to our children, Kaveh and Tobias, that this book is dedicated.

Masud Mansuripur
Tucson, December 2000

Introduction

The common threads that run through this book are the classical phenomena of diffraction, interference, and polarization. Although the reader is expected to be generally familiar with these electromagnetic phenomena, the book does cover some of the principles of classical optics in the early chapters. The basic ideas of diffraction and Fourier optics are introduced in chapters 1 through 4; this introduction is followed by a detailed discussion of spatial and temporal coherence and of partial polarization in chapters 5 through 8. These concepts are then used throughout the book to explain phenomena that are either of technological import or significant in their own right as natural occurrences that deserve attention.

Each chapter is concerned with a single topic (e.g., surface plasmons, diffraction gratings, evanescent coupling, photolithography) and attempts to develop an understanding of this subject through the use of pictures, examples, numerical simulations, and logical argument. The reader already familiar with a particular topic is likely to learn more about its applications, to appreciate better the physics behind some of the formulas he or she may have previously encountered, and perhaps even learn a thing or two about the nuances of the subject. For the reader who is new to the field, our presentation is aimed to provide an introduction, an intuitive feel for the physical and/or technological issues involved, and, hopefully, motivation for digging deeper by consulting the cited references. For the most part, this book avoids repeating what is already in the open literature, aiming instead to expose concepts and ideas, ask critical questions, and provide answers by appealing to the reader's intuition rather than to his or her mathematical skills.

Some of the chapters address fundamental problems that historically have been crucial to our modern understanding of optics; conical refraction, the Talbot effect, the principle of holography, and the Ewald–Oseen extinction theorem are representatives of this class. Other chapters introduce devices and phenomena of great scientific and technological importance; Fabry–Pérot etalons, the magneto-optical Faraday and Kerr effects, and the phenomenon of total internal reflection fall into this second category. Many of the remaining chapters single out a tool or an instrument that not only is of immense technological value but also has its unique principles of operation, worthy of detailed understanding; examples include various microscopes and telescopes, lithographic systems, ellipsometers, and so on. Occasionally a theoretical concept or a numerical method is found that has a wide range of applications; we have devoted a few chapters to these topics, such as the method of Fox and Li, the beam propagation method, and the concept of reciprocity in classical optics.

The majority of the computer simulations reported in this book were performed with the software packages DIFFRACT™, MULTILAYER™, and TEMPROFILE™, which I have written in the course of the past twenty years and which are now commercially available. These programs in turn are based on theoretical methods and numerical algorithms that are fully documented in several of my publications.[1–6] In a few chapters, I have collaborated with Professor Lifeng Li (now at the Tsinghua University in China). Here, we have used Professor Li's program DELTA™, also commercially available, for calculations pertaining to diffraction gratings. The theoretical foundations of DELTA™ are described in Professor Li's publications.[7]

Throughout the book, black-and-white pictures will be used to display the various properties of an optical beam; these include cross-sectional distributions of intensity, phase, polarization, and the Poynting vector. A unified scheme for the gray-scale encoding of real-valued functions of two variables is used in all the chapters, and it is helpful to review these methods at the outset. In the convention adopted the beam always propagates along the Z-axis, and its cross-sectional plane is XY. The Cartesian XYZ coordinate system is right-handed, the polar angles are measured from the positive Z-axis, and the azimuthal angles are measured from the positive X-axis towards the positive Y-axis. In general, the beam has three components of polarization along the X-, Y-, and Z-axes of the coordinate system, that is, its electric field E has components $E_x(x, y)$, $E_y(x, y)$, and $E_z(x, y)$ at any given cross-sectional plane, say, at $z = z_0$. Since the E-field components are complex-valued, their complete specification requires two distributions for each component, namely, amplitude and phase. The following paragraphs describe in some detail the encoding scheme used for displaying different cross-sectional properties of the beam and also provide a few examples.

Plots of intensity distribution

The electric-field intensity is the square of the field amplitude at any given location in space. Thus, for example, the intensity distribution in the cross-sectional XY-plane for the E-field component along the X-axis is denoted by $I_x(x, y) = |E_x(x, y)|^2$. Figure 0.1 shows plots of intensity distribution for the three components of polarization of a Laguerre–Gaussian beam propagating along the Z-axis. The black pixels represent locations where the intensity is at its minimum (zero in the present case), the white pixels correspond to the locations of maximum intensity within the corresponding frame, and the gray pixels linearly interpolate between these minimum and maximum values. In the case of Figure 0.1, the beam was taken to be linearly polarized at 45° to the X-axis, leading to identical distributions for the X- and Y-components of polarization.

The much weaker Z-component is computed to ensure that the Maxwell equations will be satisfied for the assumed distributions of the X- and Y-polarization components. In general, one may assume arbitrary distributions for E_x and E_y within a given cross-sectional XY-plane. To determine E_z in a self-consistent manner, one must break up the E_x and E_y distributions into their plane-wave constituents and proceed to determine E_z for each plane wave that propagates along the unit vector $\boldsymbol{\sigma} = (\sigma_x, \sigma_y, \sigma_z)$ by requiring the inner product of $\boldsymbol{E}$ and σ to vanish (i.e., $E_x\sigma_x + E_y\sigma_y + E_z\sigma_z = 0$). One must then superimpose the Z-components of all the plane waves thus obtained to arrive at the total distribution of E_z. In Figure 0.1 the peak intensities in the three frames are in the ratios $I_x : I_y : I_z = 1.0 : 1.0 : 1.47 \times 10^{-7}$.

Logarithmic plots of intensity distribution

In order to emphasize the weaker regions of an intensity distribution, we will show on numerous occasions the distribution of the logarithm of the intensity. First, the intensity distribution is normalized by its peak value, then the base-10 logarithm of the normalized intensity is computed and all values below some cutoff point are truncated. For instance, if the cutoff is set to $-\alpha$ then all values of the normalized intensity below $10^{-\alpha}$ are reset to $10^{-\alpha}$; the range of the logarithm of normalized intensity thus becomes $(-\alpha, 0)$. The continuum of gray levels from black to bright-white is then mapped linearly

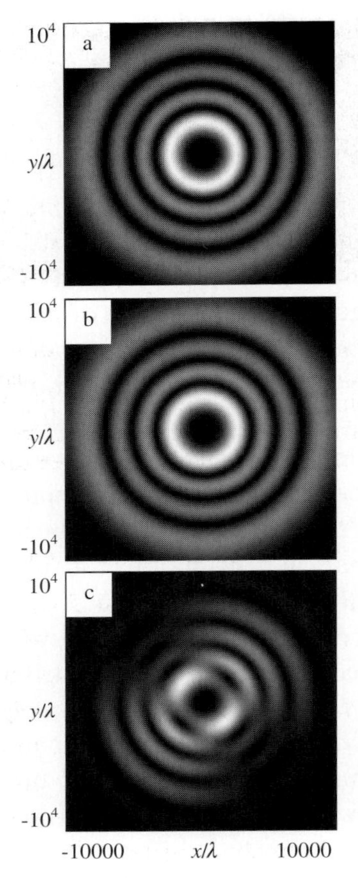

Figure 0.1 Plots of intensity distribution in the cross-sectional plane of a Laguerre–Gaussian beam for the three components of the *E*-field. In each frame the black pixels represent locations of zero intensity, while the white pixels represent locations of maximum intensity in the corresponding frame. The beam is assumed to propagate along the *Z*-axis, linearly polarized at 45° to the *X*-axis. (a) Intensity of the component of polarization along the *X*-axis, $I_x(x, y) = |E_x(x, y)|^2$, (b) $I_y(x, y) = |E_y(x, y)|^2$, (c) $I_z(x, y) = |E_z(x, y)|^2$. The peak intensities in (a), (b), (c) are in the ratios $1.0 : 1.0 : 1.47 \times 10^{-7}$, respectively.

onto this interval and used to display plots of normalized intensity on the logarithmic scale. When it is deemed useful or necessary, the corresponding value of α will be indicated in a figure's caption.

Figure 0.2 shows two plots of the same intensity distribution at the focal plane of a comatic lens. In (a) the distribution is linearly mapped onto the gray-scale, whereas in (b) the logarithm of intensity with a cutoff at $\alpha = 4$ is displayed. The latter is similar to what would be obtained by over-exposing a photographic plate placed at the focal plane of the lens.

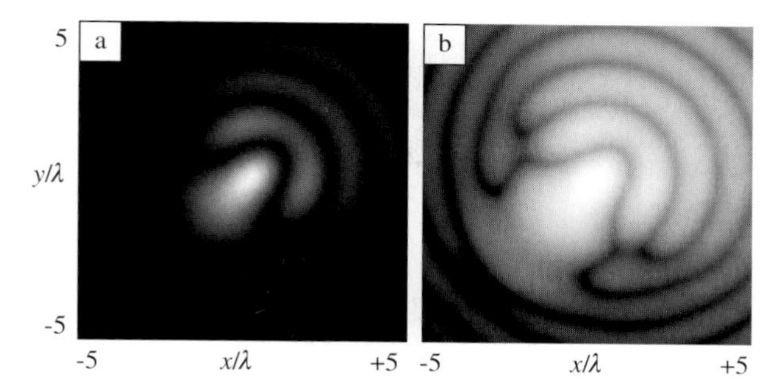

Figure 0.2 (a) Intensity distribution in the focal plane of a $0.5NA$ lens having 1.5λ of third-order coma (Seidel aberration). The uniformly distributed incident beam is assumed to be circularly polarized. In the focal plane, the X-, Y-, and Z- components of the electric field vector are added together to yield the total E-field intensity. (b) Same as (a) but on a logarithmic scale with $\alpha = 4$ (see text).

Plots of phase distribution

In several chapters we will show plots of phase distribution in a beam's cross-sectional plane. The phase, a modulo-2π entity, will always be limited to a range less than or equal to 360°. We typically divide the range of phase values for a given distribution into equal sub-intervals, assigning black to the minimum value, bright-white to the maximum value, and various gray levels to the values in between. A sharp discontinuity (from black to white or vice versa) appearing in these phase plots would be of no physical significance, since it merely indicates a 360° phase jump.

Figure 0.3 is a cross-sectional plot of the phase distribution for the Laguerre–Gaussian beam whose intensity distribution was given in Figure 0.1. The three frames of Figure 0.3 correspond to the components of polarization along the X-, Y-, and Z- axes. The black pixels represent the minimum phase, $-180°$, and the white pixels correspond to the maximum phase, $+180°$; the gray pixels cover the continuous range of values in between.

Ellipse of polarization

Consider a collimated beam of light propagating along the Z-axis. In general, the state of polarization of the beam at any given point is elliptical, as shown in Figure 0.4. So long as the electric-field vector E may be assumed to be confined to the XY-plane, it may be resolved into two orthogonal components, along the X- and Y- axes say. If E_x and E_y happen to be in phase, the polarization will be linear along some direction specified by the angle ρ. If, on the other hand, the phase difference between E_x and E_y is $\pm 90°$ then the

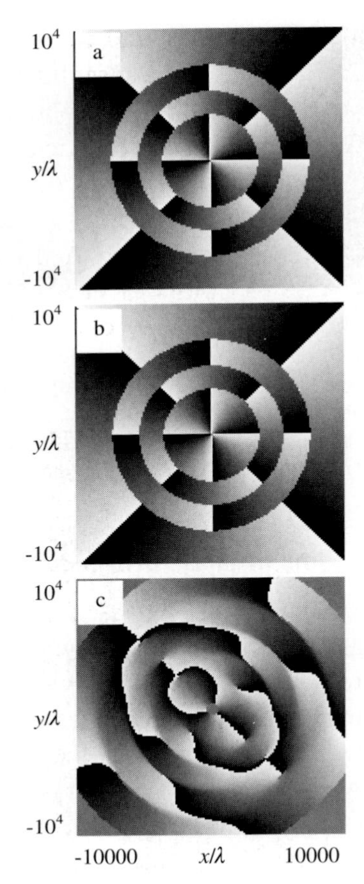

Figure 0.3 Plots of phase distribution in the cross-sectional plane of the Laguerre–Gaussian beam depicted in Figure 0.1. Frames (a), (b), and (c) correspond, respectively, to the components of the E-field along the X-, Y-, and Z- coordinate axes. In each frame the black pixels represent a phase of-180° and the white pixels correspond to a phase of $+180^\circ$; the gray pixels linearly interpolate between these two extreme values.

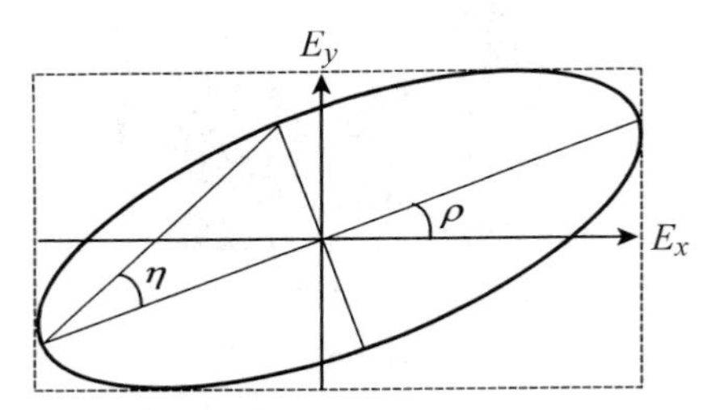

Figure 0.4 The ellipse of polarization is uniquely specified by E_x and E_y, the complex-valued electric field components along the X- and Y- axes. The major axis of the ellipse makes an angle ρ with the X-direction, and the angle η facing the minor axis represents the polarization ellipticity.

polarization will be elliptical, the major and minor axes of the ellipse lying along the X- and Y-axes. In general, the phase difference between E_x and E_y is somewhere between 0° and 360°, giving rise to an ellipse whose major axis has an angle ρ with the X-axis and whose ellipticity is given by the angle η. When the polarization is linear, $\eta = 0^\circ$; for light that is right circularly polarized (RCP), $\eta = +45^\circ$, whereas for light that is left circularly polarized (LCP), $\eta = -45^\circ$. In general, $-90^\circ < \rho \leq 90^\circ$ and $-45^\circ \leq \eta \leq 45^\circ$.

Figure 0.5 shows cross-sectional plots of intensity and polarization state for a beam with a highly non-uniform state of polarization. Frame (a) is the

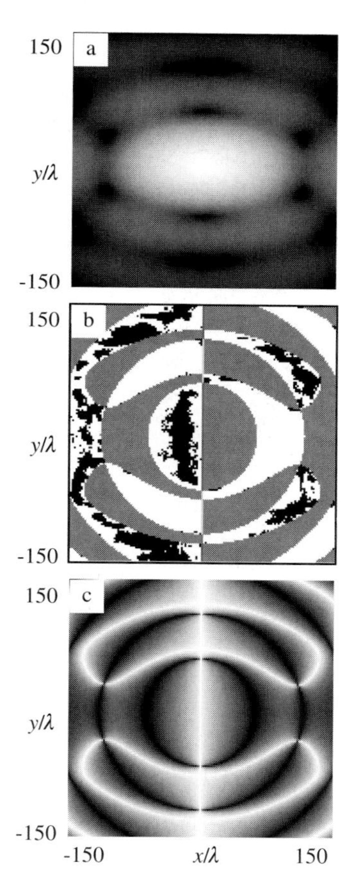

Figure 0.5 Distributions of intensity and polarization in the cross-section of a beam having a non-uniform polarization state. (a) Logarithmic plot of intensity distribution having cutoff at $\alpha = 4$. (b) Polarization rotation angle ρ; the gray-scale is linearly mapped onto ρ, from black at $\rho_{\min} = -90^\circ$ to bright-white at $\rho_{\max} = +90^\circ$. (c) Polarization ellipticity η; the gray-scale is linearly mapped onto η, from black at $\eta_{\min} = -45^\circ$ to bright-white at $\eta_{\max} = +45^\circ$.

logarithmic intensity pattern in the XY-plane. The polarization rotation angle $\rho(x, y)$ is depicted in (b), while the ellipticity $\eta(x, y)$ is shown in (c). The gray-scale in Figure 0.5(b) is a linear map of the values of ρ from -90° (black) to $+90^\circ$ (white). Similarly, the plot of η in Figure 0.5(c) is linearly encoded in gray-scale, with black representing -45° and white representing $+45^\circ$.

In the plot of ρ depicted in Figure 0.5(b), there are random-looking jumps between black and bright-white pixels. This is due to the ambiguity of the polarization rotation angle when either the E-field intensity is zero or the ellipticity η is $\pm 45^\circ$. In these regions, a small numerical error could readily cause a discrete jump between $\rho_{min} = -90^\circ$ and $\rho_{max} = +90^\circ$.

References for the Introduction

1 M. Mansuripur, *The Physical Principles of Magneto-optical Recording*, Cambridge University Press, UK, 1995.
2 M. Mansuripur, Distribution of light at and near the focus of high numerical aperture objectives, *J. Opt. Soc. Am.* **3**, 2086 (1986).
3 M. Mansuripur, Certain computational aspects of vector diffraction problems, *J. Opt. Soc. Am. A* **6**, 786 (1989). See also the erratum in *J. Opt. Soc. Am. A* **10**, 382–383 (1993).
4 M. Mansuripur, Analysis of multilayer thin film structures containing magneto-optic and anisotropic media at oblique incidence using 2 × 2 matrices, *J. Appl. Phys.* **67**, 6466–6475 (1990).
5 M. Mansuripur, G. A. N. Connell, and J. W. Goodman, Laser-induced local heating of multilayers, *Appl. Opt.* **21**, 1106 (1982).
6 M. Mansuripur and G. A. N. Connell, Laser induced local heating of moving multilayer media, *Appl. Opt.* **22**, 666 (1983).
7 Lifeng Li, Multilayer-coated diffraction gratings: differential method of Chandezon *et al.* revisited, *J. Opt. Soc. Am. A* **11**, 2816–2828 (1994).

1
Abbe's sine condition

Ernst Abbe (1840–1905), professor of physics and mathematics and director of the astronomical observatory at Jena, was also the research director of the Zeiss optical works. In 1868 he invented the apochromatic lens, thus eliminating the primary and secondary color distortion in microscopes. Abbe developed a clear theoretical understanding of limits to resolution and magnification in optical image-forming systems and discovered the sine condition for a lens to form a sharp image without the defects of coma and spherical aberration. (*Jena Review*, 1965, Zeiss Archive, Courtesy AIP Emilio Segré Visual Archives.)

Ernst Abbe (1840–1905), professor of physics and mathematics at the University of Jena, Germany, and major partner in the Carl Zeiss company, made important contributions to the theory and practice of optical microscopy.[1] His compound microscope was a superb optical design based on a theoretical understanding of diffraction and minimization of the effects of aberrations.[2] Abbe enunciated his famous sine condition regarding the axial point in the object plane of a centered image-forming system such as a microscope or a telescope. When this condition is satisfied, "aberration-free" imaging of the object points located in the vicinity of the optical axis is assured.[1–6] This chapter provides an heuristic description of the sine condition, which, in the words of Conrady, is "one of the most remarkable and labor-saving theorems in the whole realm of applied optics".[7]

As the chapter follows a rather unconventional approach towards explaining the sine condition, it is worthwhile to highlight its main features at the outset. An introduction of the necessary geometric-optical concepts provides the basis for defining the sine condition. This is followed by establishing, for an axial object point, a one-to-one mapping between the principal planes of the imaging system. The wavefront entering the system at the first principal plane (p.p.) is thus related to that emerging from the second p.p.

To describe the imaging of near-axis regions, we switch to a wave-optical viewpoint. Assuming that the axial object point is shifted to a nearby off-axis location, we derive the spatial phase modulation imparted to the emergent wavefront in consequence of this small shift. By then it should be apparent that aberration-free imaging of the off-axis point requires this spatial phase modulation to be linear in a certain coordinate system and that Abbe's sine condition is both necessary and sufficient to guarantee this linearity.

A lens that violates the sine condition

To appreciate the significance of Abbe's sine condition consider the plano-convex lens shown in Figure 1.1. A collimated beam of light propagating along the optical axis Z enters the flat facet of this lens and, upon exiting the second, hyperboloidal, surface, converges toward the focal point. The conic constant of the second surface is chosen to bring the beam to a perfect (i.e., diffraction-limited) focus at the rear focal plane of the lens. The logarithmic plot of intensity distribution at the focal plane (see Figure 1.2(a)) reveals the focused spot to be the well-known Airy pattern for this $0.75NA$ lens.

If the incident beam is tilted by a small amount, the focus shifts to an off-axis location but, more importantly, it acquires a significant amount of coma (see Figure 1.2(b)). Thus it is clear that a lens that works well for an axial object point is not necessarily suitable for the imaging of near-axis regions. The sine condition is intended to alleviate this problem. For comparison with a case to be described later, Figure 1.2(c) shows the phase distribution of the oblique beam at the front facet of the plano-convex lens; similarly Figure 1.2(d) shows the phase distribution of the emergent beam (minus the curvature) at the second p.p. Note that the clear aperture at the second p.p. is reduced in size and that the emergent phase pattern is "compressed" toward the optical axis in a nonlinear fashion. As we shall see

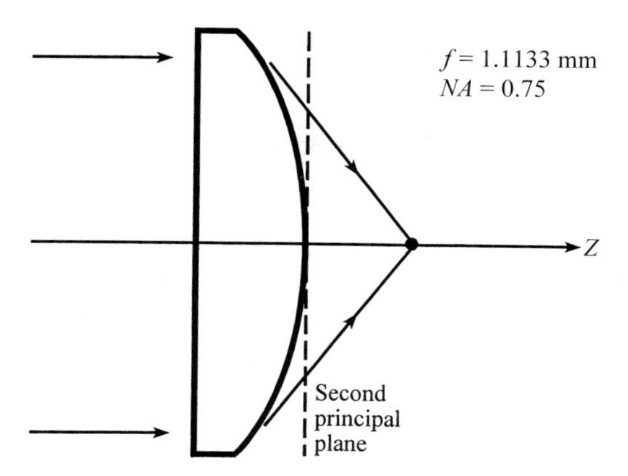

Figure 1.1 A plano-convex lens brings a collimated beam to perfect focus on an axial point. The lens is designed for $\lambda = 633$ nm; it has a 4 mm diameter clear aperture, a focal length of 1.1133 mm, and a numerical aperture of $NA = 0.75$. The refractive index of the lens glass $n = 2.5$, its thickness at the center is 1 mm, and its hyperboloidal surface has radius of curvature $R_c = 1.67$ mm and conic constant $k = -n^2 = -6.25$. The second principal plane of this lens is tangent to its curved surface at the apex. Both surfaces of the lens are assumed to be antireflection coated.

below, the emergent phase pattern is quite different for a lens that does satisfy the sine condition.

Geometric-optical concepts

The sine condition applies to a centered optical system designed for "aberration-free" imaging of a small patch within the object plane to a corresponding patch within the image plane (see Figure 1.3). The imaging system is intended for a given pair of conjugate planes, so that the distance z_0 between the object and the first p.p. of the system is fixed, as is the distance z_1 between the image and the second p.p. The lens formula $1/z_0 + 1/z_1 = 1/f$, where f is the focal length of the system, applies here.[5]

Throughout this chapter, attention is confined to systems where both the object and image are in air; extension of the results to situations where the object space and image space have differing refractive indices (e.g., immersion-oil microscopy) is straightforward but is not discussed.[4,5]

In the present context, "aberration-free" imaging means that a cone of light emanating from any point (x_0, y_0) in the small patch within the object plane, when captured by the optical system is turned into a convergent cone that – to a first approximation in the relevant parameters – comes to focus at (x_1, y_1) in the image plane (see Figure 1.4).[4,5] The point (x_1, y_1) is conjugate

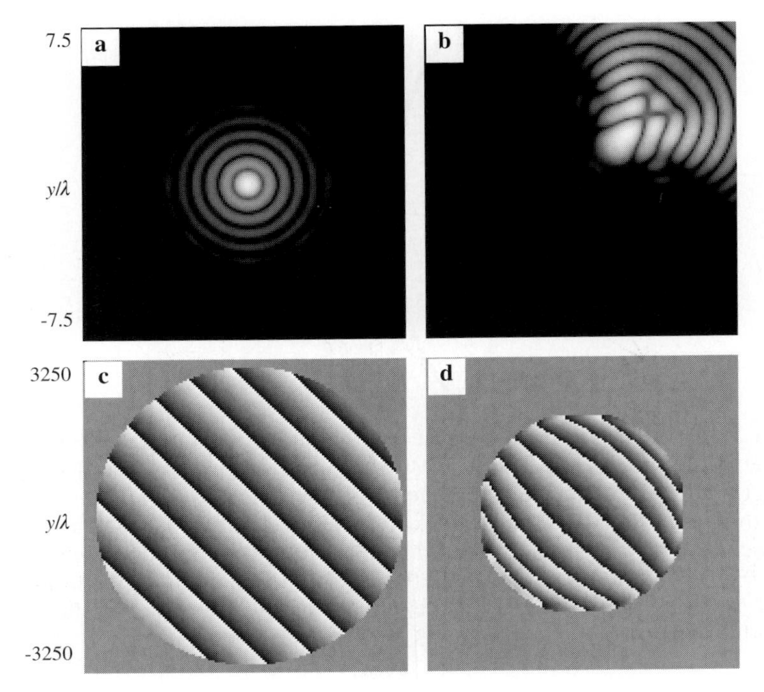

Figure 1.2 (a) Logarithmic plot of intensity distribution at the focal plane of the plano-convex lens of Figure 1.1 for a circularly polarized, collimated beam traveling along the optical axis. (b) Same as (a) but for an obliquely incident beam traveling at 0.076° relative to the optical axis. (c) Distribution of phase for the oblique beam entering the lens at its flat surface. The gray-scale covers the interval from −180° (black) to +180° (white). (d) Distribution of phase for the oblique beam emerging from the lens at its second p.p.

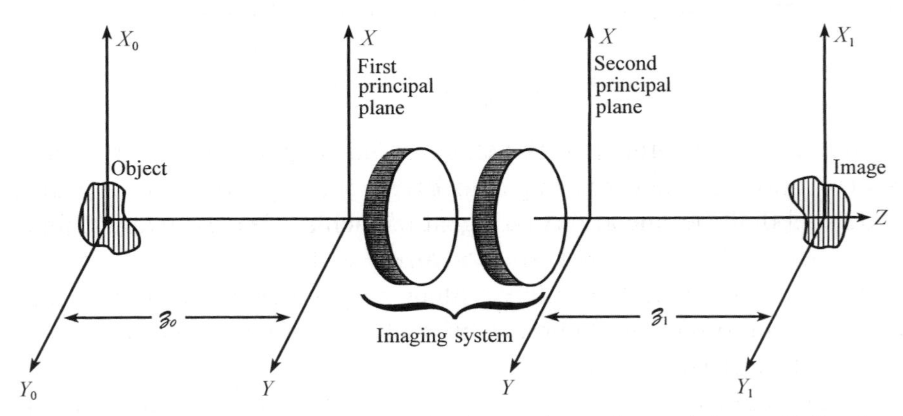

Figure 1.3 A small planar object in the vicinity of the optical axis in the $X_0 Y_0$-plane is imaged onto a small region of the $X_1 Y_1$-plane. The principal planes of the imaging system are also shown. The object and image planes are assumed to be in air, so that the refractive indices of both the object space and the image space may be set to unity.

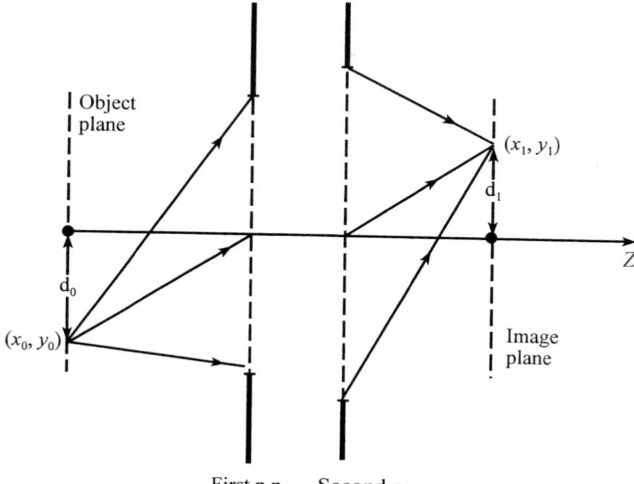

Figure 1.4 The cone of light emanating from an off-axis object point (x_0, y_0) is captured by the imaging system and brought to focus at the corresponding image point (x_1, y_1). Note that beyond the paraxial regime the rays entering the first p.p. at a given height do not necessarily emerge from the second p.p. at the same height.

to (i.e., the Gaussian image of) the point (x_0, y_0). Since the system is circularly symmetric around the optical axis, the axial point at the center of the object plane is imaged to the axial point at the center of the image plane. Denoting the distance between (x_0, y_0) and the origin of the object plane by d_0 and, similarly, the distance between (x_1, y_1) and the origin of the image plane by d_1, the transverse magnification m of the system is d_1/d_0. It is not difficult to show that m is also equal to z_1/z_0 (see Figure 1.3).

Principal planes

The concept of the principal planes is rooted in paraxial ray-tracing (i.e., Gaussian optics), where the angles between the rays and the optical axis are so small that the sine and the tangent of each angle can be approximated by the value of the angle itself, $\sin\theta \approx \tan\theta \approx \theta$. In the neighborhood of the optical axis, therefore, the entire system may be represented by a 2×2 matrix, and the principal planes are uniquely determined from this so-called ABCD matrix of the system.[5]

The principal planes are conjugate planes with unit transverse magnification. A ray entering the first p.p. at a certain height h will emerge from the second p.p. at the same height, as shown in Figure 1.5(a). Thus $h \approx z_0\theta_0 \approx z_1\theta_1$, where θ_0 and θ_1 are the angles of the incident and the emer-

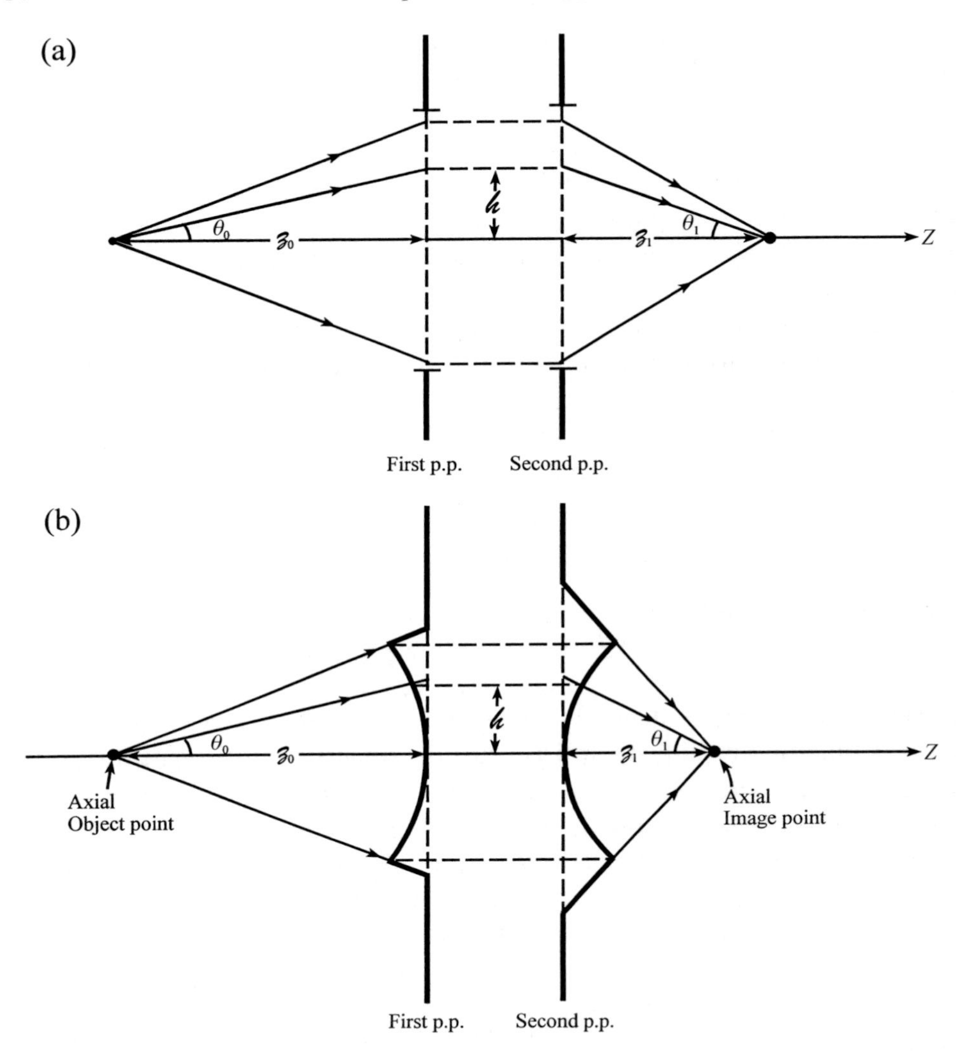

Figure 1.5 (a) In the paraxial regime the height h of a ray is measured from the optical axis in the principal planes. (b) In systems that operate beyond the paraxial regime one may define the ray height at the point where the ray crosses a reference sphere. When a system satisfies Abbe's sine condition the height of a ray thus defined remains the same upon entering and exiting the system.

gent rays with the optical axis. Note that, within the framework of the paraxial approximation, the system's entrance aperture at the first p.p. is identical in size and shape to the exit aperture located at the second p.p. (The term aperture as used here should not be confused with pupil, which has a more specific meaning in geometrical optics. The entrance and exit pupils also define the boundaries of the cones of light that enter and exit the system, but the pupils are not necessarily located at the principal planes.)

Beyond the paraxial regime, the principal planes cease to be conjugate planes. Depending on its direction, a ray entering the first p.p. at a given height h might emerge from different locations on the second p.p. One might confine attention to a specific set of rays, such as those emanating from the axial point in the object plane, in order to fix the directions of rays that enter the system. Yet there is no guarantee that the height h of a ray on entering the first p.p. will remain the same when it emerges from the second p.p. Of course one can impose this as a requirement on the system, but many other possibilities exist that are equally plausible, as long as they conform to the constraints of the paraxial regime. Abbe's sine condition is one such requirement placed on the heights of the entering and emerging rays.

The sine condition

Let us define two spherical surfaces, one in the object space, centered on the axial object point and tangent to the first p.p., and the other in the image space, centered on the axial image point and tangent to the second p.p. (see Figure 1.5(b)). Instead of assigning heights to the rays in the principal planes, the heights are assigned at the points where the rays cross these spherical surfaces. Thus, upon entering the system, $h = z_0 \sin\theta_0$. (If the height were assigned at the principal plane, the above expression would be written with tangent instead of sine.) Abbe's sine condition requires that all rays emanating from the axial object point within the incident cone must emerge in the image space, where they form a converging cone toward the axial image point, at the same height at which they entered the system.[4]

As long as the rays are close to the optical axis (where the spheres are tangent to the principal planes), the tangent and sine of a given angle are nearly the same. Thus Abbe's sine condition is consistent with the fact that, in the paraxial regime, the principal planes are unit-magnification conjugate planes. For the rays beyond the paraxial region, $\sin\theta$ deviates from $\tan\theta$ and the height of a given ray at the entrance sphere is no longer the same as its height at the first p.p. (Similarly, the height of an emergent ray at the exit sphere differs from its height at the second p.p.) In a sense, therefore, the sine condition requires the bending of the principal planes into spheres to preserve the paraxial property that a ray entering the system at a given height emerges from the system at the same height.

Whereas in the paraxial regime the angular magnification $\theta_1/\theta_0 = 1/m$, where m is the transverse magnification of the system, it is the ratio $(\sin\theta_1)/(\sin\theta_0)$

that equals $1/m$ in a system satisfying the sine condition. This turns out to be of crucial significance for the image-forming system, as will be shown below. To emphasize the point, note that in the system of Figure 1.5(a), where the entering and emerging ray heights are equal at the principal planes, the ratio $(\tan\theta_1)/(\tan\theta_0)$ equals $1/m$, whereas in the system of Figure 1.5(b), which satisfies Abbe's sine condition, the relevant ratio is $(\sin\theta_1)/(\sin\theta_0)$.

Aplanatic system

A system that yields an aberration-free image of the axial object point and satisfies Abbe's sine condition is said to be "aplanatic".[4,5] Many imaging systems in use today satisfy these conditions to a good approximation, if not exactly. Note that the clear-aperture diameter of an aplanatic system as seen on the first p.p. is no longer equal to that on the second p.p. If NA_0 is the numerical aperture of the largest cone of light emanating from the axial object point and captured by the system, the aperture radius on the entrance sphere is $z_0 NA_0$ whereas that on the first p.p. is $z_0 \tan[\sin^{-1}(NA_0)]$. Similarly, in the image space the aperture radius on the exit sphere is $z_1 NA_1$ while that on the second p.p. is $z_1 \tan[\sin^{-1}(NA_1)]$. Abbe's sine condition guarantees that $z_0 NA_0 = z_1 NA_1$ but, unless the imaging system has unit magnification, the aperture radii at the two principal planes are not equal.

What is surprising about the sine condition is that a requirement imposed solely on the cones of light corresponding to the *on-axis* points affects the quality of imaging for nearby *off-axis* points: once the sine condition has been satisfied, all near-axis points within the object plane will be imaged, essentially free of aberration, to their conjugates in the image plane. Without the sine condition, however, images of the near-axis points would be degraded by aberrations, most prominently by coma. It is this surprising property of the sine condition that we shall elucidate further.

The wave-optical viewpoint

Having secured a one-to-one mapping between the distribution of light entering the first p.p. and that exiting the second p.p, for an axial object point, we now switch to the viewpoint of wave optics and consider the perturbation of the wavefront in response to a slight off-axis shift of the axial object point.

In the diffraction analysis of lenses conducted within the paraxial approximation, it is customary to assign to the second p.p. the same complex-amplitude distribution that exists on the first p.p. This distribution is then augmented by aberrations of the lens, if any, to account for deviations of the emergent wavefront from perfect sphericity.[8] Thus if $A_1(x, y)$ represents the complex-amplitude distribution at the first p.p., the distribution at the second p.p. will be written

$$A_2(x, y) = A_1(x, y)[\mathrm{i}(2\pi/\lambda)W(x, y)] \exp[-\mathrm{i}(2\pi/\lambda)\sqrt{x^2 + y^2 + f^2}]. \quad (1.1)$$

Here λ is the wavelength of the light, $W(x, y)$ represents wavefront aberrations, and the second exponential factor corresponds to a perfect spherical wavefront converging toward the focal point in the image space.

For wide-aperture systems, Eq. (1.1) must be modified to account for deviations from the paraxial regime. For example, if the ray emerging from (x,y) in the second p.p. enters the first p.p. at (x', y'), then $A_1(x, y)$ in Eq. (1.1) must be replaced by $A_1(x', y')$, and the Jacobian of the transformation between the two principal planes must be properly taken into account, to preserve the optical energy throughput of the system.

Strictly speaking, since in non-paraxial regions the principal planes are no longer conjugate planes, it follows that a one-to-one mapping between these planes is meaningless. In practice, however, the field of view of the lens is so small that a cone of light emanating from any point within the field of view is essentially the same as the axial cone in Figure 1.5(b), but endowed with some form of phase/amplitude modulation. Thus the correspondence between a pair of points such as (x', y') on the first p.p. and (x, y) on the second p.p., established for the axial cone, remains approximately valid for all object points. Any phase/amplitude perturbation affecting the beam at (x', y') can then be transferred directly to the beam at (x, y), and the resulting distribution within the second p.p. may be used as the initial distribution for further propagation through the image space.

(Many authors prefer to use the amplitude distribution over the spherical exit surface in Figure 1.5(b) as the initial distribution, without ever referring to the principal planes. If one is interested in diffraction analysis using a plane wave spectrum, however, one should start with initial conditions that are defined on a flat surface, in which case the second p.p. provides a natural frame of reference.)

Wavefront perturbation due to off-axis shift of the object point

The distribution of the complex amplitude at the first p.p. due to a cone of light emanating from the off-axis point (x_0, y_0) may be determined by reference to Figure 1.6. The distance from (x', y') to the off-axis point differs from that to the on-axis point by

$$\Delta l \approx x_0 S'_x + y_0 S'_y. \tag{1.2}$$

To a first approximation, therefore, upon arrival at the first p.p. the cone of light that originates at (x_0, y_0) will be the same as that which originated from the axial point, albeit with a modulation by the following phase factor:

$$\exp[\mathrm{i}(2\pi/\lambda)\Delta l] \approx \exp[\mathrm{i}(2\pi/\lambda)(x_0 S'_x + y_0 S'_y)]. \tag{1.3}$$

Note that the phase in Eq. (1.3) is linear in (S'_x, S'_y) but not in (x', y'). The same phase factor will appear on the beam at (x, y) on the second p.p. Now, and this is the crux of the matter, if the sine condition is satisfied then this phase factor can be replaced by $\exp[\mathrm{i}(2\pi/\lambda)(x_1 S_x + y_1 S_y)]$, because the angular magnification between (S'_x, S'_y) and (S_x, S_y) is exactly the reverse of the transverse magnification m between (x_0, y_0) and (x_1, y_1). The distribution at the second p.p. now corresponds to a spherical wavefront, converging toward (x_1, y_1) and having no aberrations whatsoever. This is the essence of the sine

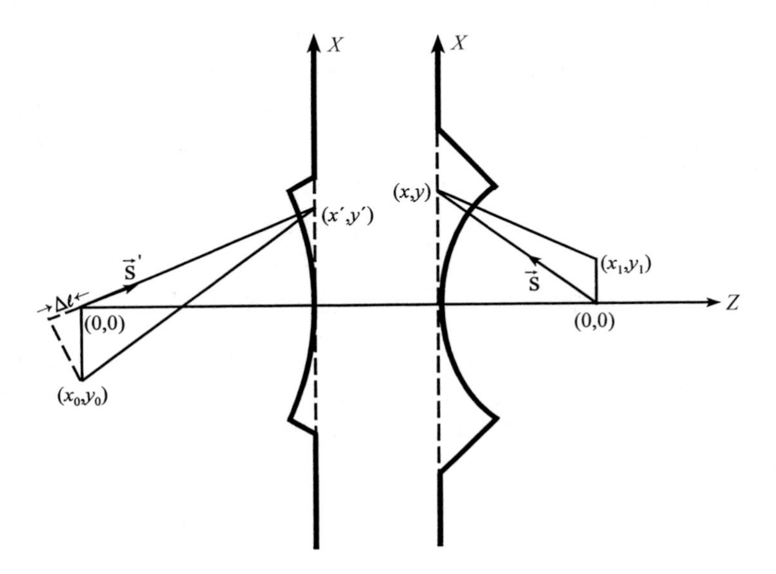

Figure 1.6 The ray leaving the off-axis point (x_0, y_0) and arriving at (x', y') will travel a slightly different distance than the ray from the axial point (0, 0) that travels along $\boldsymbol{S}'$ toward the same location. When (x_0, y_0) is sufficiently close to the optical axis, the path-length difference between these two rays can be approximated by the projection on $\boldsymbol{S}'$ of the line joining (x_0, y_0) to the point at the origin. The same argument applies to the conjugate rays in the image space.

condition, which cannot be over-emphasized; it is the reason why there is "aberration-free" imaging of near-axial points.

A wide-aperture aplanat

As an example, consider an ideal infinite-conjugate aplanatic lens having $z_0 = \infty$, $NA_0 = 0$, $z_1 = f = 4000\lambda$ and $NA_1 = 0.75$. The phase pattern of an obliquely incident plane wave at the first p.p. of this lens is shown in Figure 1.7(a). The beam has a linear phase over the entire entrance aperture, as expected of a plane wave at oblique incidence. Upon emerging from the second p.p. the phase pattern of the beam is that of Figure 1.7(b). In compliance with the sine condition the exit aperture is seen to be larger than the entrance aperture, and the phase pattern has undergone some sort of nonlinear "stretching". (The emergent phase pattern in Figure 1.7(b), however, is nonlinear because it is displayed in the x, y coordinates; in the coordinates S_x, S_y it would be perfectly linear.)

The emergent beam comes to focus at the focal plane of the lens, creating the off-axis Airy pattern shown in Figure 1.7(c). For comparison, the on-axis focused spot of the same lens is also shown in the figure. As expected, the off-axis spot is free from aberrations, and the two spots are essentially identical.

It is not difficult to design an aplanat with the characteristics of the lens in the above example; a specific design is shown in Figure 1.8. The various parameters of this meniscus, which consists of two conic surfaces, are listed in the figure caption.

Offense against the sine condition

Let us now examine the special case of a lens in which the ray heights have been made equal at the principal planes. Here $(x, y) = (x', y')$, and the difference between the actual and the ideal (i.e., aberration-free) emergent wavefronts will be

$$W(x, y) = (x_0 S'_x + y_0 S'_y) - (x_1 S_x + y_1 S_y). \tag{1.4}$$

Note that S_x and S_y are proportional to $\sin\theta$, but in the present case it is $\tan\theta$ that is magnified by $1/m$. A Taylor series expansion yields

$$\tan\theta = \sin\theta/\sqrt{1 - \sin^2\theta} = \sin\theta + \tfrac{1}{2}\sin^3\theta + \tfrac{3}{8}\sin^5\theta + \cdots. \tag{1.5}$$

To a first approximation, therefore, the difference between $\sin\theta$ and $\tan\theta$ is proportional to $\sin^3\theta$. This difference, when inserted in Eq. (1.4), produces

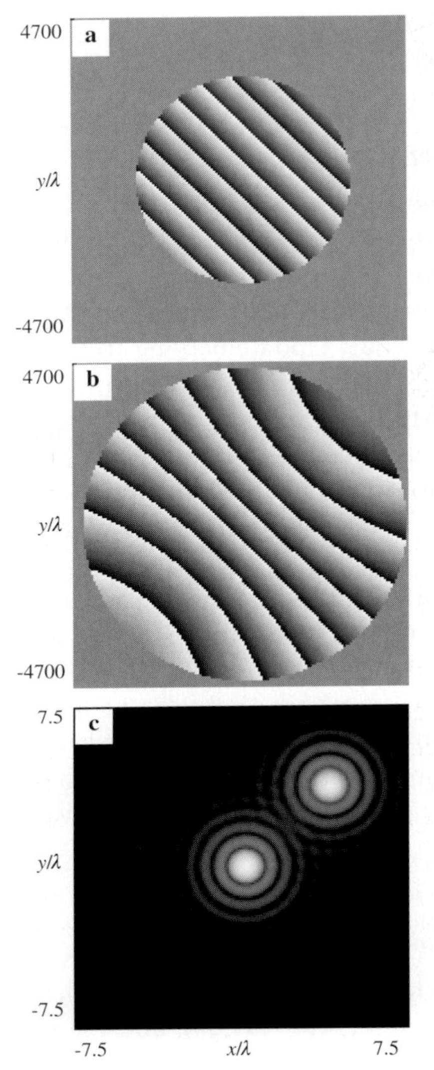

Figure 1.7 (a) Distribution of phase at the first p.p. of an infinite-conjugate lens having $NA_1 = 0.75$ and $f = 4000\lambda$. The entrance aperture radius is 3000λ, and the incident beam propagates at $\theta = 0.076^\circ$ relative to the optical axis. The gray-scale covers the interval from -180° (black) to $+180^\circ$ (white). (b) Distribution of phase at the second p.p. Since the lens satisfies Abbe's sine condition the exit aperture radius is 4536λ. (c) Logarithmic plot of intensity distribution at the focal plane showing the axial focused spot (center) and the off-axis spot corresponding to an oblique incidence angle of $\theta = 0.076^\circ$. The spots are nearly identical; both are substantially free from aberrations.

primary coma. Thus when the rays that enter at a given height on the first p.p. emerge at the same height on the second p.p., perfect imaging of the axial point results in comatic imaging of the near-axis points.

Similar arguments may be advanced for systems that violate the sine condition in ways other than described above. In general, offense against the sine condition results in primary and higher-order coma in near-axis regions of the image plane.

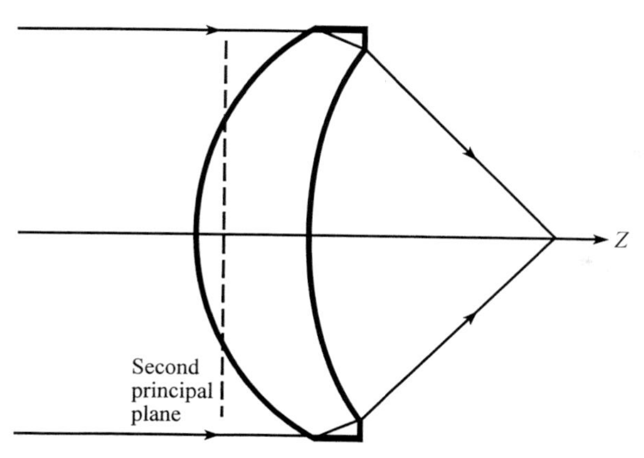

Figure 1.8 An aplanatic meniscus lens brings collimated beams to diffraction-limited focus within its focal plane in the vicinity of the optical axis. This 4 mm diameter lens has $f = 2.6733$ mm and $NA = 0.75$. The refractive index of the lens glass is $n = 2.49486$, its thickness at the center is 1 mm, and its conic surfaces have the following radii of curvature and conic constants: first surface, $R_c = 2.26875$ mm, $k = -0.20945$; second surface, $R_c = 3.87493$ mm, $k = 0.08173$. The second principal plane is 0.2894 mm to the right of the first surface's vertex.

The image of a diffraction grating

An appealing argument in favor of the sine condition involves the image of a diffraction grating.[9] Consider a small grating of period P placed perpendicular to the optical axis in the object plane of the system of Figure 1.3; the illumination is coherent, collimated, and monochromatic with wavelength λ. The nth diffraction order leaves the grating at the Bragg angle θ_n relative to the optical axis, where $\sin\theta_n = n\lambda/P$. In the image plane the grating period is mP, where m is the transverse magnification of the system. Therefore, to obtain a distortion-free image it is necessary that all the $\sin\theta_n$ be magnified by $1/m$; in other words, the sine condition must be satisfied.

References for chapter 1

1 E. Abbe, *Jenaisch. Ges. Med. Naturw.* (1879); also *Carl. Repert. Phys.* **16**, 303 (1880).
2 C. Hockin, *J. Roy. Micro. Soc.* (2) **4**, 337 (1884).
3 A. B. Porter, *Phil. Mag.* (6) **11**, 154 (1906).
4 M. Born and E. Wolf, *Principles of Optics*, sixth edition, Pergamon Press, Oxford, 1980.
5 M. V. Klein, *Optics*, Wiley, New York, 1970.
6 J. M. Stone, *Radiation and Optics*, McGraw-Hill, New York, 1963.
7 A. E. Conrady, *Applied Optics and Optical Design*, Dover, New York, 1957.
8 J. W. Goodman, *Introduction to Fourier Optics*, McGraw-Hill, New York, 1968.
9 Douglas Goodman, private communication.

Augustin Jean Fresnel

Joseph Fourier

Siméon Denis Poisson

Joseph von Fraunhofer

2
Fourier optics

The classical theory of diffraction originated in the work of the French physicist Augustin Jean Fresnel, in the first quarter of the nineteenth century. Fresnel's ideas were subsequently expanded and elaborated by, among others, William Rowan Hamilton, Gustav Kirchhoff, George Biddell Airy, John William Strutt (Lord Rayleigh), Ernst Abbe, and Arnold Sommerfeld, leading to a complete understanding of light in its wave aspects.[1]

The Fourier-transform operation occurs naturally in any formulation of the theory of diffraction, giving rise to a body of literature that has come to be known as Fourier optics.[2] The prominence of Fourier transforms in physical optics is rooted in the fact that any spatial distribution of the complex amplitude of light can be considered a superposition of plane waves.[3] (Plane waves, of course, are eigenfunctions of Maxwell's equations for the propagation of electromagnetic fields through homogeneous media.[1,4])

Many students of Fourier optics are intimidated by the approximations involved in deriving its basic formulas, but it turns out that the majority of these approximations are in fact unnecessary: by starting from a plane-wave expansion of the light amplitude distribution, rather than the traditional Huygens' principle,[1,2,4] one can readily arrive at the fundamental results of the classical theory either directly or after applying the stationary-phase approximation.[1,3] (For a detailed discussion of the stationary-phase method see the appendix to this chapter.)

The goal of the present chapter is to show how decomposition into, and subsequent superposition of, plane waves can lead straightforwardly to the near-field (Fresnel) and far-field (Fraunhofer) formulas, to elucidation of the Fourier transforming properties of a lens, and to the essence of Abbe's theory

George Biddell Airy

Gustav Robert Kirchhoff

Augustin Jean Fresnel (1788–1827). His work in optics received scant public recognition during his lifetime, but Fresnel maintained that not even acclaim from distinguished colleagues could compare with the pleasure of discovering a theoretical truth or confirming a calculation experimentally. (Photo: Smithsonian Institution, courtesy of AIP Emilio Segré Visual Archives.)

Jean Baptiste Joseph Fourier (1768–1830), began to work on the theory of heat around 1804 and by 1807 had completed a memoir, *On the Propagation of Heat in Solid Bodies*, in which periodic functions were expressed as the sum of an infinite series of sines and cosines. Lagrange and Laplace objected to Fourier's expansion on the grounds that it lacked generality and rigor. Fourier's treatise, *The Analytical Theory of Heat*, was not published until 1822. (Photo: Deutsches Museum, courtesy of AIP Emilio Segré Visual Archives.)

Siméon Denis Poisson (1781–1840). In 1818, during the judging of Fresnel's paper on diffraction at the Paris Academy, Poisson argued that the consequence of Fresnel's theory was the absurdity that the center of the shadow of an opaque disk should be illuminated. This unexpected effect was subsequently observed. (Photo: courtesy of AIP Emilio Segré Visual Archives.)

Joseph von Fraunhofer (1787–1826) German physicist who first studied the dark lines in the spectrum of the Sun. The first to use diffraction gratings, his work set the stage for the further development of spectroscopy. (Photo: Bavarian Academy of Sciences, courtesy of AIP Emilio Segré Visual Archives.)

of image formation. Along the way, several numerical examples will demonstrate the utility of the derived formulas.

Electromagnetic plane waves

A plane-wave solution of Maxwell's equations in a homogeneous environment can be expressed as

$$a(x, y, z) = \boldsymbol{A}_0 \exp[\mathrm{i}(2\pi/\lambda)(x\sigma_x + y\sigma_y + z\sigma_z)]. \tag{2.1a}$$

Here λ is the wavelength of the light, $\boldsymbol{A}_0$ is a complex vector representing the magnitude and state of polarization of the E-field at the origin of the coordinate system, and $\sigma = (\sigma_x, \sigma_y, \sigma_z)$ is a unit vector specifying the direction of propagation. In general, σ_z is related to σ_x and σ_y by

$$\sigma_z = (1 - \sigma_x^2 - \sigma_y^2)^{1/2}. \tag{2.1b}$$

On the one hand, σ_z will be real-valued if $\sigma_x^2 + \sigma_y^2 \leq 1$, in which case the plane wave is said to be homogeneous or propagating. On the other hand, if $\sigma_x^2 + \sigma_y^2 > 1$ then σ_z becomes imaginary and the plane wave is called inhomogeneous or evanescent.

In scalar diffraction theory, the state of polarization of the light is ignored and $\boldsymbol{A}_0$ is treated as a complex constant. Furthermore, if the x, y, z coordinates are normalized by the wavelength λ, then this parameter disappears from all subsequent equations. Throughout this chapter, therefore, all lengths will be assumed to be normalized by λ; a propagation distance of 1000, for example, should be understood as a distance of 1000λ.

Sir George Biddell Airy (1801–1892), became Lucasian Professor of Mathematics at Cambridge only three years after graduating from Trinity College in 1823. He was Astronomer Royal from 1835 to 1881. Airy contributed to the understanding of the rainbow by studying the effects of diffraction from raindrops. (Photo: courtesy of AIP Emilio Segré Visual Archives, E. Scott Barr Collection.)

Gustav Robert Kirchhoff (1824–1887), Professor of physics at Heidelberg, Breslau and Berlin. His discovery that a gas absorbs the same wavelengths that it emits when heated explained the numerous dark lines (Fraunhofer lines) in the Sun's spectrum, marking the beginning of a new era in astronomy. Kirchhoff placed Fresnel's ideas on a firm theoretical basis, formulating what is now referred to as the Fresnel–Kirchhoff diffraction theory. (Photo: courtesy of AIP Emilio Segré Visual Archives, W.F. Meggers Collection.)

Expansion into plane waves

Consider the complex-amplitude distribution $a(x, y, z = 0)$ in the XY-plane at $z = 0$. The Fourier transform of $a(x, y, z = 0)$ is defined as

$$A(\sigma_x, \sigma_y) = \iint_{-\infty}^{\infty} a(x, y, z = 0) \exp[-\mathrm{i}2\pi(x\sigma_x + y\sigma_y)]\, \mathrm{d}x\, \mathrm{d}y. \tag{2.2a}$$

The inverse Fourier transform may therefore be written

$$a(x, y, z = 0) = \iint_{-\infty}^{\infty} A(\sigma_x, \sigma_y) \exp[\mathrm{i}2\pi(x\sigma_x + y\sigma_y)]\, \mathrm{d}\sigma_x\, \mathrm{d}\sigma_y. \tag{2.2b}$$

Because Maxwell's equations are linear, any superposition of plane waves within homogeneous linear media is also a solution of Maxwell's equations. In general, the superposition of plane waves in Eq. (2.2b) contains both propagating and evanescent waves. At a distance $z = z_0$ from the origin, the complex-amplitude distribution of the light is thus given by

$$a(x, y, z = z_0) = \iint_{-\infty}^{\infty} A(\sigma_x, \sigma_y) \exp[\mathrm{i}2\pi(x\sigma_x + y\sigma_y + z_0\sigma_z)]\, \mathrm{d}\sigma_x\, \mathrm{d}\sigma_y. \tag{2.3}$$

Equation (2.3) is the fundamental formula of the classical theory of diffraction. It provides the following simple recipe for computing the distribution of the field at the plane $z = z_0$ given the initial distribution at $z = 0$:

(i) compute the Fourier transform $A(\sigma_x, \sigma_y)$ of the initial distribution;

(ii) multiply $A(\sigma_x, \sigma_y)$ by the phase factor, which may be written as $\exp(\mathrm{i}2\pi z_0\sigma_z) = \exp[\mathrm{i}2\pi z_0(1 - \sigma_x^2 - \sigma_y^2)^{1/2}]$;

(iii) compute the inverse Fourier transform of the resulting function.

The above recipe is applicable to many practical problems, without the need to introduce any approximations or simplifications. Some consequences of Eq. (2.3) are explored in the following examples.

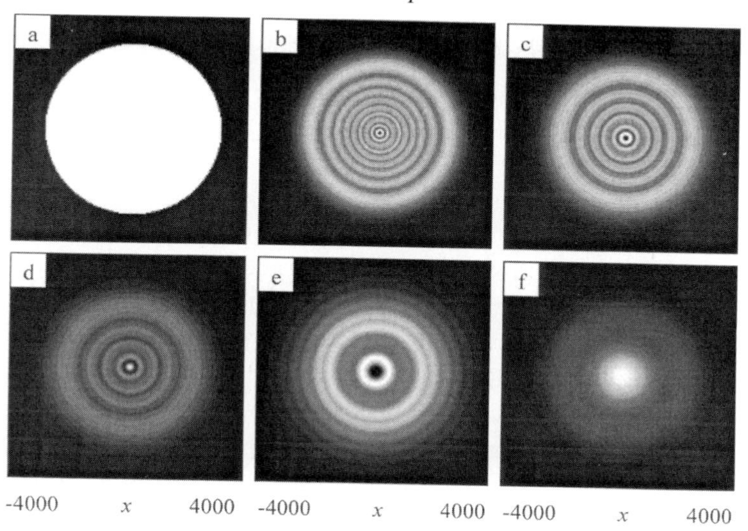

Figure 2.1 Computed intensity patterns at various distances from a circular aperture of radius $r_0 = 3000$, illuminated by a uniform plane wave. The assumed distances from the aperture are (a) $z_0 = 0$, (b) $z_0 = 0.5 \times 10^6$, (c) $z_0 = 0.75 \times 10^6$, (d) $z_0 = 10^6$, (e) $z_0 = 2.25 \times 10^6$, (f) $z_0 = 9.0 \times 10^6$. Note that the center of the diffracted beam is dark in (b), (c) and (e), while it is bright in (d) and (f).

Diffraction from a circular aperture

Figure 2.1 shows the computed intensity patterns at various distances from a circular aperture of radius $r_0 = 3000$ illuminated by a uniform plane wave. From (a) to (f) the assumed distances from the aperture are $z_0 = 0, 0.5 \times 10^6, 0.75 \times 10^6, 1.0 \times 10^6, 2.25 \times 10^6$, and 9.0×10^6. (These distances correspond to the Fresnel numbers[1] $N = r_0^2/z_0 = \infty$, 18, 12, 9, 4, and 1, respectively.) The computations were carried out by discretizing the initial distribution on a 512×512 mesh and then applying the fast Fourier transform (FFT) algorithm. On a modern personal computer the time needed for these calculations is less than a second.

Diffraction-free beams

If the propagation phase factor in Eq. (2.3) happens to be a constant then it can be taken out of the integral, in which case, aside from a multiplicative phase factor, the distribution at $z = z_0$ becomes equal to that at $z=0$. This occurs if the Fourier transform $A(\sigma_x, \sigma_y)$ of the initial distribution happens to be non-zero only over a circle of fixed radius in the Fourier plane, that is, if

$A(\sigma_x, \sigma_y) = 0$ everywhere except where $\sigma_x^2 + \sigma_y^2 = \rho_0^2$. Under these circumstances, Eq. (2.3) yields

$$a(x, y, z = z_0) = \exp\left[i2\pi z_0\left(1 - \rho_0^2\right)^{1/2}\right]a(x, y, z = 0). \quad (2.4)$$

According to Eq. (2.4), any initial distribution that is confined to a circle of radius ρ_0 in the Fourier domain will not diffract while propagating along the Z-axis.[5] A particularly simple case occurs when $A(\sigma_x, \sigma_y) = \delta(\rho - \rho_0)$, where $\delta(\cdot)$ is Dirac's delta function and $\rho = (\sigma_x^2 + \sigma_y^2)^{1/2}$. The inverse transform of this delta function is a zeroth-order Bessel function of the first kind, namely, $a(x, y, z = 0) = J_0(2\pi\rho_0 r)$, where $r = (x^2 + y^2)^{1/2}$.

Needless to say, any azimuthal variation of the amplitude and/or phase of the above delta function around the circle of radius ρ_0 in the Fourier domain yields another non-diffracting beam. Moreover, if the radius ρ_0 is less than unity then the non-diffracting beam will be a propagating beam, whereas $\rho_0 > 1$ corresponds to an exponentially attenuating, non-diffracting, evanescent beam.

Poisson's bright spot

A bright spot appearing at the center of the geometrical shadow of an opaque disk was first predicted by S. D. Poisson in an attempt to refute Fresnel's theory of diffraction. Fresnel's theory was vindicated, however, when François Arago confirmed the existence of the bright spot in an experiment.[1,4,6]

The diagram in Figure 2.2 shows a collimated beam blocked at the center by a disk of radius r_0. The complex-amplitude distribution immediately after the disk is denoted by $a(x, y, z = 0)$. The volume under the Fourier transform $A(\sigma_x, \sigma_y)$ of this distribution over the $\sigma_x\sigma_y$-plane is zero, because the central value $a(0, 0, 0)$ of the initial distribution is zero. However, the volume under the Fourier transform of the beam's cross-section at $z = z_0$ is not zero, because $A(\sigma_x, \sigma_y)$ is multiplied by the phase-factor $\exp[i2\pi z_0(1 - \rho^2)^{1/2}]$, which changes the phase of the Fourier transform as a function of the radius ρ in the $\sigma_x\sigma_y$-plane. A non-zero volume in the Fourier domain implies that the central value of the distribution $a(x = 0, y = 0, z = z_0)$ is also non-zero, that is, the center of the distribution at $z = z_0$ is no longer dark. For a disk of radius $r_0 = 2500$, Figure 2.3 shows the computed intensity distributions (a) immediately after the disk, (b) at $z_0 = 2.0 \times 10^6$, and (c) at $z_0 = 4.0 \times 10^6$.

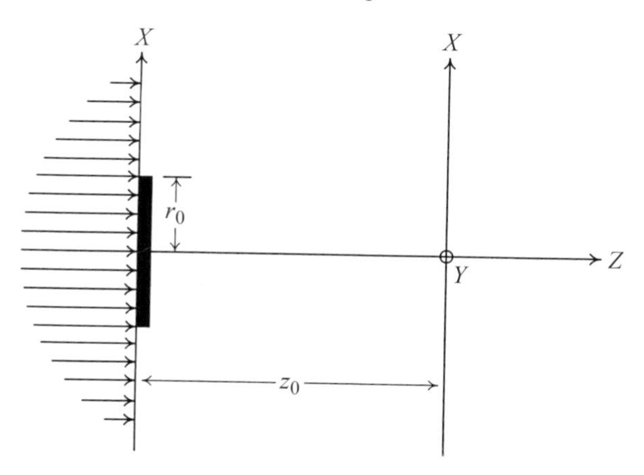

Figure 2.2 A collimated beam illuminates an opaque circular disk of radius r_0. At a distance z_0 from the disk the intensity distribution in the XY-plane contains a bright spot at the center of the geometrical shadow of the disk.

Poisson's bright spot may be considered as the focus of a collimated beam produced by an opaque disk. The disk, therefore, behaves as a lens, albeit a dark one;[6] an illuminated object placed before the disk forms an image through the dark lens, as shown in Figure 2.4. In this particular example, the object, shown in Figure 2.4(a), is a circular aperture partially covered by four small obstacles. The object is back-illuminated incoherently, by an extended quasi-monochromatic source, through a $0.005NA$ condenser lens. A dark lens of radius $r_0 = 2500$ at a distance of 10^6 from the object produces the real image shown in Figure 2.4(b) at a distance of 2.0×10^6 behind the dark lens.

We mention in passing that the incoherence of the illumination is essential for the success of this imaging process; interference effects totally obscure the image when the object is coherently illuminated.

Distribution of light in the far field

As the value of z_0 increases, Eq. (2.3) becomes exceedingly difficult to compute, because the rapid oscillations of the exponential phase factor require dense sampling of the functions in the $\sigma_x\sigma_y$-plane. In this regime, however, the stationary-phase approximation[1] becomes applicable.

For a fixed value of (x, y, z_0), the exponent under the integral in Eq. (2.3) may be considered to be a function of (σ_x, σ_y). This function has a single stationary point at $(\sigma_{x0}, \sigma_{y0}) = (x, y)/(x^2 + y^2 + z_0^2)^{1/2}$. At all other points in the $\sigma_x\sigma_y$-plane the complex exponential oscillates so rapidly that the local

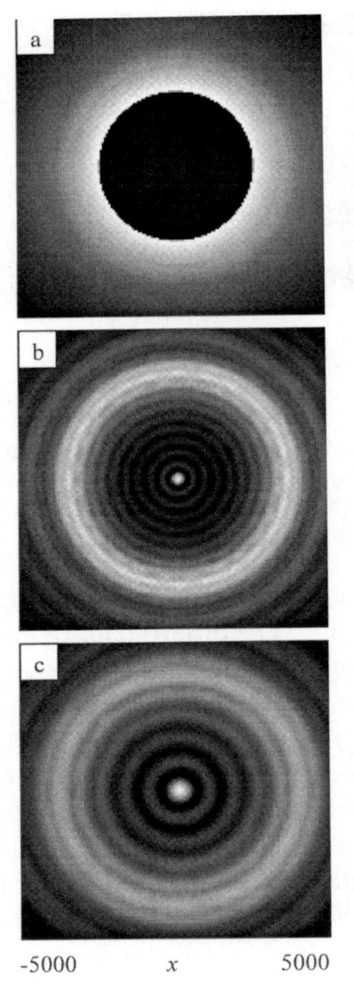

Figure 2.3 Computed intensity patterns at various distances from an opaque circular disk of radius $r_0 = 2500$, illuminated by a collimated Gaussian beam having a 1/e (amplitude) radius of 5000. The distances from the disk are (a) $z_0 = 0$, (b) $z_0 = 2.0 \times 10^6$, (c) $z_0 = 4.0 \times 10^6$.

integral effectively vanishes; only at the stationary point does the integral yield a non-zero value. At this point the exponent can be replaced by the first few terms in its Taylor-series expansion around the stationary point, namely,

$$x\sigma_x + y\sigma_y + z_0\sigma_z \approx (x^2 + y^2 + z_0^2)^{1/2}\big[1 - \tfrac{1}{2}(1 + x^2/z_0^2)(\sigma_x - \sigma_{x0})^2 - (xy/z_0^2)(\sigma_x - \sigma_{x0})(\sigma_y - \sigma_{y0}) - \tfrac{1}{2}(1 + y^2/z_0^2)(\sigma_y - \sigma_{y0})^2\big]. \tag{2.5}$$

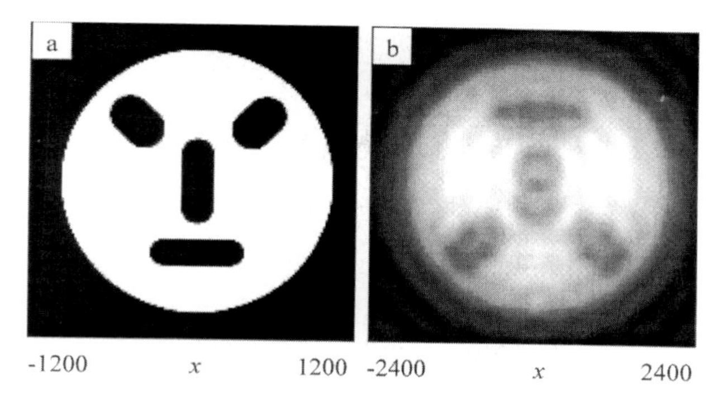

Figure 2.4 Incoherent imaging by means of a dark lens. The object in (a) is illuminated by an extended quasi-monochromatic source through a $0.005NA$ condenser of focal length $f = 6.0 \times 10^5$. The source consists of 529 mutually incoherent point sources, imaged by the condenser at a distance of $\Delta z = 10^5$ before the object. The dark lens is an opaque circular disk of radius $r_0 = 2500$, placed a distance of $\Delta z = 10^6$ from the object. The image in (b) is computed at a distance of $z_0 = 2.0 \times 10^6$ behind the dark lens.

The integral in Eq. (2.3) is then readily computed, without further approximations, yielding

$$a(x, y, z = z_0) \approx -\left[\mathrm{i}/(x^2 + y^2 + z_0^2)^{1/2}\right] \exp\left[\mathrm{i}2\pi(x^2 + y^2 + z_0^2)^{1/2}\right] \times A(\sigma_{x0}, \sigma_{y0})/\left[1 + (x/z_0)^2 + (y/z_0)^2\right]^{1/2}. \tag{2.6}$$

This is the so-called Fraunhofer (or far-field) distribution arising from the initial distribution $a(x, y, z = 0)$. The far field is expressed in terms of the Fourier transform $A(\sigma_x, \sigma_y)$ of the initial distribution evaluated at $(\sigma_{x0}, \sigma_{y0}) = (x, y)/(x^2 + y^2 + z_0^2)^{1/2}$. Note how the obliquity factor $\cos\theta = 1/[1 + (x/z_0)^2 + (y/z_0)^2]^{1/2}$ enters the above equation (see Figure 2.5).

If the far field is observed on a spherical surface of radius z_0 centered on the object (see Figure 2.5) then the curvature phase factor becomes a constant and $(\sigma_{x0}, \sigma_{y0})$ reduces to $(x/z_0, y/z_0)$, yielding the following simple formula for the far-field pattern on a spherical surface of radius z_0:

$$a(x, y, z) \approx -(\mathrm{i}/z_0) \exp(\mathrm{i}2\pi z_0)\, A(x/z_0, y/z_0) \cos\theta. \tag{2.7}$$

The conservation of optical power passing through any cross-section of the beam may be verified by integrating the squared modulus of the functions appearing in Eqs. (2.6) and (2.7) over their respective domains.

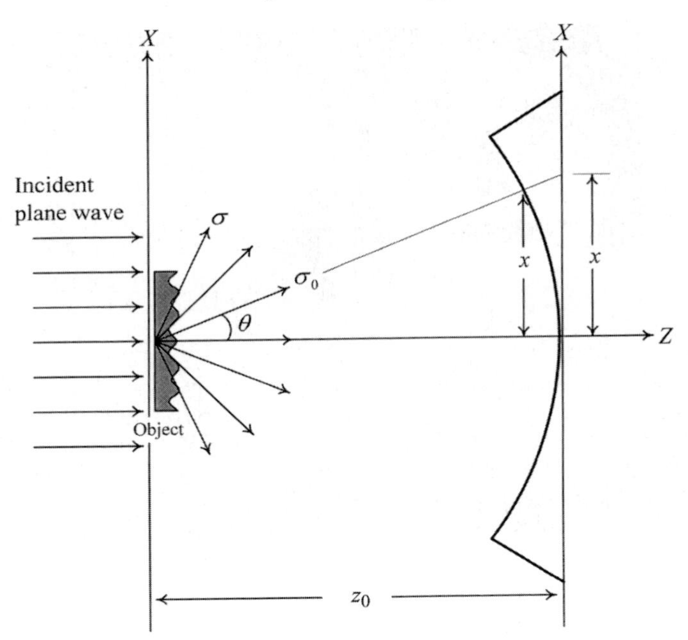

Figure 2.5 A phase/amplitude object is illuminated by a plane wave propagating along the Z-axis. The diffracted beam is a superposition of plane waves of differing amplitudes, propagating along directions indicated by the unit vectors σ. The far-field pattern appears at a sufficiently large distance z_0 from the object. Whether the field is observed on the XY-plane at $z = z_0$ or on the spherical surface of radius z_0 centered on the object, the x, y coordinates of a given point are the usual coordinates measured along the X- and Y- axes. In either case, the far-field amplitude is proportional to the complex amplitude of the plane wave whose propagation direction σ_0 is directly aimed at the observation point.

Far field of an annular aperture

To demonstrate the utility of Eq. (2.6), we use as the initial distribution the narrow ring of light transmitted through an annular aperture (of width 100 and average radius 1000), shown in Figure 2.6(a). After propagating a distance of 10^6, the far-field pattern of Figure 2.6(b) is obtained. (To enhance the weak rings of this distribution, a gray-scale plot of the logarithm of intensity is displayed.) The far field is essentially a Bessel beam with a curvature phase factor. To eliminate the curvature, we use a 0.0075NA lens of focal length $f = 10^6$ to collimate the beam in the far field of the annular aperture. The emerging truncated and collimated Bessel beam at the exit pupil of the lens is shown in Figure 2.6(c). This beam is not completely

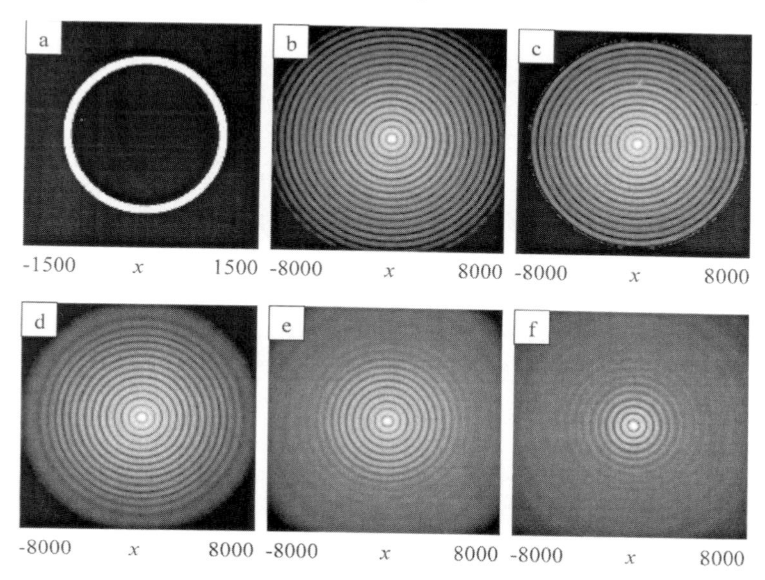

Figure 2.6 Logarithmic plots of intensity distribution at various cross-sections of a beam. (a) A transparent ring (radius 1000, width 100), illuminated with a collimated uniform beam propagating along the Z-axis. (b) Far-field pattern of the ring in the XY-plane at $z_0 = 10^6$. (c) The beam in (b) after collimation by a $0.0075NA$ lens of focal length $f = 10^6$. (d) The collimated beam in (c) after propagating in free space a distance of 10^6. (e) The beam in (d) after propagating a distance of 2.0×10^6. (f) The beam in (e) after propagating a distance of 5.0×10^6.

diffraction-free because it has a finite diameter. For instance, after it has propagated a distance of 10^6 from the exit pupil of the collimating lens one observes the intensity pattern of Figure 2.6(d). The intensity distribution after propagating another distance of 2×10^6 is shown in Figure 2.6(e). Finally, Figure 2.6(f) shows the intensity distribution observed at a distance of 5×10^6. Note how the decay of this truncated Bessel beam starts from the outer rings and moves toward the center as the beam propagates.

The Airy pattern at the focal plane of a lens

Consider the infinite-conjugate aplanatic lens of focal length f shown in Figure 2.7. (For a discussion of aplanatism see Chapter 1, Abbe's sine condition.) To determine the light amplitude distribution around the focal point F, we need the distribution in the second principal plane, which is given by

$$a(x, y, z = 0) = a_0(x_1, y_1) \cos^{3/2} \theta \exp\left[- i2\pi(x^2 + y^2 + f^2)^{1/2}\right]. \tag{2.8}$$

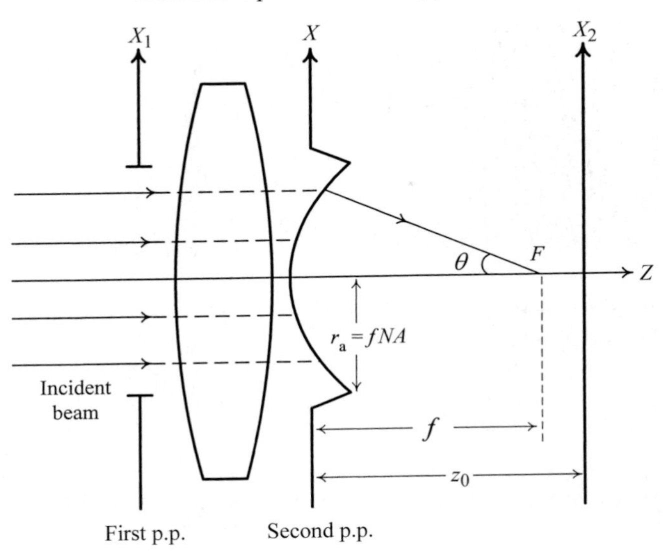

Figure 2.7 A collimated beam of light enters an infinity-corrected, aplanatic lens of focal length f and numerical aperture NA. The entrance and exit pupils are at the first and second principal planes. A ray entering at a height (x_1, y_1) on the first principal plane appears at the same height on the spherical surface centered at the rear focal point F and tangent to the second principal plane. In the absence of aberrations, all emergent rays converge to the focal point F. The distribution in the XY-plane at $z = z_0$ is given by Eq. (2.11).

The amplitude distribution at the entrance pupil (assumed to coincide with the 1st principal plane) is denoted by $a_0(x_1, y_1)$. The coordinates at the 1st and 2nd principal planes are related as follows: $(x_1, y_1) = (fx, fy)/(x^2 + y^2 + f^2)^{1/2}$. The corresponding infinitesimal areas in the two principal planes are in the ratio $\cos^3 \theta$, where $\cos\theta = f/(x^2 + y^2 + f^2)^{1/2}$; the amplitude in Eq. (2.8) is therefore scaled by $\cos^{3/2}\theta$ to conserve optical power between the entrance and exit pupils. The exponential phase factor in Eq. (2.8) is the curvature imparted by a perfect lens to the emergent beam.

To determine, in accordance with Eq. (2.2a), the Fourier transform of the initial distribution given by Eq. (2.8), we invoke the stationary-phase approximation.[1] The exponent of the integrand under the Fourier integral may be expanded in a Taylor series around its stationary point,

$$(x_0, y_0) = -(f\sigma_x, f\sigma_y)/(1 - \sigma_x^2 - \sigma_y^2)^{1/2},$$

yielding

$$x\sigma_x + y\sigma_y + (x^2 + y^2 + f^2)^{1/2} \approx (1 - \sigma_x^2 - \sigma_y^2)^{1/2}\Big\{f + \frac{1}{f}\big[\tfrac{1}{2}(1 - \sigma_x^2)(x - x_0)^2 - \sigma_x\sigma_y(x - x_0)(y - y_0) + \tfrac{1}{2}(1 - \sigma_y^2)(y - y_0)^2\big]\Big\}. \tag{2.9}$$

Without any other approximations, the Fourier transform of the initial distribution is found to be

$$A(\sigma_x, \sigma_y) \approx -\mathrm{i}fa_0(-f\sigma_x, -f\sigma_y)\exp\big[-\mathrm{i}2\pi f(1 - \sigma_x^2 - \sigma_y^2)^{1/2}\big]/(1 - \sigma_x^2 - \sigma_y^2)^{1/4}. \tag{2.10}$$

When the above function is substituted in Eq. (2.3) we obtain

$$a(x_2, y_2, z = z_0) \approx -\mathrm{i}f\iint\big[a_0(-f\sigma_x, -f\sigma_y)/(1 - \sigma_x^2 - \sigma_y^2)^{1/4}\big] \times \exp\big[\mathrm{i}2\pi(z_0 - f)(1 - \sigma_x^2 - \sigma_y^2)^{1/2}\big] \times \exp[\mathrm{i}2\pi(x_2\sigma_x + y_2\sigma_y)]\,\mathrm{d}\sigma_x\mathrm{d}\sigma_y. \tag{2.11}$$

For a given distribution $a_0(x_1, y_1)$ at the entrance pupil, Eq. (2.11) gives the distribution at and near the focal plane of the aplanatic lens of Figure 2.7. If the final distribution is sought in the focal plane (i.e., $z_0 = f$) and if the factor $\cos^{1/2}\theta = (1 - \sigma_x^2 - \sigma_y^2)^{1/4}$ is ignored (i.e., the paraxial approximation), then the focal-plane distribution becomes simply the Fourier transform of the entrance-pupil distribution. For an aberration-free lens having a circular aperture of radius $r_a = fNA$, and for a uniform incident beam, the focal-plane distribution is thus proportional to $J_1(2\pi NAr)/r$, where $J_1(\cdot)$ is the first-order Bessel function of the first kind and $r = (x_2^2 + y_2^2)^{1/2}$. This is known as the Airy pattern, a plot of which appears in Figure 2.8.

Fourier-transforming property of a lens

An infinity-corrected lens produces in its focal plane the Fourier transform $A_0(\sigma_x, \sigma_y)$ of a complex-amplitude distribution $a_0(x_1, y_1)$ placed before the lens. This behavior is readily understood if one recognizes that the input distribution is a superposition of plane waves, each propagating in a different direction. The lens captures these plane waves and brings them to focus within its focal plane. The amplitude of each focused spot is thus proportional to the corresponding plane-wave amplitude. The finite aperture of the

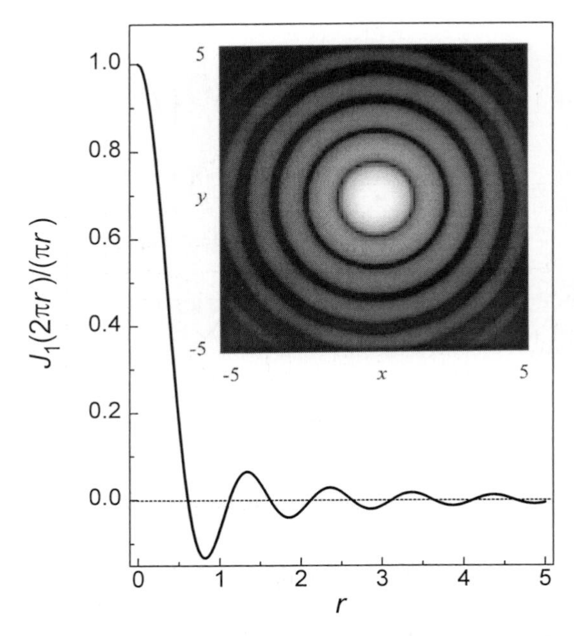

Figure 2.8 Plot of the Airy function $J_1(2\pi r)/\pi r$ versus the radial distance r from the focal point. The first zero of the Airy function is at $r \approx 0.61$. The inset shows a logarithmic plot of the intensity distribution at the focal plane of a $0.5NA$ diffraction-limited lens. This Airy pattern, being the result of a scalar calculation, shows circular symmetry. In practice, both unpolarized and circularly polarized incident beams produce circularly symmetric Airy patterns. However, for linearly polarized light the Airy pattern tends to be slightly elongated along the direction of the incident E-field.

lens spreads each focused spot into an Airy function, giving rise to a focal plane distribution that is the convolution between the object's Fourier transform and the lens's Airy pattern.

To study in some detail the properties of an aplanatic, infinite-conjugate lens, consider Figure 2.9. Here a plane wave propagating at angle θ relative to the Z-axis enters the lens at its first principal plane. At the entrance pupil (which is assumed to coincide with the 1st principal plane) the ray heights are the same as those at the exit pupil, which is a spherical cap of radius f centered at the focal point F. A ray entering at height x_1 has phase $2\pi x_1 \sigma_x$, which it retains as it emerges from the exit pupil. The ray then acquires an additional phase in propagating from the exit pupil to the focus at $x_2 = f\sigma_x$. The total phase at this focus (relative to that at F) is thus

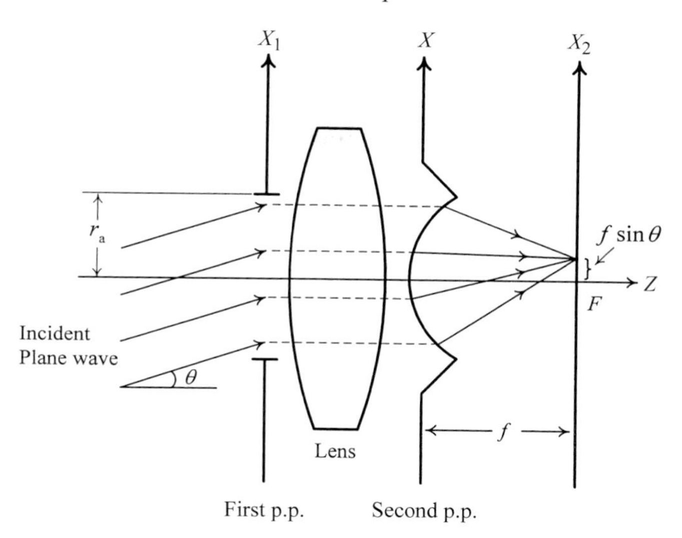

Figure 2.9 Fourier transform lens having focal length f and aperture radius $r_a = fNA$. The incident plane wave makes an angle θ with the Z-axis in the XZ-plane, that is, $(\sigma_x, \sigma_y) = (\sin\theta, 0)$. The beam emerging from the lens converges to the point $(x_2, y_2) = (f \sin\theta, 0)$ within the focal plane. The height of a ray entering the lens at the first principal plane is the same as that of the emergent ray measured on a spherical surface of radius f centered at the rear focal point F.

given by

$$\begin{aligned}\phi(x_1, \sigma_x) &= 2\pi\Big\{x_1\sigma_x + \big[(x_1 - f\sigma_x)^2 + (f^2 - x_1^2)\big]^{1/2} - f\Big\}\\ &= 2\pi f\Big\{(x_1/f)\sigma_x + \big[1 + \sigma_x^2 - 2(x_1/f)\sigma_x\big]^{1/2} - 1\Big\}. \qquad (2.12)\end{aligned}$$

For small values of both x_1/f and σ_x, the above expression may be approximated as $\phi(x_1, \sigma_x) \approx \pi f \sigma_x^2$, which is independent of x_1. The various rays of the plane wave, having thus acquired the common phase factor $\exp(i\pi f \sigma_x^2)$, converge to a common focus in the vicinity of the optical axis. Further away from the axis, of course, higher-order terms will cause aberration. Unless the lens is properly designed to correct these aberrations, the acceptable values of NA and σ_x will indeed be very small. For example, Figure 2.10 shows plots of $\phi(x_1, \sigma_x) - \pi f \sigma_x^2$ versus x_1/f for several values of σ_x, for a lens having $NA = 0.05$ and $f = 25\,000$. Note that to keep the maximum phase deviation at the edge of the pupil below 90° one must restrict the aperture radius to $r_a \approx 0.05f$ and the values of σ_x to the range within ± 0.055.

We conclude that, under appropriate conditions, a plane wave entering the lens at $\sigma_x = \sin\theta$ comes to diffraction-limited focus at $x_2 = f\sigma_x$, with a phase $\phi \approx \pi f \sigma_x^2 = \pi x_2^2/f$. Because of the finite aperture of the lens, the focused spot

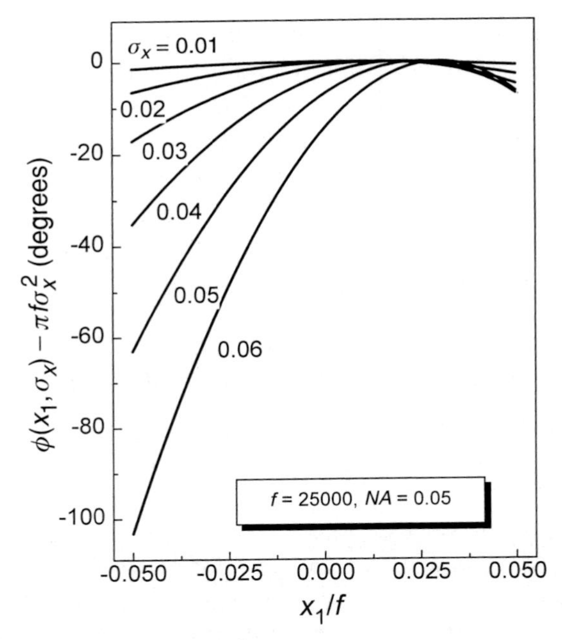

Figure 2.10 Plots of $\phi(x_1, \sigma_x) - \pi f \sigma_x^2$ versus x_1/f for several values of $\sigma_x = \sin\theta$ from 0.01 to 0.06 in the system of Figure 2.9. The function ϕ is given by Eq. (2.12), and the specific values of the lens parameters used in the calculations are $NA = 0.05$, $f = 25\,000$.

will be not a geometric point but an Airy pattern of diameter $\sim 1/NA$. Therefore, for an object $a_0(x_1, y_1)$ at the entrance pupil the focal-plane distribution is related to the Fourier transform $A_0(\sigma_x, \sigma_y)$ of the object as follows:

$$a(x_2, y_2) \approx \left\{ \exp\left[i\pi(x_2^2 + y_2^2)/f\right] A_0(x_2/f, y_2/f) \right\} * \text{Airy}(x_2, y_2). \tag{2.13}$$

Needless to say, the range of (x_2, y_2) in Eq. (2.13) is limited to the region for which the lens is properly designed to focus the incident plane waves into diffraction-limited spots. In the absence of aberrations, the angular resolution of such a lens is solely dependent on the lens-aperture radius r_a and is given by $\Delta\sigma_x = \Delta\sigma_y \approx 0.61/r_a$. (Like all other spatial dimensions in this chapter, r_a is assumed to be normalized by the wavelength λ of the light.)

Similar considerations apply when the object is placed a distance z_1 before the first principal plane. In this case each plane wave leaving the object must travel a different distance to reach the entrance pupil. By the time it reaches the entrance pupil, a plane wave traveling along the direction $(\sigma_x, \sigma_y, \sigma_z)$ will have acquired a phase $2\pi z_1 \sigma_z$, which may be approximated as $-\pi z_1(\sigma_x^2 + \sigma_y^2)$.

Under these circumstances, Eq. (2.13) remains valid provided the exponent of the first term on the right-hand side is multiplied by $(1 - z_1/f)$. In the special case where $z_1 = f$, the quadratic phase factor in Eq. (2.13) disappears altogether, leaving a simple Fourier-transform relation between the distributions in the front and rear focal planes. As an example, Figure 2.11 shows a phase/amplitude object placed in the front focal plane of a 0.05*NA* lens (see frames (a) and (b)), and the corresponding Fourier transform as observed in the rear focal plane (frames (c) and (d)).[5]

Abbe's theory of image-formation

Figure 2.12 is a diagram of the basic image-forming system. Both the entrance and exit pupils are assumed to be at the principal planes of the lens; in

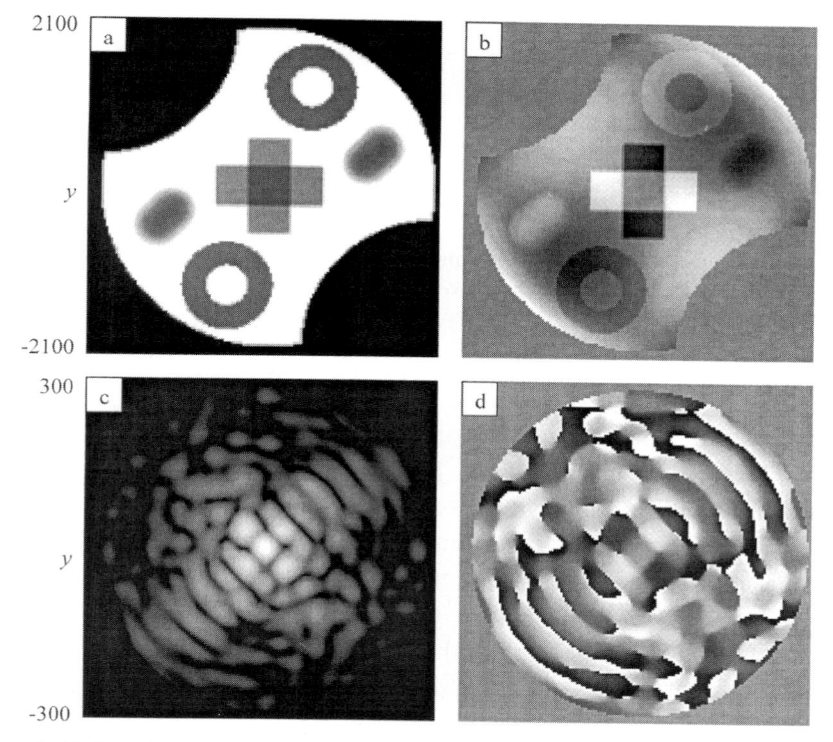

Figure 2.11 (a), (b) Intensity and phase distributions in the *XY*-plane for an object and (c), (d) for its Fourier transform. The object is in the front focal plane of a 0.05*NA* lens having $f = 10^5$, illuminated with a plane wave propagating along the *Z*-axis; the Fourier transform is observed in the rear focal plane. The intensity distribution in the Fourier-transform plane, (c), is displayed on a logarithmic scale to enhance its weak regions. The phase plots in (b) and (d) are encoded in gray-scale (black represents -180°, white represents $+180^\circ$).

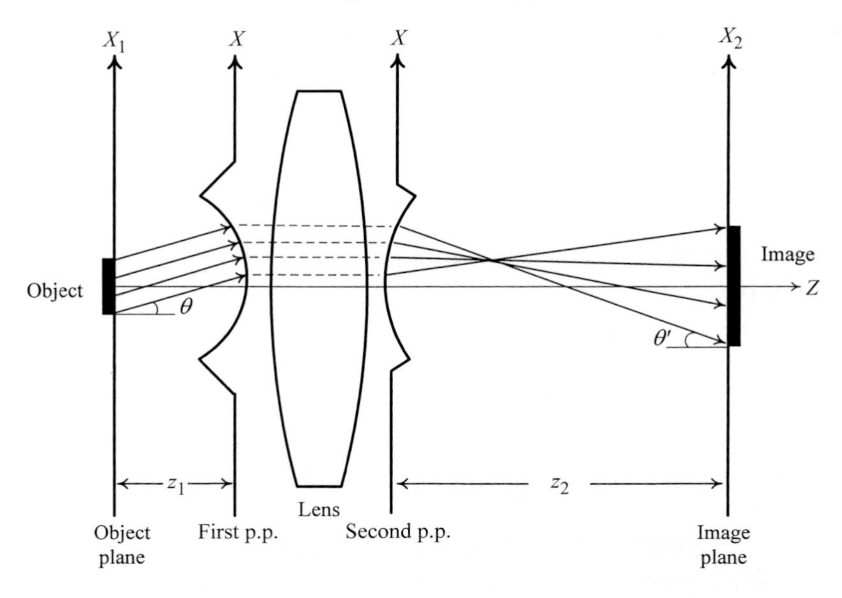

Figure 2.12 Diagram of a simple imaging system. The object and image distances from the respective principal planes are z_1 and z_2. The height of a ray entering the lens is measured on a spherical surface of radius z_1 centered at the axial object point. Similarly, the height of a ray exiting the system is measured on the spherical surface of radius z_2 centered on the axial image point. For any given ray, the entering and exiting heights are equal. Only one plane wave (leaving the object at an angle θ) is shown. The various rays of this plane wave converge to a focus in the image space, then continue to propagate to the image plane.

compliance with Abbe's sine condition, the pupils are spherical caps centered at the axial object and image points. The distance between the object and the first principal plane is z_1 and that between the second principal plane and the image is z_2. The lateral magnification of the system, therefore, is $M = z_2/z_1$.

A plane wave leaving the object at an angle θ relative to the Z-axis emerges from the exit pupil, each one of its rays having the same height and the same optical phase as at the entrance pupil. Confining attention to the two-dimensional XZ-plane, and denoting the direction cosine of a ray in the object space by $\sigma_{x1} = \sin\theta$, the ray height x at the entrance pupil is found from simple geometry to be

$$x = x_1(1 - \sigma_{x1}^2) + \sigma_{x1}\left[z_1^2 - x_1^2(1 - \sigma_{x1}^2)\right]^{1/2}. \tag{2.14}$$

A ray leaving the object at x_1 intersects the image at $x_2 = -Mx_1$. Obviously, the ray fan reaching the image plane in Figure 2.12 is not a plane wave.

However, it will be seen that this bundle of rays has a phase distribution that can be expressed as the sum of a linear term ϕ_1 and a nearly quadratic term ϕ_2. The linear term is identical with that of the plane wave leaving the object, namely,

$$\phi_1(x_2, \sigma_{x2}) = 2\pi x_2 \sigma_{x2} = 2\pi x_1 \sigma_{x1}. \tag{2.15}$$

Since x_2 is a version of x_1 magnified by a factor of M, σ_{x2} must be a version of σ_{x1} demagnified by a factor of $1/M$, so the above equality is exactly satisfied. Note in Figure 2.12 that although θ is the same for all the rays that leave the object within a given plane wave, the corresponding angle θ' in the image plane varies from ray to ray. Therefore σ_{x2}, which is defined here as $-\sigma_{x1}/M$, equals $\sin\theta'$ only for the ray that goes through the center of the image at $x_2 = 0$.

The quadratic phase ϕ_2 is acquired while covering the path from x_1 at the object plane to x_2 in the image plane. A ray leaving the object at x_1 enters the lens at a height x given by Eq. (2.14), emerges from the exit pupil at the same height and with the same optical phase as at the entrance pupil, and then proceeds to x_2 in the image plane. The phase acquired in going from x_1 to x_2 relative to that at the image center may thus be written

$$\phi_2(x_1, \sigma_{x1}) = 2\pi\left\{(x - x_1)/\sigma_{x1} + \left[(x - x_2)^2 + (z_2^2 - x^2)\right]^{1/2} - (z_1 + z_2)\right\}. \tag{2.16}$$

Noting that $x_2 = -Mx_1$, ϕ_2 may also be considered a function of x_2 and σ_{x1}. Equation (2.16) yields a nearly quadratic phase factor in x_2, which may be plotted for different values of $\sigma_{x1} = \sin\theta$. Figure 2.13 shows a set of such plots within a field of view $|x_2| < 250$ for a system in which $z_1 = 10^4$ and $z_2 = 10^5$. Different curves correspond to different values of θ. Note that if the slight differences between these curves are ignored (and the maximum difference in ϕ_2 is only about 10° in the present example), then the quadratic phase factor ϕ_2 is essentially independent of θ.

When ϕ_2 as a function of x_2 is expanded in a Taylor series, the lowest-order term is found to be

$$\phi_2 \approx (\pi/z_2)\left[1 + (z_1/z_2) - (z_1/z_2)\sigma_{x1}^2\right]x_2^2. \tag{2.17}$$

If $z_2 \gg 1$, this quadratic phase can be ignored, yielding a plane-wave output for a plane-wave input. However, when ϕ_2 is too large to be ignored its dependence on σ_{x1} may be insignificant. This happens when the magnification z_2/z_1 is either very large or very small. The case $z_2/z_1 \gg 1$ is obvious when one considers the coefficient of σ_{x1}^2 in Eq. (2.17). In the case of large

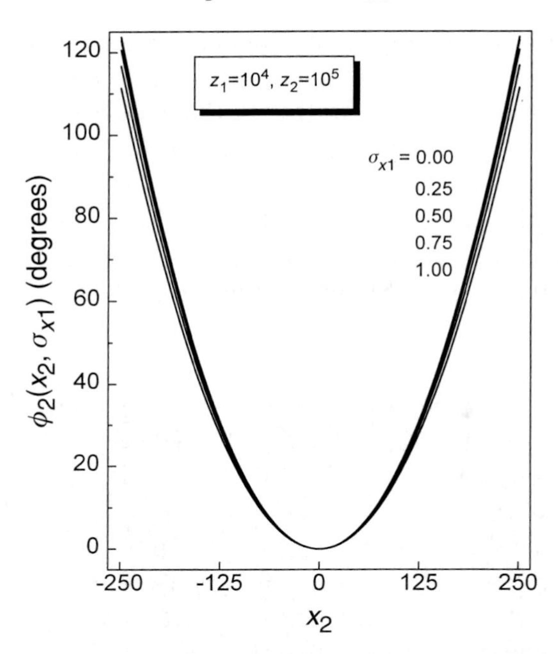

Figure 2.13 Plots of the function $\phi_2(x_2, \sigma_{x1})$ versus x_2 for several values of $\sigma_{x1} = \sin\theta$ equal to (top to bottom) 0.00, 0.25, 0.50, 0.75, 1.00. (See Figure 2.12 and Eq. (2.16); x_2 is related to x_1 through $x_2 = -Mx_1$.) The assumed system parameters are $z_1 = 10^4$, $z_2 = 10^5$. The field of view in the image plane is confined to the region $|x_2| < 250$.

demagnification, $z_2/z_1 \ll 1$, the range of σ_{x1} is limited to $|\sigma_{x1}| < z_2/z_1$, rendering ϕ_2 essentially independent of σ_{x1} once again.

The quadratic phase factor $\exp(\mathrm{i}\phi_2)$, being more or less independent of θ, can thus be factored out. This means that those plane waves that leave the object and manage to get through the lens to the image plane have the requisite uniform amplitude and linear phase expected of a plane wave. These plane waves, when superimposed upon each other, produce in the image plane a magnified (or demagnified) image of the object. Thus the differences between object and image are: (i) the image is multiplied by a nearly quadratic phase factor, $\exp(\mathrm{i}\phi_2)$; (ii) the plane waves having a large angle θ miss the lens and, therefore, do not contribute to the image. This truncation by a circular aperture in the Fourier domain is equivalent to convolution with an Airy function in the image plane. The amplitude distribution in the image plane is thus given by

$$a_{\text{image}}(x_2, y_2) = \exp(\mathrm{i}\phi_2)\left[a_{\text{object}}(-x_2/M, -y_2/M) * \mathrm{Airy}(x_2, y_2)\right]. \tag{2.18}$$

Figure 2.14 shows two examples of coherent imaging through a diffraction-limited lens. The object's intensity and phase are shown in Figures 2.14(a), (b). This object has several fine features which, being smaller than a wavelength, are below the resolution of any optical imaging system. A coherent and uniform beam propagating along the Z-axis illuminates the object. The entrance pupil of the imaging lens, located at $z_1 = 10^4$, is in the far field of the object. Figures 2.14(c), (d) show the intensity and phase patterns at the image plane of a $10\times$, $0.6NA$ lens. Similarly, Figures 2.14(e), (f) show the intensity

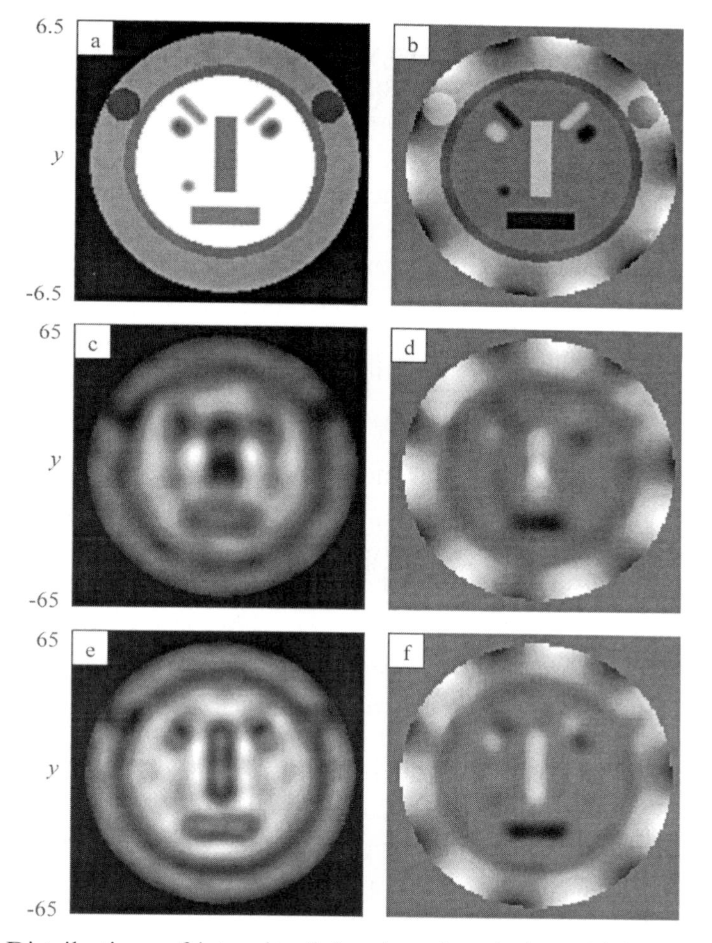

Figure 2.14 Distributions of intensity (left column) and phase (right column) at the object and image planes of a coherent imaging system. The phase plots are encoded in gray scale: black represents $-180°$, white represents $+180°$. (a), (b) Distributions in the plane of the object. (c), (d) Image obtained with a $10\times$, $0.6NA$ objective lens. (e), (f) Image obtained with a $10\times$, $0.95NA$ objective lens

and phase distributions in the image plane of a $10\times$, $0.95NA$ lens. The higher-NA lens, capturing more of the high-frequency Fourier components of the object, yields a superior image. Both lenses, however, fail to reproduce the very fine features of the object.

Appendix to Chapter 2: The stationary-phase approximation

Consider the two-dimensional integral

$$I = \iint f(x, y) \exp[\mathrm{i}\eta g(x, y)]\, \mathrm{d}x\, \mathrm{d}y, \tag{A2.1}$$

where, in general, $f(x, y)$ is a complex function, $g(x, y)$ is a real function, η is a large real number, and the domain of integration is a subset of the XY-plane. In the neighborhood of an arbitrary point (x_0, y_0), within the domain of integration, small variations in $g(x, y)$ will be amplified by η; this will result in rapid oscillations of the phase factor $\exp[\mathrm{i}\eta g(x, y)]$. Assuming that $f(x, y)$ in the neighborhood of (x_0, y_0) is a slowly varying function, the oscillations result in a negligible contribution from this neighborhood to the integral. The main contributions to the integral then come from the regions in which $g(x, y)$ is nearly constant. These regions are in the vicinity of stationary points (x_0, y_0), which are defined by the following relation:

$$\partial g(x, y)/\partial x = \partial g(x, y)/\partial y = 0. \tag{A2.2}$$

Around each stationary point one may expand $g(x, y)$ in a Taylor series up to the second-order term to obtain

$$\begin{aligned} g(x, y) \approx g(x_0, y_0) &+ \tfrac{1}{2} g_{xx}(x_0, y_0)(x - x_0)^2 \\ &+ g_{xy}(x_0, y_0)(x - x_0)(y - y_0) + \tfrac{1}{2} g_{yy}(x_0, y_0)(y - y_0)^2. \end{aligned} \tag{A2.3}$$

Replacing the expression for $g(x, y)$ in Eq. (A2.1) with that in Eq. (A2.3), and taking $f(x, y)$ outside the integral, yields

$$\begin{aligned} I \approx \sum f(x_0, y_0) &\exp[\mathrm{i}\eta g(x_0, y_0)] \\ &\times \iint_{-\infty}^{\infty} \exp\Big\{\mathrm{i}(\eta/2)\big[g_{xx}(x - x_0)^2 + 2g_{xy}(x - x_0)(y - y_0) \\ &+ g_{yy}(y - y_0)^2\big]\Big\}\, \mathrm{d}x\, \mathrm{d}y, \end{aligned} \tag{A2.4}$$

where the summation is over all stationary points (x_0, y_0). Notice that the domain of integration is now extended to the entire plane, since the contribution to the integral from regions outside the immediate neighborhood of the stationary points is, in any event, negligible. The double integral in Eq. (A2.4) can be readily carried out, yielding

$$I \approx (2\pi i/\eta) \sum \nu |g_{xx}g_{yy} - g_{xy}^2|^{-1/2} \exp[i\eta g(x_0, y_0)] f(x_0, y_0), \qquad \text{(A2.5)}$$

where the summation is again over all stationary points (x_0, y_0) and the coefficient ν is given by

$$\nu = \begin{cases} -\mathrm{i} & \text{if} \quad g_{xx}g_{yy} < g_{xy}^2 \\ \pm 1 & \text{if} \quad g_{xx}g_{yy} > g_{xy}^2 \text{ and } g_{xx} \gtrless 0. \end{cases}$$

Equation (A2.5) is the final result of this appendix. If the numerical value of $g_{xx}g_{yy} - g_{xy}^2$ happens to be exactly zero at a particular stationary point or if a stationary point occurs on the boundary of the domain of integration in Eq. (A2.1) then Eq. (A2.5) no longer applies. In our analysis of diffraction problems, however, these special cases will not be encountered.

References for chapter 2

1 M. Born and E. Wolf, *Principles of Optics*, sixth edition, Pergamon Press, Oxford, 1980.
2 J. W. Goodman, *Introduction to Fourier Optics*, second edition, McGraw-Hill, New York, 1996.
3 L. Mandel and E. Wolf, *Optical Coherence and Quantum Optics*, Cambridge University Press, UK, 1995.
4 M. V. Klein, *Optics*, Wiley, New York, 1970.
5 J. Durnin, J. J. Miceli, and J. H. Eberly, Diffraction-free beams, *Phys. Rev. Lett.* **58**, 1499–1501 (1987).
6 F. A. Jenkins and H. E. White, *Fundamentals of Optics*, fourth edition, McGraw-Hill, New York, 1976.

3

Effect of polarization on diffraction in systems of high numerical aperture

The classical theory of diffraction, according to which the distribution of light at the focal plane of a lens is the Fourier transform of the distribution at its entrance pupil, is applicable to lenses of moderate numerical aperture (*NA*). The incident beam, of course, must be monochromatic and coherent, but its polarization state is irrelevant since the classical theory is a scalar theory (see chapter 2, "Fourier optics"). If the incident beam happens to be a plane wave and the lens is free from aberrations then the focused spot will have the well-known Airy pattern. When the incident beam is Gaussian the focused spot will also be Gaussian, since this particular profile is preserved under Fourier transformation. In general, arbitrary distributions of the incident beam, with or without aberrations and defocus, can be transformed numerically, using the fast Fourier transform (FFT) algorithm, to yield the distribution in the vicinity of the focus.

There are two basic reasons for the applicability of the classical scalar theory to systems of moderate *NA*. The first is that bending of the rays by the focusing element(s) is fairly small, causing the electromagnetic field vectors (***E*** and ***B***) before and after the lens to have more or less the same orientations. A scalar amplitude assigned to each point on the emergent wavefront from a system having low to moderate values of *NA* is sufficient to describe its electromagnetic state, whereas in the high-*NA* regime one can no longer ignore the vectorial nature of light. The second reason for the success of the classical scalar theory (within its proper limits) is that a certain integral – that which represents the decomposition of a convergent wavefront into its plane-wave constituents – submits to evaluation by the method of stationary-phase approximation. The remaining integral – that which represents the superposition of plane waves arriving at the focal plane – is then calculated with the aid of Fourier transformation. When the stationary-phase technique fails, so does the classical scalar theory, as is evidenced, for instance, in systems of very low numerical aperture: The well-known focal-shift phenomenon is but one manifestation of the failure of the stationary-phase approximation in very-low-NA systems.[1]

In the stationary-phase approximation the plane-wave spectrum of the convergent beam at the exit pupil coincides with the light amplitude distribution at that pupil, thus enabling each geometric-optical ray to represent one plane wave of the spectrum, namely, that which propagates in the direction of the ray.[2] This correspondence between rays and plane waves, which is an important feature of many diffraction problems, is therefore understood to be a direct consequence of the stationary-phase approximation. Now, let θ be the angle between a converging ray in the image space and the optical axis at the focal point. Since the projection of the wave vector $\boldsymbol{k}$ onto the exit pupil has length $k \sin \theta$, whereas the intersection of the ray with the pupil occurs at a radius $r = f \tan \theta$, then in order to convert from light amplitude distribution to the corresponding plane-wave spectrum one must compress the distribution function at the exit pupil. Aside from a trivial scaling of the aperture's radius by the focal length f, the radial compression must assign to $r = \sin \theta$ the value of the function at $r = \tan \theta$; this must be followed by proper normalization to preserve the integrated intensity. The compressed distribution is therefore confined to a disk of radius $NA = \sin \theta_{\max}$, where $\theta_{\max}$ is the angle subtended by the rim of the exit pupil at the focal point. This scaling, compression, and normalization procedure is not merely justified on heuristic grounds but, as discussed in the preceding chapter, is a rigorous consequence of the stationary-phase approximation itself.

For lenses of low to moderate numerical aperture (say, $NA < 0.2$) the difference between $\sin \theta$ and $\tan \theta$ is negligible, and the effects of compression can be ignored. At the exit pupil, the plane-wave spectrum of these lenses is usually the same as the incident distribution at the entrance pupil, modified only by the presence of aberrations. For lenses of high numerical aperture, however, it is necessary to obtain the exit-pupil distribution (from the knowledge of lens characteristics and the entrance-pupil distribution) before proceeding to the compression operation. Noteworthy in this respect is the aplanatic lens, which, by virtue of satisfying Abbe's sine condition, guarantees that the compressed exit-pupil distribution is identical with the entrance-pupil distribution.

Bending of polarization vector

To account for polarization effects at high numerical aperture, one usually ignores transmission losses at the various surfaces of a lens, assuming that a ray goes through the system unattenuated but with its polarization vector bent in accordance with the known laws of refraction.[2–5] (The assumption of losslessness is not necessary here, but it simplifies the problem by enabling the polarization state of individual rays at the exit pupil to be determined solely on the basis of their coordinates, without requiring detailed knowledge of the lens structure.) For a linearly polarized incident beam, Figure 3.1 shows the bending of the E-vector at two azimuthal positions. The ray at the top of the lens contributes both an X- and a Z-component to the distribution in the image space, whereas the ray in the YZ-plane contributes only an X-component. By the same token, rays intermediate between those shown here will contribute to the polarization along all three axes.

We present a simple treatment of polarization-related phenomena within the framework of the classical theory of diffraction. This will not be a rigorous treatment based on Maxwell's equations; rather, it will be rooted in reasonable physical arguments based on the bending of rays (or plane waves) by prisms. Our approach to vector diffraction is in keeping with the spirit of diffraction theory; it is not exact as far as Maxwell's equations are concerned but incorporates intuitive ideas about the propagation of electromagnetic waves.

With reference to Figure 3.2, consider a plane wave propagating along the unit vector $\boldsymbol{\sigma}_0 = (0, 0, 1)$, i.e., along the Z-axis, having linear polarization in the X-direction. Let a prism be placed in the path of this beam, with orientation such that the emerging beam would propagate in a direction specified by the unit vector $\boldsymbol{\sigma}_1 = (\sigma_x, \sigma_y, \sigma_z)$. Now, the incident polarization vector $\boldsymbol{E}_0 = (1, 0, 0)$ may be decomposed into two components: one, the so-called

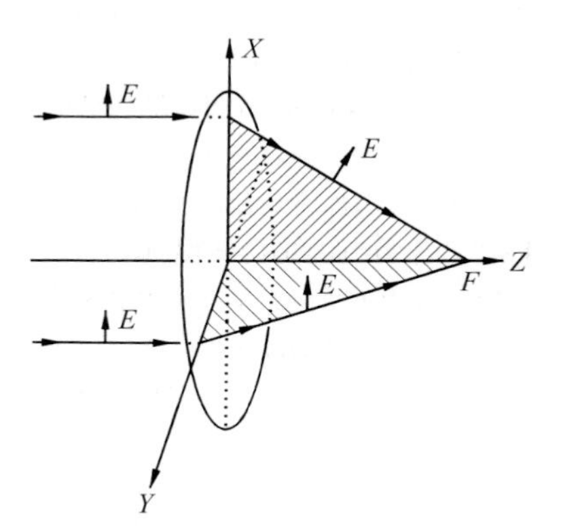

Figure 3.1 Focusing of linearly polarized light by a high-NA lens, shown in perspective, causes bending of the polarization vectors. The amount and direction of bending depend on the coordinates of the ray.

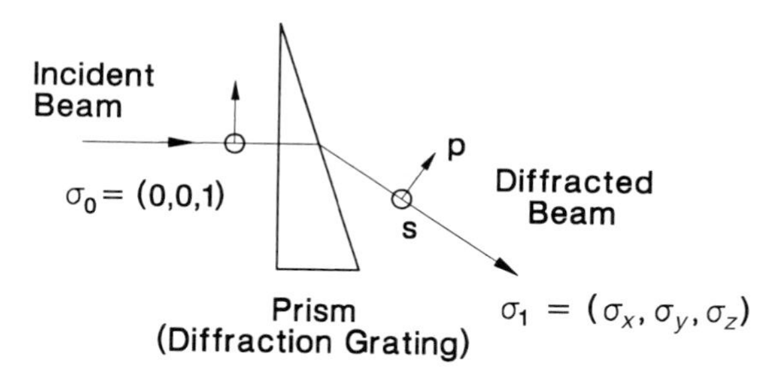

Figure 3.2 Lossless refraction of a polarized plane wave by a prism. The original direction of propagation is $\boldsymbol{\sigma}_0 = (0, 0, 1)$ and the corresponding polarization vector is $\boldsymbol{E}_0$. After refraction, the beam assumes a new direction $\boldsymbol{\sigma}_1 = (\sigma_x, \sigma_y, \sigma_z)$, and its new polarization state becomes $\boldsymbol{E}_1$. The same geometry would apply for diffraction of the beam by a grating.

p-polarization, is in the plane of $\boldsymbol{\sigma}_0$ and $\boldsymbol{\sigma}_1$; the other, known as the s-polarization, is perpendicular to this plane. As the latter component (perpendicular to the $\sigma_0\sigma_1$-plane) emerges from the prism, it will have suffered no deviation in direction. The p-component, however, will have been reoriented such that it remains perpendicular to the emergent direction. If it is further assumed that no losses, due to surface reflections or otherwise, occur in this refraction process, one can use simple geometry to determine the emerging polarization direction. A similar calculation can be performed for an incident plane wave linearly polarized along the Y-axis. Details of these calculations are left to the reader, but the final results are listed in Table 3.1. Notice that the reorientation of the polarization vector described in Table 3.1, while a consequence of the refraction of the direction of propagation, is independent of the particular mechanism responsible for refraction. Given an initial direction $\boldsymbol{\sigma}_0$ and a direction for the emerging beam $\boldsymbol{\sigma}_1$, one can use Table 3.1 to identify the emergent components of polarization for an arbitrary state of incident polarization.

In the stationary-phase approximation each ray is associated with a single plane wave, the three polarization components of which may be treated independently of each other. Therefore, for each of the components E_x, E_y, E_z of the emergent beam, a single superposition integral (i.e., Fourier transform) yields the sought-after distribution in the focal plane.

Example

The technique described in the preceding section is quite general and can be applied to arbitrary incident distributions having arbitrary polarization

Table 3.1. *Polarization $\boldsymbol{E}_1$ of a refracted beam when the incident polarization $\boldsymbol{E}_0$ is along the X- or Y- axes. The refraction (from $\boldsymbol{\sigma}_0$ to $\boldsymbol{\sigma}_1$) is lossless*

Incident polarization with $\boldsymbol{\sigma}_0 = (0,0,1)$	Emergent polarization with $\boldsymbol{\sigma}_1 = (\sigma_x, \sigma_y, \sigma_z)$
$\boldsymbol{E}_0 = (1, 0, 0)$	$\boldsymbol{E}_1 = \left(1 - [\sigma_x^2/(1+\sigma_z)], -\sigma_x\sigma_y/(1+\sigma_z), -\sigma_x\right)$
$\boldsymbol{E}_0 = (0, 1, 0)$	$\boldsymbol{E}_1 = \left(-\sigma_x\sigma_y/(1+\sigma_z), 1 - [\sigma_y^2/(1+\sigma_z)], -\sigma_y\right)$

states, while taking into account various lens aberrations (including substantial amounts of defocus). Computed results for an aberration-free, aplanatic lens having $NA = \sin 75^\circ = 0.966$ and $f = 3000\lambda$ are shown in Figure 3.3. The assumed geometry in these calculations is that depicted in Figure 3.1, where the incident beam is a uniform plane wave with linear polarization along the X-axis. Frames (a)–(c) in Figure 3.3 are intensity plots for the X-, Y-, and Z- components of polarization in the focal plane; their peak

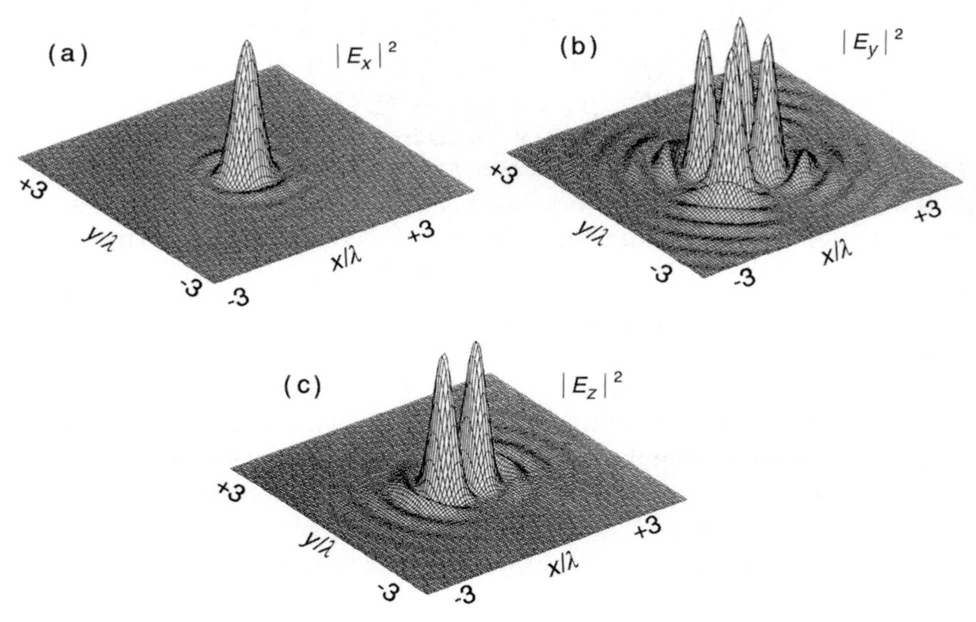

Figure 3.3 Intensity profiles of the three components of polarization at the focal plane of an aplanatic lens ($NA = 0.966$, $f = 3000\lambda$), illuminated with a linearly polarized plane wave. For best viewing, the vertical scale is chosen differently in the three cases: the peak intensities in (a), (b), (c), corresponding to the X-, Y-, Z-components of polarization, are in the ratios 1.00 : 0.0081 : 0.192.

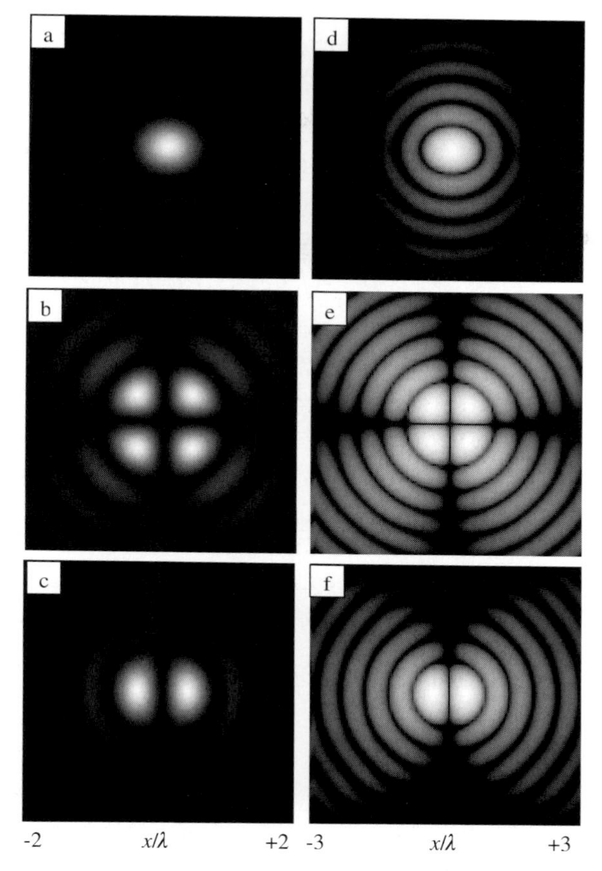

Figure 3.4 Gray-scale plots of intensity distribution at the focal plane of an aplanatic lens ($NA = 0.966$, $f = 3000\lambda$), illuminated with a linearly polarized plane wave. Frames (a)–(c) show the intensity plots, while frames (d)–(f) display the logarithm of intensity. In each column the top frame represents the X-component of polarization, the middle frame corresponds to the Y-component, and the bottom frame to the Z-component.

intensities are in the ratio 1.00 : 0.0081 : 0.192. The corresponding gray-scale plots appear in Figure 3.4; frames (a)–(c) show the intensity distributions and frames (d)–(f) display their logarithmic counterparts. The observed four-fold symmetry of the Y-component and the two-fold symmetry of the Z-component are consistent with one's expectations based on ray-bending arguments.

The contour plot in Figure 3.5 of the total E-field energy density, $|E_x|^2 + |E_y|^2 + |E_z|^2$, shows an elliptical profile, the ellipse having its major axis in the direction of the incident polarization. (Richards and Wolf[5] obtained the same result using a somewhat different formulation of the dif-

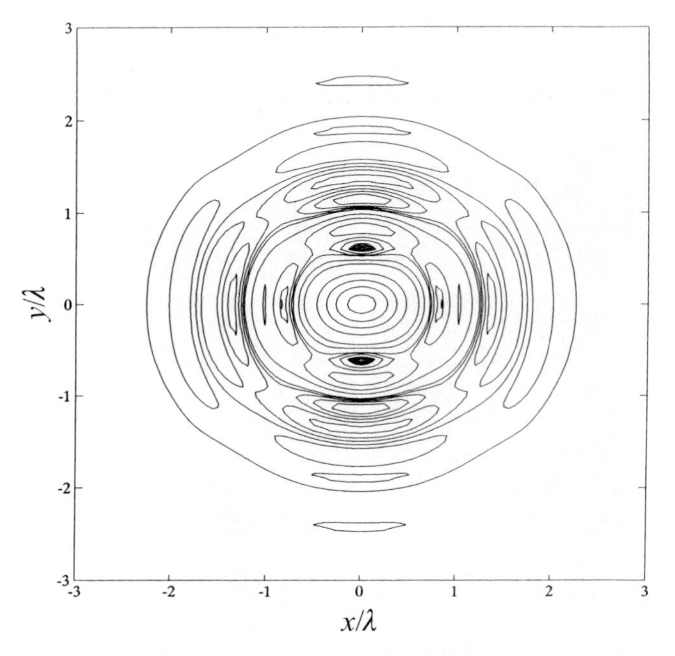

Figure 3.5 Contour plot representing the sum of the three intensity profiles shown in Figure 3.3, i.e., the total E-field energy density distribution in the focal plane of the aplanatic lens.

fraction problem.) This result indicates a slight improvement in the resolution of a microscope or telescope that uses linearly polarized light, as long as the feature that needs to be resolved is oriented along the minor axis of the ellipse, namely, in the direction perpendicular to that of the incident polarization.

The computations reported here required no more than three seconds on a modern pentium-based personal computer using a 512×512 square mesh.

References for chapter 3

1 V. N. Mahajan, Axial irradiance and optimum focusing of laser beams, *Appl. Opt.* **22**, 3042–3053 (1983).
2 J. J. Stamnes, *Waves in Focal Regions*, Adam Hilger, Bristol, 1986.
3 M. Mansuripur, Certain computational aspects of vector diffraction problems, *J. Opt. Soc. Am. A* **6**, 786–805 (1989).
4 H. H. Hopkins, The Airy disk formula for systems of high relative aperture, *Proc. Phys. Soc. London* **55**, 116–128 (1943).
5 B. Richards and E. Wolf, Electromagnetic diffraction in optical systems: structure of the image field in an aplanatic system, *Proc. Roy. Soc. Ser. A* **253**, 358–379 (1959).

4

Gaussian beam optics

A Gaussian beam is perhaps the simplest possible waveform that shows many of the effects of diffraction. Using Gaussian beams one can study diffraction in the near field and the far field, examine beam divergence upon propagation, investigate diffraction-limited focusing through a lens, observe the Gouy phase shift, and analyze many other interesting properties of electromagnetic waves.

Although Gaussian beams have been thoroughly analyzed in the literature,[1,2] it is worthwhile to examine them in the Fourier domain from a less well-known perspective. The need for the paraxial approximation (inherent in all treatments of Gaussian beams) becomes particularly clear when employing the Fourier method of analysis. There is also the issue of separability of the x- and y- dependences of the Gaussian beam profile (assuming propagation along the Z-axis), which is often assumed but not properly explained in the literature. It turns out that separability is neither necessary nor desirable and that the two-dimensional analysis of a non-separable beam is quite straightforward. It must be emphasized that separability is not always achievable by rotating the coordinate axes. When the real and imaginary parts of the Gaussian exponent require different rotations to become separable, the x- and y- dependences remain entangled, thus necessitating a two-dimensional analysis.

Cross-sectional amplitude profile

For a generalized Gaussian beam propagating along the Z-axis, the complex amplitude distribution in the cross-sectional XY-plane is given by

$$\hat{a}(x, y, z = 0) = \hat{a}_0 \exp\left[-\pi(ax^2 + 2bxy + cy^2\right]. \qquad (4.1a)$$

Here the complex constant $\hat{a}_0$ is the amplitude at the origin of the coordinate system and the coefficients $a = (a_1 + ia_2)$, $b = (b_1 + ib_2)$, $c = (c_1 + ic_2)$ are fixed complex numbers. The only constraints on these parameters are $a_1 \geq 0, c_1 \geq 0$, and $a_1 c_1 \geq b_1^2$, lest the amplitude diverges to infinity. The power content of the beam (i.e., the integrated intensity over the XY-plane) is readily found to be

$$P = \tfrac{1}{2}|\hat{a}_0|^2 / \sqrt{a_1 c_1 - b_1^2}. \tag{4.1b}$$

The real parts of the a, b, c parameters determine the profile of the beam's magnitude in the XY-plane at $z = 0$, while their imaginary parts determine the beam's phase profile. The contours of constant magnitude are ellipses oriented at θ_1 relative to X, where $\tan 2\theta_1 = 2b_1/(a_1 - c_1)$; the major and minor diameters of these ellipses are proportional to $\left[(a_1 + c_1) \pm \sqrt{(a_1 - c_1)^2 + 4b_1^2}\right]^{-1/2}$. The phase contours are ellipses or hyperbolas whose axes are oriented at θ_2 relative to X, where $\tan 2\theta_2 = 2b_2/(a_2 - c_2)$. In general $\theta_1 \neq \theta_2$ and therefore coordinate rotations cannot separate the x- and y- dependences of the Gaussian beam profile. When $a_2 c_2 > b_2^2$ the contours of constant phase are ellipses; otherwise, they are hyperbolas. Figure 4.1 shows two examples of amplitude and phase distributions for Gaussian beams having different sets of the a, b, c parameters.

Propagation in free space

The Fourier transform of the Gaussian profile in Eq. (4.1a) is given by

$$\begin{aligned} \hat{A}(\sigma_x, \sigma_y) &= \mathscr{F}\{\hat{a}(x, y, z = 0)\} \\ &= \left(\hat{a}_0/\sqrt{ac - b^2}\right) \exp\left[-\pi(\alpha\sigma_x^2 + 2\beta\sigma_x\sigma_y + \gamma\sigma_y^2)\right]. \end{aligned} \tag{4.2a}$$

Here $\alpha = c/(ac - b^2)$, $\beta = -b/(ac - b^2)$, and $\gamma = a/(ac - b^2)$. In matrix notation,

$$\begin{pmatrix} \alpha & \beta \\ \beta & \gamma \end{pmatrix} = \begin{pmatrix} a & b \\ b & c \end{pmatrix}^{-1}. \tag{4.2b}$$

When a beam travels a distance z_0 in free space, its Fourier transform is multiplied by the transfer function of propagation, $\exp\left(i2\pi z_0\sqrt{1 - \sigma_x^2 - \sigma_y^2}\right)$ (see chapter 2, "Fourier optics"). This is true irrespective of whether z_0 is positive or negative; in other words, both forward and backward propaga-

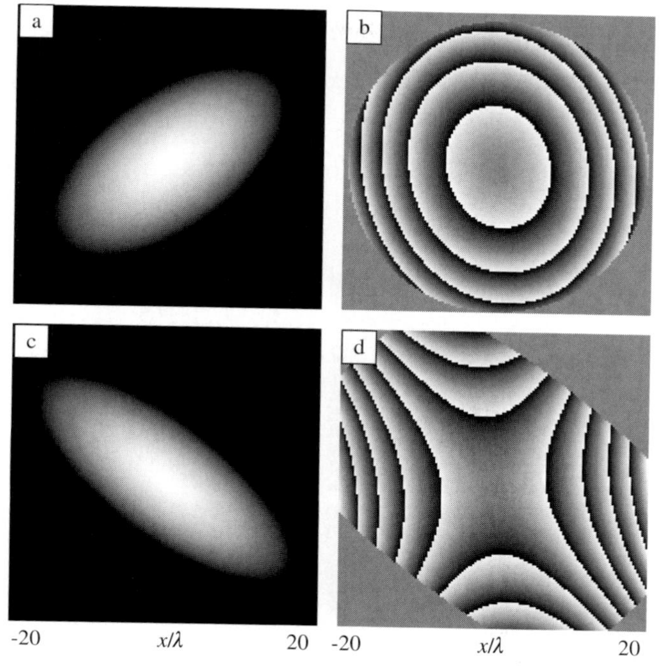

Figure 4.1 Distributions of intensity (left) and phase (right) in the cross-sections of two Gaussian beams having different a, b, c parameters. The phase plots are encoded in gray-scale, black representing -180° and white representing $+180^\circ$. (a), (b) $a = 0.009 - 0.023\mathrm{i}$, $b = -0.006 - 0.002\mathrm{i}$, $c = 0.012 - 0.016\mathrm{i}$, (c), (d) $a = 0.011 - 0.023\mathrm{i}$, $b = -0.01 - 0.003\mathrm{i}$, $c = 0.016 + 0.012\mathrm{i}$.

tion can be treated by the same formalism. The reason that the wavelength λ of the light does not appear in these equations is that all spatial coordinates are assumed to be normalized by λ, that is, x, y, z_0 are dimensionless quantities.

Invoking the standard paraxial approximation, the above transfer function is replaced by $\exp(\mathrm{i}2\pi z_0)\exp[-\mathrm{i}\pi z_0(\sigma_x^2 + \sigma_y^2)]$. Multiplying this transfer function into $\hat{A}(\sigma_x, \sigma_y)$ of Eq. (4.2a) converts $\hat{a}_0$ to $\hat{a}_0 \exp(\mathrm{i}2\pi z_0)$, α to $\alpha + \mathrm{i}z_0$, and γ to $\gamma + \mathrm{i}z_0$, while keeping β unchanged. The beam's Fourier transform thus retains its Gaussian form and, consequently, the profile of the beam at $z = z_0$ remains Gaussian, albeit with different a, b, c parameters and with a different value for $\hat{a}_0$. It is readily verified that the new parameters of the beam at $z = z_0$ are given by

$$\begin{pmatrix} a' & b' \\ b' & c' \end{pmatrix} = \begin{pmatrix} \alpha + \mathrm{i}z_0 & \beta \\ \beta & \gamma + \mathrm{i}z_0 \end{pmatrix}^{-1}, \tag{4.3a}$$

$$\hat{a}'_0 = \hat{a}_0 \exp(i2\pi z_0)\sqrt{(a'c' - b'^2)/(ac - b^2)}$$
$$= \hat{a}_0 \exp(i2\pi z_0)/\sqrt{1 - (ac - b^2)z_0^2 + i(a + c)z_0}. \tag{4.3b}$$

Thus the beam remains Gaussian as it propagates along Z, but its magnitude and phase profiles change continuously. Figure 4.2 shows computed cross-

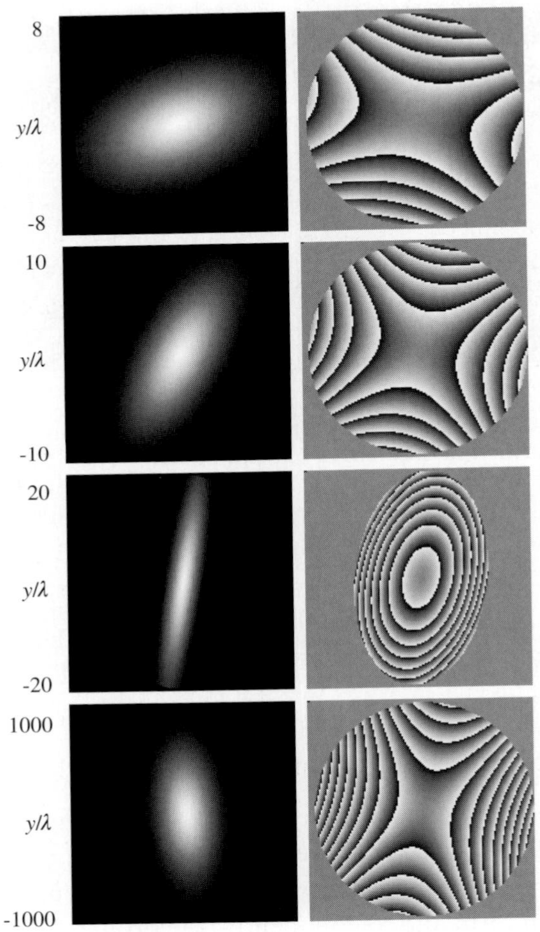

Figure 4.2 Distributions of intensity (left) and phase (right) in the cross-sectional planes of a Gaussian beam propagating along the Z-axis. The parameters of the beam at $z = 0$ are $a = 0.01 + 0.05i$, $b = -0.005 - 0.04i$, $c = 0.02 - 0.12i$. The phase plots are encoded in gray-scale, black representing -180° and white representing $+180^\circ$. From top to bottom, the propagation distances along Z are 0, 5λ, 25λ, and 1000λ. In the bottom right-hand frame the far-field curvature phase factor (corresponding to $a_2 = c_2 = 0.001$) has been subtracted.

sectional profiles for a beam at several locations along the Z-axis. Note how the elliptical cross-section of the intensity profile rotates with increasing z_0 and also how the phase contours change from hyperbolas to ellipses and vice versa.

The beam waist

The waist is the cross-section of the beam at which the phase is uniform, i.e., it is independent of x and y. In general, a Gaussian beam does not have to have a waist but if a waist exists then the a, b, c parameters in that cross-section will be real. A question arises as to when an arbitrary Gaussian beam (for which the a, b, c parameters at a given cross-section are complex) can be said to have a waist. In other words, does a value of z_0 (positive or negative) exist at which a', b', c' are real? According to Eq. (4.3a), this requirement is met if β is real and the imaginary parts of α and γ are identical, so that iz_0 will end up canceling their imaginary parts. This is equivalent to requiring both β and $\alpha - \gamma$ to be real-valued.

Considering the relationship between α, β, γ and a, b, c in Eq. (4.2b), it is not difficult to show that the necessary and sufficient condition for an arbitrary Gaussian beam to have a waist is that, in the complex plane, the three vectors b, a – c, and ac – b^2 must be parallel (or antiparallel) to each other. In other words, these three complex numbers must lie along a straight line that goes through the origin of the complex plane. This requirement, of course, is in addition to the other Gaussian beam requirements, namely, $a_1 \geq 0, c_1 \geq 0, a_1c_1 \geq b_1^2$. When the a, b, c parameters satisfy all the above constraints, the beam will have a waist at a specific location along the Z-axis. The waist is unique, because there is only one value of z_0 that can cancel the imaginary parts of both α and γ in Eq. (4.3a).

When a waist exists, there is symmetry between the locations before and after the waist. Let the waist be at $z = 0$. Then the a, b, c parameters at this location will be real, which means that the corresponding α, β, γ are real as well. Now, any value of z_0 will make α and γ complex, while $-z_0$ will yield the conjugates of the same α and γ. Therefore, the a, b, c parameters on opposite sides of the waist will be complex conjugates of each other. This means that the intensity profiles on opposite sides are identical, while the phase profiles differ by a minus sign. The beam is always convergent before, and divergent after, the waist.

The Gouy phase shift

Aside from the usual linear phase factor $\exp(i2\pi z_0)$, Eq. 4.3(b) contains an additional phase whose value depends non-linearly on z_0. This is the phase associated with the square-root factor on the right-hand side of the equation. Consider the special case when a, b, c are all real-valued, i.e., when the beam waist is at $z = 0$. As z_0 increases from zero and acquires positive values, the real part under the square root, $1 - (ac - b^2)z_0^2$, decreases while the imaginary part, $(a + c)z_0$, increases. Thus, the phase of $\hat{a}_0'$ associated with the square root, namely,

$$\psi = -\tfrac{1}{2}\tan^{-1}\left\{(a+c)z_0/[1-(ac-b^2)z_0^2]\right\}, \tag{4.4}$$

approaches $-90°$ for sufficiently large z_0. Similarly, when z_0 goes from 0 to negative values, ψ moves toward $+90°$. It is thus seen that, in crossing the waist, the beam undergoes a 180° phase shift. This phase shift, which is particularly rapid near the focus of a lens, was first observed experimentally by the French physicist L. Georges Gouy in 1890.[2–4]

To demonstrate an observable effect of the Gouy phase, consider the experiment depicted in Figure 4.3. Here an aberration-free lens is split into two identical halves, and the upper half-lens is translated forward by $\Delta z = 300\lambda$. A collimated uniform beam of light is directed at the split lens, and the distribution of intensity in the region between the two foci, F_1 and

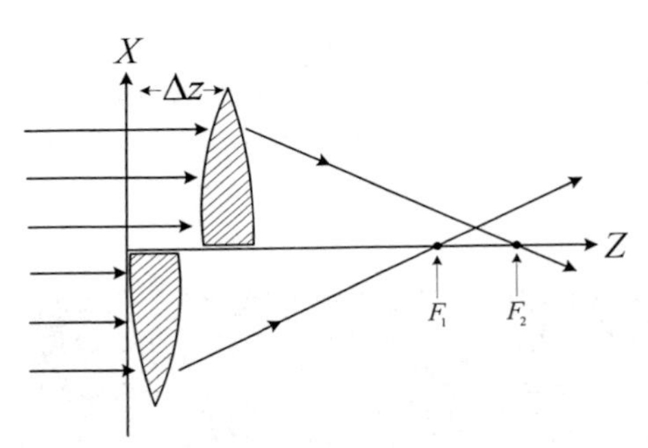

Figure 4.3 A split lens brings a collimated uniform beam of wavelength λ to two different foci, F_1 and F_2, along the Z-axis. The region of interest is between the two focal planes and above the Z-axis. At first glance the rays going through each half-lens are expected to have the same phase when they arrive in the vicinity of the Z-axis. However, because of the Gouy effect, the beam going through the upper lens and arriving at an observation point before focus will be phase-shifted by about 180° relative to the beam going through the lower lens and arriving at the same observation point after focus. In our numerical example, the lens (before splitting) has $NA = 0.1$ and focal length $= 30\,000\lambda$, and the separation between the half-lenses is $\Delta z = 300\lambda$.

F_2, is monitored. Figure 4.4 shows computed intensity patterns in a vertical plane half-way between F_1 and F_2 when (a) the upper half-lens is blocked, (b) the lower half-lens is blocked, and (c) the light is allowed to go through both half-lenses. For locations near the Z-axis, where the optical path lengths are nearly identical, the light amplitudes contributed by the two half-lenses are expected to be in phase, resulting in constructive interference. However, as Figure 4.4(c) clearly demonstrates, the vicinity of the optical axis is dark. This destructive interference is caused by the nearly -180° Gouy phase shift between the "before-focus" and "after-focus" beams arriving from the two half-lenses. Figure 4.5 shows several cross-sectional plots of intensity distribution in the region between F_1 and F_2, starting at F_1 and moving in steps of 37.5λ to F_2. The intensity at and near the Z-axis is seen to diminish as the mid-plane between the foci is approached from either side.

The Rayleigh range

For the generalized Gaussian beam of Eq. (4.1a), there is only one way to define the Rayleigh range,[2] and that is in terms of the Gouy phase shift. To admit a Rayleigh range the beam must have a waist, which we assume to be at $z = 0$, so the a, b, c parameters at this location are real-valued. With reference to Eq. (4.4), the Rayleigh range is the distance z_0 at which the Gouy phase ψ is $\pm 45^\circ$, i.e., $z_0 = \pm 1/\sqrt{ac - b^2}$. In the special case when $a = c$ and $b = 0$ (i.e., when the beam is circularly symmetric) the Rayleigh range z_0 is $\pm 1/a$. In this case the beam diameter at the Rayleigh range is a factor of $\sqrt{2}$ larger than that at the waist; also the beam curvature can be shown to attain its maximum value at the Rayleigh range.

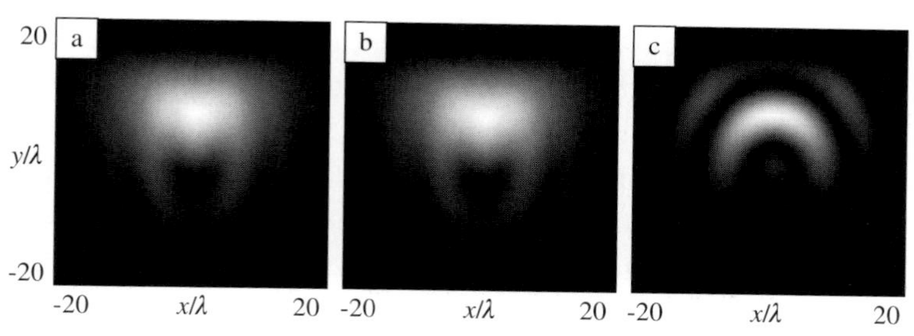

Figure 4.4 Computed intensity patterns in the XY-plane at the mid-point between the two foci in the system of Figure 4.3. (a) Upper half-lens blocked; (b) lower half-lens blocked; (c) both half-lenses transmitting the incident beam, and the two emergent beams interfering at the observation plane. Note that the central region of the distribution in (c), corresponding to points near the Z-axis, is dark.

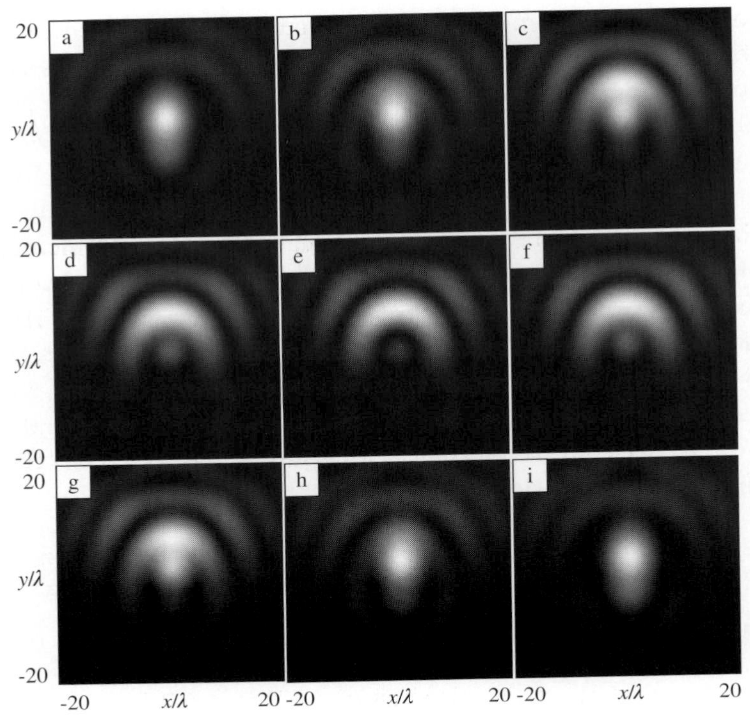

Figure 4.5 Plots of intensity distribution in the XY-plane at various locations along the Z-axis in the system of Figure 4.3. From (a) to (i) the observation plane moves in steps of 37.5λ from the first focus, at F_1, to the second focus, at F_2.

Effect of lens on Gaussian beam

In the paraxial approximation a lens imparts a quadratic phase factor to the incident beam. If the lens happens to be astigmatic, the phase factors along the X- and Y- axes will have different curvatures, and if the astigmatic lens happens to have rotated within the XY-plane then the quadratic phase factor will have an xy term as well. We assume that the lens aperture is large enough to transmit the beam without significant truncation and, therefore, to affect negligibly its amplitude profile. All in all, the effect of a lens on a Gaussian beam is to multiply its complex amplitude by the following transmission function:

$$t(x, y) = \exp[-i\pi(px^2 + 2qxy + ry^2)]. \tag{4.5}$$

Here p, q, r are real-valued constants related to the principal radii of curvature of the lens. Thus when the Gaussian beam of Eq. (4.1a) passes through the lens described by Eq. (4.5), a_2 will be augmented by p, b_2 by q, and c_2 by r.

The beam can then be propagated in the free-space region beyond the lens using the aforementioned analytical tools.

Higher-order Gaussian beams

We confine the discussion of higher-order beams to the one-dimensional case only, as the extension to two dimensions is straightforward. Consider the Gaussian function $\exp(-\pi a x^2)$, where a is a complex constant. The nth derivative of this function with respect to x may be used to define an initial amplitude distribution as follows:

$$\hat{a}(x, z=0) = \hat{a}_0 H_n(\sqrt{\pi a}\, x)\exp(-\pi a x^2) = \hat{a}_0(-1)^n(\pi a)^{-n/2}\frac{\mathrm{d}^n}{\mathrm{d}x^n}[\exp(-\pi a x^2)]. \tag{4.6a}$$

Here the nth-order Hermite polynomial $H_n(x)$ is defined as

$$H_n(x)\exp(-x^2) = (-1)^n\frac{\mathrm{d}^n}{\mathrm{d}x^n}[\exp(-x^2)]. \tag{4.6b}$$

The Fourier transform of the distribution in Eq. (4.6a) is readily evaluated using the differentiation theorem,[5]

$$\begin{aligned}\mathscr{F}\left\{\frac{\mathrm{d}^n}{\mathrm{d}x^n}[\exp(-\pi a x^2)]\right\} &= (\mathrm{i}2\pi\sigma)^n\mathscr{F}[\exp(-\pi a x^2)]\\ &= (\mathrm{i}2\pi\sigma)^n a^{-1/2}\exp(-\pi\sigma^2/a).\end{aligned} \tag{4.7}$$

To account for propagation by a distance z_0 along the Z-axis, the Fourier transform of the initial distribution in Eq. (4.6a) is multiplied by the transfer function of free-space propagation, which, in the paraxial approximation, is $\exp(\mathrm{i}2\pi z_0)\exp(-\mathrm{i}\pi z_0\sigma^2)$. This means that the coefficient $1/a$ in the exponent of the Gaussian function on the right-hand side of Eq. (4.7) is augmented by $\mathrm{i}z_0$, yielding

$$\frac{1}{a'} = \frac{1}{a} + \mathrm{i}z_0. \tag{4.8}$$

The light amplitude distribution at $z = z_0$ is then obtained by an inverse Fourier transform, yielding

$$\hat{a}(x, z=z_0) = \hat{a}_0\exp(\mathrm{i}2\pi z_0)(a'/a)^{(n+1)/2}H_n(\sqrt{\pi a'}\,x)\exp(-\pi a' x^2). \tag{4.9}$$

Since in general a' is complex, the above eigenfunctions of propagation in free space contain Hermite polynomials with a complex argument.

Siegman[2,6] refers to these as the "elegant" solutions of the wave equation in free space. The elegant solutions are substantially different from the so-called "standard" solutions, whose argument of the Hermite polynomial is real.

Assuming that the Hermite–Gaussian beam of Eq. (4.9) has its waist at $z = 0$, the parameter a of the beam will be real-valued. The normalization factor $(a'/a)^{(n+1)/2} = (1 + \mathrm{i}az_0)^{-(n+1)/2}$ thus contributes its phase angle, $-\frac{1}{2}(n+1)\tan^{-1}(az_0)$, to the Gouy phase. Note that the complex-argument Hermite polynomial $H_n(\sqrt{\pi a'}\,x)$ also has a z-dependent phase, which contributes to the overall phase pattern in the beam's cross-section.

References for chapter 4

1 H. Kogelnik and T. Li, Laser beams and resonators, *Appl. Opt.* **5**, 1150–1167 (1966).
2 A. E. Siegman, *Lasers*, University Science Books, California (1986).
3 L. G. Gouy, *Compt. Rend. Acad. Sci. Paris* **110**, 1251 (1890).
4 M. Born and E. Wolf, *Principles of Optics*, sixth edition, Pergamon Press, New York, 1980.
5 R. N. Bracewell, *The Fourier Transform and its Applications*, McGraw-Hill, New York, 1978.
6 A. E. Siegman, Hermite-gaussian functions of complex argument as optical-beam eigenfunctions, *J. Opt. Soc. Am.* **63**, 1093–1094 (1973).

5

Coherent and incoherent imaging

The basic elements of an imaging system are shown in Figure 5.1. The light from a source, either coherent (e.g., a laser) or incoherent (e.g., an incandescent lamp or an arc lamp), is collected by the illumination optics (e.g., a condenser lens) and projected onto the object. An image is then formed by an objective lens upon a screen, a photographic plate, a CCD camera, the retina of an eye, etc. Assuming that the objective lens is free from aberrations, the resolution and the contrast of the image are determined not only by the numerical aperture of the objective lens but also by the properties of the light source and the illumination optics.

The source and the illumination optics

Three types of illumination will be considered. For collimated and coherent illumination we assume a monochromatic laser beam brought to focus at the plane of the object with a condenser lens having a very small numerical aperture (NA). Figure 5.2(a) is the logarithmic intensity distribution at the object plane, produced by a $0.03NA$ condenser. This distribution has the shape of an Airy pattern, with a central lobe diameter of $1.22\lambda/NA \approx 41\lambda$, where λ is the wavelength of the light source. Since the objects of interest will be small compared to the Airy disk diameter, and since they will be placed near the center of the Airy disk, this illumination qualifies as coherent, fairly uniform, and nearly collimated.

The second type of illumination is also produced by a coherent monochromatic laser beam, but with a high-NA condenser. This time we place the focal point of the condenser somewhat before the object in order to produce within the object plane a bright spot large enough to cover the field of view of the objective lens. Figure 5.2(b) is the logarithmic intensity distribution at the object plane, produced by a coherent beam brought to focus by a $0.25NA$

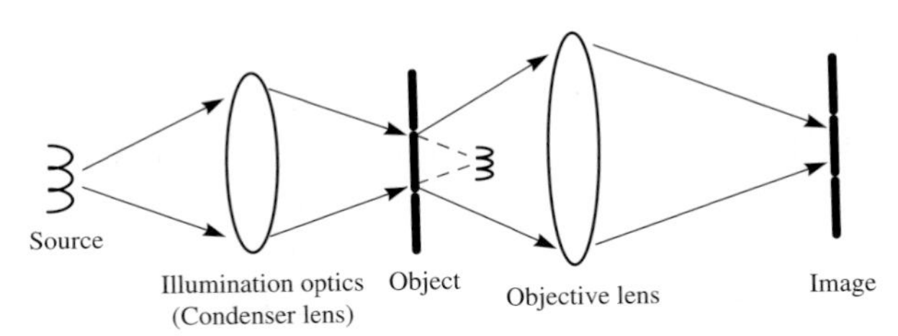

Figure 5.1 Schematic diagram of a simple imaging system. The light source is projected by the illumination optics onto an object, allowing the objective lens to form an image of this object at the image plane.

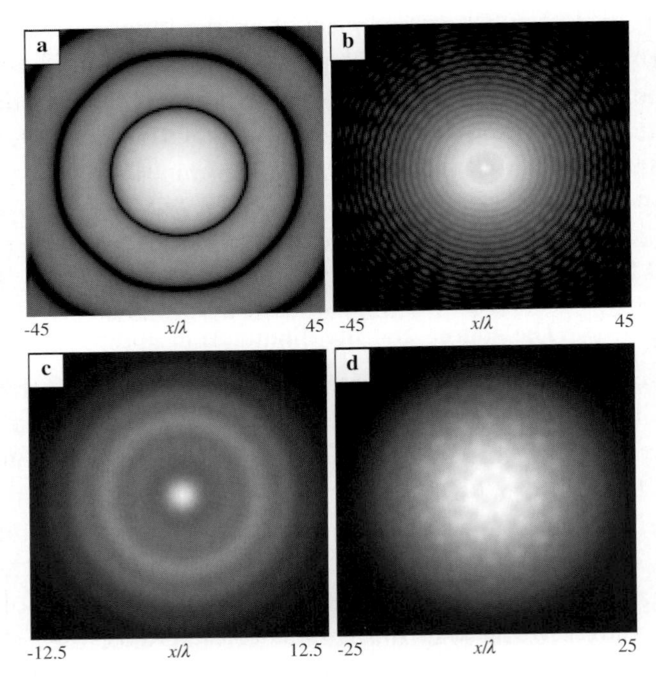

Figure 5.2 Computed intensity patterns at the plane of the object corresponding to various types of illumination. (a) Logarithmic plot ($\alpha = 4$) of the intensity distribution obtained from a coherent source with a $0.03NA$ condenser lens. (b) Logarithmic plot ($\alpha = 4$) of the intensity distribution obtained from a coherent source with a $0.25NA$ condenser lens. The beam is focused to a plane located just 50λ before the plane of the object. (c) Same as (b) but showing the intensity distribution rather than its logarithm. (d) Intensity distribution corresponding to an incoherent light source consisting of 37 independent point sources obtained with a $0.25NA$ condenser lens. Again the source is imaged to a plane located 50λ before the plane of the object.

condenser at a distance of 50λ before the object. The beam incident on the object is, therefore, divergent and, although it covers the area of interest, its intensity distribution is not very uniform. This nonuniformity may be better appreciated by considering the corresponding plot of intensity distribution in Figure 5.2(c). (Note the different scales of Figures 5.2(b), (c).)

The third type of illumination to be examined is incoherent illumination. We emphasize at the outset that our concern here is solely with spatial incoherence and, as such, we will assume that the source is quasi-monochromatic. (Departure from monochromaticity is a requirement for any source that is to exhibit spatial incoherence; the bandwidth of the source can nonetheless be narrow enough to give its light a long coherence time, making it in effect a temporally coherent source.) To simulate an incoherent source we assumed that the quasi-monochromatic light emerging from a fiber bundle consisting of 37 fibers is imaged with a $0.25NA$ condenser lens to a plane located a distance of 50λ before the plane of the object in Figure 5.1. Each fiber within the bundle acts as a coherent point source whose projected intensity distribution at the object plane will be the same as that shown in Figure 5.2(c). When these fibers are properly arranged in space and their intensity distributions added together, we obtain the intensity pattern displayed in Figure 5.2(d). This is a fairly uniform distribution over its central region, which is where the objects of interest will be placed. Although the source could have been imaged directly onto the object plane in this case, the 50λ defocus helps to create a more uniform illumination. With this type of illumination, in order to compute the intensity distribution at the image plane, we treat the 37 fibers as independent point sources – each a coherent point source in its own right. We then compute the image obtained with each source independently, and add the *intensities* of the resulting 37 images together to obtain the final image.

The imaging optics

The objective lens used in the simulations described below is free from aberrations and, therefore, its performance is diffraction-limited. The objective is a finite-conjugate lens with a numerical aperture of 0.25 (on the side of the object), a focal length of 5000λ, and a magnification of 10.

Two types of object will be used in these simulations. The first is an amplitude grating with a period of 3λ and a 50% duty cycle, shown in Figure 5.3(a). According to the classical optics textbooks,[1–3] the spatial frequency of this grating is higher than the cutoff frequency of the modulation transfer function (MTF) of the objective lens for coherent illumination,

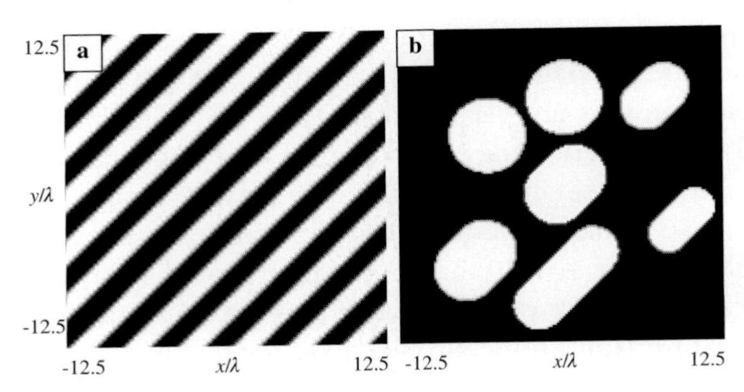

Figure 5.3 (a) Amplitude grating with a period of 3λ and a 50% duty cycle, used as the object in some of the simulations. (b) Pattern of marks with different sizes and separations on a uniform background. In some cases these marks will be black on a transparent background, in other cases they will be transparent marks on a black background, in yet other cases they will be phase objects with 100% transmissivity, imparting a 180° phase shift to the incident beam.

$f_c = NA/\lambda = (4\lambda)^{-1}$, but less than that for incoherent illumination, $f_c = 2NA/\lambda = (2\lambda)^{-1}$. We will examine the images of this grating under both coherent and incoherent illumination and draw certain conclusions about the classical treatment of this problem.

The second type of object with which we will be concerned is a mask imprinted with seven marks of various sizes and shapes, shown in Figure 5.3(b). The largest mark is 10λ long, and the smallest mark is 3λ wide. These marks are large enough to yield a reasonably clear image with both coherent and incoherent illumination. In one case the marks will be assumed to be bright objects on a dark background, in another case they will be dark objects on a bright background, in yet a third case they will be 180° phase objects having the same amplitude transmissivity as the background.

Resolution of the imaging system

Let the grating of Figure 5.3(a) be the object in the system of Figure 5.1. If the collimated beam of Figure 5.2(a) is used to illuminate this object then no image will be formed, because all diffracted orders (except the zeroth order) will miss the entrance pupil of the objective lens; the situation is depicted schematically in Figure 5.4. Denoting the period of the grating by P, the deviation angle θ of the first diffracted order will be given by $\sin\theta = \lambda/P$. This is the origin of the well-known assertion that the MTF cutoff frequency of a coherent imaging system is $f_c = NA/\lambda$.

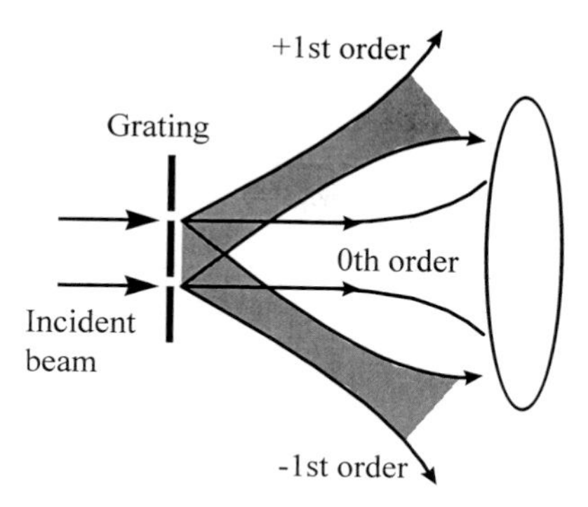

Figure 5.4 A collimated coherent beam (wavelength λ) illuminates a grating of period P at normal incidence. The ±first diffracted orders will miss the objective lens if the lens's numerical aperture NA is less than λ/P. (Some provision must be made for the expansion of the beam diameter at long propagation distances.)

If, however, the coherent illuminating beam is not collimated but is in the form of a cone of light, as in the case of the distribution shown in Figure 5.2(c), then an "image" of the grating will be formed. Figure 5.5 shows computed plots of intensity distribution (a) at the exit pupil of the objective lens, where the overlap between the zeroth-order and the ±first-order beams is clearly visible, and (b) at the image plane, where an "image" of the grating is seen superimposed on a nonuniform pattern of illumination. The reason that a coherent *cone* of light produces an image of the grating whereas a *collimated* beam fails to do so may be understood by studying Figure 5.6: the diffracted ±first-order cones are captured by the objective lens as long as the lens's NA is greater than $\lambda/(2P)$. The MTF cutoff frequency for this type of illumination, therefore, is $f_c = 2NA/\lambda$.

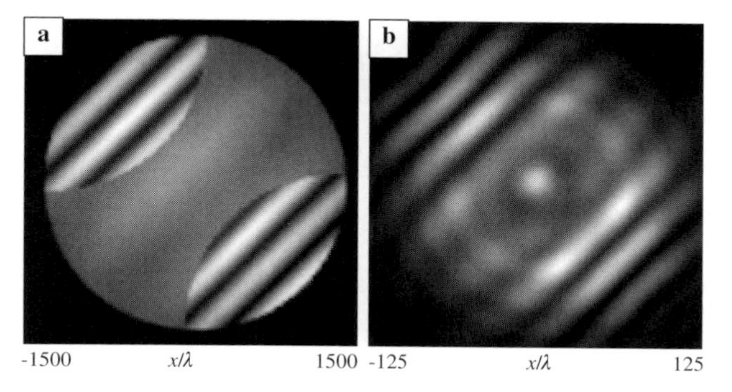

Figure 5.5 Computed intensity distribution (a) at the exit pupil of the objective lens and (b) at the image plane, corresponding to coherent illumination with the divergent beam of Figure 5.2(c). The object is the grating of Figure 5.3(a).

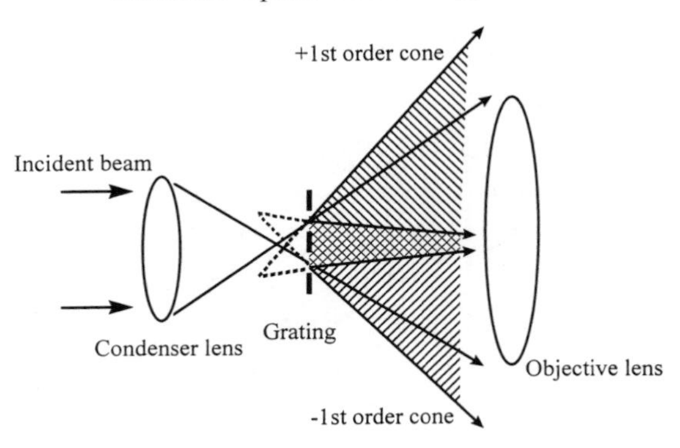

Figure 5.6 A cone of coherent light (wavelength λ), coming from a condenser lens and illuminating a grating of period P, creates several diffracted cones. If the apex angle of the incident cone is sufficiently large, the $\pm$first-order beams will be captured by the objective lens as long as $NA > \lambda/(2P)$.

The case of incoherent illumination is now easy to understand. Since the beam in Figure 5.2(d) is a superposition of 37 divergent cones similar to that of Figure 5.2(c), the grating's image will have the same resolution as that obtained with a single cone of light, but it will have a more uniform contrast because it is an average over a large number of point sources. Figure 5.7 shows the computed patterns of intensity (a) at the exit pupil of the objective lens and (b) at the image plane obtained with incoherent illumination. The MTF cutoff for this type of illumination, $f_c = 2NA/\lambda$, is the same as that for coherent illumination with a cone of light, which is twice as large as the cutoff frequency for collimated coherent illumination.

Figure 5.7 Computed intensity distribution (a) at the exit pupil of the objective lens and (b) at the image plane, corresponding to the incoherent illumination depicted in Figure 5.2(d). The object is the grating of Figure 5.3(a).

Images of non-periodic objects

The mask containing marks of different sizes shown in Figure 5.3(b) provides a good test object for comparing images obtained under coherent and incoherent illumination. Consider the case of transparent marks on a dark background (an amplitude object), imaged with a collimated coherent illumination (see Figure 5.8), and also with incoherent illumination (see Figure 5.9). The resolution of the former is obviously inferior to that of the latter, and the spurious fringes appearing in the coherent image are responsible for at least some of the image-quality degradation. (As an aside, note that the exit-pupil distribution in the case of coherent illumination displays much more structure than that obtained with incoherent light.)

Using the same object as in Figure 5.3(b) but assuming that the marks are black features on a transparent background (i.e., reversing the contrast) we

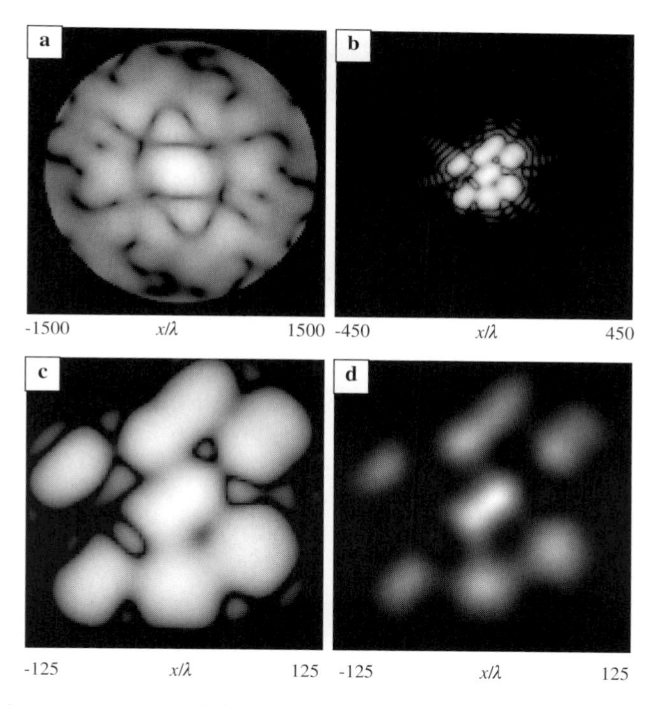

Figure 5.8 Coherent imaging of the seven transparent marks on a black background shown in Figure 5.3(b). The incident distribution is the collimated coherent beam of Figure 5.2(a). (a) Logarithmic plot ($\alpha = 4$) of intensity distribution at the exit pupil of the objective lens. (b) Logarithmic plot ($\alpha = 4$) of intensity distribution at the image plane. (c) Magnified view of the central region of the image shown in (b); in this case $\alpha = 3$. (d) Same as (c) but showing the distribution of intensity rather than its logarithm.

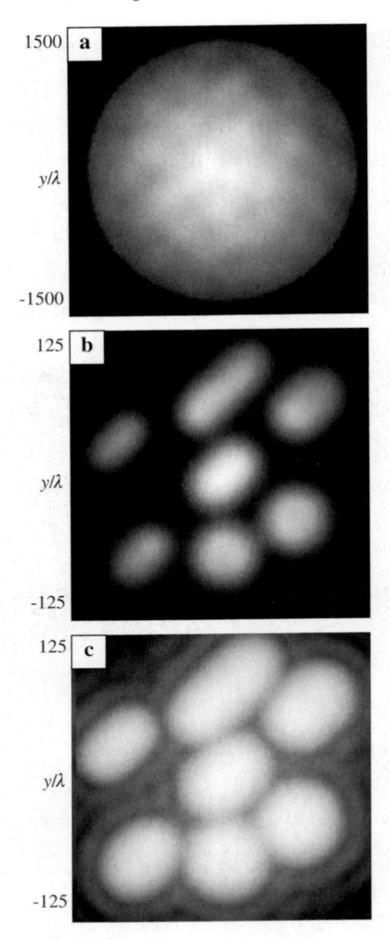

Figure 5.9 Incoherent imaging of the seven transparent marks on a black background; the incident distribution is that of Figure 5.2(d). (a) Intensity distribution at the exit pupil of the objective lens. (b) Intensity distribution at the image plane. (c) Same as (b) but on a logarithmic scale ($\alpha = 3$).

obtain the distributions of Figure 5.10 in the case of collimated coherent illumination and those of Figure 5.11 in the case of incoherent illumination. Note the similarity between the exit-pupil distributions in Figures 5.8(a) and 5.10(a), indicating that Babinet's principle is at work here.[2,3] Also note in Figure 5.10(b) that, in addition to the marks, the rings of the Airy pattern of the illuminating beam are also captured in the image. The logarithmic intensity distributions in Figures 5.10(b), (c) show gray spots in the middle of dark marks, a feature that is less prominent in the incoherent image of Figure 5.11(c).

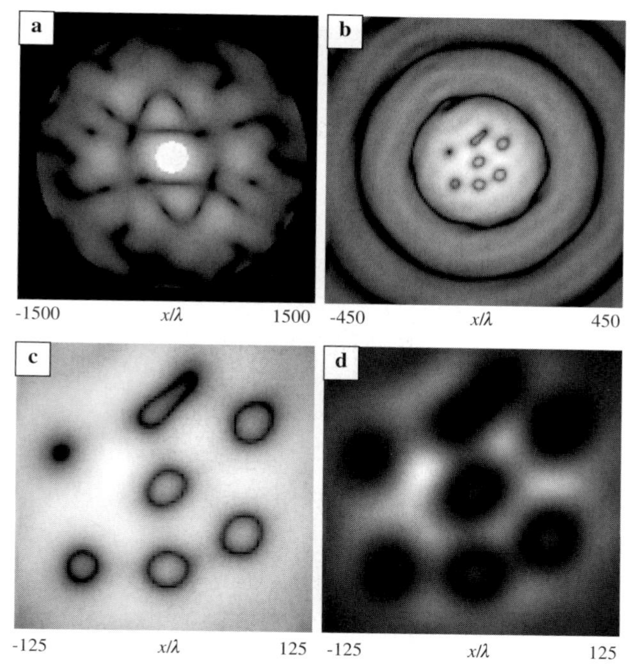

Figure 5.10 Coherent imaging of the seven black marks on a transparent background shown in Figure 5.3(b). The incident distribution is the collimated coherent beam of Figure 5.2(a). (a) Logarithmic plot ($\alpha = 4$) of intensity distribution at the exit pupil of the objective lens. (b) Logarithmic plot ($\alpha = 4$) of intensity distribution at the image plane, showing the images of the marks as well as the rings of the Airy pattern. (c) Magnified view of the central region of the image shown in (b); in this case $\alpha = 3$. (d) Same as (c) but showing the distribution of intensity rather than its logarithm.

Finally we assume that the marks on the mask of Figure 5.3(b) represent transparent phase objects that impart a phase shift of 180° (relative to the background) to the incident beam. Figure 5.12 shows the computed intensity distributions at the objective's exit pupil and at the image plane, for the case of illumination by the collimated coherent beam of Figure 5.2(a). Figure 5.13 shows the corresponding distributions for incoherent illumination. Note how diffraction from mark boundaries can create an "image" of the marks in a case where no explicit phase-contrast mechanism is present.[3] In the two simulations depicted in Figures 5.10 and 5.12, the amplitude transmission functions of the respective objects differ only by an additive constant term. Therefore, the image in Figure 5.10(b), for instance, may be derived from that in Figure 5.12(b) by the addition of the image of the incident beam, it being understood that the quantities being added are the complex amplitudes, not the intensities.

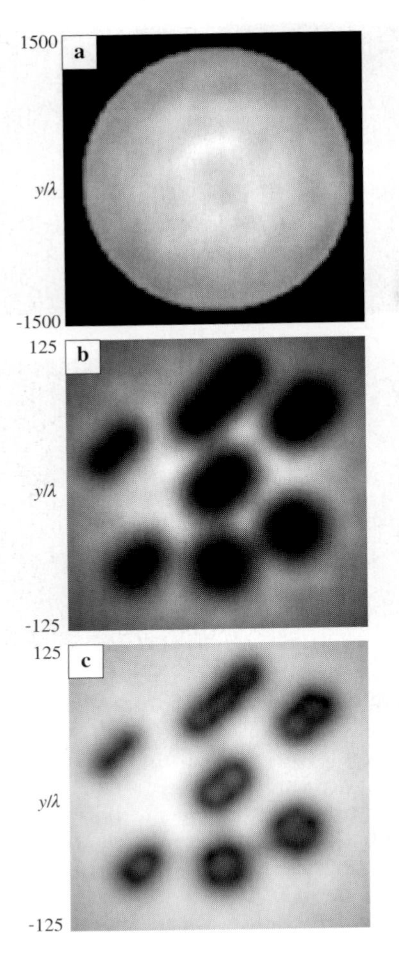

Figure 5.11 Incoherent imaging of the seven black marks on a transparent background; the incident distribution is that of Figure 5.2(d). (a) Intensity distribution at the exit pupil of the objective lens. (b) Intensity distribution at the image plane. (c) Same as (b) but on a logarithmic scale ($\alpha = 1.7$).

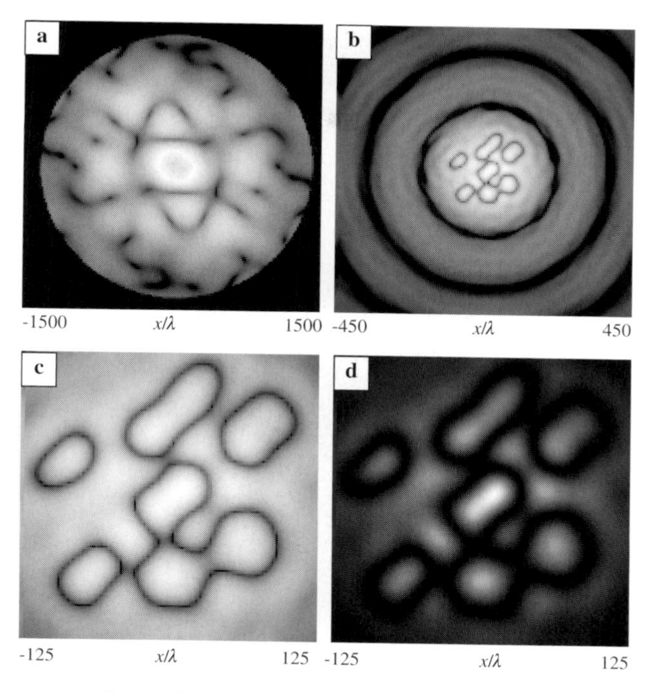

Figure 5.12 Same as Figure 5.10 but for a phase object. The assumed object in this case is the mask of Figure 5.3(b), which has uniform transmissivity everywhere; its marks impart a relative phase shift of 180° to the incident beam.

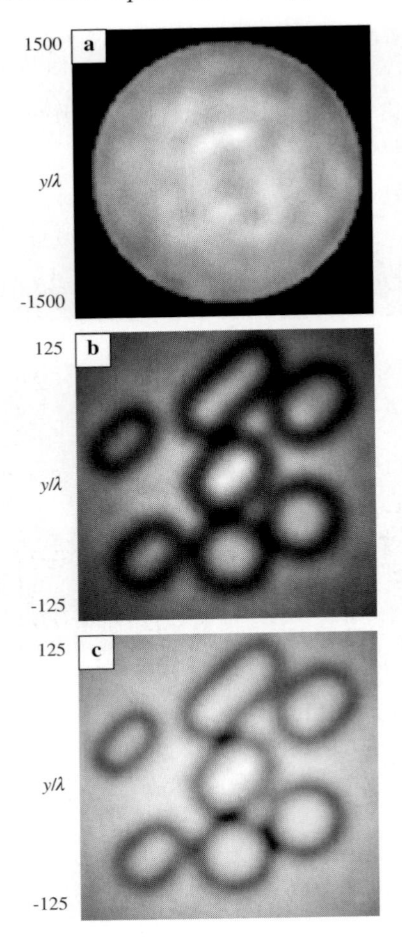

Figure 5.13 Same as Figure 5.11 but for a phase object. The assumed object in this case is the mask of Figure 5.3(b), which has uniform transmissivity everywhere; its marks impart a relative phase-shift of 180° to the incident beam. (For the logarithmic plot in (c) $\alpha = 1.4$.)

References for chapter 5

1 J. W. Goodman, *Introduction to Fourier Optics*, McGraw-Hill, New York, 1968.
2 M. Born and E. Wolf, *Principles of Optics*, 6th edition, Pergamon Press, Oxford, 1980.
3 M. V. Klein, *Optics*, Wiley, New York, 1970.

6

First-order temporal coherence in classical optics†

A truly monochromatic beam of light, if it ever existed, would be perfectly coherent. Suppose that such a beam is split into two parts and each part propagated over an arbitrary distance. When the parts are finally brought together and mixed, no matter how different the two path lengths may have been, the resulting waveform will exhibit constructive and destructive interference in the form of bright and dark fringes. The coherence length of a monochromatic beam is therefore infinite, in the sense that the path-length difference can be as large as desired without hampering one's ability to create interference patterns.

Real sources of light, of course, are never monochromatic. White light restricted to the visible range of wavelengths from 400 nm to 700 nm, for example, has a coherence length of only a couple of micrometers. A green filter passing sunlight at $\lambda_0 = 550$ nm with a 10 nm bandwidth produces a beam with a coherence length of about 50 μm. The red line of cadmium ($\lambda_0 = 643.8$ nm) has a nearly Gaussian spectrum with a 0.0013 nm width at half peak intensity, leading to a coherence length of nearly 30 cm.[1] This is similar to the coherence length of a short, inexpensive HeNe laser ($\lambda_0 = 632.8$ nm) with a few longitudinal modes and a typical bandwidth of $\Delta f \approx 1$ GHz. A stabilized HeNe laser operating in a single longitudinal mode ($\Delta f \approx 100$ kHz) has a coherence length of several kilometers. It is important therefore to understand the role of spectral bandwidth in enhancing or diminishing the performance of an optical system that, by design or by coincidence, involves interference.

The subject of temporal coherence has been covered extensively in modern and classical textbooks,[1–5] and it is not our intention here to repeat what is

† This chapter is coauthored with Ewan M. Wright, Professor of Optical Sciences at the University of Arizona.

already well known. Instead, we present an alternative viewpoint that draws on the similarities between a waveform extended over a long span of time and a compact wave packet that exists for a relatively short period. We will show that, as far as first-order temporal coherence is concerned, the wave packet can be substituted for the extended waveform in analyzing the results of interference experiments. While describing the properties of wave packets, we also mention some interesting observations concerning their reflection from, and transmission through, multilayer stacks.

Time dependence, frequency spectrum, and phase

Consider a superposition of plane waves, propagating in free space along the Z-axis and covering a range of (temporal) frequencies at and around $f = f_0$. The discrete frequencies f_n comprising the spectrum of this waveform are assumed to have a fixed spacing Δf as follows:

$$f_n = f_0 + n\Delta f = (N_0 + n)\Delta f. \tag{6.1}$$

The amplitude of the waveform is

$$a(z, t) = \sum_n A_n(\Delta f)^{1/2} \cos[2\pi f_n(z/c - t) + \phi_n], \tag{6.2}$$

where A_n and ϕ_n are the amplitude and phase of the spectral component whose frequency is f_n, and c is the speed of light in vacuum. The constant multiplier $(\Delta f)^{1/2}$ is for normalization purposes only, its significance becoming clear as the discussion proceeds. We set the central frequency $f_0 = 4.74 \times 10^{14}$ Hz (corresponding to $\lambda_0 = 632.8$ nm) and $\Delta f = 4.74 \times 10^{12}$ Hz, which leads to $N_0 = 100$. We adopt a Gaussian shape for the distribution of the amplitudes A_n, as shown in Figure 6.1(a), and let the values of n in Eq. (6.2) range from -15 to $+14$, for a total of 30 discrete wavelengths in the spectrum. To a large extent these choices are arbitrary, but the points that we seek to clarify by way of examples based on these choices are quite general in nature.

Throughout this chapter the same amplitude coefficients $\{A_n\}$ are assumed for all realizations of the waveform $a(z, t)$, but the phase angles $\{\phi_n\}$, although fixed for any particular waveform, differ for different realizations. The statistical properties of $a(z, t)$ are thus uniquely determined by the joint probability distribution over $\{\phi_n\}$. Furthermore, we consider stationary processes for which the ensemble average over different phase-angle realizations coincides with the time average derived from a single realization. This restric-

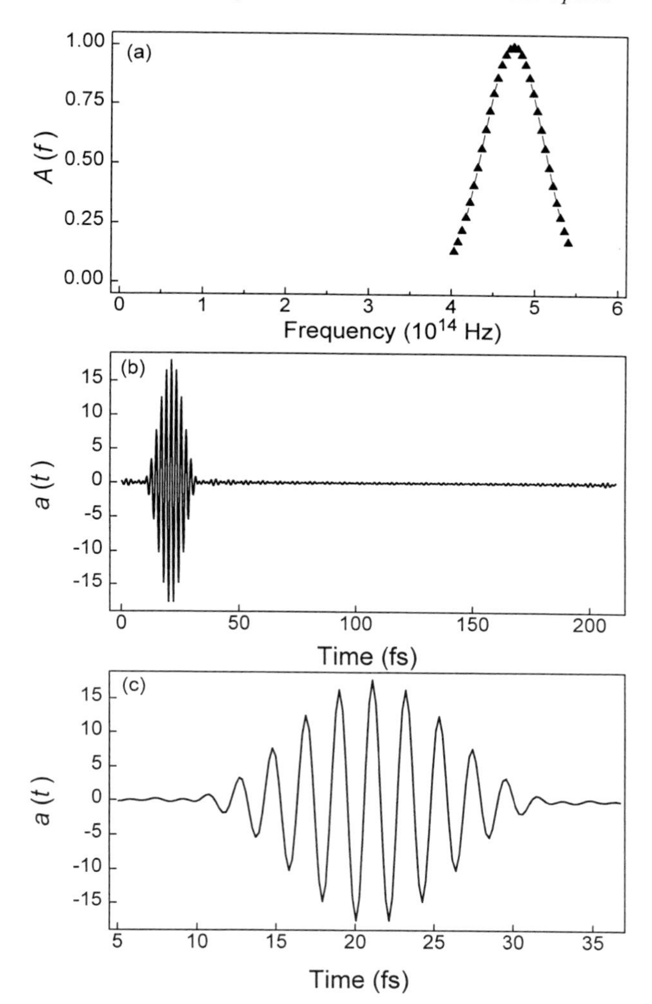

Figure 6.1 (a) A truncated Gaussian function sampled at regular intervals represents the frequency spectrum of a waveform. (b) The waveform as a function of time obtained by Fourier-transforming the spectrum in (a), assuming that the phase is a linear function of frequency. Since the spectrum is sampled at $\Delta f = 4.74 \times 10^{12}$ Hz, the waveform is repeated with a period of 211 fs; only one period of the wave packet is shown. (c) Close-up of the wave packet.

tion of randomness to spectral phase simplifies the discussion without affecting the validity of the final results.

Since the spectrum in Figure 6.1(a) is a discrete function of frequency, the corresponding amplitude $a(z, t)$ considered either as a function of time at a fixed point z, or as a function of z at a given instant of time t, will be periodic. With z fixed, for example, the period of the function in the time domain will be $T = 1/\Delta f = 211$ femtoseconds. A plot of $a(z = 0, t)$ over a full period T is shown in Figure 6.1(b), and a close-up of the wave packet appears in Figure 6.1(c). (This is reminiscent of the pulse train emerging from a mode-locked laser.) The width of the packet in Figure 6.1(b) is ≈ 20 fs, which is of the same order of magnitude as the inverse of the spectral width ($29\Delta f = 1.27 \times 10^{14}$ Hz). To increase the period T without changing the overall shape of the wave packet one must increase the rate of sampling of the spectrum of Figure 6.1(a), by selecting additional frequencies in between those that are already chosen. In this way, both the spectrum and the wave packet retain their shapes but Δf becomes smaller while T becomes larger. In the limit $\Delta f \to 0$ the separation T between adjacent wave packets approaches infinity.

Where the first-order coherence of a given waveform is concerned, the phase distribution over its spectral range is irrelevant, even though the shape of the waveform as a function of time is significantly affected by this phase distribution. For example, in Figure 6.1(b) the phase ϕ_n is assumed to be a linear function of frequency, whereas if ϕ_n is picked randomly for each f_n then an extended function such as that in Figure 6. 2 is obtained. (The latter might, for example, be the output of a multi-longitudinal-mode laser.) There are many possible choices for $\{\phi_n\}$ and each choice yields a more or less extended function of time. Only in rare occasions do we find a compact wave packet similar to that in Figure 6. 1(b). However, all functions obtained by different choices of $\{\phi_n\}$ are identical in their first-order coherence attributes. In other words, the compact packet of Figure 6.1(b) has the same degree of first-order coherence as the extended waveform of Figure 6. 2.

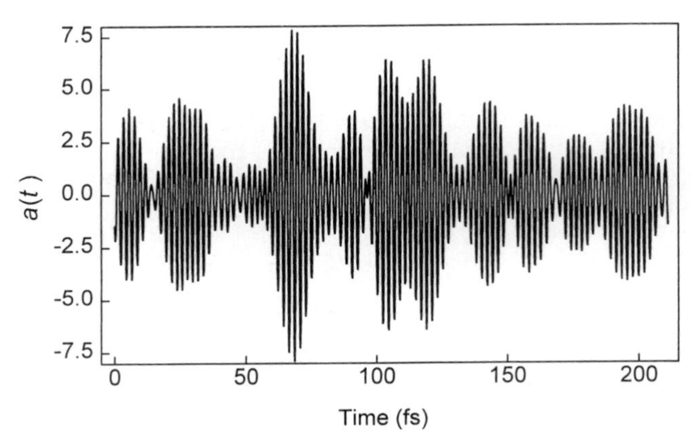

Figure 6.2 Waveform obtained by Fourier-transforming the frequency spectrum of Figure 6.1(a) after assigning it a randomly selected phase at each frequency.

The time-averaged intensity of the waveform at an arbitrary point $z = z_0$ is readily computed from Eq. (6.2) as follows:

$$\langle I(z = z_0) \rangle = \frac{1}{T} \int_0^T a^2(z = z_0, t)\, \mathrm{d}t = \frac{1}{2} \sum_n A_n^2 \Delta f. \tag{6.3}$$

Note that the right-hand side of Eq. (6.3), being the area under the square of the spectral distribution of Figure 6.1(a), remains constant as the sampling rate increases. Thus reducing Δf in order to increase the period T does not affect the average intensity of the waveform.

The Mach–Zehnder interferometer

Temporal coherence is usually measured with a Michelson interferometer. For our present purposes, however, we will consider a slightly modified version of the Mach–Zehnder interferometer, shown in Figure 6.3. The collimated beam of light entering the device is split equally between its two arms

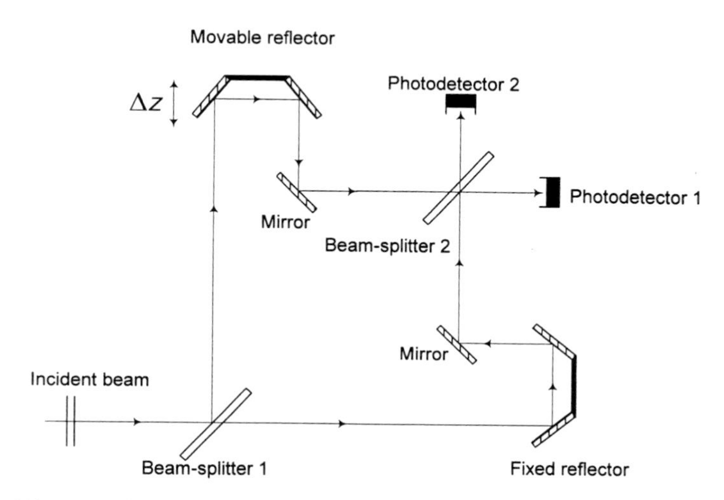

Figure 6.3 The Mach–Zehnder interferometer is used in analyzing the temporal coherence of a collimated beam of light. The incoming beam is split equally between the two arms of the device at the first BS. The two arms are identical except that the end-reflector is fixed in one arm and movable in the other. After traveling along these separate arms the beams are recombined at the second BS. When the optical path lengths of the two arms are identical, the beams interfere constructively in channel 1 and deliver their entire energy to photodetector 1. Deviations from path-length equality can send to channel 2 either the entire beam or a fraction of it. The movable reflector is used to adjust the optical path-length difference between the arms.

at the first beam splitter (BS). The two beams are reflected by the mirrors at the end of each arm, then recombined at the second BS. If, on the one hand, the two beams happen to be perfectly in phase when they arrive at the second BS, they interfere constructively in channel 1 (see Figure 6.3) and deliver their combined total optical energy to photodetector 1; photodetector 2 in this case receives no light at all. If, on the other hand, the two beams are relatively phase-shifted by $\Delta\phi = 180°$, they appear collectively at detector 2, leaving detector 1 in the dark. For intermediate values of $\Delta\phi$ the energy of the beams is split between the two detectors, the splitting ratio being 50/50 when $\Delta\phi = +90°$.

Now suppose the relative phase between the two beams can be varied continuously by adjusting the length of one of the interferometer's arms. Then S_1, the output of detector 1, reaches its maximum when the two arms become identical in length. As the length of the adjustable arm then increases by a quarter of a wavelength, $\Delta\phi$ becomes 180° and S_1 reaches a minimum. As long as the two beams remain coherent (or partially coherent) this behavior is periodically repeated, the output of each detector oscillating between a maximum and a minimum. Once the arm lengths differ by more than the coherence length of the beam, the oscillations die down and both channels receive equal amounts of light, irrespective of the path-length difference between the arms. For the wave packet of Figure 6.1, we show in Figure 6.4 the computed output of detector 1 as a function of $\Delta z = \frac{1}{2}c\tau$, where τ is the time delay between the two arms of the interferometer.

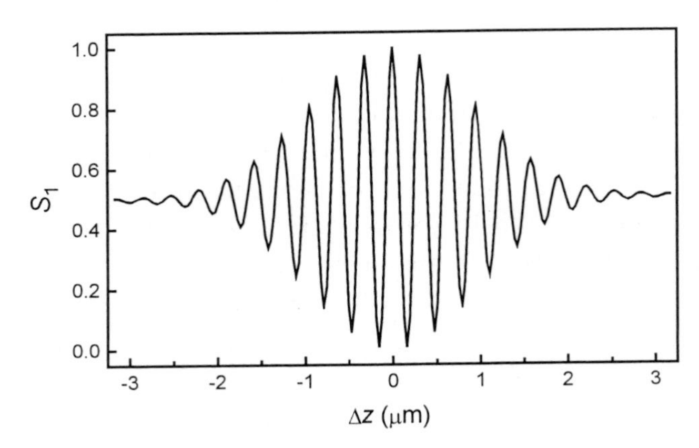

Figure 6.4 The signal S_1 of detector 1 as a function of the extension Δz of the movable end-reflector of the interferometer. The assumed incoming beam is the packet shown in Figure 6.1.

The time-averaged detector outputs may be written

$$S_{1,2}(\tau) = \frac{1}{T}\int_0^T \left\{\frac{1}{2}[a(t) \pm a(t-\tau)]\right\}^2 \mathrm{d}t = \frac{1}{2T}\int_0^T a^2(t)\,\mathrm{d}t \pm \frac{1}{2T}\int_0^T a(t)\,a(t-\tau)\,\mathrm{d}t$$

$$= \frac{1}{2}\left[\langle I\rangle \pm \frac{1}{2}\sum_n A_n^2 \Delta f \cos(2\pi f_n \tau)\right]. \tag{6.4}$$

The first term on the right-hand side of this equation is a constant, independent of τ, while the second term is the autocorrelation function of the waveform $a(t)$ and coincides with the first-order field coherence function in the case of a stationary process. The Fourier series coefficients of this autocorrelation function are $\{A_n^2\}$ and are independent of $\{\phi_n\}$. It is thus clear that the signals $S_1(\tau)$ and $S_2(\tau)$, and hence the first-order temporal coherence of the waveform, depend only on the magnitude – and not the phase – of the spectral distribution, as was asserted earlier.

Coherence length

Figure 6.5 shows the waveforms arriving in channels 1 and 2 when the wave packet of Figure 6.1(b) is sent through the interferometer, with its movable arm extended by $\Delta z = cT/8 = 7.91$ μm. The time delay between the packets traveling in the two arms is therefore $\tau = \frac{1}{4}T$. Since this delay is longer than the duration of each packet, the two packets upon arriving at the second BS do not overlap and, therefore, appear separately in both channels. Obviously no interference takes place in this case and each channel receives an equal share from each packet, each with one-half of the original amplitude.

In the above example, where the delay τ between the two arms of the interferometer is $\frac{1}{4}T$, one can divide the frequency content of the wave packet into four categories. The first category consists of the frequencies $f = 85\Delta f, 89\Delta f, 93\Delta f, \ldots, 113\Delta f$. All these terms are phase-shifted by 90° and, when combined at the second BS, are equally split between channels 1 and 2. The output of channel 1 for these frequency components is shown in Figure 6.6(a). The second category consists of the frequencies $f = 86\Delta f, 90\Delta f, 94\Delta f, \ldots, 114\Delta f$, which are phase-shifted by 180° and, therefore, appear exclusively in channel 2. The third category, consisting of the frequencies $f = 87\Delta f, 91\Delta f, 95\Delta f, \ldots, 111\Delta f$, is phase-shifted by −90° and is, once again, equally split between the two channels; the output of channel 1 for these components is shown in Figure 6.6(b). The fourth and last category consists of frequencies $f = 88\Delta f, 92\Delta f, 96\Delta f, \ldots, 112\Delta f$, which are not phase-shifted at all and appear in their entirety in channel 1;

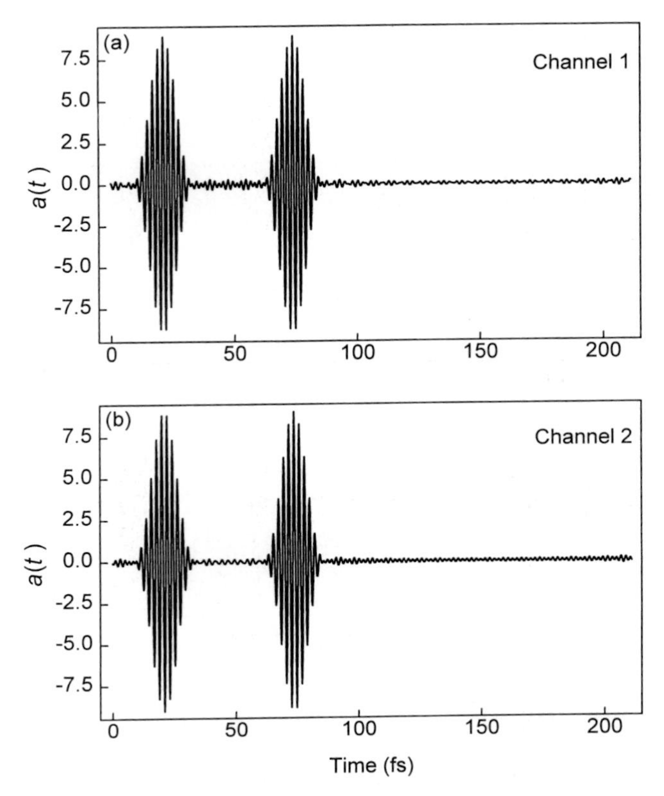

Figure 6.5 Waveforms arriving at (a) channel 1 and (b) channel 2 of the Mach–Zehnder interferometer. The assumed incoming beam is the packet of Figure 6.1, and the movable arm of the interferometer has been extended by $\Delta z = cT/8 = 7.91\ \mu\text{m}$. Because the delay is longer than the width of the packet no interference takes place. The two packets act independently and appear in both channels, albeit at half the original magnitude of the incoming wave. Note that the first packet in channel 2, having been transmitted through both beam-splitters, is flipped relative to the second packet, which has been reflected at both beam-splitters. In contrast, each packet arriving in channel 1 has been reflected at one and transmitted at the other beam-splitter. As a result, there is no relative phase shift between the two packets in channel 1.

these are shown in Figure 6.6(c). Now if the three sets of signals in Figure 6.6 are added together the twin packet of Figure 6.5(a) will be obtained.

It is clear that the behavior of individual frequency components (or groups of such components that acquire the same phase shift) is independent of all the other components; this is simply a statement of the principle of superposition for the linear system under consideration. Furthermore, the fraction of each component appearing in a given channel is only a function of the phase delay acquired by that component between arms 1 and 2, independent

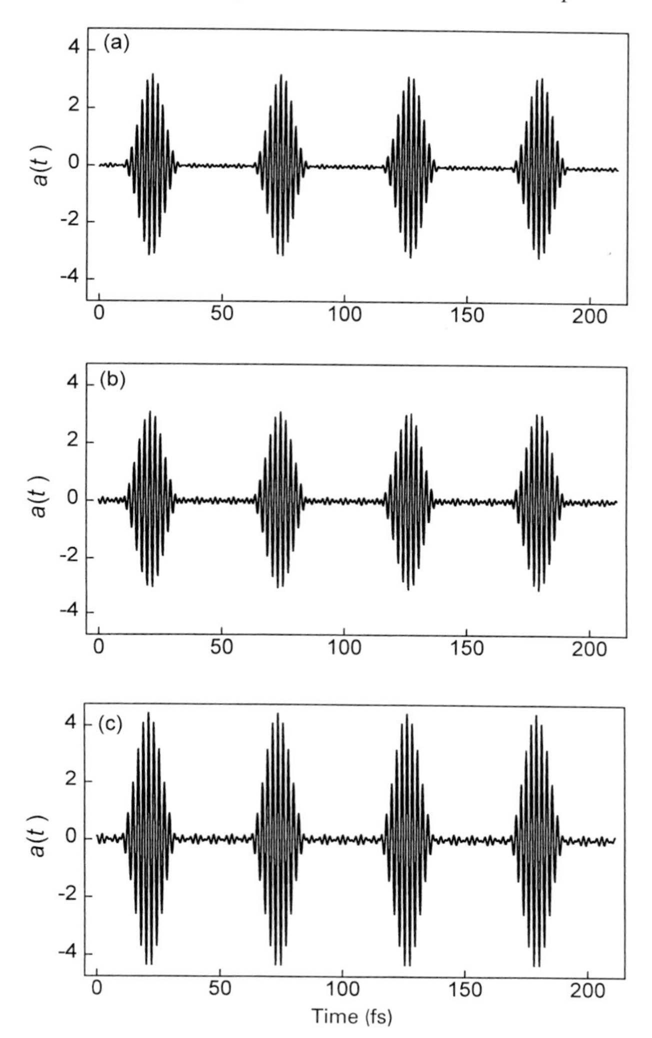

Figure 6.6 The spectrum of the wave packet in Figure 6.1(a) can be considered as the superposition of four groups of frequencies. One of these groups appears exclusively in channel 2. The other three groups appear in channel 1 either fully or partially. The waveforms shown here are those that would have appeared in channel 1 had the other groups been absent. When these three waveforms are added together they reconstruct the pair of wave packets shown in Figure 6.5(b).

of the original phase of that component. Remembering that the various frequency terms are orthogonal to each other, the behavior of the overall waveform within the interferometer must be independent of the initial phase of its individual components. Thus we see that the analysis of the packet of Figure 6.1(b) applies equally to the extended waveform of Figure 6.2. These different-looking functions share the same spectrum but have differing phase distributions over their common range of frequencies. In particular, the coherence length is equal to the width of the wave packet obtained by setting all ϕ_n equal to zero. The width of the packet, of course, is roughly equal to the inverse of its spectral bandwidth.

In addition to the phase angles ϕ_n initially present in, and those acquired during propagation of, a given wave packet, the field may accumulate further phase shifts due to dispersive elements (such as mirrors and prisms) in its path. These phase shifts manifest themselves as delays or distortions of the packet. It is of some interest, therefore, to study reflection and transmission delays caused by dispersive elements in order to evaluate their impact on interferometric measurements.

Delay upon reflection

As an example consider a 12-layer dielectric stack, Figure 6.7, consisting of alternating layers of quartz and strontium titanate. At the central wavelength of $\lambda_0 = 632.8$ nm the refractive indices of these materials are 1.46 and 2.39, respectively.[6] (The indices vary somewhat within the wavelength range of interest, and the corresponding dispersion is taken into account in the following calculations.) The thickness of the quartz layer is 108 nm and that of $SrTiO_3$ is 66 nm, each being a quarter-wave thick at λ_0. The stack is grown on a substrate whose central region has been subsequently removed. The hole thus created in the substrate is of no consequence for our analysis of reflec-

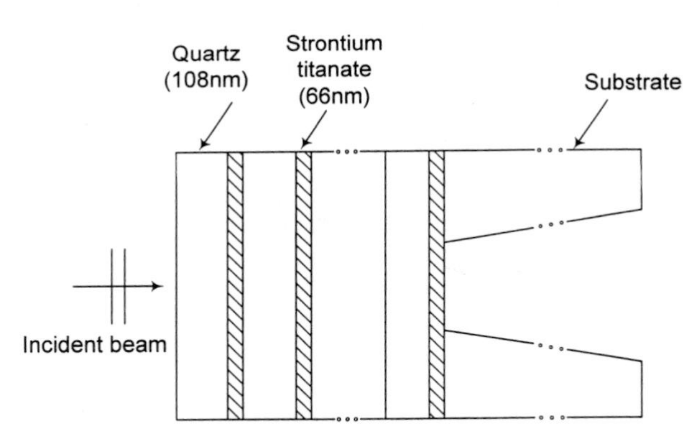

Figure 6.7 Schematic diagram of a quarter-wave stack consisting of six pairs of SiO_2/$SrTiO_3$ layers; the entire stack is 1044 nm thick. To simplify the analysis of the transmitted beam, the central region of the substrate is assumed to have been etched away. In calculating the reflection and transmission coefficients of the stack the wavelength dependence of the refractive indices of both types of layer has been taken into consideration.

tion, but it simplifies the discussion in the following section concerning transmission through the stack.

Figure 6.8 shows computed plots of amplitude and phase for the reflection and transmission coefficients of the stack in the frequency range covered by the wave packet of Figure 6.1. Note that, within the bandwidth of interest, the phase ϕ_r of the reflection coefficient is essentially a linear function of frequency with a slope of 1.5° per THz. This slope represents a 4.2 fs delay for the packet upon reflection from the stack. It might therefore be argued that, upon arrival at the surface, the packet spends 4.2 fs in exploring the

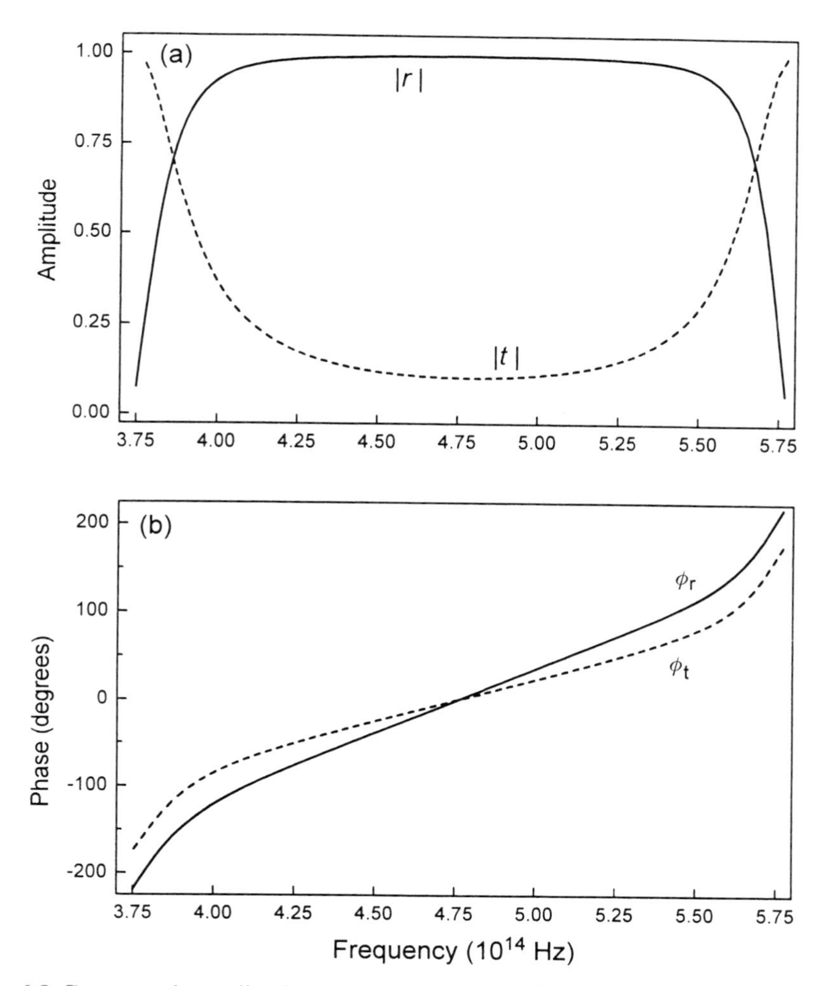

Figure 6.8 Computed amplitude and phase of the reflection and transmission coefficients r and t of the multilayer stack of Figure 6.7. The depicted range of frequencies covers the entire bandwidth of the wave packet shown in Figure 6.1.

stack before bouncing back. Roughly speaking, the delay may be associated with a penetration depth of 625 nm for this stack 1044 nm thick. (For an aluminum mirror the corresponding slope is found to be 0.03° per THz, leading to a reflection delay of 0.083 fs and an estimated penetration depth of only 12.5 nm.)

Delay upon transmission

For the wave packet transmitted through the stack of Figure 6.7 the slope of the phase $\phi_t(f)$ in Figure 6.8(b) is about 0.95° per THz, which amounts to a delay of $\Delta t = 2.6$ fs. Note, however, that the total thickness of the stack is 1044 nm, requiring 3.5 fs for the light to cover this distance at its vacuum speed c. It appears therefore that in passing through the stack the packet has exceeded the speed of light.[7–10] Since the special theory of relativity appears to have been violated, we take a closer look at the transmitted beam.

Note in Figure 6.8(a) that the transmitted amplitude $|t|$ is not constant over the range of frequencies of the wave packet but rises at both ends. This means that the actual transmitted spectrum is somewhat broadened (see Figure 6.9(a)). Taking into account the actual amplitude and phase of the transmission coefficient, we find the transmitted packet to be that of Figure 6.9(b). The peak of this packet is in fact delayed by about 2.6 fs, implying its faster-than-light propagation, but the entire packet is also compressed, which means that its starting point is about 5 fs behind that of the incoming packet (compare Figure 6.9(b) with Figure 6.1(c)). This delay of the starting point ensures that special relativity is not violated. Had we ignored the broadening of the spectrum and only included the phase shift $\phi_t(f)$ in our transmission calculations, we would have obtained the packet of Figure 6.9(c), which is only delayed relative to the incoming packet by 2.6 fs, in obvious violation of special relativity. Spectral broadening caused by the transmission curve of the stack thus results in a compression that ultimately delays the emergence of the packet, and in so doing reaffirms the impossibility of communication beyond the speed of light.

It is interesting to note that a measurement of the transmission delay by the interferometer of Figure 6.3 also leads to an apparent violation of special relativity. Such a measurement ostensibly determines the delay by measuring the peak of S_1 (the output of detector 1) when the multilayer stack is inserted in the fixed arm of the device and the movable arm is extended to maximize S_1. The corresponding signal (see Figure 6.10) is obtained by cross-correlat-

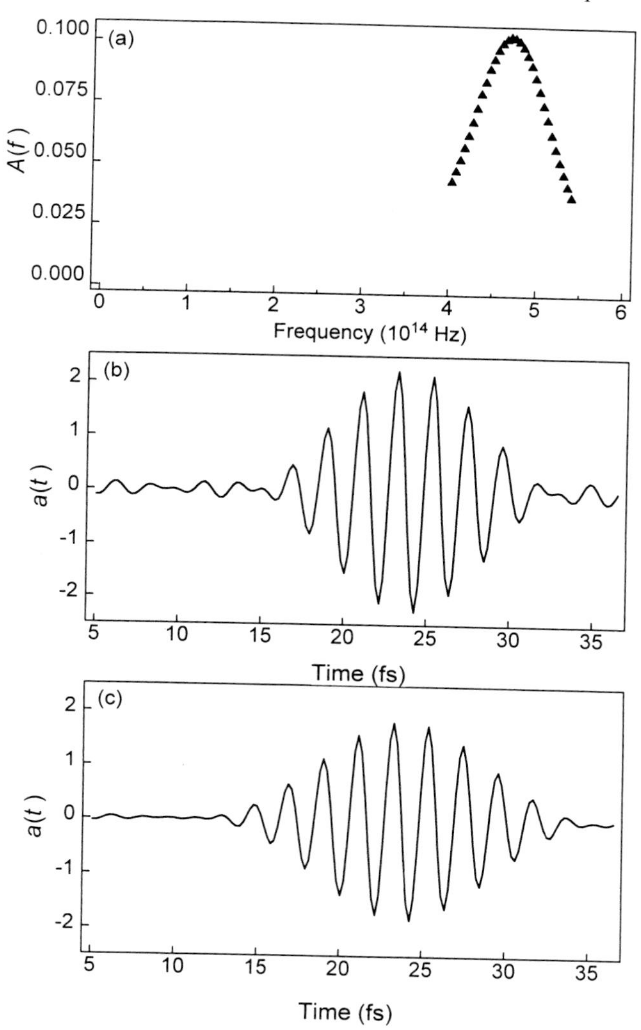

Figure 6.9 The wave packet transmitted through the stack of Figure 6.7 has a broadened spectrum as shown in (a). This spectral broadening, together with the linear phase shift $\phi_t(f)$ depicted in Figure 6.8(b), results in the compressed and delayed packet shown in (b). Had the spectral broadening been ignored and only the phase shift $\phi_t(f)$ taken into account, the transmitted packet would have resembled that in (c).

ing the wave packets of Figures 6.1(c) and 6.9(b). The peak of this curve occurs at $\Delta z = 0.4\,\mu\text{m}$, which is in agreement with the 2.6 fs delay calculated earlier. One must bear in mind, of course, that the interferometer measures the average delay of the packet upon transmission through the stack, and not the delay of its starting point.

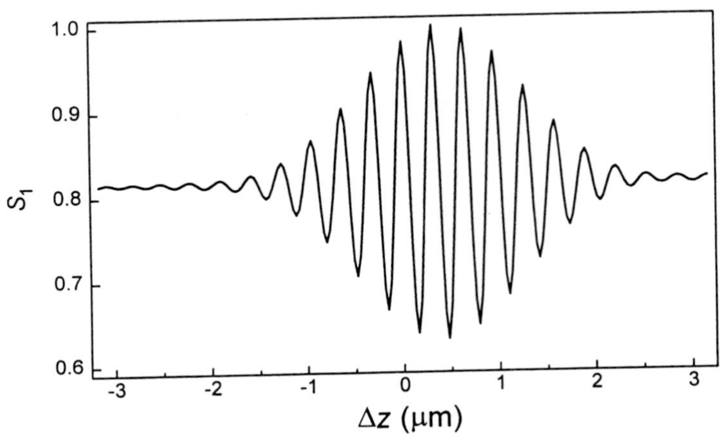

Figure 6.10 The signal S_1 of detector 1 versus the extension Δz of the movable end-reflector of the interferometer. The stack of Figure 6.7 is installed in the fixed arm while the adjustable arm is extended to compensate for the transmission delay through the stack. The incoming beam is assumed to be the wave packet of Figure 6.1(b)

References for chapter 6

1 M. Born and E. Wolf, *Principles of Optics*, sixth edition, Pergamon Press, Oxford, 1980.
2 L. Mandel and E. Wolf, *Optical Coherence and Quantum Optics*, Cambridge University Press, UK, 1995.
3 M. V. Klein, *Optics*, Wiley, New York, 1970.
4 R. Loudon, *The Quantum Theory of Light*, second edition, Clarendon Press, Oxford, 1992.
5 P. Meystre and M. Sargent III, *Elements of Quantum Optics*, Springer-Verlag, Berlin, 1990.
6 The refractive indices of strontium titanate in the wavelength range (500 nm, 800 nm) are taken from W. J. Tropf, M. E. Thomas, and T. J. Harris, *Handbook of Optics*, Vol. II, Michael Bass, editor, chapter 33, p. 33.72. Those of quartz and aluminum are taken from the *Handbook of Chemistry and Physics*, 67th edition, R. C. Weast, editor.
7 C. G. B. Garrett and D. E. McCumber, *Phys. Rev. A* **1**, 305, 1970.
8 S. Chu and S. Wong, *Phys. Rev. Lett.* **49**, 1293, 1982.
9 R. Y. Chiao, P. G. Kwiat, and A. M. Steinberg, Faster than light? *Scientific American*, 52–60, August 1993.
10 R. Y. Chiao and A. M. Steinberg, Tunneling times and superluminality, in *Progress in Optics*, ed. by E. Wolf, Vol. **37**, 347–406, Elsevier, Amsterdam, 1997.

7

The van Cittert–Zernike theorem

The beam of light emanating from a quasi-monochromatic point source (or a sufficiently distant extended source) is said to be spatially coherent: the reason is that, at any two points on a given cross-section of the beam, the oscillating electromagnetic fields maintain their relative phase at all times. If an opaque screen with two pinholes is placed at such a cross-section, Young's interference fringes will form, and the observed fringe contrast will be 100% (at and around the center of the fringe pattern). This is the sense in which the fields at two points are said to be spatially coherent relative to each other. If the relative phase of the fields at the two points varies randomly with time, the pair of point sources will fail to produce Young's fringes and, therefore, the fields are considered to be incoherent. In practice there is a continuum of possibilities between the aforementioned extremes, and the resulting fringe contrast may fall anywhere between zero and 100%. The fields at the two points are then said to be partially coherent with respect to one another, and the properly defined fringe contrast in Young's experiment is used as the measure of their degree of coherence.

Optical systems involving partially coherent illumination are explored in several other chapters of this book; see, for example, "Coherent and incoherent imaging" (chapter 5), "Michelson's stellar interferometer" (chapter 24), "Zernike's method of phase contrast" (chapter 27), and "Polarization microscopy" (chapter 28). Chapter 6 described first-order temporal coherence using a simple analytical method. A similar approach will be employed here to study first-order spatial coherence.

Coherence theory has been treated extensively in modern and classical textbooks, and it is not our intention here to repeat what is already well known.[1–4] Our goal is to present a simple derivation of the van Cittert–Zernike theorem without invoking the theories of probability and stochastic processes. This is possible because most sources of practical interest are ergodic, meaning that time-averaging over a typical waveform emanated by the source yields statistical information about the source's inherently random radiation processes. We shall make exclusive use of time-averaging to derive the degree of coherence of a pair of points within the field of an extended, quasi-monochromatic, incoherent source.

Time dependence, frequency spectrum, and phase

Consider a point source P, radiating into free space with a range of temporal frequencies at and around $f = f_0$. The discrete frequencies f_n are assumed to have a fixed spacing Δf as follows:

$$f_n = f_0 + n\Delta f = (N_0 + n)\Delta f. \tag{7.1}$$

At a given point in space, the (scalar) amplitude of the radiated waveform may be written

$$a(t) = \sum_n A_n(\Delta f)^{1/2}\cos(2\pi f_n t - \phi_n), \tag{7.2}$$

where A_n and ϕ_n are the amplitude and phase of the component whose frequency is f_n. The significance of the constant multiplier $(\Delta f)^{1/2}$, which is there for normalization purposes only, becomes clear shortly. We set the central frequency $f_0 = 5.454 \times 10^{14}$ Hz (corresponding to yellow light of wavelength $\lambda = 550$ nm) and choose $\Delta f = 5.454 \times 10^{12}$ Hz, which leads to $N_0 = 100$. We adopt a Gaussian shape for the distribution of the amplitudes A_n, as shown in Figure 7.1(a), and let the value of n in Eq. (7.2) range from -4 to $+4$, for a total of nine discrete wavelengths in the spectrum. To a large extent these choices are arbitrary but, as before, the points that we seek to clarify by way of examples based on these choices are quite general in nature.

Since in the Fourier domain the spectrum in Figure 7.1(a) is a discrete function of frequency, the corresponding amplitude $a(t)$ must be periodic, with a period of $T = 1/\Delta f \approx 183$ fs. A plot of $a(t)$ over a full period T is shown in Figure 7.1(b), where the values of ϕ_n at each frequency are chosen randomly and independently of each other. To increase the period T without changing the overall shape of the spectrum one must increase the rate of spectral sampling in Figure 7.1(a), by selecting additional frequencies in between those that are already chosen. In this way the spectrum retains its

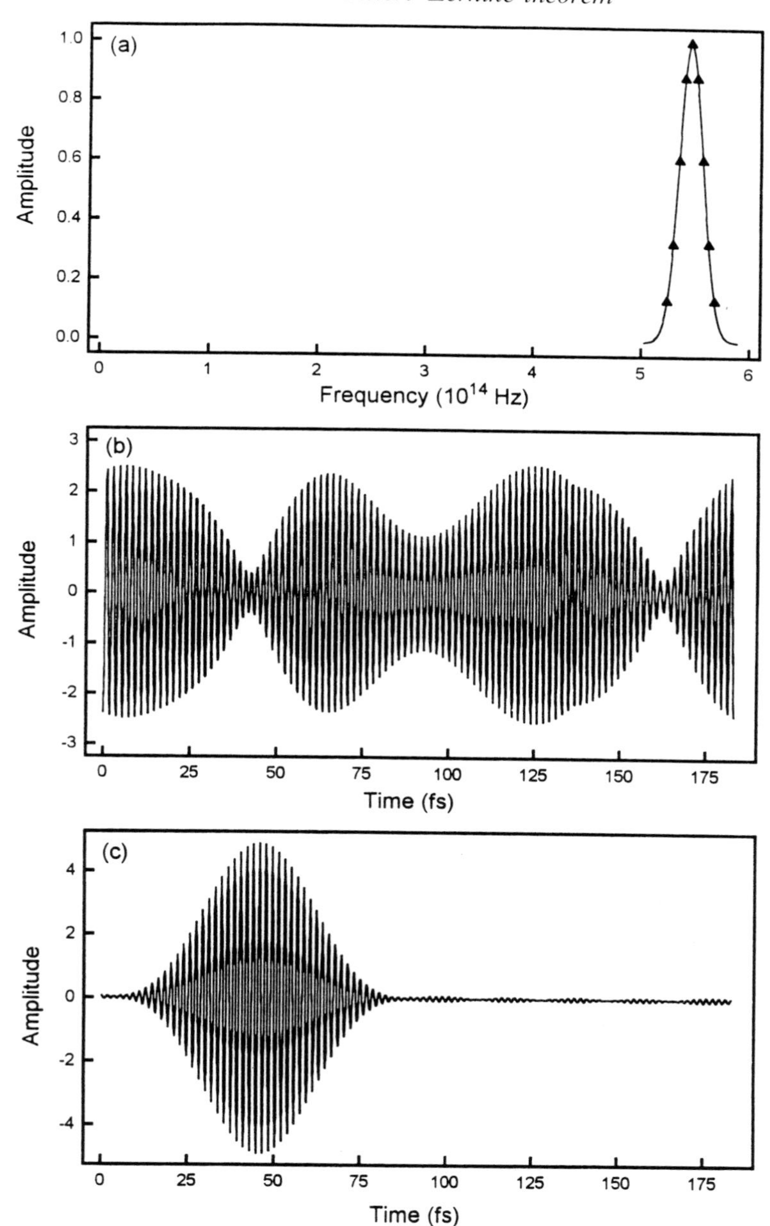

Figure 7.1 (a) A truncated Gaussian function sampled at regular intervals represents the frequency spectrum of a waveform. (b) The waveform obtained by Fourier-transforming the spectrum in (a), after assigning it a randomly selected phase at each frequency. Since the spectrum is sampled at $\Delta f = 5.454 \times 10^{12}$ Hz, its Fourier transform is repeated with a period of 183 fs. Only one period of the waveform is shown. (c) The waveform as a function of time derived by Fourier-transforming the spectrum in (a), assuming that its phase is a linear function of frequency.

shape, but Δf becomes smaller while T becomes larger. In the limit $\Delta f \to 0$ the period T of the waveform approaches infinity.

As far as the first-order coherence of a given waveform is concerned, the specific phase distribution over its spectral range is irrelevant, even though the shape of the function $a(t)$ may be significantly affected by this phase distribution. For example, in Figure 7.1(b) the value of ϕ_n at each frequency is chosen randomly, whereas if ϕ_n were chosen as a linear function of frequency then a waveform such as that of Figure 7.1(c) would have been obtained. There are many possible choices for $\{\phi_n\}$, and each one yields a more or less extended function of time such as that of Figure 7.1(b). Only in rare occasions does one find a compact wave packet similar to that of Figure 7.1(c). However, the compact packet has the same first-order coherence properties as the extended waveform.

Intensity

The average intensity of the waveform $a(t)$ in Eq. (7.2) is readily computed as follows:

$$\langle I \rangle = \frac{1}{T}\int_0^T a^2(t)\,\mathrm{d}t = \frac{1}{2}\sum_n A_n^2 \Delta f. \tag{7.3}$$

Note that the right-hand side of Eq. (7.3), being the area under the square of the spectral distribution function of Figure 7.1(a), remains constant as the sampling rate increases. Thus reducing Δf in order to increase the period T does not affect the average intensity.

Although the average intensity does not depend on $\{\phi_n\}$, the fluctuations in intensity are most definitely affected by this phase distribution. A thermal source (such as an incandescent lamp) tends to "assign" the values of ϕ_n randomly and independently of each other, thus resulting in significant fluctuations in $I(t)$. This behavior may be observed by examining the typical waveform in Figure 7.1(b). In fact, it can be shown that, for a thermal source, $\langle [I(t) - \langle I \rangle]^2 \rangle = \langle I \rangle^2$. However, it is possible to assign the phase angles in such a way as to minimize the intensity fluctuations. In a well-stabilized single-mode laser, for instance, the locking of the phase angles renders the root-mean-square fluctuations of intensity negligible, that is, $\langle I^2(t) \rangle = \langle I \rangle^2$. These considerations, however, pertain to higher-order statistics and, as far as first-order coherence is concerned, one could as well ignore the specific phase distribution.

The cross-correlation function

Consider two points P and P' on an extended source. The oscillations at these points are independent of each other. Thus the set of values $\{\phi_n\}$ and $\{\phi'_n\}$ assigned to $a_P(t)$ and $a_{P'}(t)$ may be considered independent. The cross-correlation function between the amplitudes at P and P' is given by

$$C_{PP'}(\tau) = \frac{1}{T}\int_0^T a_P(t)a_{P'}(t-\tau)\,\mathrm{d}t = \frac{1}{2}\sum_n A_n^2 \Delta f \cos(2\pi f_n \tau + \phi_n - \phi'_n). \quad (7.4)$$

Since ϕ_n and ϕ'_n are randomly selected with a uniform distribution over $[0, 2\pi]$, it follows that their difference $\phi_n - \phi'_n$ is also a random variable with the same distribution. The function $C_{PP'}(\tau)$ thus resembles $a(t)$ of Eq. (7.2) (with random phase angles), depicted in Figure 7.1(b). There is, however, a major difference between these functions: whereas Δf in Eq. (7.2) appears with a power $\frac{1}{2}$, the corresponding power of Δf in Eq. (7.4) is unity. This means that the average of $C^2_{PP'}(\tau)$ is inversely proportional to T, namely,

$$\frac{1}{T}\int_0^T C^2_{PP'}(\tau)\,\mathrm{d}\tau = \frac{1}{8T}\sum_n A_n^4 \Delta f. \quad (7.5)$$

Thus when $T \to \infty$ the magnitude of $C_{PP'}(\tau)$ for essentially all τ goes to zero, whereas in the same limit the average intensity $\langle I \rangle$ given by Eq. (7.3) remains non-zero. If the fields from P and P' are brought together in an attempt to create interference fringes, their combined intensity will be the sum of their individual intensities plus the cross-correlation term $C_{PP'}(\tau)$. Since $C_{PP'}(\tau) \to 0$ for sufficiently long T, the intensity of the sum will be the sum of individual intensities and, therefore, no fringes will be observed.

Interpretation

One may think of the radiation emanating from the two point sources P and P' in terms of two finite-duration wave packets (see Figure 7.1(c)). However, since the wave packets do not have a random relative phase it is impossible to get their cross-correlation to vanish. Nonetheless, we can assume that the packets are separated in time by an interval much longer than their individual widths and also much longer than any time delay that might occur in a system under consideration. In other words, as far as first-order coherence is concerned, an extended incoherent source emitting continuous radiation from its various points is equivalent to an identical source that emits relatively short bursts of light separated by long intervals. In this model of an incoherent, quasi-monochromatic, extended source each point emits only one pulse, no two points emit overlapping pulses, and all pulses from the various source locations have the same duration and shape.

As an example, consider an imaging system where a quasi-monochromatic spatially incoherent light source illuminates a sample, of which an image is formed on a photographic plate. One may imagine the individual points of the source as being independent coherent point sources, each creating a coherent image of the sample on the photographic plate. Because different points radiate at different times, there will be no interference among the various images. The photographic plate duly records the intensity pattern produced by each point source, automatically adding these images together as they arrive sequentially. The final image is thus the sum of the intensity distributions of all the coherent images produced by the various point sources.

Double pinhole interference

Figure 7.2 shows a quasi-monochromatic point source P, illuminating a screen located at $z = 0$. The screen is pierced with two small pinholes,

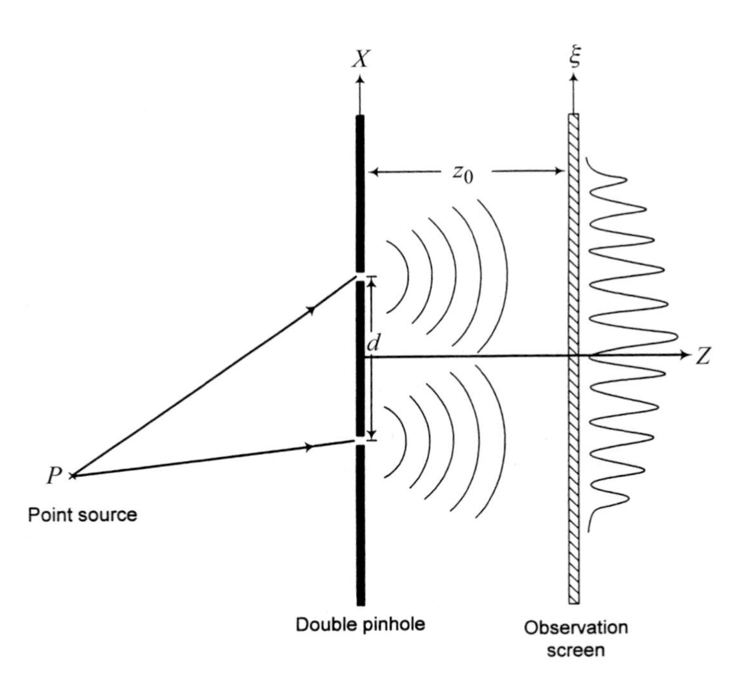

Figure 7.2 A pair of pinholes in the XY-plane at $z = 0$ is illuminated by a relatively distant point source at P. The resulting interference fringes are observed at the $\zeta\eta$-plane located at $z = z_0$.

which are separated by a distance d along the X-axis; their interference fringes are observed on the $\xi\eta$-plane at $z = z_0$. We assume that P is far enough away to yield equal intensities at the pinholes. Also the path-length difference $\Delta\ell$ from P to the pinholes must be short compared to the coherence length of the source, in order to give the pinhole radiations a high degree of temporal coherence. The relative phase between the fields at the pinholes, however, is not negligible and is given by $\Delta\phi = 2\pi\Delta\ell/\lambda$. This phase difference causes a translation of the fringe pattern along the ξ-axis. In the neighborhood of the origin at the observation plane the intensity distribution is given by

$$I(\xi) = \alpha I(P)\{1 + \cos[(2\pi/\lambda)(d/z_0)\xi + \Delta\phi]\}. \tag{7.6a}$$

Here α is an inconsequential proportionality constant and $I(P)$ is the intensity of the point source at P. Note that the fringe periodicity is independent of the location of the source P; it is determined solely by the values of d, z_0, and λ. The shift of the fringe pattern along the ξ-axis, however, is a function of $\Delta\phi$, which does depend on the location of the source.

For future reference we rewrite Eq. (7.6a) using complex notation as follows:

$$I(\xi) = \alpha I(P) + \alpha \operatorname{Re}\{I(P)\exp(\mathrm{i}\Delta\phi)\exp[\mathrm{i}2\pi d\xi/(\lambda z_0)]\}. \tag{7.6b}$$

The van Cittert–Zernike theorem

This theorem, which was first discovered by van Cittert[5] and later in a simpler form by Zernike,[6] relates the intensity distribution of an extended, quasi-monochromatic, planar source to the degree of spatial coherence observed on a parallel plane located at a relatively large distance from the source. Figure 7.3 shows a spatially incoherent, quasi-monochromatic source of wavelength λ in the $X'Y'$-plane at $z = -z_s$. The distance z_s between the source and the XY-plane at $z = 0$, on which we seek to determine the degree of coherence, is large enough that all the simplifying assumptions invoked in the previous sections still apply. We wish to determine the first-order coherence properties of the light that reaches the XY-plane at $z = 0$. We select two points (x_1, y_1) and (x_2, y_2) on this plane and assume that two pinholes are placed at these points. The light reaching the pinholes from a point source at $(x, y, -z_s)$ will have nearly the same amplitude but different phase. The phase difference at the pinholes is given by

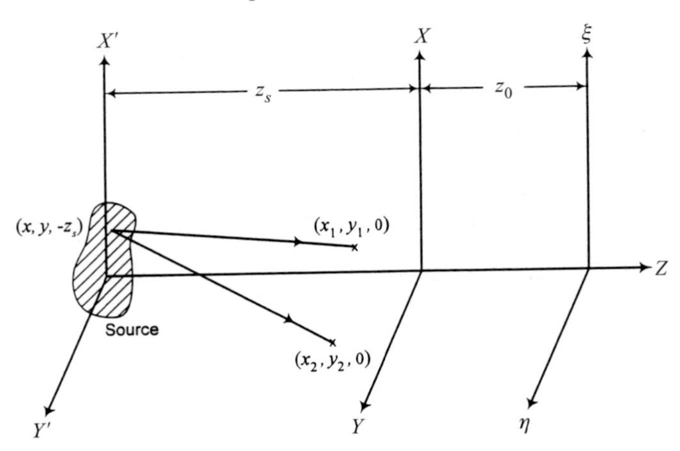

Figure 7.3 An extended, quasi-monochromatic, spatially incoherent, planar source is placed in the $X'Y'$-plane at $z = -z_s$. The light from each point $(x, y, -z_s)$ of this source reaches the points $(x_1, y_1, 0)$ and $(x_2, y_2, 0)$ on the XY-plane at $z = 0$. A pair of pinholes placed at the latter locations produces a fringe pattern in the $\zeta\eta$ observation plane at $z = z_0$. The superposition of all intensity distributions thus produced by the various points of the source yields the final intensity pattern at the observation plane.

$$\Delta\phi = 2\pi\Delta\ell/\lambda \approx \frac{2\pi}{\lambda z_s}\left\{\tfrac{1}{2}(x_1^2 + y_1^2) - \tfrac{1}{2}(x_2^2 + y_2^2)\right.$$
$$\left. - [(x_1 - x_2)x + (y_1 - y_2)y]\right\}. \qquad (7.7)$$

Consider what happens when all the point sources are active. They all act independently, each creating its own fringe pattern at the observation screen. All fringes thus produced will have the same period but different strengths and are shifted by different amounts along the ξ-axis. Because the point sources are completely incoherent, their overlapping fringe patterns must simply be added together. In other words, the final intensity distribution is the sum of Eq. (7.6) over all points P. We assume α to be the same for all the point sources. The fringe period $\lambda z_0/d$ is also the same. Therefore, the sum of Eq. (7.6b) over all point sources may be written as follows:

$$I(\xi) = \alpha \int_{\text{source}} I(x, y)\,\mathrm{d}x\,\mathrm{d}y$$
$$+ \alpha\,\mathrm{Re}\left(\left\{\int_{\text{source}} I(x, y)\exp[\mathrm{i}\Delta\phi(x, y)]\,\mathrm{d}x\,\mathrm{d}y\right\}\exp[\mathrm{i}2\pi d\xi/(\lambda z_0)]\right). \qquad (7.8)$$

To simplify the notation we define the following parameters:

$$I_0 = \alpha \int_{\text{source}} I(x, y)\,\mathrm{d}x\,\mathrm{d}y \tag{7.9a}$$

$$\hat{I}(x, y) = I(x, y) \Big/ \int_{\text{source}} I(x, y)\,\mathrm{d}x\,\mathrm{d}y \tag{7.9b}$$

$$\gamma(x_1, y_1; x_2, y_2) = \int_{\text{source}} \hat{I}(x, y) \exp[\mathrm{i}\Delta\phi(x, y)]\,\mathrm{d}x\,\mathrm{d}y. \tag{7.9c}$$

Equation (7.8) may then be rewritten as

$$I(\xi) = I_0\big(1 + \mathrm{Re}\big\{\gamma(x_1, y_1; x_2, y_2) \exp[\mathrm{i}2\pi d\xi/(\lambda z_0)]\big\}\big). \tag{7.10}$$

A comparison of Eqs. (7.10) and (7.6) reveals that the fringe contrast produced by the pinholes at (x_1, y_1) and (x_2, y_2) is equal to $|\gamma|$ and that the phase of γ determines the shift of these fringes from the center. The function γ is thus described as the complex degree of spatial coherence between (x_1, y_1) and (x_2, y_2). Substituting expression (7.7) for $\Delta\phi$ in Eq. (7.9c) yields

$$\begin{aligned}\gamma(x_1, y_1; x_2, y_2) = {}& \exp\left\{\mathrm{i}\pi[(x_1^2 + y_1^2) - (x_2^2 + y_2^2)]/(\lambda z_s)\right\} \\ & \times \int_{\text{source}} \hat{I}(x, y) \exp\{-\mathrm{i}2\pi[(x_1 - x_2)x \\ & \qquad + (y_1 - y_2)y]/(\lambda z_s)\}\,\mathrm{d}x\,\mathrm{d}y.\end{aligned} \tag{7.11}$$

Equation (7.11) is a compact statement of the van Cittert–Zernike theorem: aside from a phase factor, the complex degree of spatial coherence is the Fourier transform of the (normalized) intensity distribution at the incoherent source.

Example

Consider the uniform, quasi-monochromatic, incoherent source depicted in Figure 7.4(a). The source's central wavelength is λ, and its linear dimensions are 3250λ on each side. A square array of 13×13 independent point sources on a rectangular mesh (with spacing 250λ) is used to simulate this source. A pair of pinholes in an otherwise opaque screen is located at $z_s = 10^7\lambda$ from the source. The square pinholes shown in Figure 7.4(b) are each of side-length 350λ and separated by a distance d along the X-axis. The light from the source, having gone through the pinholes, arrives at the observation plane located at $z_0 = 10^6\lambda$.

Figure 7.5 shows the computed fringe patterns at the observation plane for four different values of d. Note that with increasing d the fringe period decreases. The fringe contrast also declines at first, going to zero when $d = 3333\lambda$. Subsequently, however, the contrast increases as d continues to

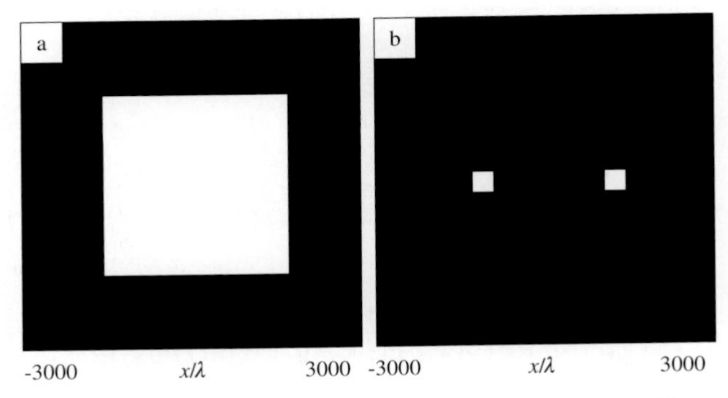

Figure 7.4 (a) Intensity distribution over the surface area of a uniform, quasi-monochromatic, incoherent source. The linear dimensions of the source are 3250λ along each side, where λ is the wavelength of its radiation. (b) A pair of square pinholes each measuring 350λ along each side. The center-to-center spacing d between the pinholes is an adjustable parameter of the simulations.

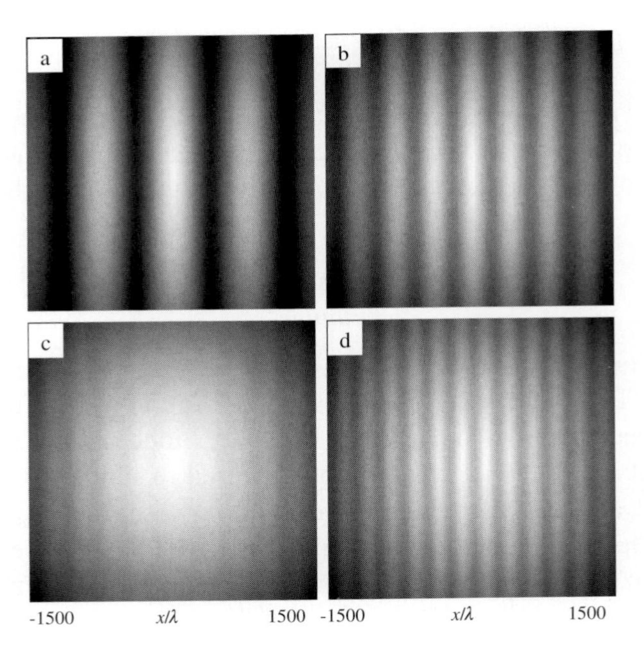

Figure 7.5 Computed intensity distributions in the vicinity of the optical axis at the observation plane of Figure 7.3 for the source and pinholes of Figure 7.4. The distance between the source and the plane of the pinholes is $z_s = 10^7\lambda$, while the distance between the pinholes and the observation screen is $z_0 = 10^6\lambda$. Each frame corresponds to a different spacing d between the pinholes: (a) $d = 1250\lambda$; (b) $d = 2500\lambda$; (c) $d = 3333\lambda$; (d) $d = 4000\lambda$.

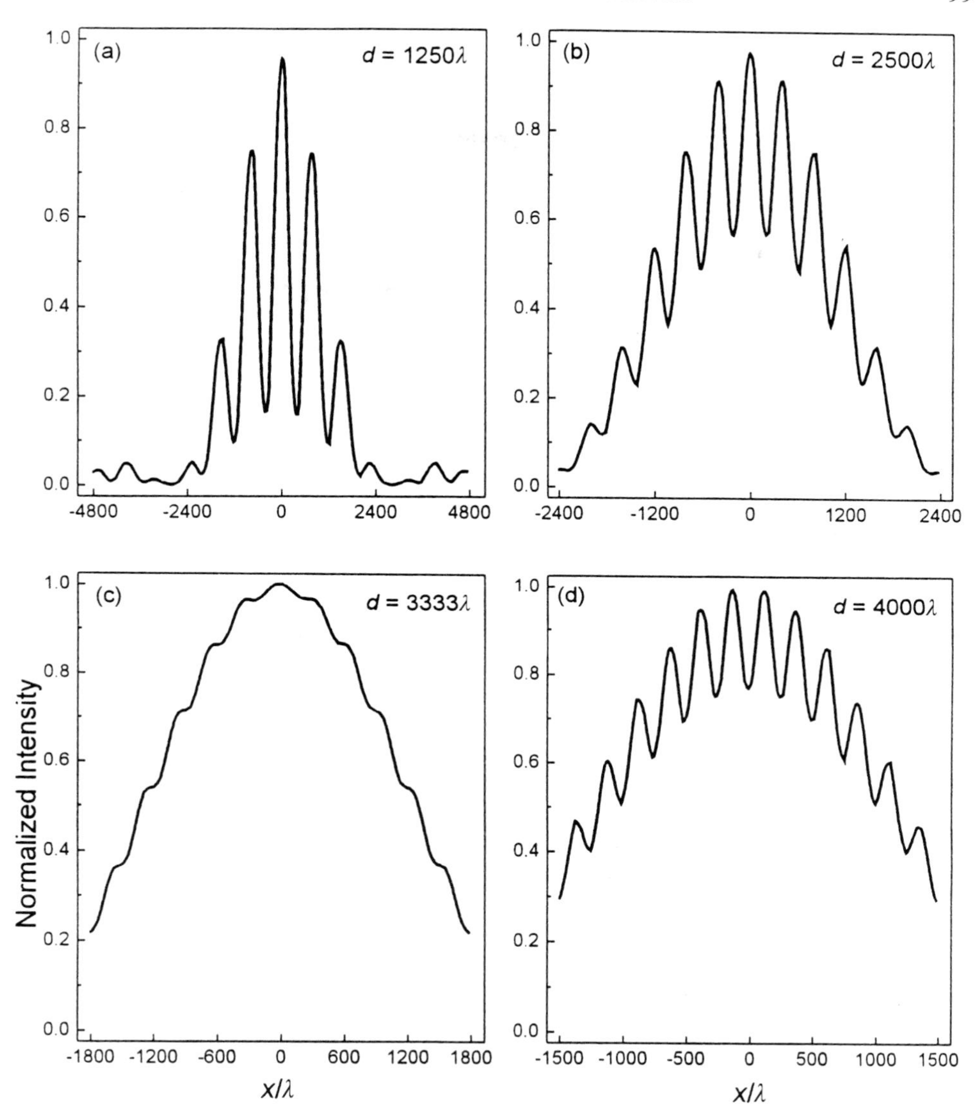

Figure 7.6 Cross-sectional view of the intensity distributions of Figure 7.5. (Note: the scale of the horizontal axis is different for each plot.) The fringe contrast, which is ≈ 0.75 in (a), drops to ≈ 0.3 in (b), and is effectively zero in (c). With the increasing of the pinhole separation the contrast climbs up once again to ≈ 0.15 in (d). Whereas the central fringe in (a) and (b) is bright, it is dark in (d).

increase. Whereas in frames (a) and (b) the central fringe is bright, in frame (d) corresponding to $d = 4000\lambda$ the central fringe becomes dark. This is equivalent to a half-period shift of the pattern upon crossing the point of zero contrast.

Figure 7.6 shows cross-sections of the fringe patterns of Figure 7.5. The contrast calculated from these plots can be shown to be in good agreement with the values predicted by the van Cittert–Zernike theorem.

References for chapter 7

1 M. Born and E. Wolf, *Principles of Optics*, sixth edition, Pergamon Press, Oxford, 1980.
2 L. Mandel and E. Wolf, *Optical Coherence and Quantum Optics*, Cambridge University Press, UK, 1995.
3 M. V. Klein, *Optics*, Wiley, New York, 1970.
4 R. Loudon, *The Quantum Theory of Light*, second edition, Clarendon Press, Oxford, 1992.
5 P. H. van Cittert, *Physica* **1**, 201 (1934).
6 F. Zernike, *Physica* **5**, 785 (1938).

8

Partial polarization, Stokes parameters, and the Poincaré sphere

A strictly monochromatic plane wave is fully polarized; to obtain partial polarization one must consider a superposition of two or more plane waves of differing wavelengths. A collimated beam of light is considered to be fully polarized if a quarter-wave plate followed by an ideal polarizer can be used to extinguish the beam. Failure at extinction reveals the beam as either fully or partially unpolarized.

In the classical literature it is customary to analyze the degree of polarization of a beam of light in terms of the cross-correlation function between two orthogonal components of the beam's E-field.[1–3] It is somewhat easier, however, to carry out the same calculations in the frequency domain and so to derive the relevant parameters as integrals over the frequency spectrum of the beam. One advantage of the latter approach is that it applies to beams of arbitrary bandwidth, thus removing from the results the restriction to quasi-monochromaticity. Another advantage is that it avoids the use of mutual coherence, which, at times, tends to confuse discussion of the subject. In the following sections we show how a frequency-domain analysis leads to a compact expression for the degree of polarization of a polychromatic beam of light in terms of its Stokes parameters.

Orthogonal polarization components

Consider a polychromatic beam of light propagating along the Z-axis and possessing polarization components in both the X- and the Y-direction:

$$E_x(z, t) = \sum_n A_n(\Delta f)^{1/2} \cos[2\pi f_n(t - z/c) + \phi_n], \tag{8.1a}$$

$$E_y(z, t) = \sum_n B_n(\Delta f)^{1/2} \cos[2\pi f_n(t - z/c) + \psi_n]. \tag{8.1b}$$

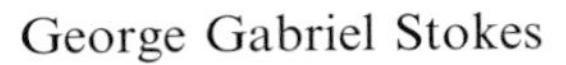

George Gabriel Stokes

Jules Henri Poincaré

Sir George Gabriel Stokes (1819–1903). Irish-born mathematical physicist; he spent most of his adult life at Cambridge, where he held the Lucasian chair for over half a century. He was an intimate friend of Lord Kelvin and James Clerk Maxwell. Stokes' most important researches were concerned with hydrodynamics, optics, and geodesy. In optics he was mainly responsible for the explanation of fluorescence, and made significant contributions to the theory of diffraction. He was generous in sharing his ideas with colleagues and students and readily gave credit to others when there were any priority disputes. A few days after his death, *The Times* of London wrote in an obituary that "Sir G. Stokes was remarkable . . . for his freedom from all personal ambitions and petty jealousies." (Photo: courtesy of AIP Emilio Segré Visual Archives, E. Scott Barr Collection.)

Jules Henri Poincaré (1854–1912), received his doctorate in mathematics from the University of Paris in 1879, and was appointed, in 1886, to the chair of mathematical physics at the Sorbonne and to a chair at the Ecole Polytechnique. Having made significant contributions to many aspects of mathematics, physics, and philosophy, Poincaré is often described as one of the great geniuses of all time and as the last universalist in mathematics. In applied mathematics he studied optics, electricity, telegraphy, capillarity, elasticity, thermodynamics, potential theory, the theory of relativity, and cosmology. His studies of the three-body problem in celestial mechanics mark the beginning of modern chaos theory. He is acknowledged as a co-discoverer, with Albert Einstein and Hendrik Lorentz, of the special theory of relativity. (Photo: Percy Bridgman Collection, courtesy of AIP Emilio Segré Visual Archives.)

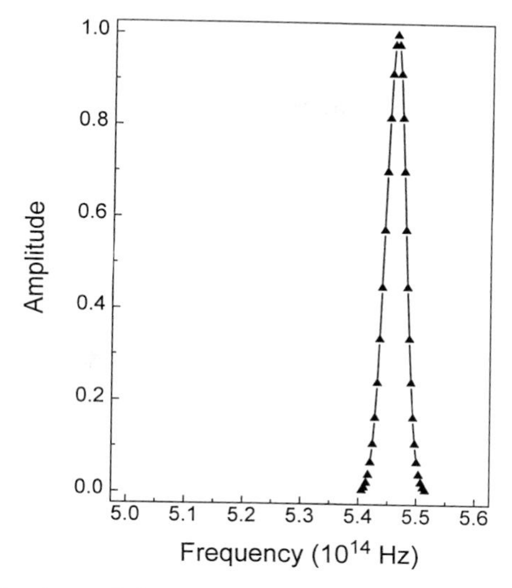

Figure 8.1 Distribution of the amplitudes of a polychromatic beam in the frequency range $(5.4054-5.5146) \times 10^{14}$Hz. The arrowheads represent the sampled values of the amplitude spectrum at intervals of $\Delta f = 3.64 \times 10^{11}$Hz. The central frequency $f_0 = 1500\Delta f = 5.46 \times 10^{14}$ Hz corresponds to the wavelength $\lambda = 549.5$ nm; the upper and lower bounds of the spectrum are at $\lambda = 544$ nm and 555 nm, respectively. The sampled amplitudes are representative of A_n and/or B_n in Eqs. (8.1). Associated with each sample is a corresponding phase ϕ_n or ψ_n (not shown).

This beam's spectrum consists of individual frequencies $f_n = (N_0 + n)\Delta f$ within a finite bandwidth around the central frequency $f_0 = N_0\Delta f$, as depicted in Figure 8.1. The term associated with f_n along the X-axis has amplitude A_n and phase ϕ_n; the corresponding term along the Y-axis has amplitude B_n and phase ψ_n; c is the speed of light in vacuum, and the constant multiplier $(\Delta f)^{1/2}$ is for normalization purposes only, its significance becoming clear as the discussion proceeds.

As described by Eqs. (8.1), the contribution to the beam of each frequency term f_n is a fully polarized plane wave. For this plane wave, which is elliptically polarized in general, one can determine the ellipticity and the orientation of the ellipse of polarization in terms of A_n, B_n, and $\phi_n - \psi_n$. However, the superposition of different frequency terms, each having a different state of polarization, results in partially polarized light.

Ideal phase-retarder and polarizer; transmitted power

To determine the degree of polarization of the beam described by Eqs. (8.1), we assume a perfect retarder and a perfect polarizer placed in the path of the

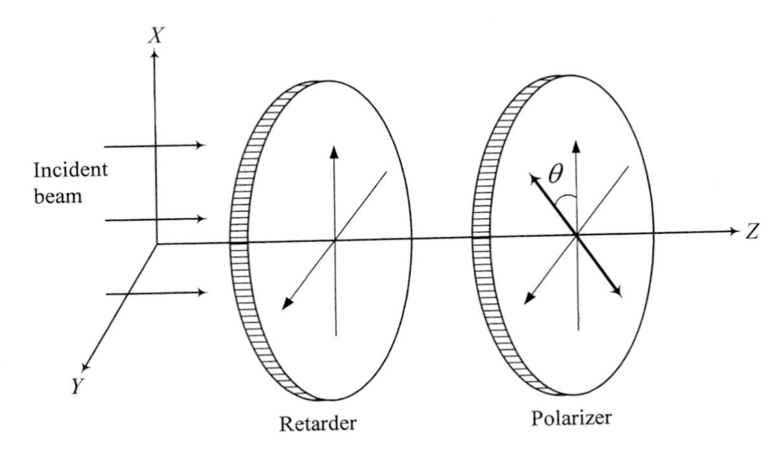

Figure 8.2 A polychromatic beam of light propagating along the Z-axis is sent through a variable retarder and a polarizer. The retarder's fast and slow axes are fixed along the X- and Y- directions, but its phase shift χ as well as the polarizer's orientation angle θ may be adjusted to minimize the amount of light that is transmitted through the system. χ must be the same for all the wavelengths contained in the incident beam.

beam, orthogonal to the propagation direction, as in Figure 8.2. The variable retarder induces a phase delay χ between E_x and E_y, the value of χ being adjustable in the range $\pm 180°$. It is imperative for the following analysis that χ be independent of optical frequency within the bandwidth of interest; in other words, for the beam described by Eqs. (8.1) the phase delay χ must be the same for all f_n contained in the spectrum. (A retarder based on total internal reflection provides a good approximation to an ideal retarder; this is the same principle of operation as used in a Fresnel rhomb.[1]) Such a retarder can modify the shape of the cross-correlation function between E_x and E_y, but, contrary to what has been asserted in the literature, it cannot destroy the mutual coherence between these components of polarization. The confusion is perhaps rooted in the fact that a time delay between E_x and E_y can destroy the mutual coherence, whereas a frequency-independent phase shift χ leaves the mutual coherence essentially intact.

The beam is subsequently passed through the polarizer, which may be rotated around the Z-axis until the transmitted optical power is minimized. The retardation χ is then adjusted and the orientation θ of the polarizer is changed accordingly until the transmitted power reaches its absolute minimum. The optimum retardation χ_0 together with the optimum orientation θ_0 of the polarizer thus obtained determine the state of polarization of that fraction of the beam which is fully polarized. The minimum transmitted power is a measure of the unpolarized content of the beam.

In the system of Figure 8.2 the amplitude of the light emerging from the polarizer is

$$E(z=0,t) = \sum_n \big[A_n \cos(2\pi f_n t + \phi_n)\cos\theta + B_n \cos(2\pi f_n t + \psi_n + \chi)\sin\theta\big](\Delta f)^{1/2}. \tag{8.2}$$

Because all frequencies f_n in Eq. (8.2) are integer multiples of Δf, namely, $f_n = (N_0 + n)\Delta f$, the transmitted amplitude $E(z=0,t)$ is a periodic function of time, with period $T = 1/\Delta f$. The time-averaged transmitted intensity as a function of χ and θ is thus given by

$$I(\chi,\theta) = \frac{1}{T}\int_0^T E^2(z=0,t)\,\mathrm{d}t = \frac{1}{2}\sum_n \big[A_n^2\cos^2\theta + B_n^2\sin^2\theta + A_n B_n \sin(2\theta)\cos(\phi_n - \psi_n - \chi)\big]\Delta f. \tag{8.3}$$

The presence of Δf in the above expression allows a smooth transition from the discrete sum to a continuous integral in the limit $\Delta f \to 0$; this, of course, is the same limit in which $T \to \infty$.

Stokes parameters

To streamline the calculation of the values of χ and θ that minimize $I(\chi,\theta)$, we follow Sir George Gabriel Stokes (1819–1903) in defining the four parameters that now bear his name:[4]

$$S_0 = \frac{1}{2}\sum_n (A_n^2 + B_n^2)\Delta f, \tag{8.4a}$$

$$S_1 = \frac{1}{2}\sum_n (A_n^2 - B_n^2)\Delta f, \tag{8.4b}$$

$$S_2 = \sum_n A_n B_n \cos(\phi_n - \psi_n)\Delta f, \tag{8.4c}$$

$$S_3 = \sum_n A_n B_n \sin(\phi_n - \psi_n)\Delta f. \tag{8.4d}$$

To minimize the transmitted intensity in Eq. (8.3) we first set the derivative of $I(\chi,\theta)$ with respect to χ equal to zero. This yields χ_0, independently of the value of θ, as follows:

$$\chi_0 = \arctan(S_3/S_2). \tag{8.5a}$$

Substituting χ_0 for χ in Eq. (8.3) and differentiating with respect to θ, we find the optimum θ_0 as

$$\theta_0 = \tfrac{1}{2}\arctan\left[(S_2/S_1)\cos\chi_0 + (S_3/S_1)\sin\chi_0\right]. \quad (8.5b)$$

The transmitted intensity thus turns out to have a minimum at (χ_0, θ_0) and a maximum at $(\chi_0, \theta_0 + 90^\circ)$, or vice versa. These values are given by

$$I_{\text{min}} = \tfrac{1}{2}S_0 - \tfrac{1}{2}(S_1^2 + S_2^2 + S_3^2)^{1/2}, \quad (8.6a)$$

$$I_{\text{max}} = \tfrac{1}{2}S_0 + \tfrac{1}{2}(S_1^2 + S_2^2 + S_3^2)^{1/2}. \quad (8.6b)$$

Degree of polarization

The minimum transmitted intensity I_{min} in Eq. (8.6a), being that part of the beam which cannot be extinguished with a retarder and a polarizer, represents the depolarized content of the beam. This, of course, is only half the total amount of depolarized light, because the same amount must also be contained in I_{max}. The total amount of depolarized light, therefore, is $2I_{\text{min}}$, while the remaining part, $I_{\text{max}} - I_{\text{min}}$, is fully polarized. The degree of polarization P of the beam may thus be defined as

$$P = (I_{\text{max}} - I_{\text{min}})/(I_{\text{max}} + I_{\text{min}}) = \left[(S_1/S_0)^2 + (S_2/S_0)^2 + (S_3/S_0)^2\right]^{1/2}. \quad (8.7)$$

Using the Schwartz inequality,[5] it is not difficult to show that $S_1^2 + S_2^2 + S_3^2 \le S_0^2$; consequently, $0 \le P \le 1$. (See Note 1 at the end of the chapter.)

One may question the generality of the above result because, in deriving it, the fast and slow axes of the wave-plate were fixed along the X- and Y- axes. In other words, one wonders if the result would have been different had the axes of the wave-plate been allowed to rotate around the Z-axis. The result can be shown to be quite general, however, because P of Eq. (8.7) remains invariant under a rotation of the XY-plane around Z. The value of S_0, being the total power of the beam, obviously remains the same for arbitrary orientations of the coordinate system. Moreover, with some elementary algebra, the quantity $S_1^2 + S_2^2 + S_3^2$ may also be shown to be invariant under coordinate rotation. (See Note 2 at the end of the chapter.)

In retrospect the variable retarder of Figure 8.2 could have been replaced by an achromatic quarter-wave plate (e.g., a Fresnel rhomb) in a rotary mount. The axes of the quarter-wave plate could then be made to coincide with the axes of the ellipse of polarization in order to linearize that part of the beam which is fully polarized. This is precisely what the variable retarder accomplishes in that it adjusts the retardation χ while maintaining a fixed orientation in the XY-plane.

The Poincaré sphere

In general, the fraction of the beam that is fully polarized has elliptical polarization, with ellipticity η and orientation angle ρ (this is the angle between the major axis of the ellipse and the X-axis). These parameters may be readily expressed in terms of the Stokes parameters:

$$\sin(2\eta) = S_3/(S_1^2 + S_2^2 + S_3^2)^{1/2}, \tag{8.8a}$$

$$\tan(2\rho) = S_2/S_1. \tag{8.8b}$$

Using the above relations, the French mathematical physicist Henri Poincaré (1854–1912) represented the state of polarization as a point S on the surface of a sphere, as shown in Figure 8.3. In this representation the three Cartesian coordinates of S are S_1, S_2, and S_3. Thus, according to Eq. (8.7), the radius of the Poincaré sphere is PS_0, the power of that fraction of the beam which is fully polarized. The latitude of S is twice the ellipticity η of the polarized component, in accordance with Eq. (8.8a), while the longitude of S represents twice the orientation angle ρ of the major axis of the ellipse of polarization, as prescribed by Eq. (8.8b).

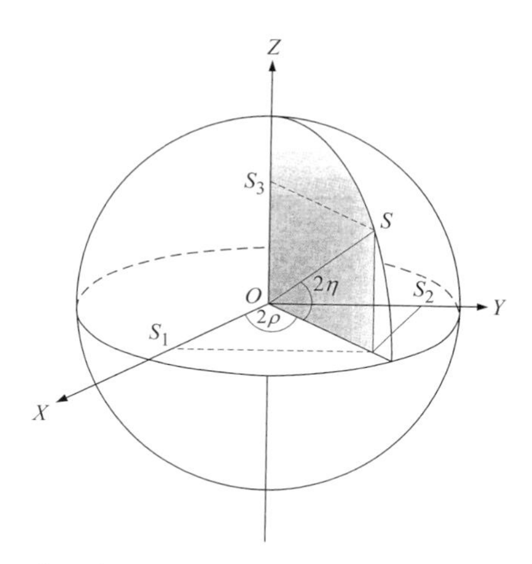

Figure 8.3 The Poincaré sphere is the location of all points S with coordinates $(x, y, z) = (S_1, S_2, S_3)$. The radius of the sphere is PS_0, and the latitude and longitude of S specify the ellipticity η and orientation angle ρ of the polarized component of the beam.

Unpolarized light

A completely unpolarized beam of light cannot be altered by the wave-plate and polarizer of Figure 8.2. No matter what the phase shift χ of the retarder and the orientation θ of the polarizer may be, the output power will be one-half the input power. For this light S_0 will be the total power of the beam, but $S_1 = S_2 = S_3 = 0$. Since $S_1 = 0$, the relation $\sum A_n^2 \Delta f = \sum B_n^2 \Delta f$ implies that the power along the X-axis equals that along the Y-axis. For natural light, where the polarization components along the X- and Y- axes are independent of each other, the relative phase angles $\phi_n - \psi_n$ are uniformly distributed over $(0, 2\pi)$ and tend to be a random function of n. Hence, in the limit $\Delta f \to 0$, the Stokes parameters S_2 and S_3 approach zero as well. However, there exist other combinations of ϕ_n and ψ_n that yield totally unpolarized light. For example, a superposition of two equal-magnitude beams of frequencies f_1 and f_2, where one beam is right- and the other left-circularly polarized, can be readily shown to be fully unpolarized.

Partial depolarization by a glass slab upon reflection or transmission

Figure 8.4 shows a glass slab 100 μm thick and of refractive index $n = 1.5$, upon which a linearly polarized beam is incident at an oblique angle $\gamma = 75°$. The incident beam has equal amounts of p- and s-polarization with equal phase, giving its linear polarization a 45° angle relative to both p- and s-directions. The spectral content of the beam is that depicted in Figure 8.1.

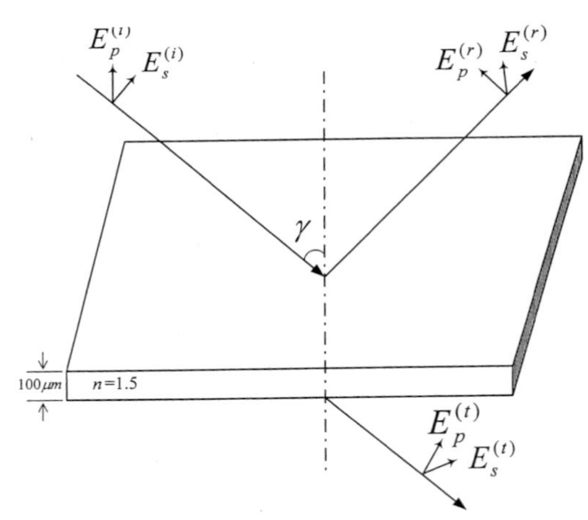

Figure 8.4 A polychromatic plane wave is incident on a glass slab 100 μm thick at $\gamma = 75°$. The index of refraction of the glass, $n = 1.5$, is independent of the wavelength. The incident beam is linearly polarized at 45° to the plane of incidence, that is, it has equal amounts of p- and s-polarization. The reflected and transmitted beams are slightly depolarized.

Upon reflection from the slab the computed amplitudes of the p- and s-components of the beam as functions of λ are depicted in Figure 8.5(a). Multiple reflections at the two facets of the slab interfere with each other to produce the fine structure seen in the spectra of Figure 8.5(a). The phase angles of the reflected p- and s- components are shown in Figure 8.5(b), and the resulting polarization rotation angle ρ and ellipticity η appear in Figure 8.5(c). The knowledge of these quantities allows one to compute the Stokes parameters from Eqs. (8.4), yielding $S_1/S_0 = -0.495$, $S_2/S_0 = -0.844$, $S_3/S_0 = -0.89 \times 10^{-6}$. Thus the degree of polarization of the reflected beam is $P = 0.978$, the wave-plate's required phase shift χ_0 is very small, 0.00006°, and the polarizer's angle for minimum transmission must be set to $\theta_0 = 29.8°$. It is seen that the polarized content of the reflected beam is essentially linear ($\eta = -0.000026°$) and is oriented at $\rho = -60.2°$ relative to the p-direction.

Similar results may be obtained for the beam transmitted through the slab. The corresponding amplitudes and phases are shown in Figure 8.6,

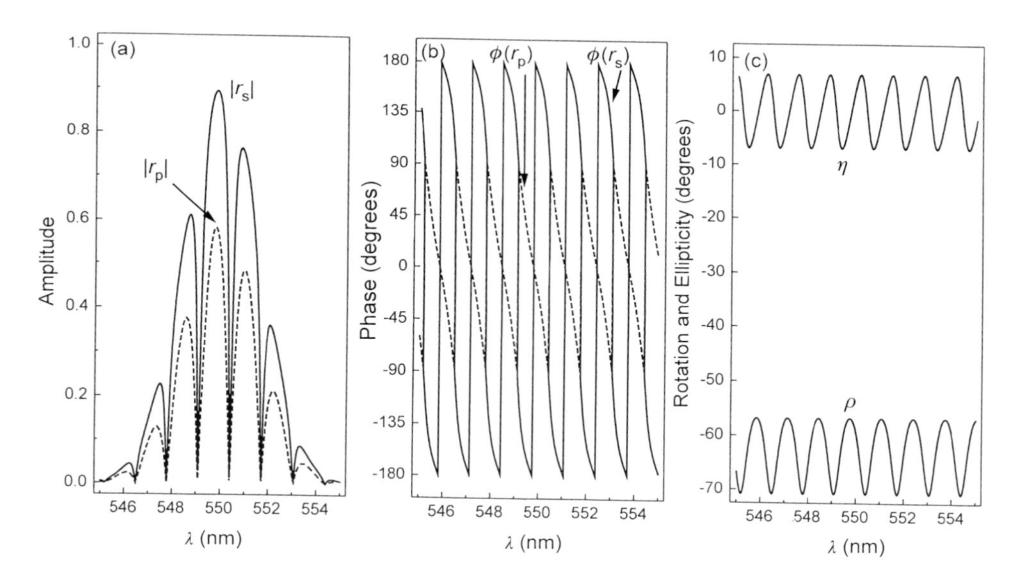

Figure 8.5 A polychromatic plane wave, having the spectrum of Figure 8.1 and a linear polarization at 45° to the plane of incidence, is reflected from a glass slab at $\gamma = 75°$ (see Figure 8.4). Shown as functions of λ: (a) the reflected amplitudes $|r_p|$ (broken line) and $|r_s|$ (solid line); (b) the phase angles of r_p (broken line) and r_s (solid line); (c) the reflected polarization state, defined by the rotation angle ρ and ellipticity η. For the reflected beam the computed degree of polarization is $P = 0.978$, the polarized component is essentially linear ($\eta = -0.000026°$), and the polarization vector makes an angle $\rho = -60.2°$ with the p-direction.

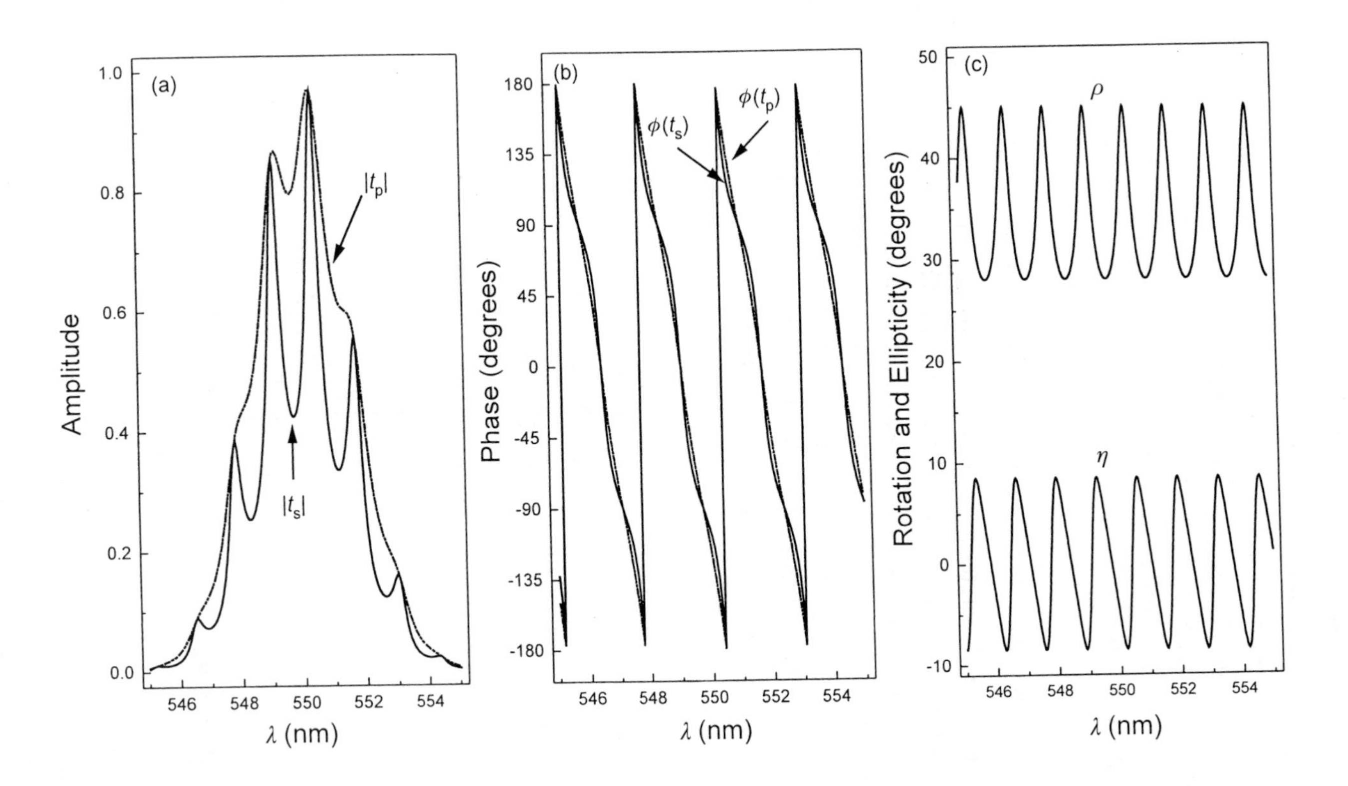

Figure 8.6 The counterpart of Figure 8.5 for the case of transmission through the glass slab 100 μm thick. The computed degree of polarization of the transmitted beam is $P = 0.957$, the polarized component is essentially linear ($\eta = 0.000022°$), and the polarization vector makes an angle $\rho = 35.7°$ with the p-direction.

and the Stokes parameters are found to be $S_1/S_0 = 0.306$, $S_2/S_0 = 0.907$, $S_3/S_0 = 0.7 \times 10^{-6}$. Thus the degree of polarization is $P = 0.957$, the wave-plate's required phase shift is $\chi_0 = 0.000047°$, and the polarizer's angle for minimum transmission is $\theta_0 = -54.3°$. Therefore, the polarized content of the transmitted beam, oriented at $\rho = 35.7°$ relative to the p-direction, is essentially linear ($\eta = 0.000022°$).

Partial depolarization upon transmission through a birefringent slab

Figure 8.7 shows the characteristics of a polychromatic beam of light upon transmission through a birefringent slab of calcite. The thickness of the slab is 85 μm and its ordinary and extraordinary refractive indices are $n_o = 1.6613$ and $n_e = 1.488$. The normally incident beam, which is linearly polarized at 45° to the crystal axes, has the spectrum of Figure 8.1. The computed Stokes parameters of the transmitted beam are: $S_1/S_0 = -0.023$, $S_2/S_0 = 0.259$, $S_3/S_0 = -0.902$. Thus the degree of polarization of the

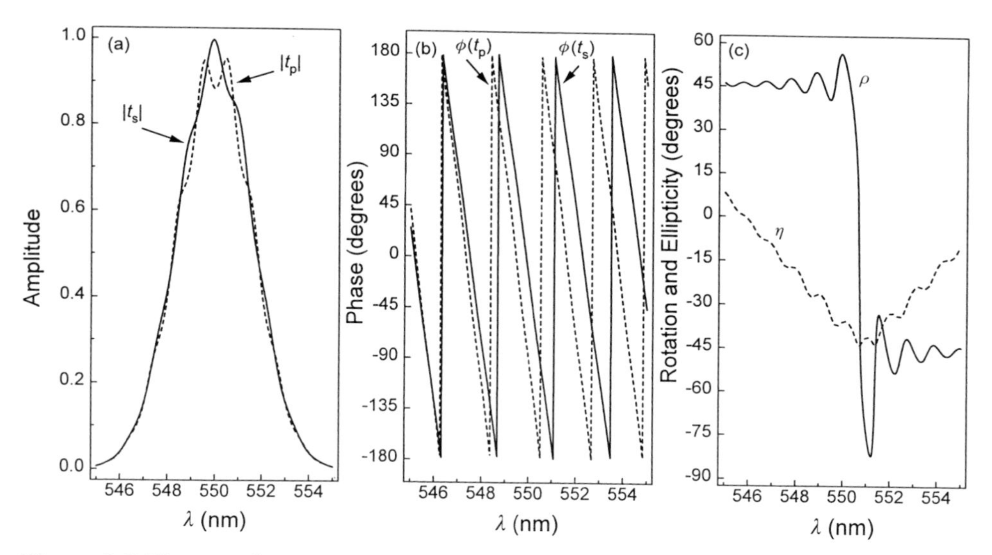

Figure 8.7 The amplitude, phase, and polarization state of a polychromatic beam, having the spectrum of Figure 8.1, upon transmission through a calcite slab 85 μm thick ($n_0 = 1.6613$, $n_e = 1.488$). The normally incident plane wave is linearly polarized at 45° to the crystal axes. (a) Transmitted amplitudes of the p-component (broken line) and s-component (solid line) versus λ. (b) Phase angles of the p-component (broken line) and s-component (solid line) versus λ. (c) Polarization rotation angle ρ and ellipticity η of the transmitted beam versus λ. The computed degree of polarization upon transmission is $P = 0.939$, and the rotation angle and ellipticity of the polarized fraction of the beam are $\rho = 47.6°$, $\eta = -37°$.

transmitted beam is $P = 0.939$, the wave-plate's required phase shift is $\chi_0 = -74°$, and the polarizer's angle of minimum transmission is $\theta_0 = -44.3°$. The polarized fraction of the transmitted beam is, therefore, elliptical with $\eta = -37°$ and $\rho = 47.6°$, relative to the p-direction.

Note 1

The Schwartz inequality,[5] which concerns the integral of the product of two complex functions of the real variable x, is written as follows:

$$\left| \int f(x) g^*(x) \, \mathrm{d}x \right|^2 \leq \int |f(x)|^2 \, \mathrm{d}x \int |g(x)|^2 \, \mathrm{d}x.$$

Defining the complex vectors $\boldsymbol{A}$ and $\boldsymbol{B}$ as

$$\boldsymbol{A} = [A_1 \exp(\mathrm{i}\phi_1), A_2 \exp(\mathrm{i}\phi_2), \ldots, A_\mathrm{N} \exp(\mathrm{i}\phi_N)]$$

$$\boldsymbol{B} = [B_1 \exp(\mathrm{i}\psi_1), B_2 \exp(\mathrm{i}\psi_2), \ldots, B_N \exp(\mathrm{i}\psi_N)]$$

we find $S_2^2 + S_3^2 = |S_2 + \mathrm{i}S_3|^2 = |\boldsymbol{A}\boldsymbol{B}^{*T}|^2 \leq \|\boldsymbol{A}\|^2 \|\boldsymbol{B}\|^2 = S_0^2 - S_1^2$, establishing the desired inequality.

Note 2

The 2×2 matrix

$$\boldsymbol{M} = \begin{bmatrix} \boldsymbol{A} \\ \boldsymbol{B} \end{bmatrix} \begin{bmatrix} \boldsymbol{A}^{*T} & \boldsymbol{B}^{*T} \end{bmatrix}$$

has the following properties:

$$\tfrac{1}{2} \operatorname{Trace} \boldsymbol{M} = S_0$$

$$\tfrac{1}{4} \operatorname{Trace}^2 \boldsymbol{M} - \operatorname{Det} \boldsymbol{M} = S_1^2 + S_2^2 + S_3^2.$$

Upon rotating the XY-plane through an angle ζ, the vector $\begin{bmatrix} \boldsymbol{A} \\ \boldsymbol{B} \end{bmatrix}$ will be multiplied on the left by the rotation matrix

$$\begin{bmatrix} \cos\zeta & \sin\zeta \\ -\sin\zeta & \cos\zeta \end{bmatrix}.$$

However, under this unitary transformation both the trace and the determinant of $\boldsymbol{M}$ remain unchanged. Therefore, the beam's total power S_0 and the power of its polarized component $(S_1^2 + S_2^2 + S_3^2)^{1/2}$ are rotation invariant.

References for chapter 8

1 M. Born and E. Wolf, *Principles of Optics*, sixth edition, Pergamon Press, 1983.
2 L. Mandel and E. Wolf, *Optical Coherence and Quantum Optics*, Cambridge University Press, UK, 1995.
3 M. V. Klein, *Optics*, Wiley, New York, 1970.
4 G. G. Stokes, *Trans. Camb. Phil. Soc.* **9**, 399 (1852). Reprinted in his *Mathematical and Physical Papers*, Vol. III, p. 233, Cambridge University Press, 1901.
5 Papoulis, *Probability, Random Variables, and Stochastic Processes*, McGraw-Hill, New York, 1984.

9

What in the world are surface plasmons?†

Despite its scary name, a surface plasmon is simply an inhomogeneous plane-wave solution to Maxwell's equations. Typically, a medium with a large but negative dielectric constant ε is a good host for surface plasmons. Because in an isotropic medium having refractive index n and absorption coefficient κ we have $\varepsilon = (n + \mathrm{i}\kappa)^2$, whenever $\kappa \gg n$ the above criterion, large but negative ε, is approximately satisfied; as a result, most common metals such as aluminum, gold, and silver can exhibit resonant absorption by surface plasmon excitation. In order to excite, within a metal, a plane wave that has a large enough amplitude to carry away a significant fraction of the incident optical energy, one must create a situation whereby the metal is "forced" to accept such a wave; otherwise, as normally occurs, the wave within the metal ends up having a small amplitude, causing nearly all of the incident energy to be reflected, diffracted, or scattered from the metallic surface, depending upon the condition of that surface.

In this chapter several practical situations in which surface plasmons play a role will be presented. We begin by describing the results of an experiment that can be readily set up in any optics laboratory, and we give an explanation of the observed phenomenon by scrutinizing the well-known Fresnel's reflection formula at a metal-to-air interface. We then describe other, slightly more complicated, situations involving the excitation of surface plasmons, in an attempt to convey to the reader the generality of the phenomenon and its various manifestations.

Surface plasmons in a thin metallic film

Perhaps the simplest arrangement in which one may observe surface plasmons is that shown schematically in Figure 9.1. A thin metal film, coated on

† The coauthor of this chapter is Lifeng Li, now at the Tsinghua University in China.

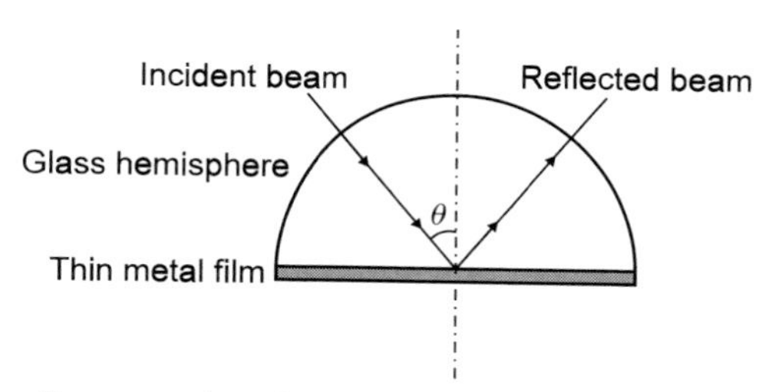

Figure 9.1 Schematic diagram showing a monochromatic plane wave incident on a thin metal film through a hemispherical glass substrate. When the film is sufficiently thin, at a specific incidence angle θ a surface plasmon is excited within the metal layer, causing a substantial fraction of the incident beam's energy to be absorbed and converted to heat within the metal layer.

the flat face of a glass hemisphere, is illuminated at oblique incidence through the hemisphere. In this example the glass will be assumed to have refractive index 1.5, and the metal film will be assumed to be aluminum, although most common metals coated on just about any type of glass will exhibit a similar behavior. A plane monochromatic beam of red HeNe light is directed at the glass–metal interface, and its reflection is monitored as a function of the angle of incidence θ. Figure 9.2(a) shows computed plots of the reflection coefficients $|r_p|$ and $|r_s|$ versus θ for the case of a very thin ($d = 5\,\text{nm}$) aluminum film. At the critical angle of total internal reflection (TIR) for a glass–air interface, $\theta_{crit} = 41.8°$, the reflection coefficients show a sudden rise, but $|r_p|$ drops sharply above θ_{crit}, attaining a minimum at $\theta \approx 45°$. This sharp reduction in the reflectivity of p-polarized light is due to the excitation of a surface plasmon in the aluminum layer.

Figure 9.2(b) shows plots of the magnitude of the Poynting vector $\boldsymbol{S}$ through the thickness of the aluminum layer for both p- and s-polarized light at the incidence angle $\theta = 45°$. Note that for the s-light, approximately 30% of the incident optical power enters the film, which then proceeds to be absorbed (rather uniformly) through the film thickness. With the p-light, however, the fraction of the incident power absorbed by the film is much higher (close to 90%). Evidently, at this particular angle of incidence the p-light has been able to excite a very strong wave in the metal layer.

Similar calculations can be done for other thicknesses of the aluminum layer; the results shown in Figure 9.3 correspond to a film thickness $d = 10$ nm. The minimum reflectivity now occurs at $\theta = 42.95°$, and the percentage of p-light absorbed by the film has climbed to over 98%. If we continue to increase the film thickness, however, the effect begins to decrease (and eventually to disappear), as demonstrated by the plots of Figure 9.4, which correspond to $d = 20\,\text{nm}$. In fact, aside from the weak, plasmon-

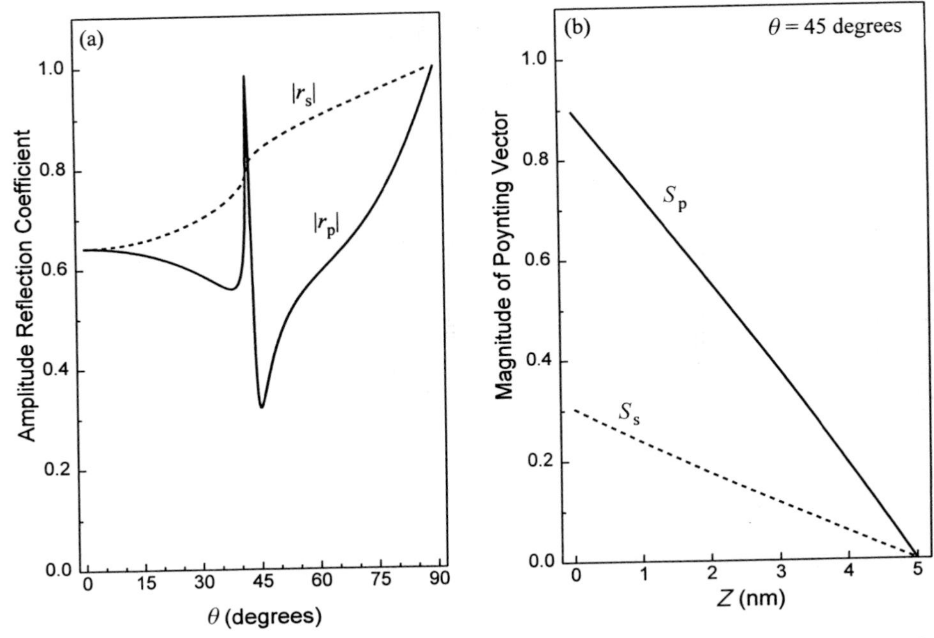

Figure 9.2 (a) Computed plots of amplitude reflection coefficients for the p- and s-components of polarization versus the angle of incidence θ, for the monochromatic plane wave ($\lambda = 633\,\text{nm}$) incident at the interface between glass and a thin aluminum layer ($d = 5\,\text{nm}$) shown in Figure 9.1. The dip in $|r_p|$ at $\theta \approx 45°$ is caused by the excitation of a surface plasmon in the aluminum film. (b) Plots of the magnitude of the Poynting vector $\boldsymbol{S}$ against the depth z within the aluminum layer, at $\theta = 45°$. Note that approximately 90% of the incident power of the p-polarized light enters the aluminum film and is absorbed fairly uniformly within the film's thickness. In contrast, only 30% of the s-polarized light is absorbed by the film.

related feature in the vicinity of θ_{crit}, the plots of $|r_p|$ and $|r_s|$ in Figure 9.4(a) already resemble those for a very thick aluminum film (i.e., one for which $d \gg$ skin depth). It is thus obvious that the lower interface, between aluminum and air, is responsible for the excitation of surface plasmons: increasing the film thickness prevents the electromagnetic field from reaching the aluminum–air interface, thus suppressing the excitation of the plasma wave. Also note in Figure 9.4(b) that the slope of S_s is greatest near the glass–aluminum interface, and the flux of optical energy contained in the s-polarized beam decays exponentially as it moves away from this interface towards the aluminum–air interface. In contrast, the slope of S_p is greatest at the aluminum–air interface, indicating that most of the energy is deposited at that site. This is yet another indication that the aluminum–air interface is responsible for the excitation of surface plasmons in the system of Figure 9.1.

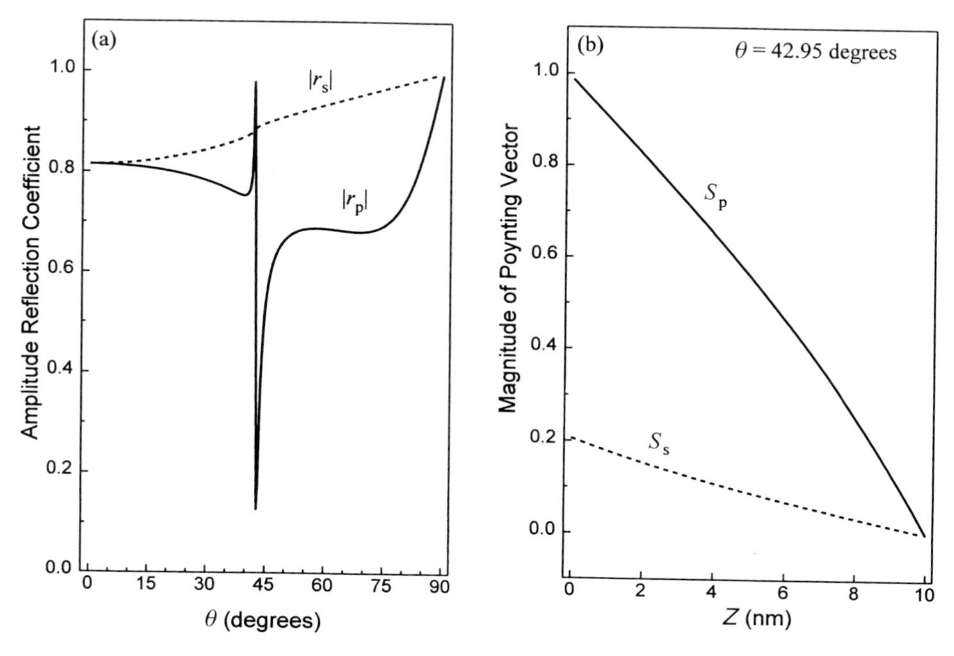

Figure 9.3 Same as Figure 9.2, except for the thickness of the aluminum film, which is now 10 nm. The resonant absorption in this case occurs at $\theta = 42.95°$, and the fraction of p-polarized light absorbed by the aluminum layer is over 98%.

A simple explanation based on Fresnel's reflection coefficients

The Fresnel reflection coefficients at the interface between air and metal provide a good starting point for an explanation of the nature of surface plasmons and the conditions under which they occur. Consider the case of a polished metal surface of dielectric constant ε, upon which a monochromatic plane wave of wavelength λ_0 is incident from air, at the oblique incidence angle of θ. The k-vector of the incident beam (in air) has magnitude $k_0 = 2\pi/\lambda_0$, and its projections parallel and perpendicular to the air–metal interface are denoted by $k_\parallel$ and $k_\perp$. The complex Fresnel reflection coefficients r_p and r_s for p- and s-polarized light are written

$$r_p = \frac{\sqrt{\varepsilon - (k_\parallel/k_0)^2} - \varepsilon k_\perp/k_0}{\sqrt{\varepsilon - (k_\parallel/k_0)^2} + \varepsilon k_\perp/k_0}, \tag{9.1}$$

$$r_s = \frac{k_\perp/k_0 - \sqrt{\varepsilon - (k_\parallel/k_0)^2}}{k_\perp/k_0 + \sqrt{\varepsilon - (k_\parallel/k_0)^2}}. \tag{9.2}$$

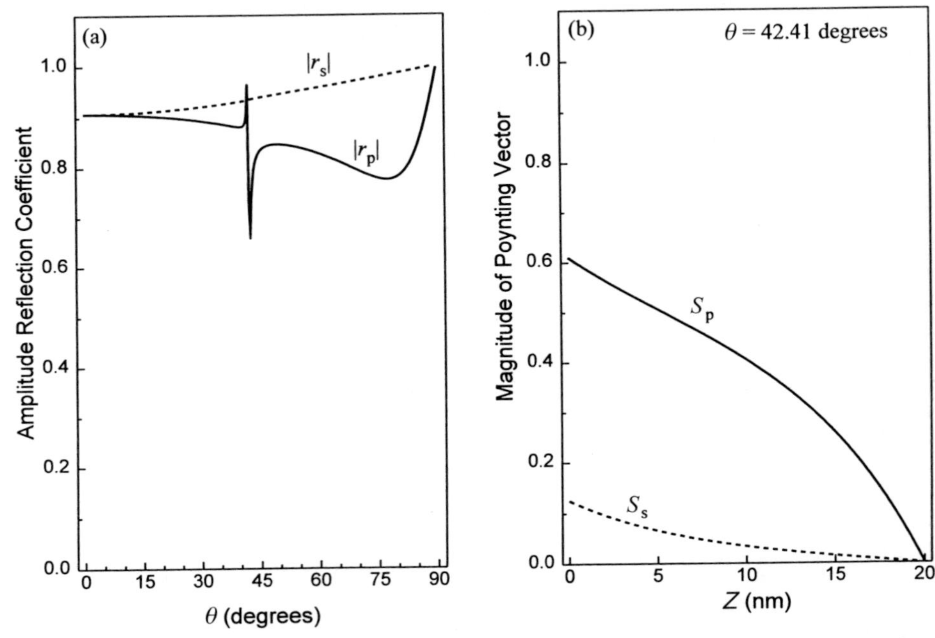

Figure 9.4 Same as Figures 9.2 and 9.3, except for the thickness of the aluminum film, which is now 20 nm. The resonant absorption in this case occurs at $\theta = 42.41^\circ$, and the fraction of p-polarized light absorbed within the aluminum layer is just over 60%.

The denominator of the expression for r_p in Eq. (9.1) goes to zero at $k_\parallel/k_0 = \sqrt{\varepsilon/(1+\varepsilon)}$, indicating that r_p has a pole at this point. No such pole, however, exists for r_s. In the case of aluminum, $n + \mathrm{i}\kappa = 1.38 + 7.6\mathrm{i}$ at $\lambda_0 = 633\,\mathrm{nm}$, yielding $\varepsilon = -55.86 + 20.98\mathrm{i}$; this results in a value $1.008 + 0.003\mathrm{i}$ for the pole of r_p. Under ordinary circumstances, when the metal surface is illuminated in air at an oblique angle θ we have $k_\parallel/k_0 = \sin\theta$, which is less than unity and, therefore, far from the pole. However, if evanescent waves are somehow created at an air–aluminum interface, then $k_\parallel/k_0$ can exceed unity and, in the neighborhood of the pole, the reflectivity r_p at that interface will approach infinity. This means that an evanescent p-polarized plane wave of very small amplitude impinging at the metal surface can excite a very strong plane wave within the metal. This plane wave, of course, is the surface plasmon, which is capable of absorbing a good fraction of the energy from the incident beam and converting it to heat within the metallic medium.

In light of the above arguments it is not difficult to see that, in the system of Figure 9.1, the creation of evanescent waves with $k_\parallel/k_0 \approx 1$ at the aluminum–air interface is responsible for the sharp decline in r_p at angles slightly greater than the critical TIR angle. Since the expression for r_s in Eq. (9.2) does not admit a pole, no such behavior could be expected from the s-polarized light.

Attenuated total internal reflection (ATIR)

Another setup in which the excitation of surface plasmons is readily observed is shown in Figure 9.5. (Some results from this type of experiment are also described in chapter 19, "Some Quirks of Total Internal Reflection".) Here the presence of an air gap between the glass hemisphere and the metal plate guarantees the creation of evanescent waves whereas in the setup of Figure 9.1 the metal layer had to be sufficiently thin to provide access for the electromagnetic waves to the metal–air interface.

For the system of Figure 9.5, computed reflection coefficients versus the gap-width are plotted in Figure 9.6 for several angles of incidence in the vicinity of θ_{crit}. Because the variations of r_s were imperceptible within the chosen range of incidence angles, 41°–43°, it was deemed pointless to label the various coinciding r_s curves. In contrast, r_p was very sensitive to changes in θ, and the various r_p curves in Figure 9.6 are clearly labeled to indicate this dependence. A dip in the plots of r_p versus the gap-width begins to appear at angles of incidence just below θ_{crit}; the dip becomes more pronounced with an increasing θ until the minimum reflectivity actually reaches zero at $\theta = 41.95°$. The dip then decreases with further increases in θ and, by the time θ reaches 43°, it has practically disappeared.

As before we observe in the plots of Figure 9.6 the salient features of absorption by surface plasmon excitation, namely, a p-polarized incident

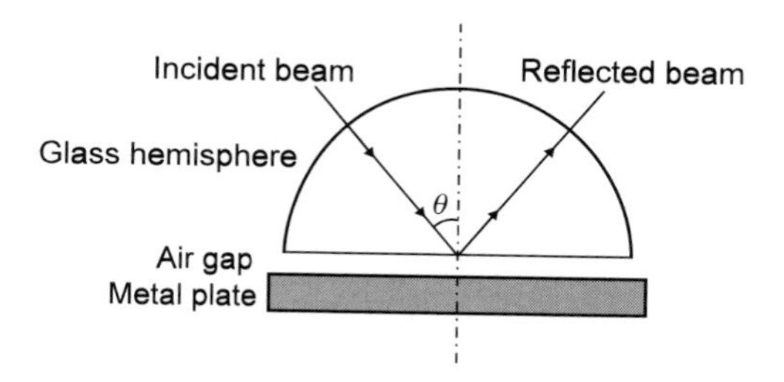

Figure 9.5 Schematic diagram showing a monochromatic plane wave at oblique incidence on the flat surface of a glass hemisphere. When the air gap separating the hemisphere and the polished metal surface is sufficiently thin, and at a specific angle of incidence θ, a substantial fraction of the incident beam will be coupled into a surface plasmon and thus absorbed by the metal plate.

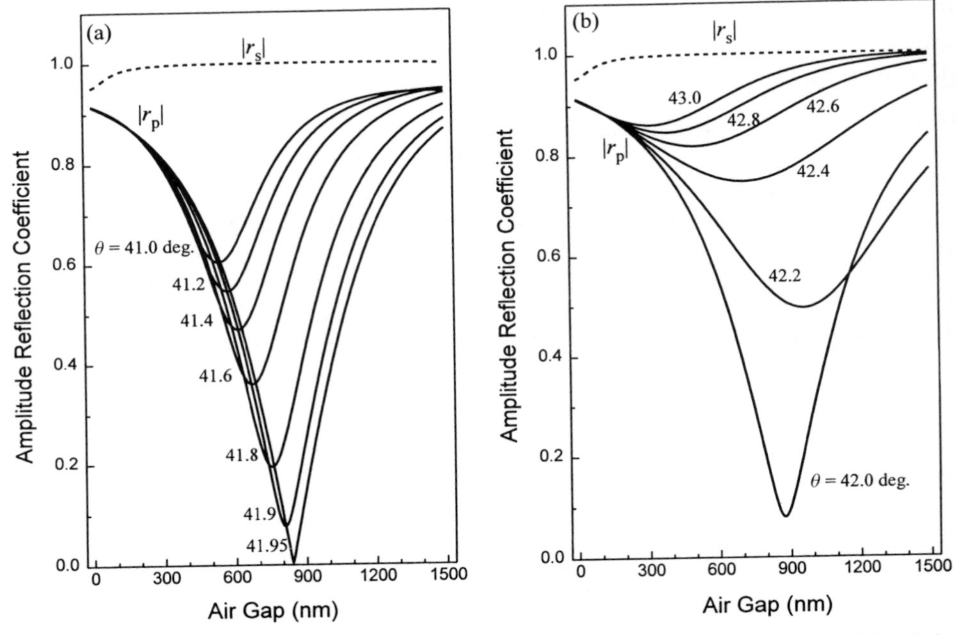

Figure 9.6 Computed plots of amplitude reflection coefficients versus the width of the air gap for the experiment depicted in Figure 9.5. Each curve represents a specific incidence angle θ in the range (41°, 43°). The reflection coefficient $|r_s|$ for the s-polarized light does not change very much in this narrow range of incident angles; thus all the $|r_s|$ curves coincide at this scale. The dip in $|r_p|$ begins to appear at angles of incidence just below the critical TIR angle ($\theta_{crit} = 41.81°$); it becomes a maximum at $\theta = 41.95°$ and then decreases again to insignificance at $\theta = 43°$. Note also that the gap-width at which $|r_p|$ reaches a minimum varies with the angle of incidence. (a) θ-values from 41.0° to 41.95°; (b) θ-values from 42.0° to 43.0°.

beam, the existence of evanescent waves at an air–metal interface, and an angle of incidence in the vicinity of the critical TIR angle for the glass–air interface, that is, when $k_{\|}/k_0 \approx 1$.

Excitation of surface plasmons in metalized diffraction gratings

A third experiment in which surface plasmons may be observed involves the reflection of light from metallized diffraction gratings. Here, as shown in Figure 9.7, the incident beam excites one or more propagating diffracted orders but also creates non-propagating evanescent waves near the surface. Whenever one of these evanescent waves happens to have the $k_{\|}$ of a surface plasmon, the conditions for its excitation are ripe, and a good fraction of the optical energy will be coupled into the grating medium. As might be guessed from the cases discussed in the preceding examples, the range of parameters over which surface plasmon excitation can be expected is very narrow and,

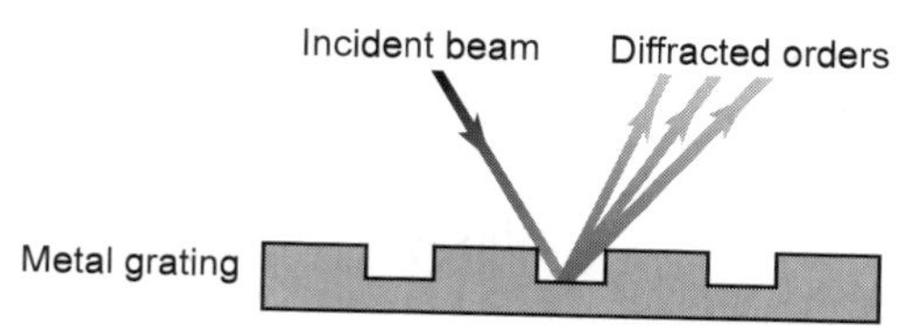

Figure 9.7 A monochromatic plane wave incident on a metal grating creates multiple diffracted orders within the reflected beam. At certain angles of incidence, when the polarization of the beam happens to have a component perpendicular to the grooves, surface plasmons are excited within the grating. These plasmons, which can convert a good fraction of the incident optical power into heat, cause a sudden and substantial drop in the diffraction efficiencies of the various orders.

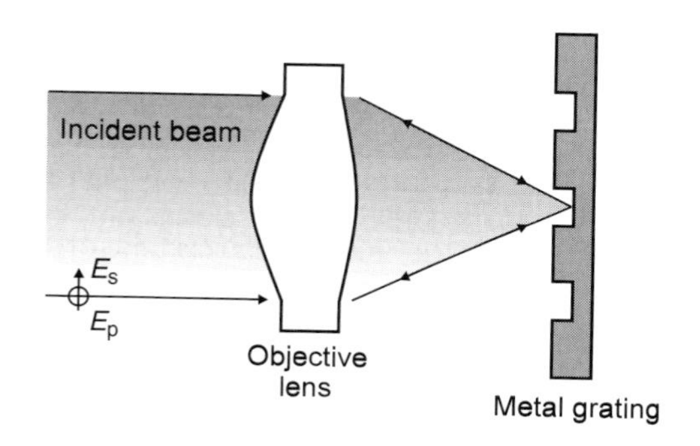

Figure 9.8 A good method of observing surface plasmons in practice involves the reflection of a focused beam of light from the grooved surface of a metal grating. Because the focused cone contains rays within a wide range of angles of incidence, upon reflection from the grating dark bands will appear in the exit pupil of the lens corresponding to those rays that have succeeded in exciting the surface plasmons. A camera, set up to photograph the reflected beam at the exit pupil of the lens, will not only reveal the narrow bands corresponding to surface plasmons but also show the superposition of the various diffracted orders captured by the lens. To observe the surface plasmon bands, one must allow the incident beam to have a component of polarization perpendicular to the grooves of the grating. In this figure, E_s is in the direction that will excite the plasmons.

therefore, the angle of incidence at which surface plasmons are excited must be sharply defined.

If one directs a focused beam onto a metal grating, as shown in Figure 9.8, then a wide angular spectrum will be present in the beam, and some of the rays will be strongly absorbed. A photograph of the reflected beam at the exit pupil of the lens will show one or more dark lines corresponding to the absorption of surface plasmons within the grating. Figure 9.9 shows a typical set of results obtained in an experiment of this type. When the polarization is

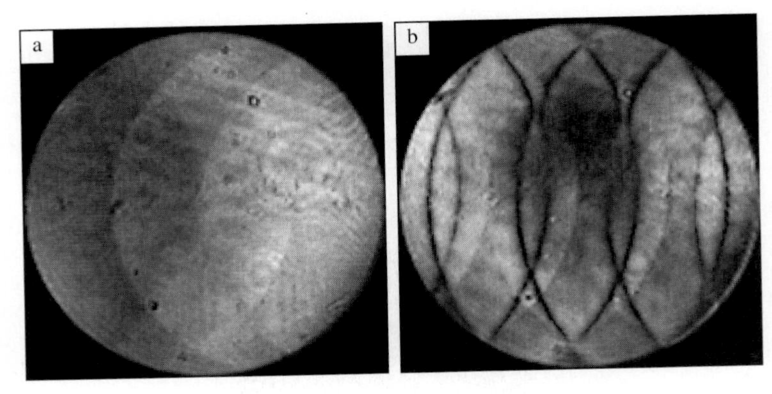

Figure 9.9 Photographs showing the intensity distribution at the exit pupil of a 0.8*NA* microscope objective lens, through which a collimated beam of laser light ($\lambda = 633$ nm) is focused on a gold-coated diffraction grating; $(n, k)_{\text{gold}} = (0.13, 3.16)$. The grooves of the grating are oriented along the Y-axis, the grating period is 1.6 μm, and the grooves, which have a trapezoidal cross-section, are 0.5 μm wide at the top and 70 nm deep. The direction of the linear polarization of the incident beam is parallel to the grooves in (a) and perpendicular to the grooves in (b). (From Ronald E. Gerber, Ph.D. dissertation, Optical Sciences Center, University of Arizona, Tucson.)

parallel to the grooves, as is the case in Figure 9.9(a), there are no surface plasmon bands. However, with the polarization vector perpendicular to the grooves, surface plasmons are clearly excited, as shown in Figure 9.9(b). Results of theoretical calculations confirming these results are shown in Figures 9.10 and 9.11. In these calculations, Maxwell's equations were solved for about 10 000 plane waves impinging on the metal grating at various angles. These results were then combined to represent the focused cone of light created by a 0.8*NA* objective lens.

In the case of Figure 9.10, where the incident polarization vector was parallel to the grooves, no plasmons were observed. We did the calculations for three different positions of the focused spot over the grooves, however, to show the so-called baseball pattern that results from superposition of the various diffracted orders. Frames (a), (b), and (c) correspond respectively to a beam focused on one groove edge, on the middle of a groove, and on an opposite groove edge. The phase differences between various diffracted orders create constructive and destructive interference among these various orders in their regions of mutual overlap, thus giving rise to black and white areas. When the polarization is perpendicular to the grooves, the

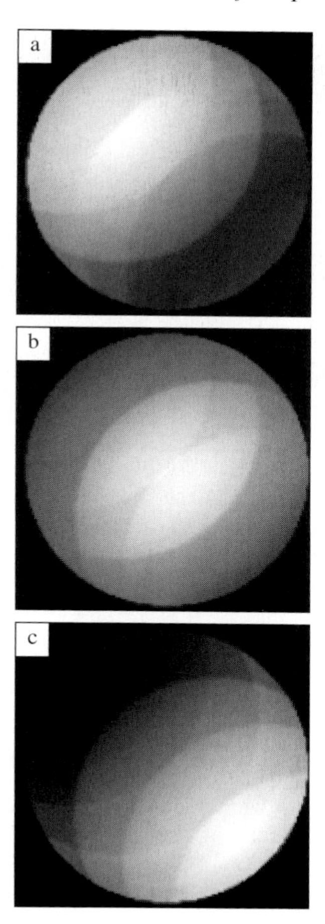

Figure 9.10 Computed plots of intensity distribution at the exit pupil of a 0.8NA objective lens through which a uniform plane wave is focused on a diffraction grating. The grooves are oriented at 45° relative to the X-axis. The parameters of the grating are the same as those used in the experiment (see the caption to Figure 9.9). The various diffraction orders are clearly visible in these so-called "baseball" patterns. The incident linear polarization is parallel to the grooves, thus explaining the absence of plasmon-related dark bands in these pictures. The center of the focused spot is (a) on a groove edge, (b) in the middle of a groove, and (c) on the opposite groove edge.

pattern in Figure 9.11 is obtained. For this computation the position of the focused spot on the grating was on a grooved edge similar to that shown in Figure 9.10(c). The dark bands of Figure 9.11, predicted by this theoretical calculation to arise from surface plasmon excitation, agree quite well with the experimental results of Figure 9.9(b).

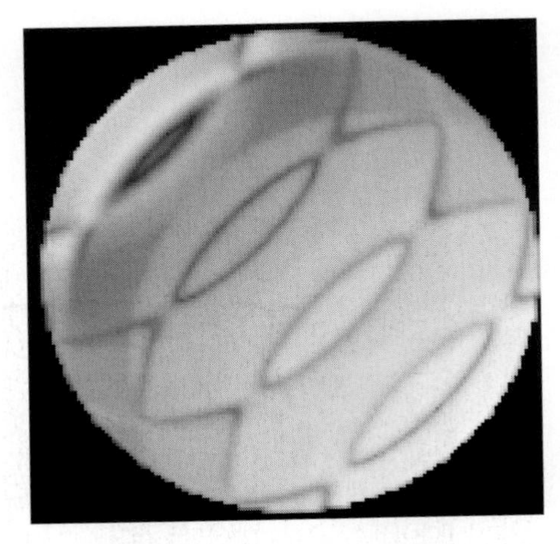

Figure 9.11 Logarithmic plot of computed intensity distribution at the exit pupil of a 0.8*NA* objective lens. The simulation parameters are the same as those used to obtain Figure 9.10(c), with the exception of the direction of incident polarization, which is perpendicular to the grooves. The grooves are oriented at 45° to the *X*-axis, and the focused spot is centered on the edge of a groove. The absorption bands caused by the excitation of surface plasmons are identical to those observed experimentally in Figure 9.9(b).

References for chapter 9

1 R. E. Gerber, Lifeng Li, and M. Mansuripur, Effects of surface plasmon excitations on the irradiance pattern of the return beam in optical disk data storage, *Appl. Opt.* **34**, 4929–4936 (1995).

2 R. W. Wood, On a remarkable case of uneven distribution of light in a diffraction grating spectrum, *Phil. Mag.* **4**, 396–402 (1902).

3 Lifeng Li, Multilayer-coated diffraction gratings: differential method of Chandezon *et al.* revisited, *J. Opt. Soc. Am. A* **11**, 2816–2828 (1994).

4 R. H. Ritchie, Plasma losses by fast electrons in thin films, *Phys. Rev.* **106**, 874–881 (1957).

5 For the computations leading to Figures 9.10 and 9.11, reflection coefficients of the grating were first computed by a vector diffraction program developed by Lifeng Li. These coefficients were subsequently imported to DIFFRACT, where they were combined to represent the effects of a focused beam.

6 J. C. Quail, J. G. Rako, and H. J. Simpson, Long-range surface plasmon modes in silver and aluminum, *Opt. Lett.* **8**, 377 (1983).

7 D. Sarid, Long-range surface-plasma waves on very thin metal films, *Phys. Rev. Lett.* **47**, 1927 (1981).

8 A. D. Boardman, ed., *Electromagnetic Surface Modes*, Wiley, New York, 1982.

9 A. E. Craig, A. Olson, and D. Sarid, Experimental observation of the long-range surface-plasmon polariton, *Opt. Lett.* **8**, 380 (1983).

10
The Faraday effect

Michael Faraday (1791–1867) (Photo: National Portrait Gallery, London, courtesy of AIP Emilio Segré Visual Archives.)

Michael Faraday (1791–1867) was born in a village near London into the family of a blacksmith. His family was too poor to keep him at school and, at the age of 13, he took a job as an errand boy in a bookshop. A year later he was apprenticed as a bookbinder for a term of seven years. Faraday was not only binding the books but was also reading many of them, which excited in him a burning interest in science.

When his term of apprenticeship in the bookshop was coming to an end, he applied for the job of assistant to Sir Humphry Davy, the celebrated chemist, whose lectures Faraday was attending during his apprenticeship. When Davy asked the advice of one of the governors of the Royal Institution of Great Britain about the employment of a young bookbinder, the man said: "Let him wash bottles! If he is any good he will accept the work; if he refuses, he is not good for anything." Faraday accepted, and remained with the Royal Institution for the next fifty years, first as Davy's assistant, then as his collaborator, and finally, after Davy's death, as his successor. It has been said that Faraday was Davy's greatest discovery.

In 1823 Faraday liquefied chlorine and in 1825 he discovered the substance known as benzene. He also did significant work in electrochemistry, discovering the laws of electrolysis. However, his greatest work was with electricity. In 1821 Faraday built two devices to produce what he called electromagnetic rotation, that is, a continuous circular motion from the circular magnetic force around a wire. Ten years later, in 1831, he began his great series of experiments in which he discovered electromagnetic induction. These experiments form the basis of modern electromagnetic technology.

Apart from numerous publications in scientific magazines, the most remarkable document pertaining to his studies is his *Diary*, which he kept continuously from the year 1820 to the year 1862. (This was published in 1932 by the Royal Institution in seven volumes containing a total of 3236 pages, with a few thousand marginal drawings.) Queen Victoria rewarded Faraday's lifetime of achievement by granting him the use of a house at Hampton Court and a knighthood. Faraday accepted the cottage but gracefully rejected the knighthood.[1]

On 13 September 1845, Faraday discovered the magneto-optical effect that bears his name. This day's entry in his *Diary* reads: "Today worked with lines of magnetic force, passing them across different bodies (transparent in different directions) and at the same time passing a polarized ray of light through them and afterwards examining the ray by a Nichol's Eyepiece or other means." After describing several negative results in which the ray of light was passed through air and several other substances, Faraday wrote in the same day's entry: "A piece of heavy glass which was 2 inches by 1.8 inches, and 0.5 of an inch thick, being silico borate of lead, and polished on the two shortest edges, was experimented with. It gave no effects when the *same magnetic poles* or the *contrary* poles were on opposite sides (as respects the course of the polarized ray) – nor when the same poles were on the same side, either with a constant or intermitting current – BUT, when contrary magnetic poles were on the same side, there *was an effect produced on the polarized ray*, and thus magnetic force and light were proved to have relation to each other. This fact will most likely prove exceedingly fertile and of great value in the investigation of both conditions of natural force."

Electromagnetic basis of the Faraday effect

Magneto-optical (MO) effects are best described in terms of the dielectric tensor $\boldsymbol{\varepsilon}$ of the medium in which the interaction between the light and the applied magnetic field (or the internal magnetization of the medium) takes place:[2]

$$\boldsymbol{\varepsilon} = \begin{pmatrix} \varepsilon_{xx} & \varepsilon_{xy} & \varepsilon_{xz} \\ \varepsilon_{yx} & \varepsilon_{yy} & \varepsilon_{yz} \\ \varepsilon_{zx} & \varepsilon_{zy} & \varepsilon_{zz} \end{pmatrix}.$$

In an isotropic material (such as ordinary glass) the three diagonal elements are identical and, in the presence of a magnetic field along the Z-axis, there is a non-zero off-diagonal element $\boldsymbol{\varepsilon}'$, which couples the x- and y- components of the optical E-field, that is,

$$\boldsymbol{\varepsilon} = \begin{pmatrix} \varepsilon & \varepsilon' & 0 \\ -\varepsilon' & \varepsilon & 0 \\ 0 & 0 & \varepsilon \end{pmatrix}.$$

In general, ε and ε' are wavelength dependent, but over a narrow range of wavelengths they might be treatable as constants. In a transparent material, where there is no optical absorption, ε is real and ε' is imaginary. However, in the most general case of an absorbing MO material both ε and ε' may be complex numbers. For diamagnetic and paramagnetic media ε' is proportional to the applied magnetic field H, while for ferromagnetic and ferrimagnetic materials spin–orbit coupling is the dominant source of the MO interaction, making ε' proportional to the magnetization $\boldsymbol{M}$ of the medium.[2] Since $\boldsymbol{B} = \boldsymbol{H} + 4\pi\boldsymbol{M}$ (in CGS units), we consider the B-field inside the medium as the source of the MO effects.

Now we discuss the basis of the MO effect. When a polarized beam of light propagates in a medium along the direction of the magnetic field $\boldsymbol{B}$, the right and left circularly polarized (RCP and LCP) components of the beam experience different refractive indices, $n^{\pm} = (\varepsilon \pm \mathrm{i}\varepsilon')^{1/2}$. For fused silica glass at a wavelength $\lambda = 550\,\mathrm{nm}$, for example, $\varepsilon \approx 2.25$, and $\varepsilon' \approx 10^{-7}\mathrm{i}$ per kOe of applied magnetic field. (Note that both n^+ and n^- in this case are real-valued and, therefore, there is no absorption.) For linearly polarized light passing through a length L of the material under the influence of a B-field, the two circular-polarization components suffer a relative phase shift $\Delta\phi = 2\pi L(n^+ - n^-)/\lambda$.[3,4] As shown in Figure 10.1, a change in the relative phase of the RCP and LCP components is equivalent to a rotation of the plane of polarization by the Faraday angle $\theta_{\mathrm{F}} = \frac{1}{2}\Delta\phi$. In the above example, $\theta_{\mathrm{F}} \sim 0.22°$ at $\lambda = 550\,\mathrm{nm}$ for a slab 1 cm thick immersed in a 1 kOe magnetic field. The figure of 0.22°/cm kOe is known as the Verdet constant of fused silica at the specified wavelength.[3]

Certain magnetic materials (e.g., magnetic garnets) are transparent enough to transmit a good fraction of the light while producing a fairly large Faraday rotation. These materials can be magnetized in a given direction and sustain

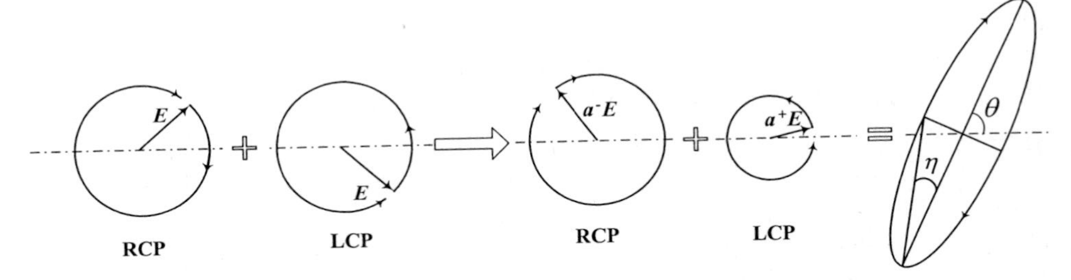

Figure 10.1 A linearly polarized beam of light may be considered as the superposition of equal amounts of right and left circularly polarized beams. In going through a perpendicularly magnetized slab of material at normal incidence, the two components of circular polarization experience different (complex) refractive indices and, therefore, each emerges from the medium with a different phase and amplitude. The amplitudes of the emergent beams may be denoted by a^+ and a^-, and their phase difference by $\Delta\phi$. The superposition of the emergent circular polarization states yields elliptical polarization. The angle of rotation of the major axis of the ellipse from the horizontal direction (which is the direction of the incident linear polarization) is given by $\theta = \frac{1}{2}\Delta\phi$, and the ellipticity η is given by $\tan\eta = (a^+ - a^-)/(a^+ + a^-)$.

their magnetization when the external field is removed. Therefore, the Faraday effect in these media may be observed in the absence of an external magnetic field. At $\lambda = 550\,\text{nm}$, for instance, a typical crystal of bismuth-substituted rare-earth iron garnet may have $\varepsilon \approx 5.5 + 0.025\text{i}$ and $\varepsilon' \approx 0.002 - 0.01\text{i}$. The complex refractive indices for RCP and LCP light are thus $(n + \text{i}k)^+ \approx 2.347 + 0.006\text{i}$ and $(n + \text{i}k)^- \approx 2.343 + 0.005\text{i}$, yielding a Faraday rotation angle $\theta_\text{F} \approx 1.3^\circ$ for a micron-thick slab of this crystal. The absorption coefficient of the material is $\alpha = 4\pi kL/\lambda$, where k is the imaginary part of the complex refractive index. For the above garnet, therefore, $\alpha \approx 0.12$ per micron, which is equivalent to 1 dB loss of light for every 2 μm of crystal thickness. In other words, this garnet delivers 2.6° of polarization rotation per dB of loss. These crystals can be grown in a range of thicknesses from a fraction of a micron to about 100 microns. Thicker crystals are useful at longer wavelengths, where the losses are small, but the Faraday rotation generally decreases with increasing wavelength as well.

Faraday rotation in a transparent slab

For the sake of simplicity we will ignore the effects of absorption in the Faraday medium and consider a transparent slab of magnetic material having a real ε and a purely imaginary ε'. Thus we consider a slab 20 μm thick having $\varepsilon = 5.5$, $\varepsilon' = 0.01\text{i}$. The material is magnetized perpendicularly to the plane of its surface, and a linearly polarized beam of light (with its E-field along the X-axis) is sent at normal incidence through the slab, as in Figure 10.2(a).[5,6] Real sources of light, of course, are never perfectly monochromatic and, therefore, we assume a finite spectral bandwidth for the light source, covering the range $\lambda = 545 - 55$ nm. Figure 10.3 shows computed plots of the transmitted amplitudes, $|t_x|$ and $|t_y|$, as well as the polarization rotation and ellipticity angles, θ_F and η_F, versus λ. Because of multiple reflections at the front and rear facets of the slab these functions vary periodically with λ. (The same interference phenomena are responsible for the non-zero values of η_F, which would otherwise be absent in a transparent medium.) The net Faraday rotation angle is the average value of θ_F over the relevant range of wavelengths, but one should also recognize that the wavelength dependence of the direction of emergent polarization produces a certain amount of depolarization in the emergent beam. The Faraday rotation combined with the spectral bandwidth of the light source thus causes partial depolarization as a direct consequence of interference among the multiple reflections.

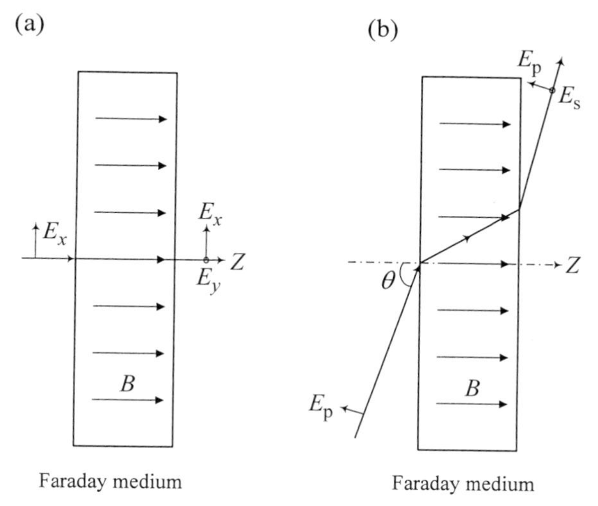

Figure 10.2 Faraday effect in the polar geometry. (a) In going normally through a slab of magnetic material, a linearly polarized beam of light with its E-field along the X-axis acquires a component of polarization along Y. The lines of B-field shown within the medium represent either an externally applied magnetic field or the intrinsic magnetization of the medium. (b) The effect is also observed at oblique incidence. Shown here is a p-polarized incident beam, which acquires a s-component upon transmission through the magnetic medium. (If the incident beam is s-polarized, the magneto-optically induced polarization is then in the p-direction.) In general, upon reversing the B-field from the $+Z$ to the $-Z$ direction the magneto-optically induced component of polarization changes sign.

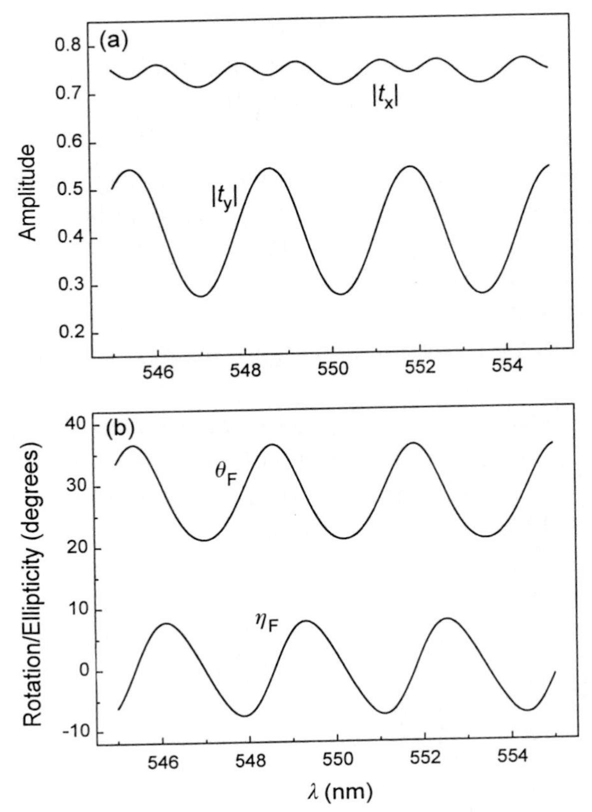

Figure 10.3 A plane wave, linearly polarized along the X-axis, is normally incident on a slab 20 μm thick, as shown in Figure 10.2(a). The slab ($\varepsilon = 5.5, \varepsilon' = 0.01\mathrm{i}$) is magnetized along the Z-axis. (a) Plots of $|t_x|$ and $|t_y|$, the transmitted polarization components along the X- and Y- axes, as functions of λ. (b) Plots of polarization rotation angle θ_{F} and ellipticity η_{F}, versus λ.

Oblique incidence

Figure 10.4 shows the transmitted amplitudes and polarization angles versus the angle of incidence θ in the case of the slab 20 μm thick magnetized along the Z-axis ($\varepsilon = 5.5, \varepsilon' = 0.01\mathrm{i}$) when, as shown in Figure 10.2(b), a p-polarized plane wave at the single wavelength of $\lambda = 550$ nm is incident on the slab. The oscillations in the transmitted amplitudes and polarization angles are caused by interference among the beams multiply reflected from the facets of the slab. Aside from these interference oscillations, however, note that the Faraday effect does not show any signs of abatement with increasing angle of incidence. The reason is that even though the direction of propagation of the beam increasingly deviates from the direction of the B-field, the propagation

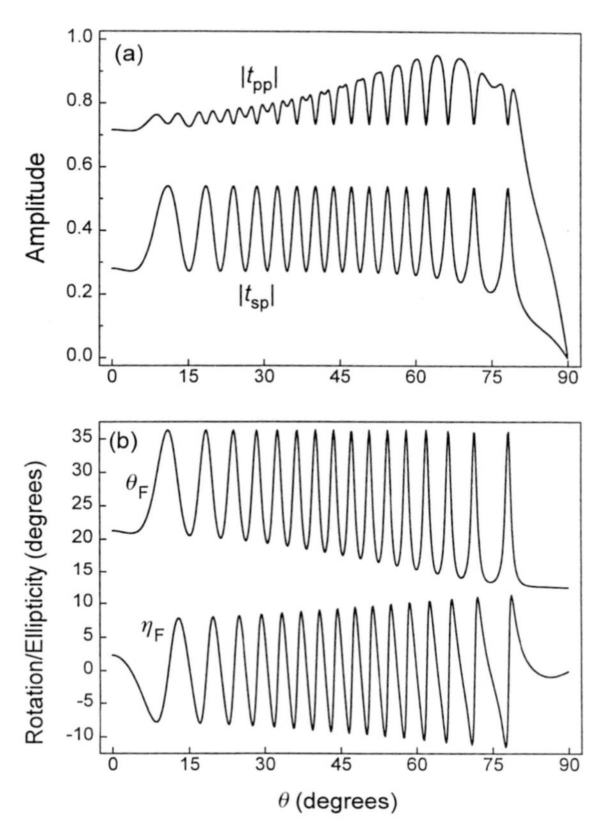

Figure 10.4 A p-polarized plane wave ($\lambda = 550\,\text{nm}$) is incident at oblique angle θ on a slab 20 μm thick, as shown in Figure 10.2(b). The slab ($\varepsilon = 5.5$, $\varepsilon' = 0.01\text{i}$) is magnetized along the Z-axis. (a) Plots of $|t_{\text{pp}}|$ and $|t_{\text{sp}}|$, the transmitted polarization components along the p- and s-directions, as functions of θ. (b) Plots of θ_{F} and η_{F} versus θ.

distance simultaneously increases, keeping the net interaction between the magnetic material and the beam of light at a constant level.

Figure 10.5 shows results for the case of oblique incidence, at $\theta = 85^\circ$, on the same slab as above in the range $\lambda = 545-55\,\text{nm}$. As in the case of normal incidence depicted in Figure 10.3, we note a significant variation of the Faraday angles and the amplitudes within this narrow range of wavelengths. Although the beam inside the slab travels at $\sim 25^\circ$ relative to the direction of magnetization of the material, the maximum Faraday effect as exemplified by $|t_{\text{sp}}|$ is the same as at normal incidence, because the propagation distance is correspondingly adjusted. The wavelength-averaged Faraday rotation may be lower at larger angles of incidence, but this is just a consequence of interference; it is not caused by any reduction in the intrinsic optical activity

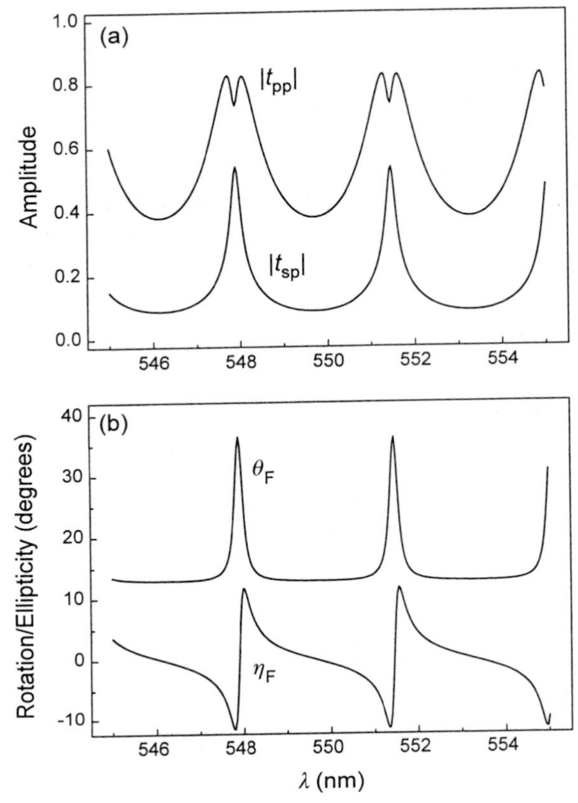

Figure 10.5 A p-polarized plane wave is incident at $\theta = 85^\circ$ on the slab described in Figures 10.2–10.4. (a) Plots of $|t_{pp}|$ and $|t_{sp}|$, the transmitted p- and s-components of polarization, as functions of λ. (b) Plots of θ_F and η_F versus λ.

of the slab. If, for instance, the facets are antireflection coated, or if the beam enters and exits through index-matched spherical surfaces, then multiple reflections would be eliminated and the Faraday rotation becomes independent of the incidence angle.

The above discussions were confined to the case of a p-polarized incident beam, but the conclusions remain valid for s-polarized light as well. For example, Figure 10.6 is the counterpart of Figure 10.4, showing the transmitted amplitudes and polarization angles versus the angle of incidence for a s-polarized incident beam. Note that the magneto-optically generated component of polarization t_{ps} in Figure 10.6 is identical to t_{sp} in Figure 10.4. This is an important and completely general result, indicating that the amount of light converted from one polarization state to another is independent of the incident polarization state.

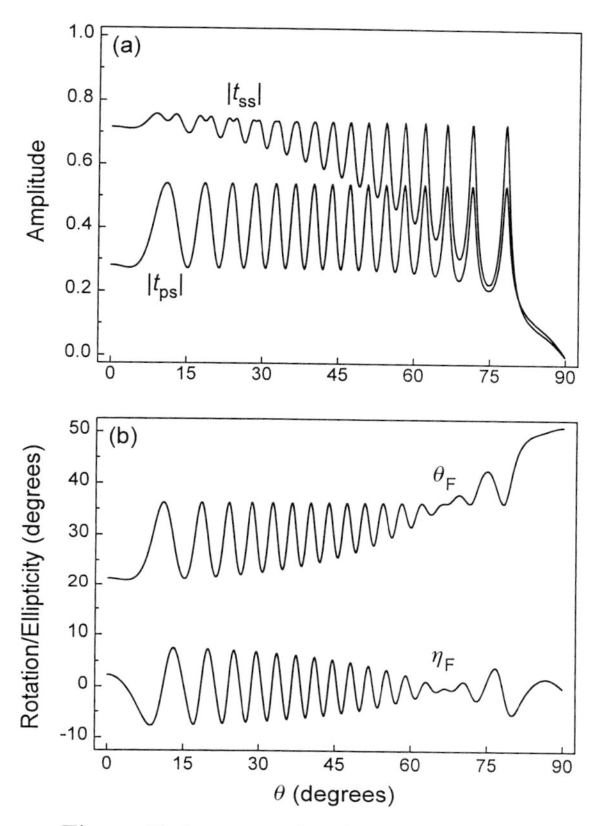

Figure 10.6 Same as Figure 10.4, except that here the incident beam is s-polarized.

Faraday medium in a Fabry–Pérot resonator

Because the Faraday effect is amplified when the beam propagates back and forth within a magnetized medium, it is interesting to observe the enhancement of the Faraday effect in a Fabry–Pérot resonator. Figure 10.7 shows a system that may be used to monitor such enhancement over a range of angles of incidence. The first objective lens ($NA = 0.8$) focuses a linearly polarized beam of light onto the Fabry–Pérot resonator, and the second, identical, lens collimates the transmitted beam, thus allowing observation at the exit pupil. For a slab of transparent magnetic material 20 μm thick sandwiched between a pair of dielectric mirrors, Figure 10.8 shows the computed patterns of intensity and polarization angle at the exit pupil of the collimator. This figure indicates that the rings of maximum transmission also correspond to locations of maximum polarization rotation. The maximum and minimum rotation angles in Figure 10.8(c) are $+63^\circ$ and -23°, respectively, well in excess of

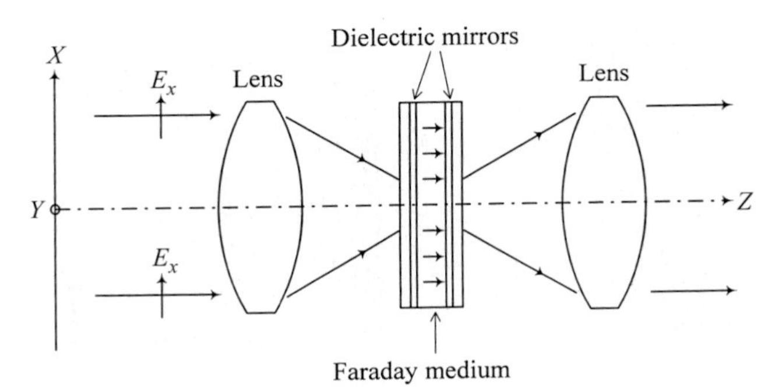

Figure 10.7 A Faraday medium in a Fabry–Pérot resonator is placed in a convergent cone of light. The incident plane wave is linearly polarized along the X-axis, and the $0.8NA$ focusing lens is free from aberrations. The Faraday medium, 20 μm thick, and with $\varepsilon = 5.5, \varepsilon' = 0.01i$, is uniformly magnetized along the Z-axis. The mirrors coated on the front and back facets of the Faraday slab each consist of 10 alternating layers of high-index ($n = 2.0$) and low-index ($n = 1.5$) quarter-wave-thick dielectrics. The collimating lens is identical to the focusing objective, and the emergent beam is observed at the exit pupil of the collimator.

the rotations obtained from the bare slab. Also note in Figures 10.8(c), (d) the asymmetrical nature of the polarization angles in the first and third quadrants, on the one hand, and in the second and fourth quadrants on the other hand.

Longitudinal and transverse geometries

When the direction of the B-field is in the plane of the slab as well as in the plane of incidence, as in Figure 10.9(a), one observes the longitudinal Faraday effect. In this case ε' occupies the position of ε_{yz} in the dielectric tensor. The transverse effect occurs when the B-field, while in the plane of the sample, is perpendicular to the incidence plane, as in Figure 10.9(b). In this case ε' occupies the position of ε_{xz}.

In the longitudinal case at normal incidence no polarization rotation occurs, but the effect begins to show with increasing angle of incidence. For a p-polarized plane wave ($\lambda = 550$ nm) obliquely incident on a slab of magnetic material 20 μm thick ($\varepsilon = 5.5, \varepsilon' = 0.01i$), Figure 10.10 shows the computed amplitudes of the transmitted p- and s-polarized light as well as the angles of rotation and ellipticity versus the incidence angle θ. One could readily compute similar results for a s-polarized incident beam as well. In both cases the MO effect is bipolar, meaning that a reversal of the direction of the B-field reverses the signs of θ_F and η_F. Moreover, as in the polar case discussed earlier, the magneto-optically generated component of polarization

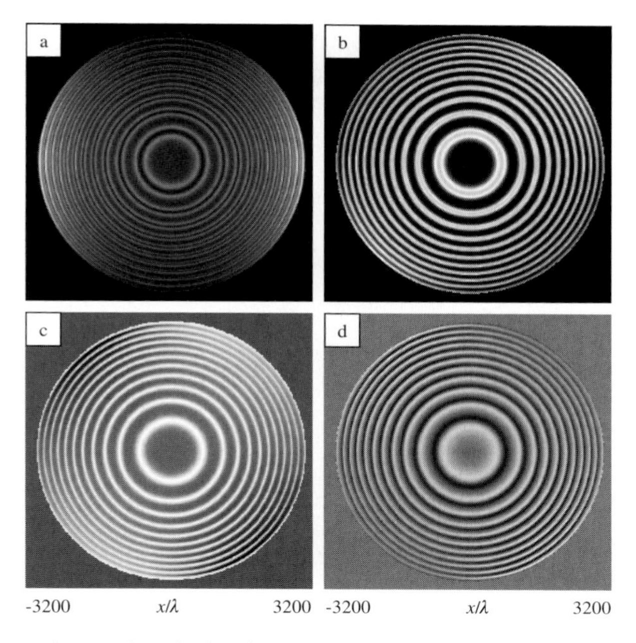

Figure 10.8 Intensity and polarization patterns in the exit pupil of the collimating lens of Figure 10.7. (a) The intensity distribution of the emergent X-polarized component. The bright rings indicate the regions where the conditions of resonance are met and the light passes through the resonator. (b) The intensity distribution of the emergent Y-polarized component. The bright rings coincide with those in (a), indicating that the conditions of resonance for the incident polarization are the same as those for the magneto-optically induced polarization. (c) Polarization rotation angle θ_F of the emergent beam encoded in gray-scale. The range of values of θ_F is -23° (black) to $+63^\circ$ (white). (d) The polarization ellipticity η_F of the emergent beam encoded in gray-scale. The range of values of η_F is -32° (black) to $+42^\circ$ (white).

turns out to be the same for both directions of incident polarization; that is, $t_{sp} = t_{ps}$.

The transverse effect is very different from both the polar and the longitudinal effects. With s-polarized incident light, where the optical E-field is parallel to the direction of the B-field in the slab, there is no MO effect whatsoever, but for the p-polarized light the medium exhibits an effective refractive index $n = [\varepsilon + (\varepsilon'^2/\varepsilon)]^{1/2}$. Thus in the transverse case neither s- nor p-polarized beams undergo polarization rotation, but the magnitude of the transmitted p-light shows a weak dependence on magnetization, that is, $T_p = |t_p|^2$ becomes a function of the strength of the B-field. The transverse effect is not bipolar, so that changing the direction of the B-field from $+Y$ to $-Y$ does not alter the magnitude of T_p. For a slab of transparent material 20 μm thick and with a fairly large MO coefficient

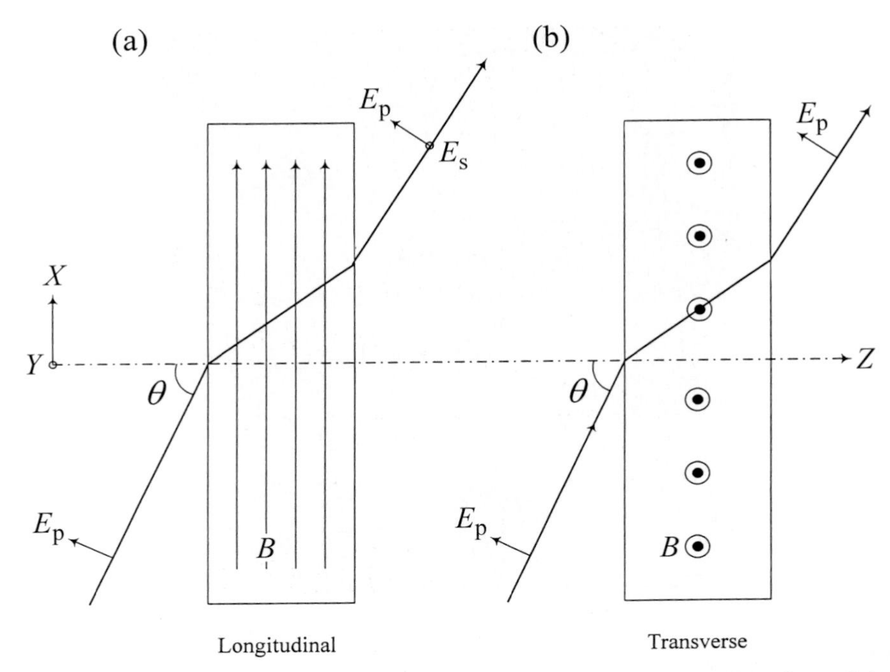

Figure 10.9 (a) Longitudinal Faraday effect is observed when the direction of the B-field within the slab of material is parallel both to the surface of the slab and to the plane of incidence. The rotation of polarization in this case occurs only at oblique incidence, where, upon transmission, a p-polarized beam acquires a s-component and vice versa. If the direction of $\boldsymbol{B}$ is reversed, the magneto-optically induced component of polarization will change sign. (b) The transverse effect occurs when the B-field lies in the plane of the sample perpendicular to the plane of incidence. The MO interaction in this case occurs only when the incident beam is p-polarized. Even then there is no polarization rotation; the only effect is that a change in the magnitude of the B-field causes a slight change in the magnitude of the transmitted p-light. The transverse effect is small and is not bipolar, meaning that reversing the direction of $\boldsymbol{B}$ does not affect the emergent beam.

($\varepsilon = 5.5, \varepsilon' = 0.1\mathrm{i}$), Figure 10.11 shows computed plots of $T_\mathrm{p}^{(0)}$ (i.e., transmission in the absence of a B-field, when $\varepsilon' = 0$) and $\Delta T_\mathrm{p} = T_\mathrm{p} - T_\mathrm{p}^{(0)}$ versus the angle of incidence θ. Note, in particular, that $\Delta T_\mathrm{p} \approx 0$ around the Brewster angle $\theta_\mathrm{B} = 66.9^\circ$, where a vanishing surface reflectivity results in minimal interference effects.

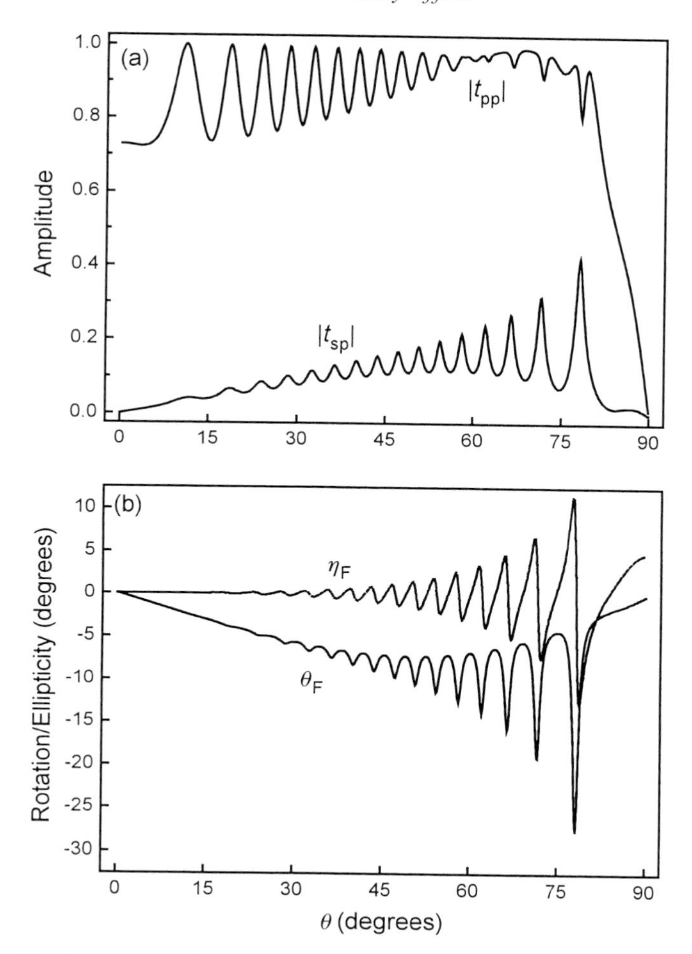

Figure 10.10 The longitudinal Faraday effect arising when a p-polarized plane wave ($\lambda = 550\,\text{nm}$) is incident at oblique angle θ on a slab 20 μm thick. The slab ($\varepsilon = 5.5$, $\varepsilon' = 0.01\text{i}$) is magnetized along the X-axis, as depicted in Figure 10.9(a). (a) The transmitted amplitudes $|t_{pp}|$ and $|t_{sp}|$ versus θ. (b) The polarization rotation angle θ_F and the ellipticity η_F versus θ.

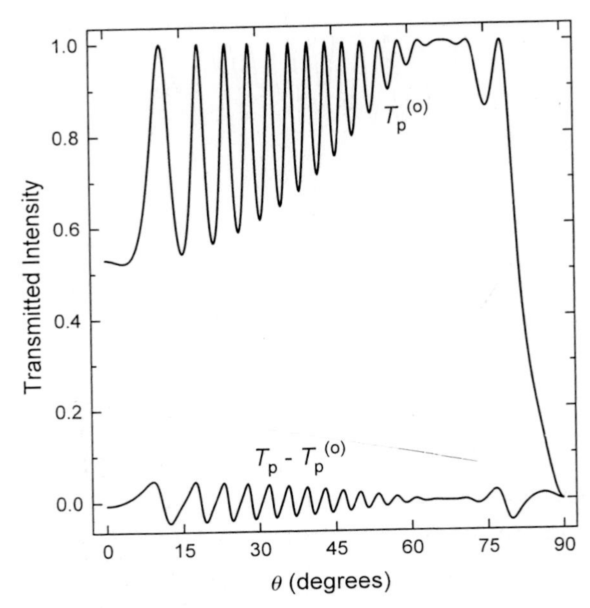

Figure 10.11 The transverse Faraday effect arising when a p-polarized plane wave ($\lambda = 550$ nm) is incident at oblique angle θ on a slab 20 μm thick. The slab ($\varepsilon = 5.5$) is magnetized along the Y-axis, as shown in Figure 10.9(b). In the absence of the B-field, $\varepsilon' = 0$, and the transmission of the slab for a p-polarized incident beam is denoted by $T_p^{(0)}$. When a strong B-field is introduced (corresponding to $\varepsilon' = 0.1\mathrm{i}$ in this case), the transmission changes to T_p. Shown here is the transmission differential $\Delta T_p = T_p - T_p^{(0)}$ as a function of θ.

References for chapter 10

1 Adapted from George Gamow, *The Great Physicists from Galileo to Einstein*, Dover Publications, New York, 1961. Some of the historical anecdotes have been compiled from information available on the worldwide web; see, for example, www.phy.uct.ac.za, www.iee.org.uk, www.woodrow.org.
2 P. S. Pershan, Magneto-optical effects, *J. Appl. Phys.* **38**, 1482–1490 (1967).
3 F. A. Jenkins and H. E. White, *Fundamentals of Optics*, fourth edition, McGraw-Hill, New York, 1976.
4 R. W. Wood, *Physical Optics*, third edition, Optical Society of America, Washington DC, 1988.
5 D. O. Smith, Magneto-optical scattering from multilayer magnetic and dielectric films, *Opt. Acta* **12**, 13 (1965).
6 M. Mansuripur, *The Physical Principles of Magneto-optical Recording*, Cambridge University Press, UK, 1995.

11

The magneto-optical Kerr effect

The Scottish physicist John Kerr (1824–1907) discovered the magneto-optical effect named after him in 1888. When linearly polarized light is reflected from the polished surface of a magnetized medium its polarization vector rotates and becomes somewhat elliptical. The direction of rotation and the sense of ellipticity are reversed when the direction of magnetization $\boldsymbol{M}$ of the sample is reversed, thus providing a powerful tool for optically monitoring the state of magnetization of the sample under investigation.[1–3]

The physical mechanism of the Kerr effect is identical to that of the Faraday effect and, in fact, the same theoretical model can be used to describe both phenomena, one in reflection, the other in transmission (see chapter 10, "The Faraday effect").

The Kerr effect can be analyzed under quite general conditions, with the direction of magnetization of the sample oriented arbitrarily relative to the plane of incidence of the light beam. However, the three geometries shown in Figure 11.1 are of particular importance and will be analyzed separately in the present chapter. When the magnetization $\boldsymbol{M}$ is perpendicular to the sample's surface, the observed phenomenon is referred to as the *polar* Kerr effect. When $\boldsymbol{M}$ is parallel to the surface and in the plane of incidence, the Kerr effect is *longitudinal*. Finally, when $\boldsymbol{M}$ is parallel to the surface but perpendicular to the plane of incidence, the observed phenomenon is known as the *transverse* Kerr effect.[4,5]

Electromagnetic basis of the Kerr effect

For convenience, we repeat in this short section the relevant text from chapter 10. Magneto-optical (MO) effects are best described in terms of the dielectric tensor $\boldsymbol{\varepsilon}$ of the medium in which the interaction between the light

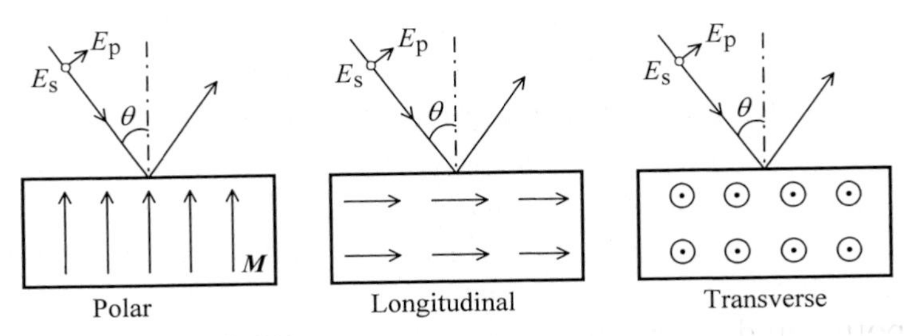

Figure 11.1 The MO Kerr effect is polar, longitudinal, or transverse, depending on the orientation of the magnetic moment $\boldsymbol{M}$ relative to the sample's surface and to the plane of incidence. The incident beam is p- or s-polarized according to whether its E-field is in the plane of incidence (E_p) or perpendicular to it (E_s).

and the applied magnetic field (or the internal magnetization of the medium) takes place:[1]

$$\boldsymbol{\varepsilon} = \begin{pmatrix} \varepsilon_{xx} & \varepsilon_{xy} & \varepsilon_{xz} \\ \varepsilon_{yx} & \varepsilon_{yy} & \varepsilon_{yz} \\ \varepsilon_{zx} & \varepsilon_{zy} & \varepsilon_{zz} \end{pmatrix}.$$

In an isotropic material the three diagonal elements are identical and, in the presence of a magnetic field along the Z-axis, there is a non-zero off-diagonal element ε', which couples the x- and y- components of the optical E-field:

$$\boldsymbol{\varepsilon} = \begin{pmatrix} \varepsilon & \varepsilon' & 0 \\ -\varepsilon' & \varepsilon & 0 \\ 0 & 0 & \varepsilon \end{pmatrix}.$$

In general, ε and ε' are wavelength-dependent, but over a narrow range of wavelengths they might be treatable as constants. In a transparent material, where there is no optical absorption, ε is real and ε' is imaginary. However, in the most general case of an absorbing MO material both ε and ε' would be complex numbers. For diamagnetic and paramagnetic media ε' is proportional to the applied magnetic field H, while for ferromagnetic and ferrimagnetic materials spin–orbit coupling is the dominant source of the MO interaction, making ε' proportional to the magnetization $\boldsymbol{M}$ of the medium.[1] Since $\boldsymbol{B} = \boldsymbol{H} + 4\pi\boldsymbol{M}$ (in CGS units), in general the B-field inside the medium may be considered the source of the MO effects.

When a polarized beam of light propagates in a medium along the direction of the magnetic field $\boldsymbol{B}$, the right and left circularly polarized (RCP and LCP) components of the beam experience different refractive indices $n^{\pm} = (\varepsilon \pm \mathrm{i}\varepsilon')^{1/2}$. Since the Fresnel reflection coefficients depend on the refractive index, the two circular polarizations are reflected with different reflectivities, r^+ and r^-, say. When r^+ and r^- happen to have a phase difference, the reflected beam exhibits a polarization rotation, and if the magnitudes $|r^+|$ and $|r^-|$ differ from each other, then there will be some degree of ellipticity. When the medium is transparent, $n^{\pm}$ are real and, therefore, there is no phase difference between r^+ and r^-, although their magnitudes will be different. In this case the reflected light exhibits polarization ellipticity only. However, in the general case of reflection from the surface of an absorbing medium (both ε and ε' complex), the reflected light exhibits elliptical polarization, with the major axis of the ellipse rotated relative to the direction of incident polarization.

For concreteness, we will confine our attention throughout this chapter to a metallic magnetic material having $\varepsilon = -8 + 27\mathrm{i}$ and $\varepsilon' = -0.6 + 0.2\mathrm{i}$ at the red HeNe wavelength, $\lambda_0 = 633\,\mathrm{nm}$. This is typical of the TbFeCo amorphous alloys used in magneto-optical disks for data storage. The discussion, however, will be kept quite general in nature, and the conclusions drawn from specific examples should be applicable to a wide variety of magnetic materials.

The polar effect

Figure 11.2(a) shows computed plots of the various reflection coefficients versus the angle of incidence θ for the case of a perpendicularly-magnetized sample. The conventional reflection coefficients for p- and s-light, r_{pp} and r_{ss}, show the behavior expected for a metallic surface. We denote by r_{ps} the cross-polarization factor from incident p to reflected s, and by r_{sp} that from incident s to reflected p. These coefficients represent the ability of the magnetic medium to convert, upon reflection, p-polarized light into s, and vice versa. It can be shown quite generally that $r_{\mathrm{ps}} = r_{\mathrm{sp}}$ at all angles of incidence. Thus the power of the magnetic medium to "rotate" the polarization is independent of whether the incident beam is p- or s-polarized. However, the polarization rotation and ellipticity angles, ρ and η, which depend on r_{pp} and r_{ss} as well as r_{ps}, exhibit differing behaviors for p- and s-light (see Figures 11.2(b), (c)). Note also in Figure 11.2(a) that r_{ps} remains more or less constant up to fairly large angles of incidence.

The longitudinal effect

Plots of the various reflection coefficients versus the angle of incidence θ for the longitudinal geometry appear in Figure 11.3(a). As in the polar case, it turns out that $r_{\mathrm{sp}} = r_{\mathrm{ps}}$ for all values of θ. At normal incidence the interaction between the incident E-field and the magnetization of the medium cannot

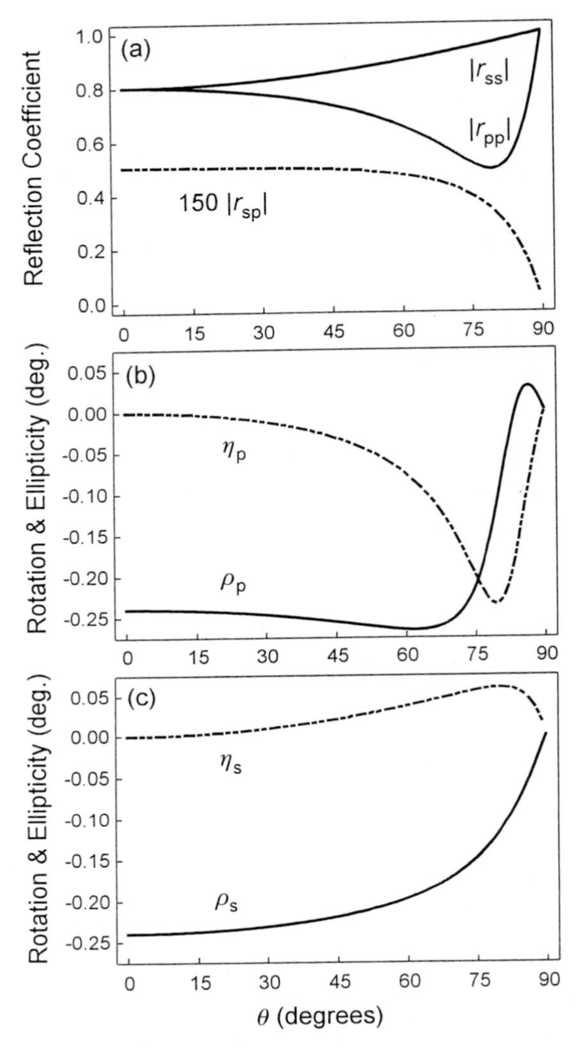

Figure 11.2 A linearly polarized plane wave is reflected from the polished surface of a magnetic material having perpendicular magnetization (the polar case); $\varepsilon_{xx} = -8 + 27\mathrm{i}$, $\varepsilon_{xy} = -0.6 + 0.2\mathrm{i}$. (a) Plots of $|r_{\mathrm{pp}}|$, $|r_{\mathrm{ss}}|$, and $|r_{\mathrm{sp}}| = |r_{\mathrm{ps}}|$ versus the angle of incidence θ. (b) The polarization rotation angle ρ and the ellipticity η versus θ for p-polarized incident beam. (c) Same as (b) for s-polarized beam.

produce any polarization rotation; therefore, $r_{\mathrm{sp}} = 0$ at $\theta = 0$. As θ increases, however, the MO signal gains strength, peaking at $\theta = 65°$. Again, ρ and η depend on whether the incident polarization is p or s (see Figures 11.3(b), (c)), but the effective MO signal, r_{sp}, is independent of the incident polarization. The longitudinal MO signal is typically weaker than its polar counterpart by almost one order of magnitude.

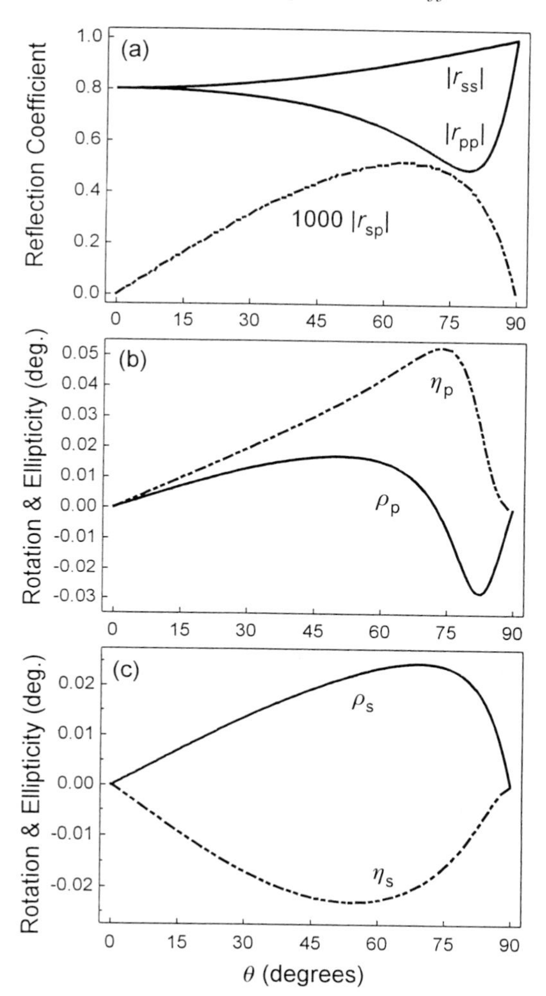

Figure 11.3 Same as Figure 11.2 but here for the longitudinal Kerr effect. Again $r_{sp} = r_{ps}$ at all angles of incidence. The MO effect is zero at normal incidence, reaching its peak at a fairly large angle. Note that $|r_{ps}|$, ρ, η are about an order of magnitude smaller than their counterparts for the polar geometry case. Both polar and longitudinal effects are bipolar, in the sense that a reversal in the direction of $\boldsymbol{M}$ results in a π phase shift of r_{ps}, leading to a reversal in the signs of both ρ and η.

The transverse effect

The behavior of the reflected light in this case differs fundamentally from that in the other two cases. First, there is no interaction whatsoever between the magnetic moment of the sample and s-polarized light. Here the optical E-field is parallel to $\boldsymbol{M}$ and, therefore, does not "see" the magnetization of the sample. When the incident beam is p-polarized, the interaction is confined to the plane of incidence, creating an extra E-field component within the same plane. Unlike the polar and longitudinal effects, no E-fields are generated perpendicular to the plane of incidence. Therefore, there are no polarization rotations in the transverse geometry. What is interesting, however, is that the reflectivity of the sample, $R_p = |r_{pp}|^2$, depends on the magnitude and direction of the magnetic moment $\boldsymbol{M}$.

In Figure 11.4(a) the reflectivity in the absence of $\boldsymbol{M}$ is denoted by $R_p^{(0)}$ (i.e., ε' is set to zero). With $\boldsymbol{M}$ pointing along $+Y$ the reflectivity changes slightly, becoming $R_p^{(+)}$; the difference is shown as the solid curve at the bottom of Figure 11.4(a). Similarly, when $\boldsymbol{M}$ is reversed to point along $-Y$, the corresponding change in R_p is given by the broken curve. The change in R_p is thus seen to depend on the direction of $\boldsymbol{M}$. This behavior is rather curious and, at first sight, appears to violate the principles of symmetry, although a careful analysis shows it to be correct.[1] It is noteworthy that this bipolar nature of R_p critically depends on the magnetic medium being absorptive; for transparent magnetic media (where ε is purely real and ε' purely imaginary), the depen-

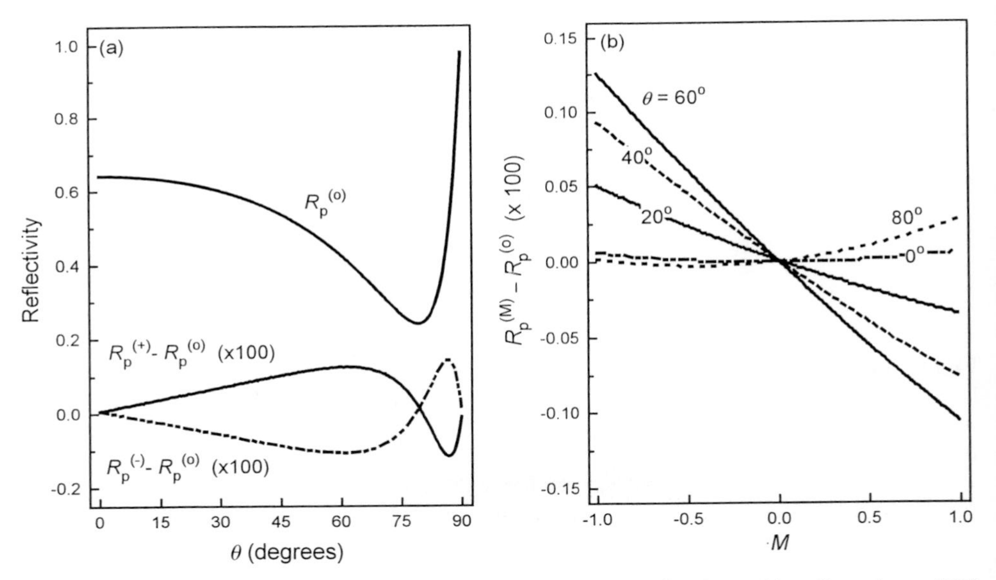

Figure 11.4 Variation of the reflectivity R_p with the magnitude and/or direction of $\boldsymbol{M}$ in the transverse geometry. The incident beam is p-polarized in all cases; there are no transverse effects for s-polarized light. (a) The dependence of R_p on the angle of incidence θ; the superscript zero indicates that $M = 0$. When the medium is fully magnetized in the $\pm Y$ direction, the reflectivity is denoted by $R_p^{(\pm)}$. (b) The variation of R_p with M at various angles of incidence. At $\theta = 0°$ the dependence on M is quadratic, while at $\theta = 20°, 40°, 60°$ it is nearly linear. (The off-diagonal element ε' of the dielectric tensor is assumed to be directly proportional to M.)

dence of R_p on $\boldsymbol{M}$ is quadratic, showing no change with the reversal of the direction of magnetization.

Figure 11.4(b) shows the variations in the reflectivity difference $R_p^{(M)} - R_p^{(0)}$ with the magnitude of $\boldsymbol{M}$, as $\boldsymbol{M}$ varies continuously from a maximum value along $+Y$ to zero and then reverses direction and reaches a maximum in the opposite direction. At normal incidence the dependence on $\boldsymbol{M}$ is quadratic, but at larger angles (20°, 40°, 60°) it is almost (but not quite) linear. Like the longitudinal effect, the transverse effect in this case is about an order of magnitude weaker than the polar effect.

Localized probe of the state of magnetization

It is sometimes desirable to probe the local state of a magnetic surface. This can be done by focusing onto the surface a polarized laser beam through a high-NA objective, as shown in Figure 11.5. The lens focuses the beam to a diffraction-limited spot (diameter $\sim \lambda_0$), providing access to the sample's magnetization within a tiny region. The focused beam, of course, contains many rays arriving at the sample from different directions, making the analysis of the resulting Kerr signal somewhat tedious.

To begin with, even in the absence of a magnetic moment $\boldsymbol{M}$ the reflected polarization state is complicated. Figure 11.6 shows the various distributions at the exit pupil of the objective when $\boldsymbol{M}$ is set to zero. The intensity of the x-

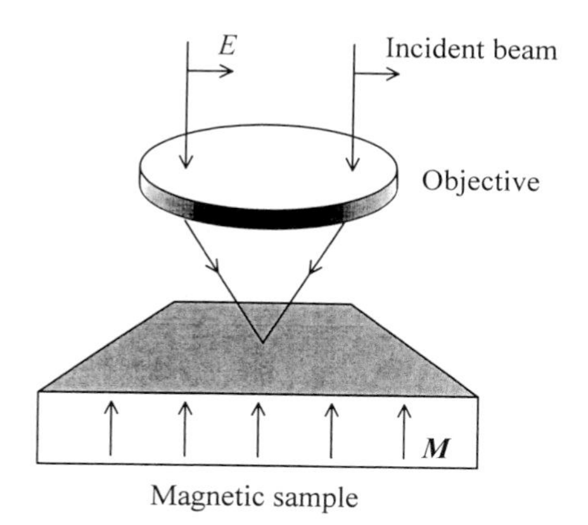

Figure 11.5 A linearly polarized beam of light having its E-field parallel to the X-axis is focused onto the flat surface of a magnetic medium through a diffraction-limited microscope objective lens ($NA = 0.95$, $f = 3158\lambda$). The power of the incident beam – its integrated intensity – is set to unity. The reflected light's distribution at the exit pupil has a small but important contribution from the magnetization $\boldsymbol{M}$ of the sample.

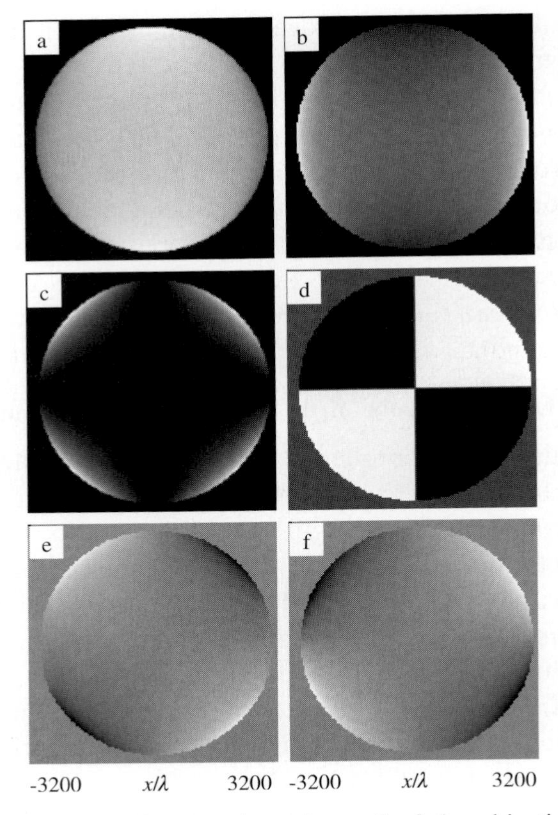

Figure 11.6 Various distributions at the exit pupil of the objective of Figure 11.5, when $\boldsymbol{M}$ is set to zero (i.e., no Kerr effect). (a) Distribution of intensity for the reflected E_x; the total power = 0.62. (b) Distribution of phase for E_x; $\phi_{\min} = 0°$, $\phi_{\max} = 55°$. (c) Distribution of intensity for the reflected E_y; the total power = 0.011. (d) Distribution of phase for E_y; $\phi_{\min} = -36°$, $\phi_{\max} = 150°$. (e) The polarization rotation angle ρ; $\rho_{\min} = -20.5°$, $\rho_{\max} = 20.5°$. (f) The polarization ellipticity η; $\eta_{\min} = -25.1°$, $\eta_{\max} = 25.1°$.

component of the reflected light, $I_x = |E_x|^2$, depicted in Figure 11.6(a), shows slight variations across the aperture, in agreement with the r_{pp} and r_{ss} curves of Figure 11.2(a). Similar variations are seen in the corresponding phase plot of Figure 11.6(b). In addition to E_x, the reflected light also contains a y-component, E_y, whose intensity and phase plots appear in Figures 11.6(c), (d). While the total power (i.e., the integrated intensity) of E_x is 62% of the incident power, that of E_y is only $\sim 1.1\%$. The reflected E_y in adjacent quadrants of the aperture exhibits a phase shift of π, indicating a sign reversal from one quadrant to the next. The presence of E_y in the reflected beam gives rise to the patterns of polarization rotation and ellipticity depicted in Figures

11.6(e), (f); note the fairly large values of ρ and η in the four corners of the aperture ($\rho_{\min}, \rho_{\max} = \pm 20.5°$; $\eta_{\min}, \eta_{\max} = \pm 25.1°$).

To determine the contribution to the reflected E-field by the sample's magnetization, we compute the complex reflected amplitudes for $\boldsymbol{M}$ up and $\boldsymbol{M}$ down, then subtract one distribution from the other. In the process the x-component of polarization disappears, indicating that E_x is indifferent to the reversal of $\boldsymbol{M}$. However, the residual y-component shows the distribution depicted in Figure 11.7. The total power of E_y contributed by the MO interaction in this case is $\sim$0.0042% of the incident power. Both the phase and intensity of this residual E_y are fairly uniform, with the intensity showing a mild decline towards the edge of the aperture, consistent with the behavior of r_{sp} in Figure 11.2(a). (Note that, even at $NA = 0.95$, the largest angle of incidence on the sample is less than 72°.)

A similar calculation for the longitudinal case yields the plots in Figure 11.8. Here the complex amplitude distributions are computed for $\boldsymbol{M}$ along $+X$ and $-X$, then subtracted from each other. Unlike the polar Kerr signal in Figure 11.7, both the reflected E_x and the reflected E_y in the longitudinal geometry contain some MO contribution. The total power of the MO contribution to E_x is 0.0000065%, which is rather small and concentrated in the four corners of the aperture. Note that the top half of the aperture containing the E_x signal has a π phase shift relative to the bottom half. In contrast, the E_y contribution to the MO signal (see Figures 11.8(c), (d)) contains 0.000054% of the incident power, equally divided between the right and left halves of the aperture with a π phase shift.

Finally, if the magnetization of the sample in Figure 11.5 is aligned with the Y-axis (perpendicular to the plane of the figure) then the MO contribu-

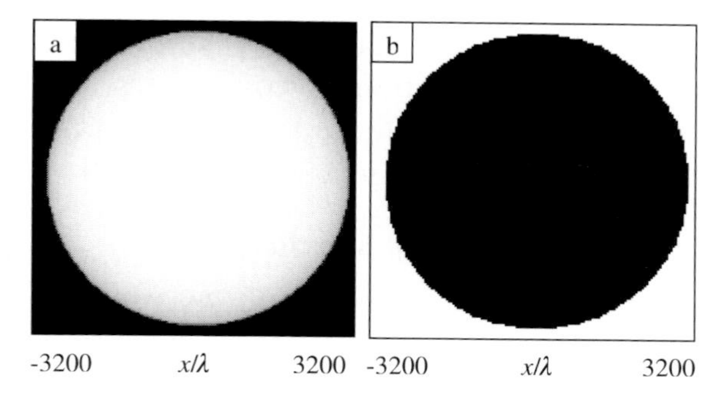

Figure 11.7 Contribution of the magnetic moment $\boldsymbol{M}$ of the sample in Figure 11.5 to the E_y distribution at the objective's exit pupil; $\boldsymbol{M}$ is assumed to be perpendicular to the sample's surface. (a), (b) Intensity and phase patterns of E_y; the total power $= 0.42 \times 10^{-4}$.

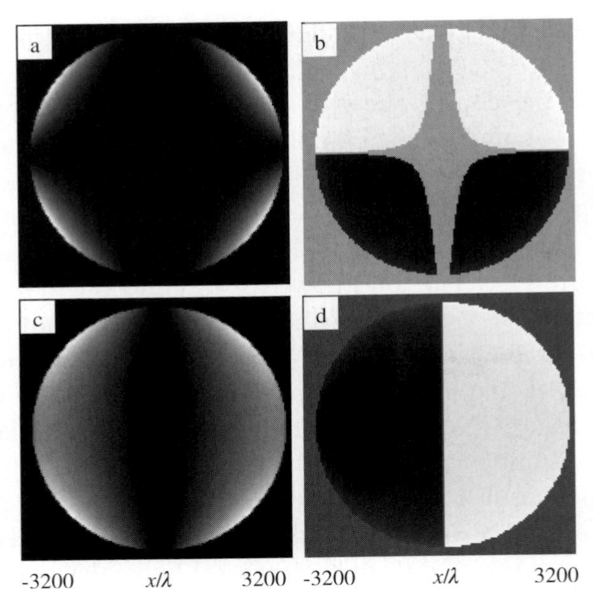

Figure 11.8 Contribution of the magnetic moment of the sample to the E-field distribution at the exit pupil of the objective of Figure 11.5; $\boldsymbol{M}$ is assumed to be aligned with the X-axis. (a), (b) Intensity and phase patterns of E_x; total power 0.65×10^{-7}. The top and bottom halves of the aperture have a relative phase of π. (c), (d) Intensity and phase patterns of E_y; the total power $= 0.54 \times 10^{-6}$. Note the π phase difference between the right and left halves of the aperture.

tions to the reflected beam will be those shown in Figure 11.9. As before, we obtain these distributions by computing the complex amplitudes at the exit pupil with $\boldsymbol{M}$ along $+Y$ and $-Y$ and then subtracting one from the other. The MO contribution to E_x, having 0.00026% of the incident power, is fairly strong. The contribution to E_y contains 0.000054% of the incident power, exactly as in the longitudinal case depicted in Figures 11.8(c), (d). Note that, with the exception of a 90° rotation of coordinates, the distributions in Figures 11.9(c), (d) are identical to those in Figures 11.8(c), (d).

Signal detection

The MO contribution to the reflected polarization state can be converted to an electronic signal with the aid of polarization-sensitive optics and photodetectors. For instance, to detect the polar Kerr signal shown in Figure 11.7, one can employ the differential scheme shown in Figure 11.10. Here the reflected beam is directed toward a quarter-wave plate, which helps to eliminate the phase shift between E_x and E_y. The quarter-wave plate is followed

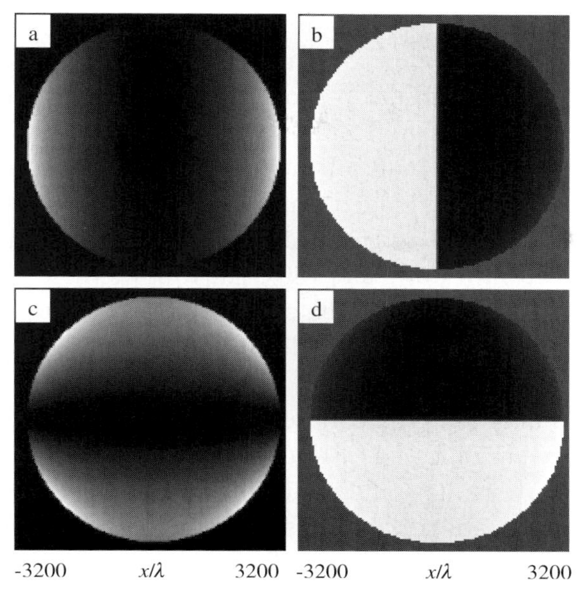

Figure 11.9 Same as Figure 11.8, but for the transverse geometry, where $\boldsymbol{M}$ is switched between $+Y$ and $-Y$ directions. E_x, depicted in (a), (b), has total power 0.26×10^{-5}. E_y, depicted in (c), (d), has total power 0.54×10^{-6}.

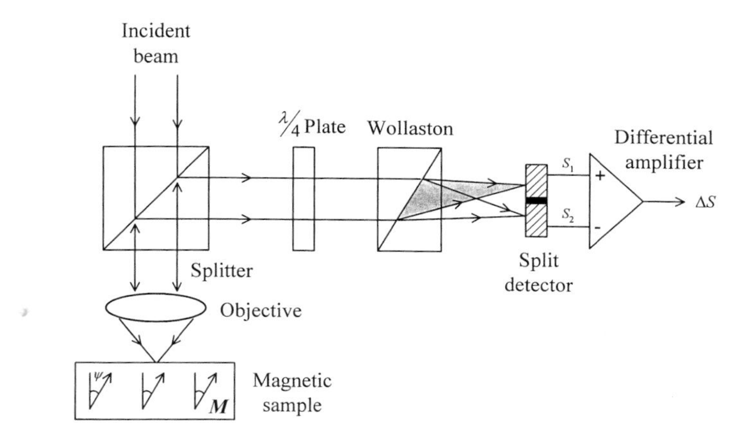

Figure 11.10 A differential detection scheme is used to probe the direction of $\boldsymbol{M}$ via the state of polarization of the reflected beam. To attain high spatial resolution, the laser beam is focused on the sample surface. The reflected beam goes through a quarter-wave plate, whose fast and slow axes are at 45° to the direction of incident polarization. The Wollaston prism divides the beam between two photodetectors, and the difference ΔS between the outputs of these detectors is monitored. To maximize the swing of ΔS one must adjust the orientation of the Wollaston around the optical axis.

by a Wollaston prism, which mixes the MO component of polarization contained in E_y with the reflected x-component of polarization, E_x. The two mixed beams emerging from the Wollaston are detected by a pair of photodetectors whose difference signal ΔS conveys information about the sample's magnetic state. A computed plot of the normalized differential signal versus the orientation angle ψ of $\boldsymbol{M}$ is given in Figure 11.11. As $\boldsymbol{M}$ moves away from its initial orientation at $\psi = 0°$ toward the plane of the sample at $\psi = 90°$, and continues downward until $\psi = 180°$, ΔS follows these changes continuously. (We mention in passing that, as $\boldsymbol{M}$ rotates, the sum signal $S_1 + S_2$ undergoes slight variations, but, for all practical purposes, it remains a constant.)

Similar systems may be designed to extract the longitudinal and transverse MO signals depicted in Figures 11.8 and 11.9. However, because in these cases the E-field contributions have different signs in opposite halves of the aperture, any viable detection scheme must extract the signals from these half-apertures separately, before combining them with the proper sign at the end.

Enhancing the Kerr signal

To enhance the MO signal one should force the magnetic sample to absorb a greater fraction of the incident beam. As an example of how this can be done, consider the system of Figure 11.12, which consists of a magnetic sample

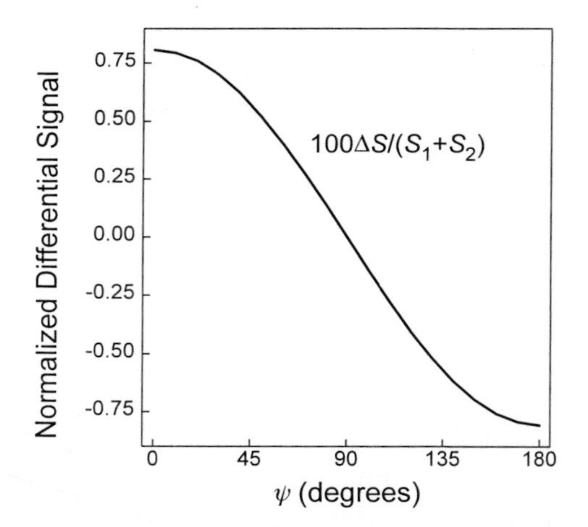

Figure 11.11 The normalized differential signal as a function of the orientation angle ψ of $\boldsymbol{M}$ (see Figure 11.10). The detection module has been adjusted for maximum swing of ΔS. This signal is bipolar, in the sense that it switches sign when $\boldsymbol{M}$ is reversed. ΔS is zero at $\psi = 90°$ (i.e., $\boldsymbol{M}$ in the plane of the sample).

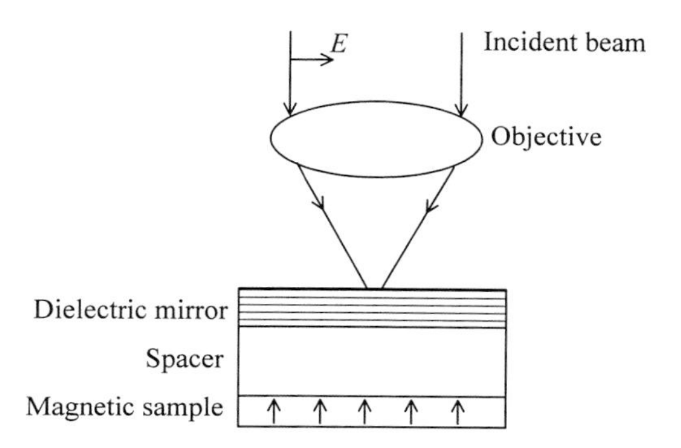

Figure 11.12 A collimated linearly polarized laser beam ($\lambda = 633\,\text{nm}$) is focused through a $0.75NA$, $f = 4000\lambda$ objective onto a high-reflectivity dielectric mirror sitting on top of a magnetic sample ($\varepsilon = -8 + 27\text{i}$, $\varepsilon' = -0.6 + 0.2\text{i}$). The mirror, consisting of seven pairs of high- and low-index quarter-wave layers ($n_{\text{H}} = 2$, $n_{\text{L}} = 1.5$), is deposited on a glass substrate 10μm thick and of index $n = 1.5$. The substrate is in direct contact with the magnetic surface. The magnetization $\boldsymbol{M}$ is uniform and perpendicular to the plane of the sample's surface, and the beam entering the lens is polarized along the X-axis. Upon reflection from the sample, the light collected by the objective is photographed at the exit pupil. Most rays within the focused beam are reflected at the mirror, without ever reaching the magnetic sample. At certain angles of incidence, however, where the cavity becomes resonant, the light passes through to the magnetic medium and is absorbed by it. It is only for these resonant rays that the MO effect is observed at the exit pupil.

placed under a high-reflectivity dielectric mirror. The majority of the rays in the focused beam are reflected from the mirror without ever reaching the magnetic sample. However, when the direction of the ray is such that the cavity between the mirror and the sample becomes resonant, the ray is strongly absorbed by the magnetic sample. This strong absorption produces in the reflected beam a rather large polarization component perpendicular to the incident E-field, which can then be detected at the exit pupil of the objective.

Figure 11.13 shows the computed distributions at the exit pupil of a $0.75NA$ lens. The intensity plot for E_x in Figure 11.13(a) shows absorption bands in the angular spectrum of the incident beam. The reflected E_y in Figure 11.13(b) is strong in certain regions of the aperture, but these contributions mostly come from spurious light reflected from the mirror, not from the magnetic sample.

To determine the MO signal at the exit pupil, we once again compute the reflected complex amplitudes with $\boldsymbol{M}$ up and $\boldsymbol{M}$ down, then subtract the

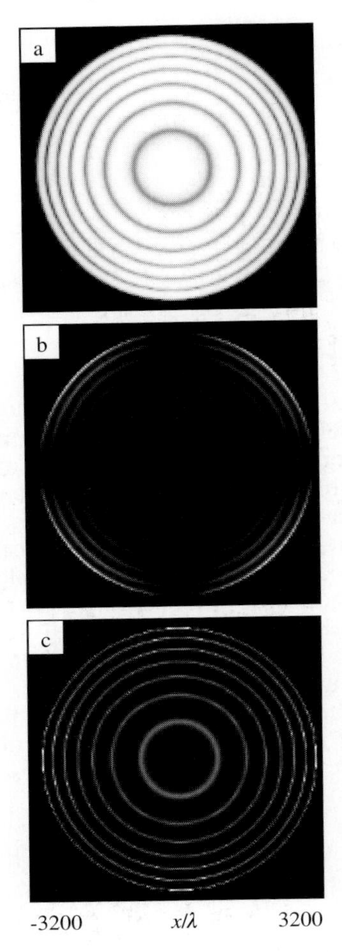

Figure 11.13 Various distributions at the exit pupil of the objective of Figure 11.12. (a) Intensity of the reflected E_x; the total power = 87% of the incident power. (b) Intensity of E_y; the total power = 0.3% of the incident power. Most of this E_y, which is primarily produced by oblique reflections from the dielectric mirror, serves only to obscure the MO-generated component of polarization. (c) The true MO signal obtained by subtracting the distributions produced with $\boldsymbol{M}$ up and $\boldsymbol{M}$ down; the total power $\approx$ 0.0007% of the incident power.

corresponding distributions. Figure 11.13(c) is the result of this calculation, showing the intensity of the residual E_y contributed by the MO interaction. The peak value of this MO signal is nearly twice that shown in Figure 11.7(a).

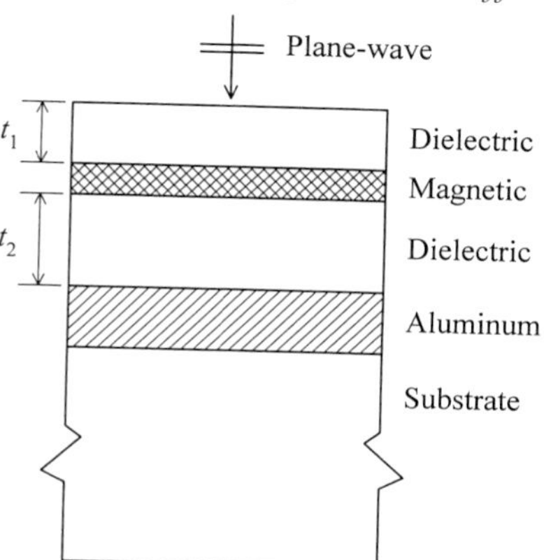

Figure 11.14 A quadrilayer MO stack consists of an aluminum reflector, an intermediate dielectric layer, a thin magneto-optic film, and an overcoating dielectric layer. The thicknesses of the various layers may be adjusted to maximize the MO signal $|r_{sp}|$ obtained upon reflection.

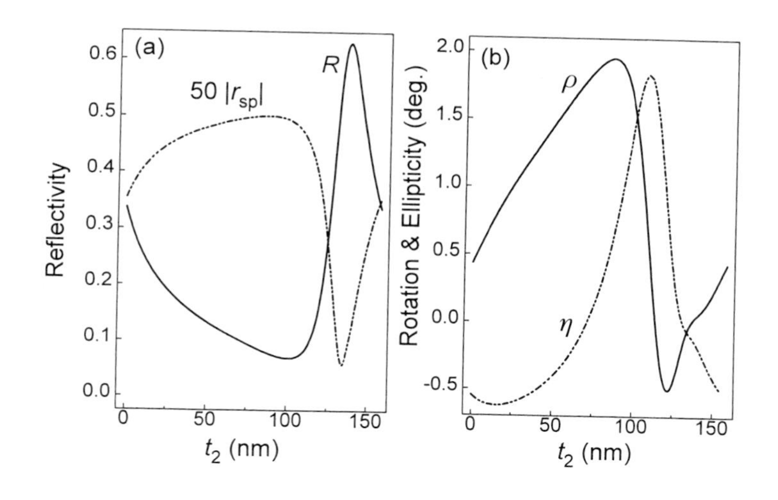

Figure 11.15 The dependence of reflected signals from a quadrilayer MO stack on the thickness of the dielectric underlayer ($\lambda_0 = 633$ nm, normal incidence). The aluminum layer ($n + ik = 1.4 + 7.6i$) is 50 nm thick, the MO film is 20 nm thick, and the overcoat layer ($n = 2$) is 80 nm thick ($t_1 = \lambda_0/(4n)$). The underlayer's index of refraction is $n = 2$, but its thickness t_2 is adjustable. (a) Computed plots of the reflectivity R and the MO signal $|r_{sp}|$ versus t_2. (b) The Kerr rotation angle ρ and the ellipticity η versus t_2. The maximum of ρ occurs when η is nearly zero, and vice versa.

Quadrilayer stack

A practical method of enhancing the MO Kerr effect involves the incorporation of a thin magnetic film in a quadrilayer stack structure. Figure 11.14 shows one such stack, consisting of an aluminum reflector, a dielectric underlayer, a thin magnetic film, and a dielectric overlayer. By optimizing the thicknesses of these layers it is possible to improve the MO signal substantially. In the following example, we will fix the thicknesses of three of the layers and optimize the thickness of the remaining one. This results in a significant gain in the performance of the stack. (It is possible to achieve further improvement by optimizing the other layers as well.)

Figure 11.15 shows plots of the reflectivity $R = |r_{pp}|^2 = |r_{ss}|^2$, the MO Kerr signal $|r_{sp}|$, and the polarization rotation and ellipticity all versus the thickness t_2 of the dielectric underlayer. (Since the dependence on t_2 is periodic, only one period, ranging from zero to $\lambda_0/(2n)$, is shown.) Note that $|r_{sp}|$ peaks when R is at a minimum, and vice versa. The maximum value of $|r_{sp}|$ in this example is about three times greater than that of the bare magnetic sample shown in Figure 11.2.

References for chapter 11

1 P. S. Pershan, Magneto-optical effects, *J. Appl. Phys.* **38**, 1482–1490 (1967).
2 F. A. Jenkins and H. E. White, *Fundamentals of Optics*, fourth edition, McGraw-Hill, New York, 1976.
3 R. W. Wood, *Physical Optics*, third edition, Optical Society of America, Washington DC, 1988.
4 D. O. Smith, Magneto-optical scattering from multilayer magnetic and dielectric films, *Optica Acta* **12**, 13 (1965).
5 M. Mansuripur, *The Physical Principles of Magneto-optical Recording*, Cambridge University Press, UK, 1995.

12

Fabry–Pérot etalons in polarized light

The principles of operation of Fabry–Pérot interferometers are well known, and their application in spectroscopy has established their status as one of the most sensitive instruments ever invented.[1,2] However, the behavior of a Fabry–Pérot device in polarized light, especially when birefringence and optical activity are present within the mirrors or in the cavity, is less well known. We devote this chapter to a description of some of these phenomena, in the hope of clarifying their physical origins and perhaps suggesting some new applications.

The dielectric mirror

A multilayer stack mirror is shown schematically in Figure 12.1. The substrate is a transparent slab of glass, and the layers are made of high- and low-index dielectric materials.[3] In the examples used in this chapter the low-index layers will have $(n, k) = (1.5, 0)$ and thickness $d = 105.5$ nm, and the high-index layers will have $(n, k) = (2, 0)$ and $d = 79.125$ nm. (At the operating wavelength, 633 nm, both these layers will be one quarter-wave thick.) Figure 12.2 shows computed plots of amplitude and phase for the reflection coefficients of a 10-layer mirror. At normal incidence ($\theta = 0$) both p- and s-components of polarization have an amplitude reflectivity $|\rho| = 0.844$. The mirror, therefore, reflects about 71% of the incident optical power and transmits the remaining 29%. Ignoring any loss of light at the uncoated facet of the substrate, the amplitude transmission coefficient (outside the substrate) turns out to be $|\tau| = 0.536$. At larger angles of incidence both the amplitude and phase for p- and s-light begin to deviate from their normal-incidence values and from each other, but we are not concerned with these variations here. What is important is to note that, at small angles of incidence (say up to 30°), the reflectivity

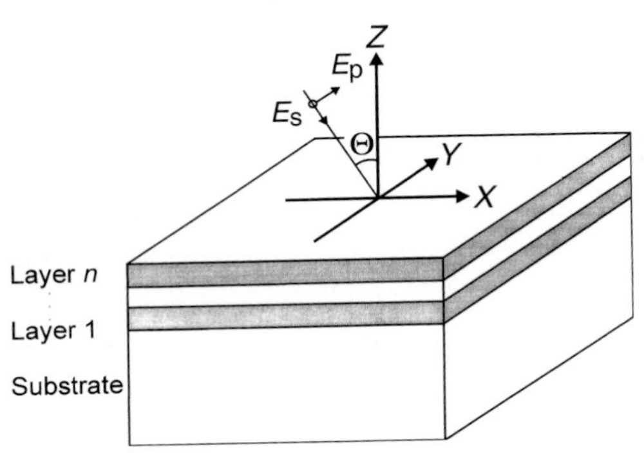

Figure 12.1 A multilayer dielectric mirror and a plane wave at oblique incidence. In the examples used in this chapter, the substrate refractive index n is 1.5, the odd-numbered layers have an index of 2 and are 79.125 nm thick, and the even-numbered layers have an index of 1.5 and are 105.5 nm thick. At the design wavelength, $\lambda = 633$ nm, these layer thicknesses correspond to one-quarter of the wavelength.

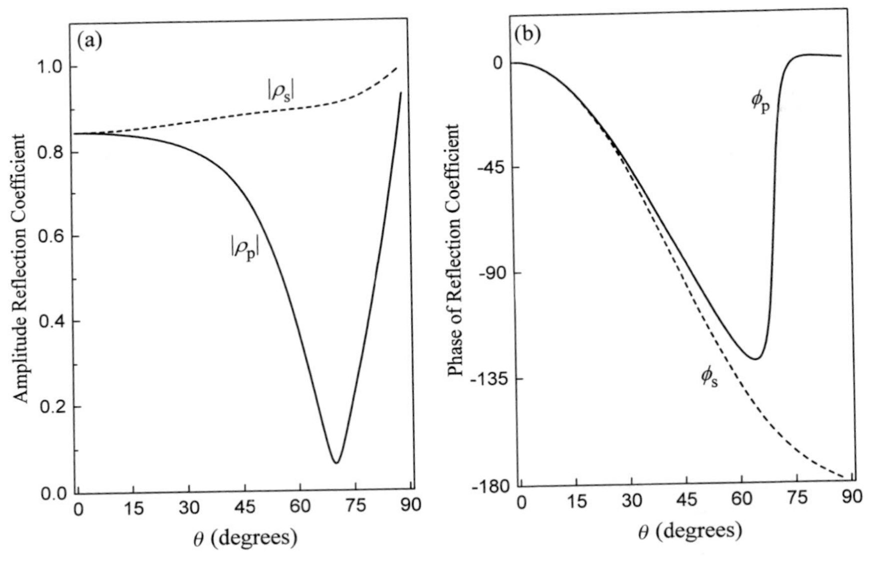

Figure 12.2 Computed plots of (a) amplitude and (b) phase of the reflection coefficients of a dielectric mirror for p- and s-polarized beams versus the angle of incidence. The assumed mirror is as shown in Figure 12.1, with a total of 10 layers; the medium of incidence is air.

remains high. Let us also mention in passing that, when light shines on a dielectric mirror through its substrate, the reflection and transmission coefficients ρ' and τ' generally retain the same amplitudes as above, but their phases will differ from those of ρ and τ.

The Fabry–Pérot etalon

Figure 12.3 shows the schematic of a Fabry–Pérot etalon. Two dielectric mirrors, separated by an air gap, are placed face-to-face and parallel to each other. A plane monochromatic beam is shown at oblique incidence θ on one of the mirrors. (In practice the uncoated facets of both substrates are given a slight wedge to eliminate spurious reflections.) For the system of Figure 12.3 the computed plots of reflection amplitude and phase versus θ are shown in Figure 12.4, for mirrors with 10 dielectric layers each, and an air gap 8.229 μm wide, which is exactly 13λ. Note that, within the 0° to 30° range of angles of incidence depicted, the p- and s- reflectivities are nearly the same. Sharp drops in the etalon's reflectivity occur at $\theta = 0°$, 15.37°, 21.83°, and 26.85°; at these angles $(d/\lambda)\cos\theta = 13$, 12.53, 12.07, and 11.6, respectively. In other words, when the effective gap-width is an integer multiple of a half-wavelength, the etalon becomes transparent to the incident light. To be sure, there are slight deviations from exact half-wavelength multiplicity here, which have to do with the θ-dependence of the phase of the individual mirror reflectivities (see Figure 12.2 (b)), but, for our purposes, these differences are small and may be ignored.

Next, we study the setup of Figure 12.5, which is designed to send a focused beam of light onto an etalon and to analyze the resulting reflection. The setup includes a path for a reference beam, so that Twyman–Green interferometry may be used to reveal the reflected phase pattern. It also includes a polarizer before the observation plane to allow selection of the

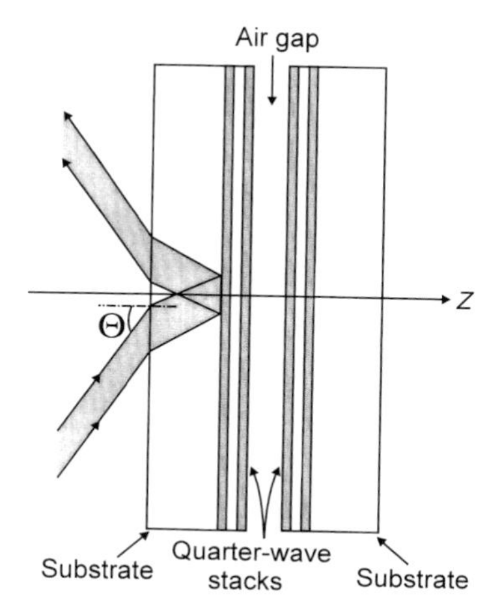

Figure 12.3 A Fabry–Pérot etalon consists of two face-to-face dielectric mirrors separated by an air-gap. Also shown is a plane wave at an oblique incidence angle θ.

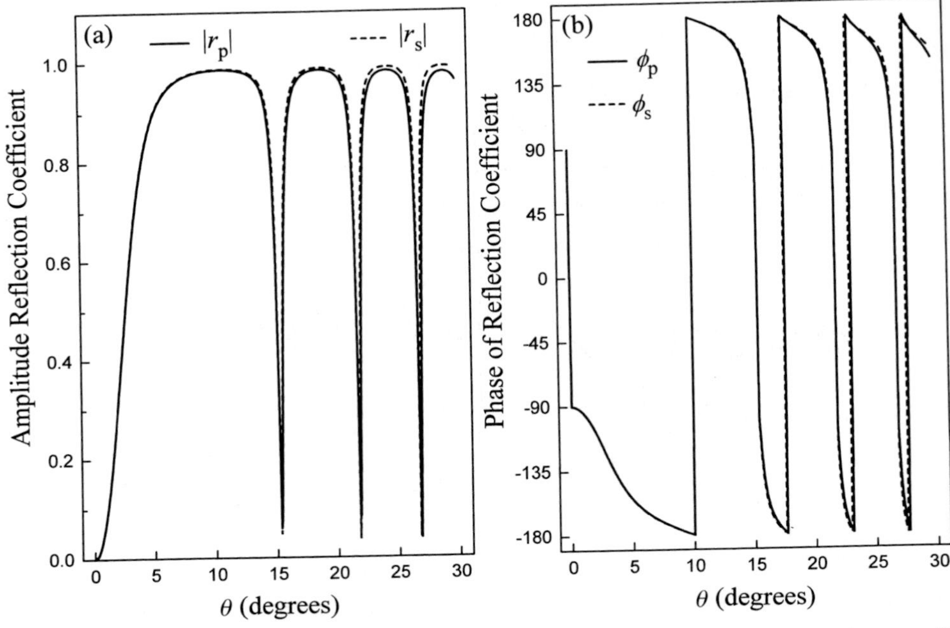

Figure 12.4 Computed plots of (a) amplitude and (b) phase of the reflection coefficients of a Fabry–Pérot etalon for p- and s-polarized beams versus the angle of incidence. The assumed etalon is that shown in Figure 12.3 with 10-layer mirrors and a 8.229 μm gap.

polarization direction of interest. Figure 12.6 shows computed plots of intensity at the observation plane obtained under various conditions. Frames (a) and (b) are obtained when the reference beam is blocked, whereas frames (c) and (d) are interferograms obtained in the presence of the reference beam. The circular area in each frame represents the aperture of the objective lens ($NA = 0.5$).

In (a) the polarizer is taken to transmit the same direction of polarization as that of the incident beam. The dark rings correspond to the angles of incidence at which the reflectivity plots of Figure 12.4(a) exhibit their minima. In frame (b) the polarizer is turned by 90° so that only a small fraction of the light (about 3×10^{-4} of the original incident power) passes through to the observation plane. The four corners of this distribution correspond to the four corners of the focused cone of light, which have a mix of p- and s-polarization. Here, the rays incident on the etalon are subject to slightly different reflectivities in their p- and s-components (see Figure 12.4(a)), which gives rise to a small rotation of polarization from its original direction. It is this polarization rotation in the four corners of the

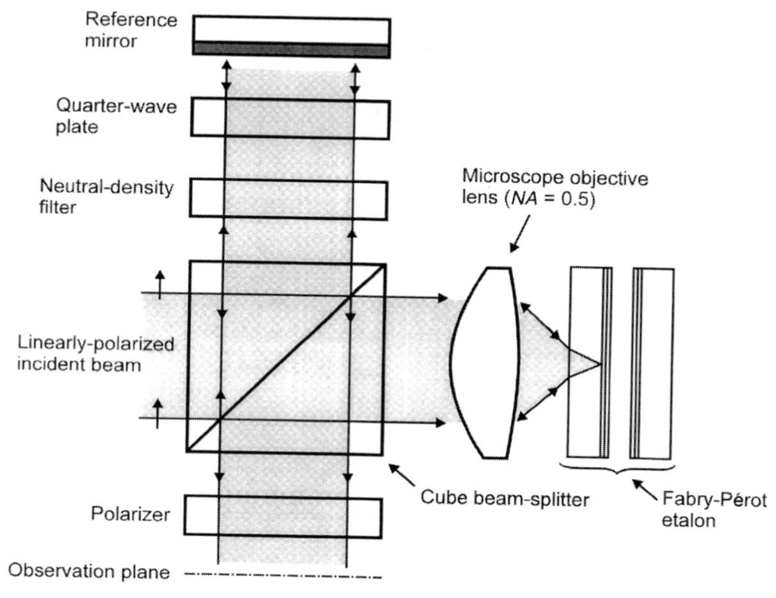

Figure 12.5 Schematic diagram of a system used for observing Fabry–Pérot fringes. A linearly polarized beam of light is focused on an etalon, which reflects the beam and sends it back through the system. The recollimated beam may then be viewed after passing through a polarizer. By setting the transmission axis of the polarizer perpendicular to the direction of incident polarization, one may observe regions of the beam which have suffered a small (but measurable) polarization rotation. A Twyman–Green interferometer is also incorporated into the system for observing the reflected phase pattern. The neutral-density filter is needed to adjust the amplitude of the reference beam in order to obtain high-contrast interferograms. A 45° rotation of the quarter-wave plate around the optical axis causes a 90° rotation of the reference beam's polarization; this is needed when the polarizer's transmission axis is set perpendicular to the direction of incident polarization.

lens that is responsible for the four corners of the intensity distribution in Figure 12.6(b).

Frames (c) and (d) of Figure 12.6 are obtained by unblocking the reference arm in the system of Figure 12.5, thus allowing the interference pattern (between the beam reflected from the etalon and that reflected from the reference mirror) to impinge on the observation plane. The case for the parallel component of polarization depicted in (c) shows the phase of the pattern to be more or less uniform over the entire aperture; in particular, it shows that there are no phase jumps between adjacent rings. The case for the perpendicular component of polarization depicted in (d) shows a 180° phase shift between adjacent corners. This is caused

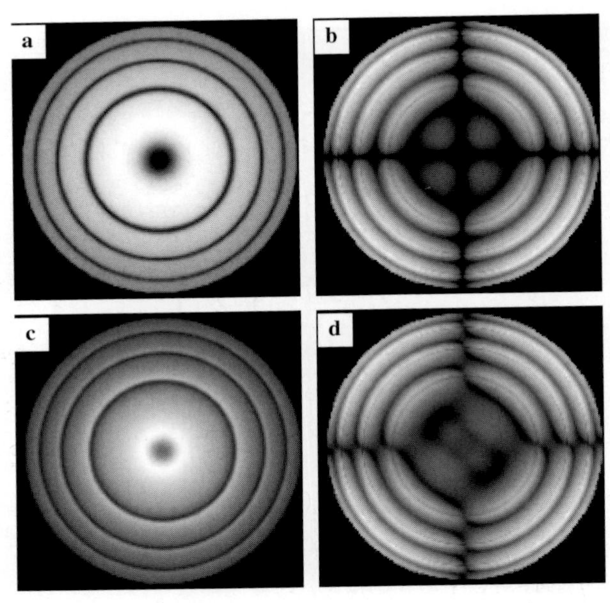

Figure 12.6 Computed plots of intensity distribution at the observation plane of the system of Figure 12.5. Frame (a) is obtained when the reference beam is blocked and the polarizer is set to transmit the direction of incident polarization. In the case of frame (b), the reference beam is still blocked, but the polarizer is rotated by 90°. The logarithm of the intensity distribution is plotted here in order to enhance the weak regions of the pattern; this is similar to over-exposing a photographic film placed at the observation plane. Frames (c) and (d) show the corresponding interference patterns obtained when the reference beam is unblocked. In the case of frame (d) the polarizer is rotated by 90° and the quarter-wave plate by 45°, a strong neutral-density filter is used to attenuate the reference beam substantially, and again the logarithm of the intensity distribution is plotted to enhance weak regions.

by the fact that, in adjacent corners of the lens, the polarization vector rotates in opposite directions.

Mirror birefringence

Next we consider the effects of birefringence in the mirrors of the Fabry–Pérot etalon.[4,5] For this analysis we assume that the mirrors have 20 layers each ($|\rho| = 0.9905$, $|\tau| = 0.1375$) and that the uppermost layer of both mirrors is slightly birefringent. We will suppose that the uppermost layer has a nominal index of 1.5, except along the Y-axis (see Figure 12.1) where the index is 1.505. We also assume a normally incident plane wave and an adjustable gap-width. For this etalon the computed transmission coefficients and the polarization state of the transmitted light versus the gap-width are plotted in Figure 12.7. Two different peaks are observed in transmission, one for the p-polarized, the other for the s-polarized incident beam. (The E-field of the p-light is parallel to X, while that of the s-light is parallel to Y.) The peak separation arises because the mirrors give a slightly different phase upon reflection to the two components of polarization. The gap-width, therefore, must be adjusted to compensate for this phase difference. If the incident beam is linearly polarized at 45° (i.e., halfway between p and s), the transmitted beam will show the rotation angle ψ and the ellipticity ξ depicted in Figure 12.7(b). The maximum ellipticity is close to 40°, which shows that the transmitted light at this point is nearly circularly polarized. A very small amount of birefringence in the mirrors can, therefore, have substantial effects on the polarization state of the transmitted (or reflected) beam.

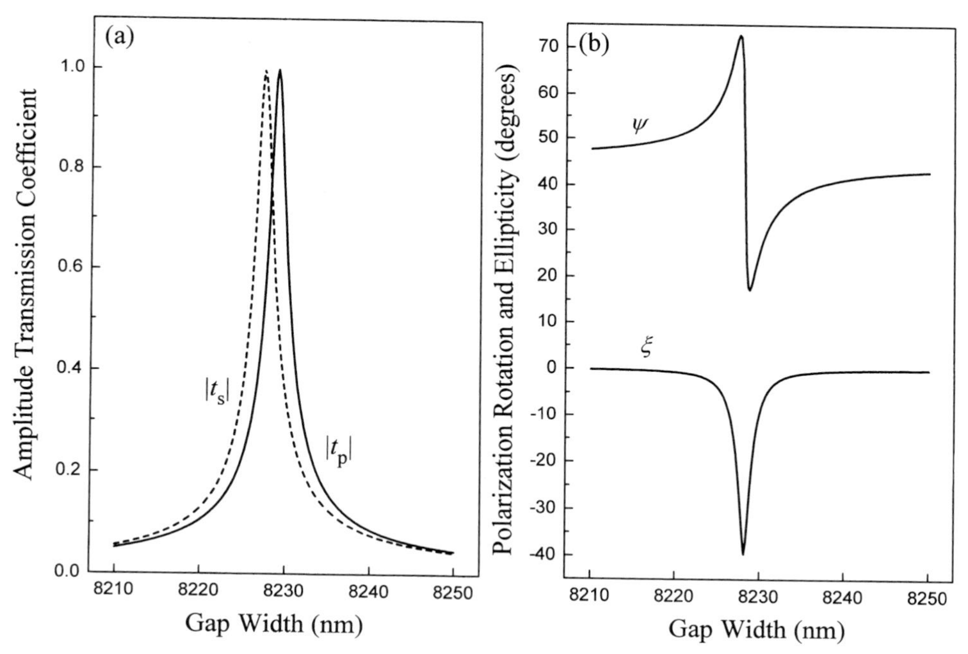

Figure 12.7 (a) Computed amplitude transmission coefficients and (b) the state of transmitted polarization plotted versus the gap-width for a normally incident beam on the Fabry–Pérot etalon of Figure 12.3. The mirrors are assumed to have 20 layers each and, for both mirrors, the uppermost layer is assumed to be birefringent. With reference to Figure 12.1, the refractive indices of the top layer along the X-, Y-, and Z- axes are 1.500, 1.505, and 1.500, respectively. The normally incident beam is linearly polarized at 45° to the X-axis. In (b) the polarization rotation angle, ψ, is also referred to the X-axis. By definition, the polarization ellipticity ξ is the arctangent of the ratio of the minor axis of the ellipse of polarization to its major axis. Thus $\varepsilon = 0°$ corresponds to linear polarization, whereas $\xi = 45°$ represents circular polarization.

In practice, if the mirrors are known to have the same amount of birefringence, these problems can be avoided by rotating one of the mirrors by 90° relative to the other. Also, it might be of some interest to note that birefringence of the top layer poses the most serious problem for the Fabry–Pérot etalons. In our calculations, the effects diminished as we moved the birefringent layer down the stack (closer to the substrate). By the time the birefringence is moved to layer 14 of both mirrors, its effects are totally negligible.

Enhancement of Faraday rotation

Figure 12.8 shows a Fabry–Pérot etalon with a Faraday rotator inserted in the gap between its mirrors. The Faraday rotator may be a slice of a transparent magnetic crystal (such as an iron garnet) or a transparent piece of glass in which an externally applied magnetic field has induced polarization rotation. In our simulations this medium was 2.11 μm thick, with refractive indices $(n, k)_{\pm} = (1.5 \pm 0.333 \times 10^{-4}, 0)$ for the states of right and left circular polarization (RCP and LCP). This small amount of optical activity would produce only 0.04° of polarization rotation in a single pass of the beam through the medium. However, as we shall see shortly, the etalon enhances

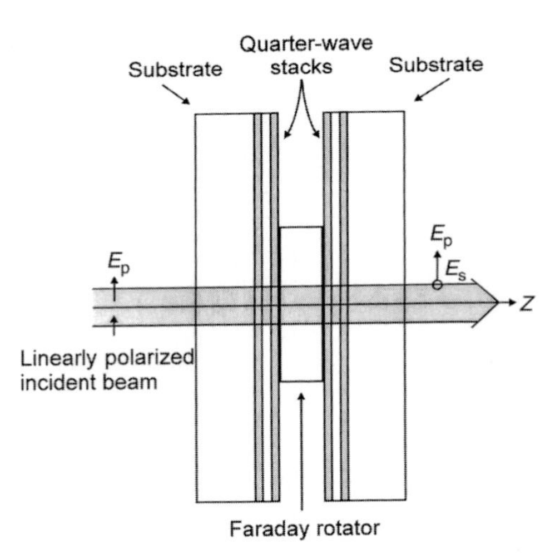

Figure 12.8 Schematic diagram showing a Fabry–Pérot etalon with a Faraday rotator placed in the gap. The normally incident plane wave is linearly polarized along the p-direction, but the optical activity of the Faraday rotator produces a transmitted component of polarization along the s-direction. The mirrors are taken to have 21 layers each and the Faraday medium to be 2.11 μm thick and to have refractive indices $(n, k)_{\pm} = (1.5 \pm 0.333 \times 10^{-4}, 0)$ for the states of right and left circular polarization.

the rotation because, in effect, it circulates the beam through the Faraday medium.

Figure 12.9 shows the transmitted amplitudes and the polarization state of a linearly polarized beam after going through the etalon of Figure 12.8. Since tuning of the cavity may be accomplished by varying the incident wavelength λ, we have plotted the data versus λ in the vicinity of resonance (633 nm). Note that almost all the incident beam is transmitted through the etalon and that its polarization rotation at resonance is close to 11°. (Even more rotation may be obtained if higher-reflectivity mirrors are used.)

Absorption within the Faraday medium reduces the quality factor Q of the cavity and, therefore, hampers its ability to enhance the Faraday rotation. If the same medium as above is assumed to have an absorption coefficient $k = 10^{-4}$, the characteristics of the etalon shown in Figure 12.9 will change to those in Figure 12.10. Note that the transmitted power has dropped by more than 60% and that the peak rotation angle is reduced by about 4°. The

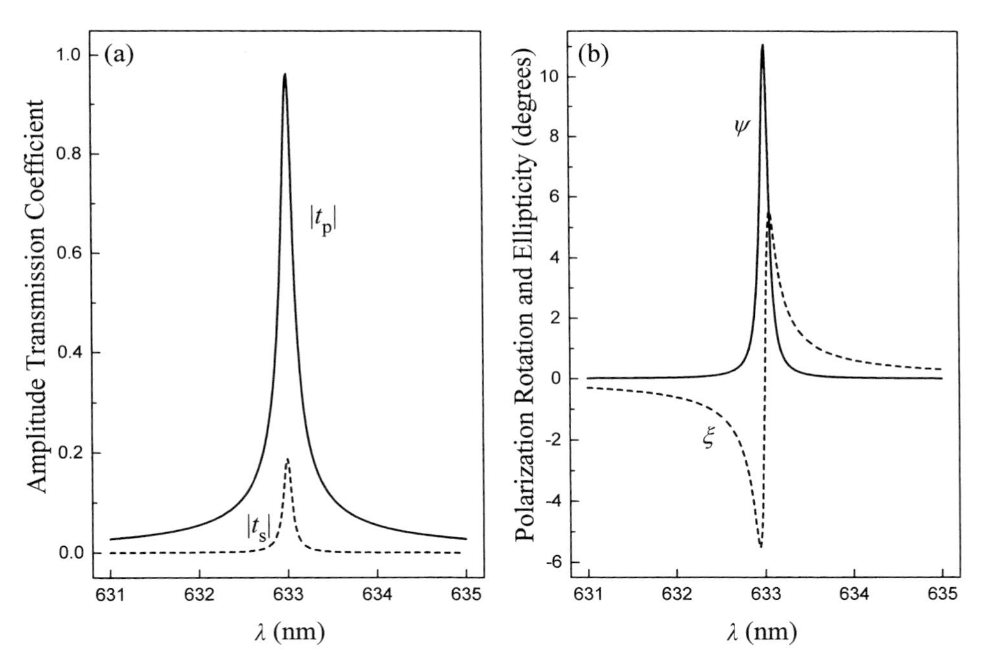

Figure 12.9 (a) Computed amplitude transmission coefficients and (b) the state of transmitted polarization plotted versus λ for a plane wave normally incident on the etalon of Figure 12.8. The assumed direction of incident polarization is p. In (a) the transmission coefficient t_s is defined as the ratio of the transmitted s-component to the incident p-component. In (b) the polarization rotation angle ψ is relative to the direction of incident polarization.

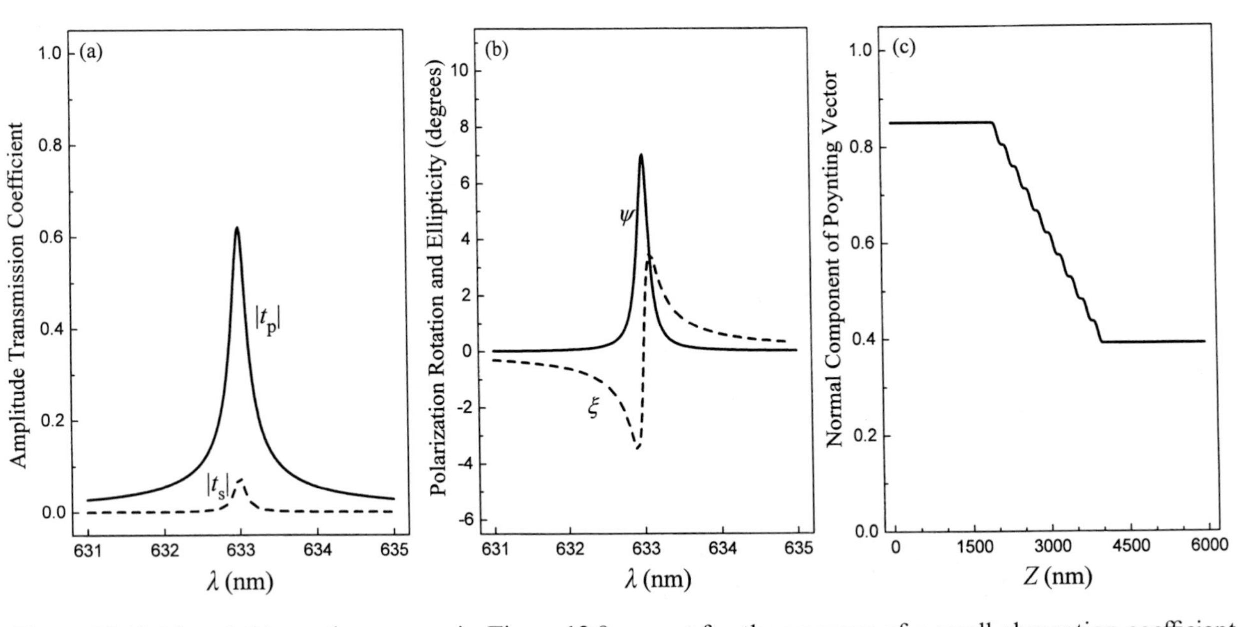

Figure 12.10 (a) and (b) are the same as in Figure 12.9, except for the presence of a small absorption coefficient ($k = 10^{-4}$) in the Faraday medium. The plot in (c) shows the magnitude of the Poynting vector as a function of position along the beam's propagation path. The flat parts of the curve indicate that optical energy passes unattenuated through the dielectric mirrors. The steep, staircase-like drop in the curve is caused by absorption within the Faraday medium.

plot in Figure 12.10(c) of the magnitude of the Poynting vector along the propagation path shows constant values in the multilayer mirrors but a rapid decline within the Faraday medium. Of the roughly 85% of the optical power that enters the etalon, 46% gets absorbed in the Faraday rotator and only 39% eventually passes out of the device. The small plateaux in the Poynting vector plot of Figure 12.10(c) are caused by the standing-wave pattern of the E-field within the Faraday medium: the absorption rate goes through minima and maxima following the E-field intensity variations. In our example, where the Faraday medium is 5λ thick, there are exactly 10 such plateaux.

A simple analysis

We now present a simple derivation of the basic properties of the Fabry–Pérot interferometer. Let us assume that the two mirrors are identical, with reflection coefficients ρ and transmission coefficients τ. If the light is incident from the substrate side of the mirror, these coefficients will be denoted by ρ' and τ', respectively. For dielectric mirrors, in general, we have $|\rho| = |\rho'|$ and $|\tau| = |\tau'|$, and there are simple relations among the corresponding phase factors, but these are not needed here. Also, since there is no absorption in the mirrors, we have $|\rho|^2 + |\tau|^2 = 1$.

Consider the case of a unit-amplitude beam normally incident on a Fabry–Pérot etalon, such as that shown in Figure 12.3 but with $\theta = 0$. Denote the gap-width by Δ, and let there be two counter-propagating beams in the cavity, one with amplitude A traveling to the right and the other with amplitude B traveling to the left. Since at the second mirror there are no incoming beams from the outside, and since the beam with amplitude A is reflected with coefficient ρ from this mirror, we must have $B = \rho A \exp(\mathrm{i}2\pi\Delta/\lambda)$. At the first mirror, the beam with amplitude B is reflected once again, and its amplitude becomes $\rho^2 A \exp(\mathrm{i}4\pi\Delta/\lambda)$. Since the incident beam has unit amplitude, its contribution to the field just inside the cavity is τ'. Therefore

$$A = \rho^2 A \exp(\mathrm{i}4\pi\Delta/\lambda) + \tau', \tag{12.1}$$

which yields

$$A = \frac{\tau'}{1 - \rho^2 \exp(\mathrm{i}4\pi\Delta/\lambda)}. \tag{12.2}$$

Resonance occurs when the phase of ρ^2 (if any) plus the phase acquired in a round trip through the cavity, $4\pi\Delta/\lambda$, becomes a multiple of 2π, at which point the denominator in Eq. (12.2) will be at a minimum and the field amplitude A within the cavity at a maximum.

The light transmitted through the device will have amplitude

$$t = \tau A \exp(\mathrm{i}2\pi\Delta/\lambda) \tag{12.3}$$

and that reflected from the device will have amplitude

$$r = \rho' + \tau\rho A \exp(\mathrm{i}4\pi\Delta/\lambda). \tag{12.4}$$

The same equations may be used at oblique incidence, provided that the gap-width Δ is multiplied by $\cos\theta$ and that ρ, τ, ρ', and τ' represent the corresponding quantities at the particular angle of incidence. When the medium of the cavity happens to be absorptive, the same type of analysis may still be used to arrive at the relevant formulas.

We now demonstrate the application of the preceding equations to some of the cases discussed earlier. In the case of the 10-layer stack, the mirror coefficients were $\rho = -\rho' = 0.844$ and $\tau = \tau' = 0.536$. From Eqs. (12.2)–(12.4) we find that, at resonance, $A = 1.863$, $t = 1$ and $r = 0$. In the case of the 20-layer stack, $\rho = -\rho' = 0.9905$ and $\tau = \tau' = 0.1375$, yielding at resonance $A = 7.27$, $t = 1$ and $r = 0$. For the component of polarization that sees the higher refractive index of the uppermost layer, ρ acquires a phase angle of $0.9°$, which is canceled when the gap-width is reduced by 1.6 nm. This is exactly the peak shift observed in Figure 12.7(a).

The 21-layer stacks used with the Faraday rotator were symmetric, in the sense that their substrate and their medium of incidence had the same refractive index, $n = 1.5$. For these mirrors $\rho = \rho' = 0.9964$ and $\tau = \tau' = 0.0843\mathrm{i}$, yielding at resonance $A = 11.73\mathrm{i}$, $t = -1$ and $r = 0$. Since the refractive indices of the Faraday medium for RCP and LCP light deviated from the nominal value by $\pm 0.0022\%$, a similar change in wavelength was needed to re-establish the conditions of resonance for each state of circular polarization. At $\lambda = 633$ nm, however, both RCP and LCP components of the incident beam were slightly off resonance. This, according to Eq. (12.2), caused a large phase shift between the values of A for RCP and LCP light, which translated into a large phase difference between the transmitted RCP and LCP components, in accordance with Eq. (12.3). The resulting Faraday rotation angle of the transmitted beam was a manifestation of this phase difference. Similar arguments can be advanced to explain the consequences of the absorption observed in Figure 12.10.

Note

In general, it is possible to eliminate from a stack non-absorbing layers whose thicknesses are multiples of $\lambda/2$. Now, if the gap happens to be an integer multiple of $\lambda/2$ its elimination will bring the top dielectric layers of the two mirrors into contact. These layers, each being a quarter-wave thick, will combine into a half-wave layer that can be subsequently eliminated, paving the way for the elimination of all the remaining layers in similar fashion. At the end, the two substrates will come into direct contact, and the incident light will be fully transmitted, as is expected of a well-tuned etalon.

References for chapter 12

1 C. Fabry and A. Perot, *Ann. Chim. Phys.* (7) **16**, p. 115 (1899).
2 R. W. Wood, *Physical Optics*, third edition, Optical Society of America, Washington, 1988.
3 H. A. Macleod, *Thin Film Optical Filters*, second edition, Macmillan, New York, 1986.
4 S. C. Johnston and S. F. Jacobs, Some problems caused by birefringence in dielectric mirrors, *Appl. Opt.* **25**, 1878 (1986).
5 C. Wood, S. C. Bennett, J. L. Roberts, D. Cho, and C. E. Wieman, Birefringence, mirrors, and parity violation, *Opt. & Phot. News* **7**, 54 (1996).

13

The Ewald–Oseen extinction theorem

When a beam of light enters a material medium, it sets in motion the resident electrons, whether these electrons are free or bound. The electronic oscillations in turn give rise to electromagnetic radiation which, in the case of linear media, possesses the frequency of the exciting beam. Because Maxwell's equations are linear, one expects the total field at any point in space to be the sum of the original (exciting) field and the radiation produced by all the oscillating electrons. However, in practice the original beam appears to be absent within the medium, as though it had been replaced by a different beam, one having a shorter wavelength and propagating in a different direction. The Ewald–Oseen theorem[1,2] resolves this paradox by showing how the oscillating electrons conspire to produce a field that exactly cancels out the original beam everywhere inside the medium. The net field is indeed the sum of the incident beam and the radiated field of the oscillating electrons, but the latter field completely masks the former.[3,4]

Although the proof of the Ewald–Oseen theorem is fairly straightforward, it involves complicated integrations over dipolar fields in three-dimensional space, making it a brute-force drill in calculus and devoid of physical insight.[5,6] It is possible, however, to prove the theorem using plane waves interacting with thin slabs of material, while invoking no physics beyond Fresnel's reflection coefficients. (These coefficients, which date back to 1823, predate Maxwell's equations.) The thin slabs represent sheets of electric dipoles, and the use of Fresnel's coefficients allows one to derive exact expressions for the electromagnetic field radiated by these dipolar sheets. The integrations involved in this approach are one-dimensional, and the underlying procedures are intuitively appealing to practitioners of optics. The goal of the present chapter is to outline a general proof of the Ewald–Oseen theorem using arguments that are based primarily on thin-film optics.

Dielectric slab

Consider the transparent slab of dielectric material of thickness d and refractive index n, shown in Figure 13.1. A normally incident plane wave of vacuum wavelength λ_0 produces overall a reflected beam of amplitude r and a transmitted beam of amplitude t. Both r and t are complex numbers in general, having a magnitude and a phase angle. Using Fresnel's coefficients at each facet of the slab and accounting for multiple reflections, it is fairly straightforward to obtain expressions for r and t. The reflection and transmission coefficients at the front facet of the slab are[5,7]

$$\rho = (1 - n)/(1 + n), \tag{13.1}$$

$$\tau = 2/(1 + n). \tag{13.2}$$

At the rear facet the corresponding entities are

$$\rho' = (n - 1)/(n + 1), \tag{13.3}$$

$$\tau' = 2n/(n + 1). \tag{13.4}$$

A single path of the beam through the slab causes a phase shift ψ, where

$$\psi = 2\pi n d/\lambda_0. \tag{13.5}$$

Adding up all partial reflections at the front facet yields an expression for the reflection coefficient r of the slab. Similarly, adding all partial transmissions at the rear facet yields the transmission coefficient t. Thus

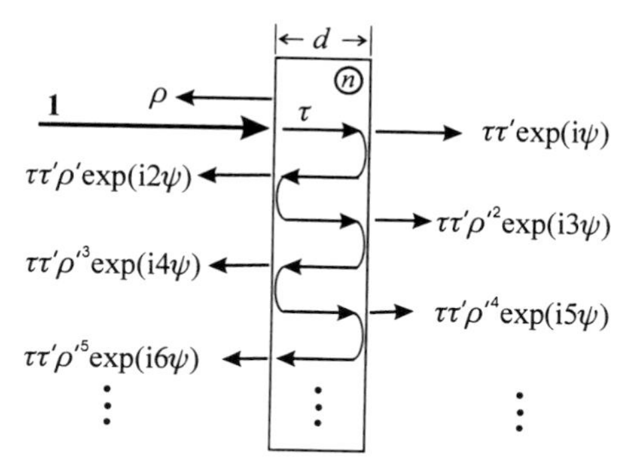

Figure 13.1 A transparent slab of homogeneous material of thickness d and refractive index n, on which is normally incident a monochromatic plane wave of wavelength λ_0. The beam suffers multiple reflections at the two facets of the slab. By adding the various reflected and transmitted amplitudes one obtains the expressions for the total r and t given in Eqs. (13.6) and (13.7).

$$r = \rho + \tau\tau'\rho' \exp(i2\psi) \sum_{m=0}^{\infty} [\rho' \exp(i\psi)]^{2m} = \rho + \frac{\tau\tau'\rho' \exp(i2\psi)}{1 - \rho'^2 \exp(i2\psi)} \tag{13.6}$$

$$t = \tau\tau' \exp(i\psi) \sum_{m=0}^{\infty} [\rho' \exp(i\psi)]^{2m} = \frac{\tau\tau' \exp(i\psi)}{1 - \rho'^2 \exp(i2\psi)}. \tag{13.7}$$

Rather than try to simplify these complicated functions of n, d and λ_0, we give numerical results in Figure 13.2 for the specific case of $n = 2$ and $\lambda_0 = 633\,\text{nm}$. The magnitudes of r and t are shown in Figure 13.2(a), and their phase angles in Figure 13.2(b), both as functions of the thickness d of the slab. For any given value of d it is possible to represent r and t as complex vectors (see Figure 13.3). Since the phase difference between r and t is always $90°$, these complex vectors are orthogonal to each other. Also, the conservation of energy requires that $|r|^2 + |t|^2 = 1$. These observations lead to the conclusion that the hypotenuse of the triangle in Figure 13.3 must have unit length, that is $|t - r| = 1$, which is also confirmed numerically in Figure 13.2(c).

Within the slab the incident beam sets the atomic dipoles in motion. These dipoles in turn radiate plane waves in both the forward and the backward directions, as shown in Figure 13.4. When the slab is sufficiently thin, symmetry requires forward- and backward-radiated waves to be identical, that is, they must both have the same amplitude r. In the forward direction, however, the incident beam continues to propagate unaltered, except for a phase-shift caused by propagation in free-space through a distance d. Thus we must have

$$t = r + \exp(i2\pi d/\lambda_0). \tag{13.8}$$

It was pointed out earlier in conjunction with the diagram of Figure 13.3 that $t - r$ has unit amplitude, which is in agreement with Eq. (13.8). It is by no means obvious, however, that the phase of $t - r$ must approach $2\pi d/\lambda_0$ as $d \to 0$. Figure 13.2(c) shows computed plots of the phase of $t - r$ normalized by $2\pi d/\lambda_0$. It is seen that in the limit $d \to 0$ the normalized phase approaches unity as well. This confirms that the slab radiates equally in the forward and backward directions, and that the incident beam, having set the dipolar oscillations in motion, continues to propagate undisturbed in free space.

Radiation from a uniform sheet of oscillating dipoles

In the limit of small d Eq. (13.6) reduces to the following simple form:

$$r \approx i\,[\pi(n^2 - 1)d/\lambda_0] \exp[i\pi(n^2 + 1)d/\lambda_0] \qquad d/\lambda_0 \ll 1. \tag{13.9}$$

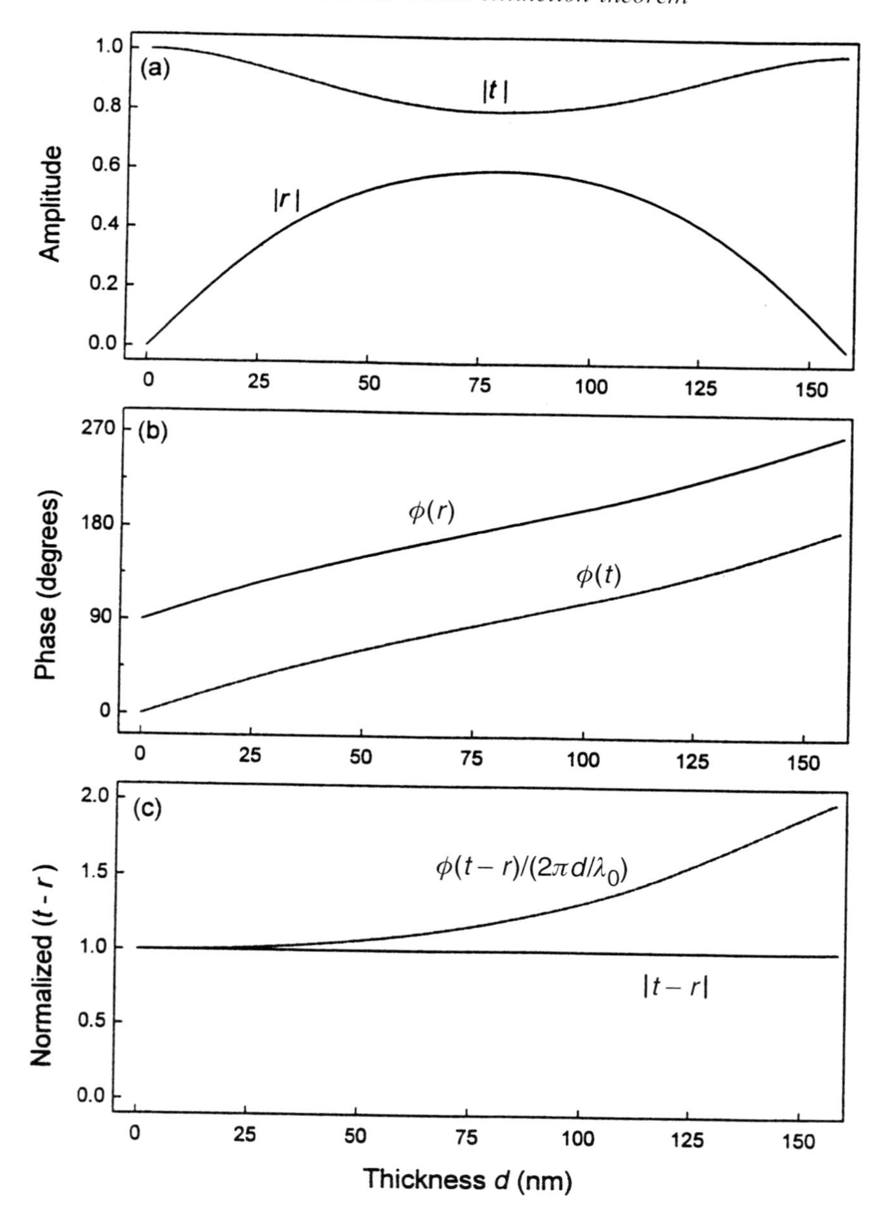

Figure 13.2 Computed plots of r and t for a slab of thickness d and refractive index $n = 2$, when a plane wave with $\lambda_0 = 633\,\text{nm}$ is normally incident on the slab. The horizontal axis covers one cycle of variations in r and t, corresponding to a half-wave thickness of the slab.

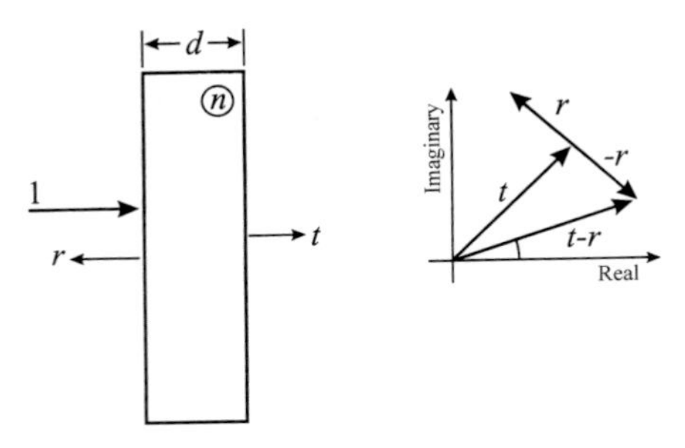

Figure 13.3 A dielectric slab of thickness d and refractive index n, reflecting the unit-amplitude incident beam with coefficient r while transmitting it with coefficient t. The complex-plane diagram on the right shows the relative orientations of r, t and their difference $t - r$. For a non-absorbing slab (i.e., one with a real-valued index n) r and t are orthogonal to each other, and $t - r$ has unit magnitude.

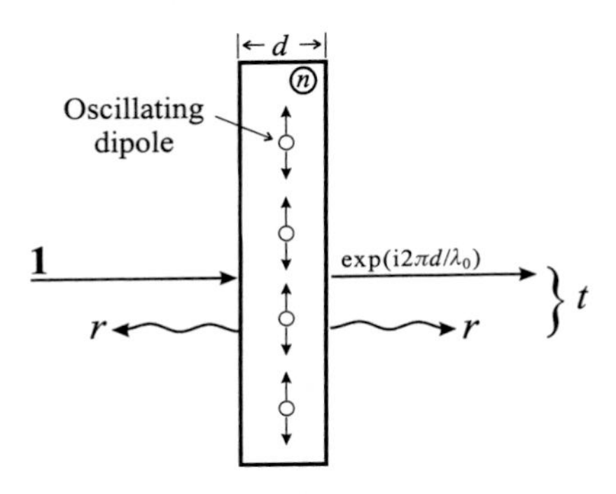

Figure 13.4 Bound electrons within a very thin dielectric slab, when set in motion by a normally incident plane wave of unit amplitude, radiate with equal strength in both the forward and backward directions. The magnitude of the radiated field is the reflection coefficient r of the slab. The incident beam continues to propagate undisturbed as in free-space, acquiring a phase shift of $2\pi d/\lambda_0$ upon crossing the slab. The sum of the incident beam and the forward-propagating part of the radiated beam constitutes the transmitted beam.

In this limit the radiated field is slightly more than 90° ahead of the incident field, while its amplitude is proportional to d/λ_0 and also proportional to $n^2 - 1$, the latter being the coefficient of polarizability of the dielectric material. Note that the small phase angle of r over and above its 90° phase, i.e., the exponential factor in Eq. (13.9), is essential for the conservation of energy among the incident, reflected, and transmitted beams (see Figure 13.4).

Equation (13.9) is in fact the exact solution of Maxwell's equations for the radiation field of a sheet of dipole oscillators. Although derived here as an aid in proving the extinction theorem, it is an important result in its own right. Note, for example, that the amplitude of the radiated field is proportional to $1/\lambda_0$ even though the field of an individual dipole radiator is known to be proportional to $1/\lambda_0^2$. The coherent addition of amplitudes over the sheet of dipoles has thus modified the wavelength dependence of the radiated field.[3]

The extinction theorem

Having derived Eq. (13.9) for the field radiated by a sheet of dipoles, we are now in a position to outline the proof of the extinction theorem. Consider a semi-infinite, homogeneous medium of refractive index n, bordering with free space at $z = 0$, as shown in Figure 13.5. A unit-magnitude plane wave of wavelength λ_0 is directed at this medium at normal incidence from the left side. To determine the reflected amplitude r at the interface, divide the medium into thin slabs of thickness Δz, then add up (coherently) the reflected fields from each of these slabs. Similarly, the field at an arbitrary plane $z = z_0$ inside the medium may be computed by adding to the incident beam the contributions of the slabs located to the left of z_0 as well as those to the right of z_0. The simplest way to proceed is by assuming that the field inside the medium has the expected form, $\tau \exp(\mathrm{i}2\pi n z/\lambda_0)$, then showing self-consistency. These calculations involve simple one-dimensional integrals, and are in fact so straightforward that there is no need to carry them out here. The interested reader may take a few minutes to evaluate the integrals and convince himself or herself of the validity of the theorem.

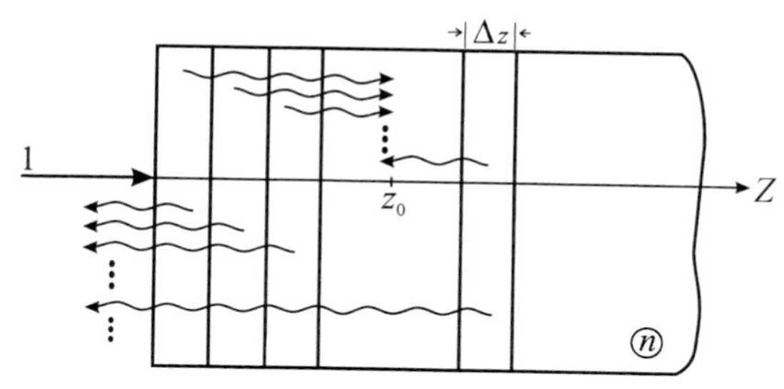

Figure 13.5 A semi-infinite medium of refractive index n is illuminated by a unit-amplitude plane wave at normal incidence. The medium may be considered as a contiguous sequence of thin slabs, each radiating with equal strength in both the forward and backward directions. Adding coherently the backward-radiated fields yields the reflection coefficient at the front facet of the medium. Similarly, the internal field at $z = z_0$ is obtained by coherent addition of the incident beam, the forward-propagating radiations from the left side of z_0, and the backward-propagating radiations from the right side of z_0.

Slab of absorbing material

When the material of the slab is absorbing, similar arguments to those above may be advanced to prove the Ewald–Oseen theorem, although the expressions for the reflection and transmission coefficients become more complicated. Numerically, however, it is still possible to describe the situation with great accuracy.

Figure 13.6(a) shows computed plots of r and t for a metal slab having complex index $n + ik = 2 + i7$. (Compare these plots with the corresponding plots for the dielectric slab in Figure 13.2.) It is seen that the reflectance drops sharply while the transmittance increases as the film thickness is reduced below about 20 nm. The phase plots in Figure 13.6(b) are quite different from those of the dielectric slab, indicating a phase difference greater than 90° between r and t. A complex-plane diagram for this type of material is given in Figure 13.7. The angle between r and t being greater than 90° implies that $|t - r|^2 > |t|^2 + |r|^2$, while the conservation of energy requires that $|t|^2 + |r|^2 < 1$ in the case of absorbing media. The fact that $|t - r|$ can approach unity is borne out by the numerical results depicted in Figure 13.6(c). In the limit $d \to 0$, not only does the magnitude of $t - r$ become unity but also its phase approaches $2\pi d/\lambda_0$. Therefore, in the limit of small d, the transmitted beam may be expressed as the sum of the reflected beam and the phase-shifted incident beam, the phase shift being due to free-space propagation over the distance d. This is all that one needs in order to prove the extinction theorem for absorbing media.

Oblique incidence on a dielectric slab

Figure 13.8 shows a s-polarized plane wave at oblique incidence on a dielectric slab of thickness d and index n. The oscillating dipoles are parallel to the s-direction of polarization and radiate with equal magnitude in the forward and backward directions. The computed plots of r_s and t_s versus d for the specific case of $\lambda_0 = 633$ nm, $n = 2$, and $\theta = 50°$ are shown in Figure 13.9. The angle of propagation inside the medium is obtained from Snell's law as $\theta' = 22.52°$, and the half-wave thickness of the slab is given by $\lambda_0/(2n\cos\theta') = 171.3$ nm. These curves are again very similar to those of Figure 13.2, showing a 90° phase difference between r_s and t_s, unit magnitude for $t_s - r_s$, and a phase for $t_s - r_s$ that approaches $2\pi(d/\lambda_0)\cos\theta$ as $d \to 0$. The Ewald–Oseen theorem for the case of s-polarized light at oblique incidence can therefore be proven along the same lines as described earlier for normal incidence.

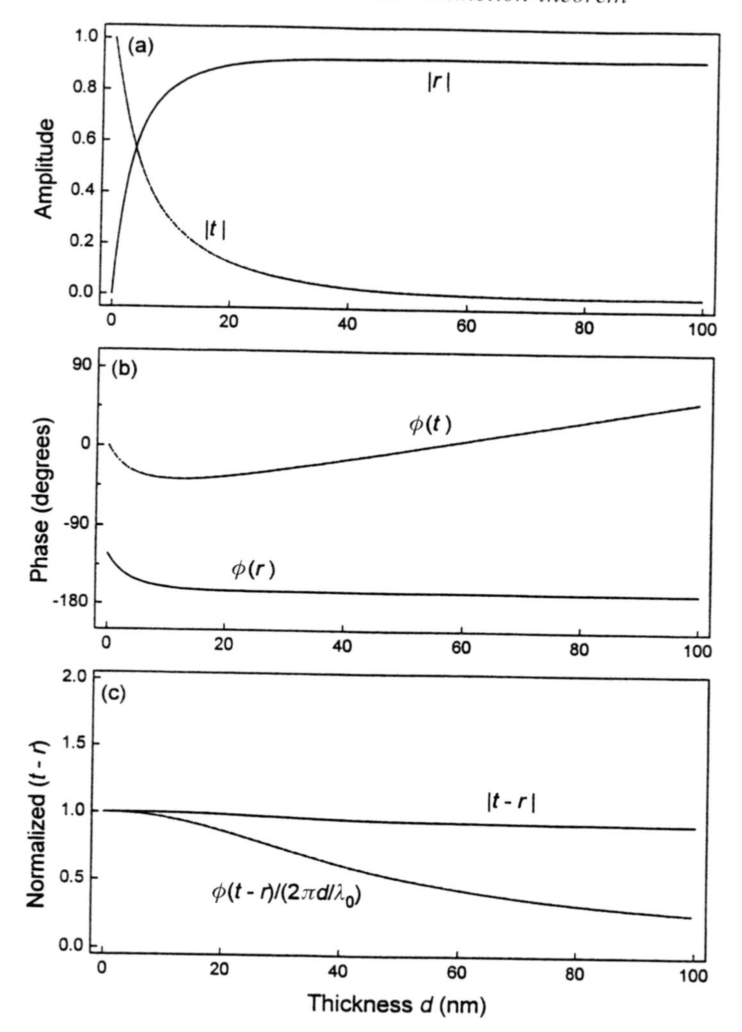

Figure 13.6 Computed plots of r and t for a slab of thickness d and complex refractive index $(n, k) = (2, 7)$, when a plane wave with $\lambda_0 = 633\,\mathrm{nm}$ is normally incident on the slab. The horizontal axis covers the penetration depth of the material.

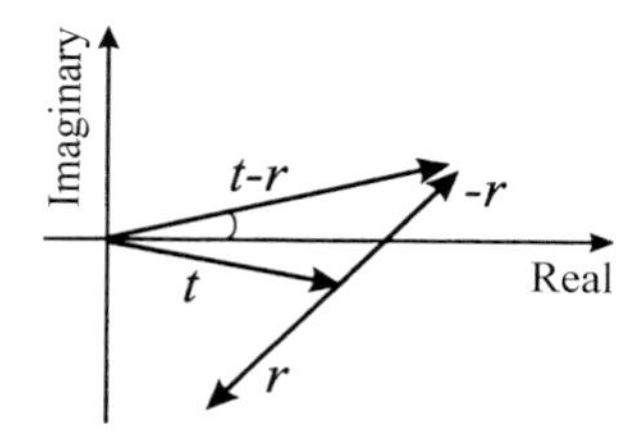

Figure 13.7 A complex-plane diagram showing the reflection coefficient r, transmission coefficient t, and their difference $t - r$ for a thin slab of an absorbing material.

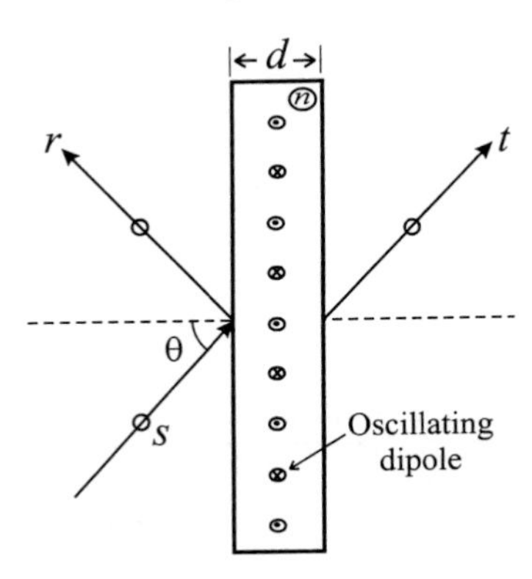

Figure 13.8 An s-polarized plane wave is obliquely incident at an angle θ on a dielectric slab of thickness d and index n. The electric dipoles of the slab oscillate in a direction perpendicular to the plane of the diagram, radiating identical fields in the forward and backward directions.

The case of p-polarized light, depicted in Figure 13.10, is somewhat different, however. Here the directionality of the dipole oscillations within the slab breaks the symmetry between the forward- and backward-radiated beams. The angle θ'' between the direction of oscillation of the dipoles and the plane of the slab may be determined by considering multiple reflections within the slab. For very thin slabs, it is possible to show that

$$\tan\theta'' = (1/n^2)\tan\theta. \tag{13.10}$$

Note that at Brewster's angle, where $\tan\theta = n$, we have $\tan\theta'' = 1/n$, that is, $\theta'' = \theta'$, where θ' is the propagation angle within the medium as given by Snell's law. At angles below the Brewster angle $\theta'' < \theta'$, while above the Brewster angle $\theta'' > \theta'$.

For the case of p-polarized light of wavelength $\lambda_0 = 633$ nm incident at $\theta = 50°$ on a slab of index $n = 2$, plots of r and t versus the slab thickness d are shown in Figure 13.11. Although the magnitude of $t_p - r_p$ can still be shown to be unity, its phase does not approach $2\pi(d/\lambda_0)\cos\theta$ as $d \to 0$. This is a manifestation of the breakdown of symmetry between the forward and backward radiations. If the magnitudes of the beams radiated in the two directions are taken into account, however, the preceding arguments can be restored. One may readily observe from Figure 13.10 that the ratio of the forward- and backward-propagating magnitudes must be given by

$$W(\theta) = \cos(\theta - \theta'')/\cos(\theta + \theta''). \tag{13.11}$$

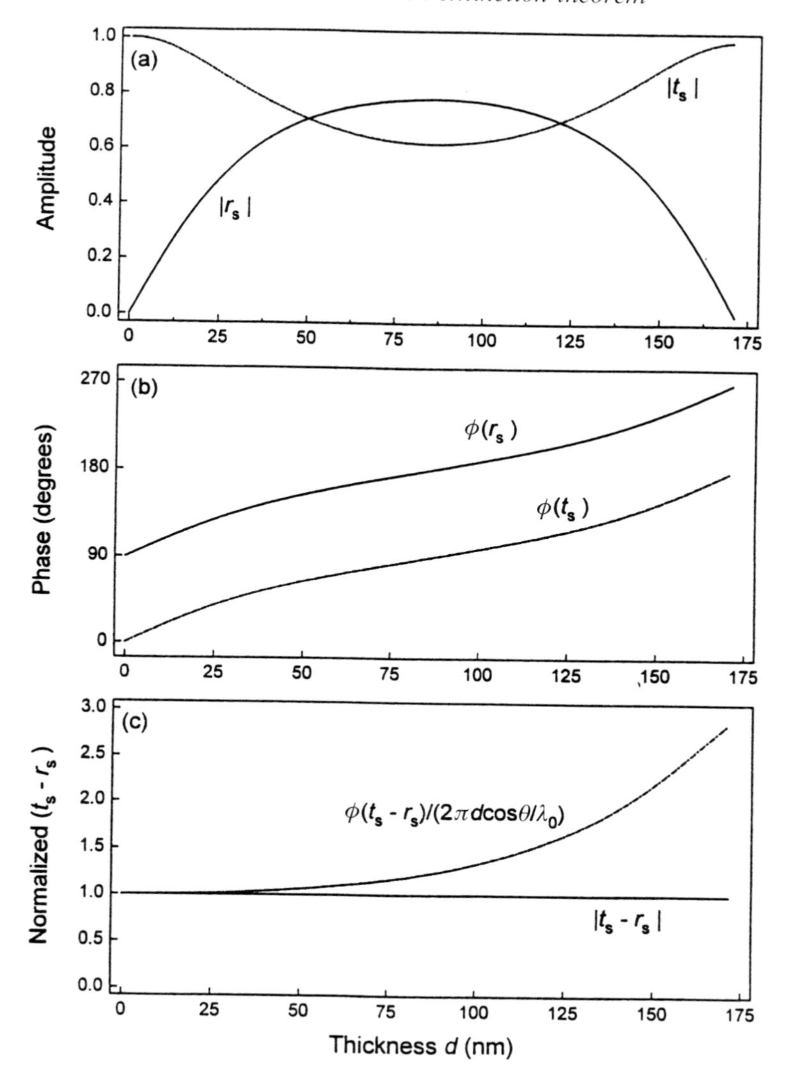

Figure 13.9 Computed plots of r and t for a slab of thickness d and index $n = 2$, when a s-polarized plane wave with $\lambda_0 = 633\,\text{nm}$ illuminates the slab at $\theta = 50^\circ$. The horizontal axis covers one cycle of variations of r and t, corresponding to a half-wave thickness of the slab at this particular angle of incidence.

Therefore, for p-polarized light at oblique incidence, it is $t - Wr$ that approaches $\exp(i2\pi d\cos\theta/\lambda_0)$ as $d \to 0$. This is seen to be verified in Figure 13.11(c).

As a further test of Eq. (13.11), we show in Figure 13.12 the computed plot versus θ of $r_p/[t_p - \exp(i2\pi d\cos\theta/\lambda_0)]$ for a slab with $d = 10\,\text{nm}$ and $n = 2$, illuminated by a plane wave with $\lambda_0 = 633\,\text{nm}$. This curve overlaps the plot of

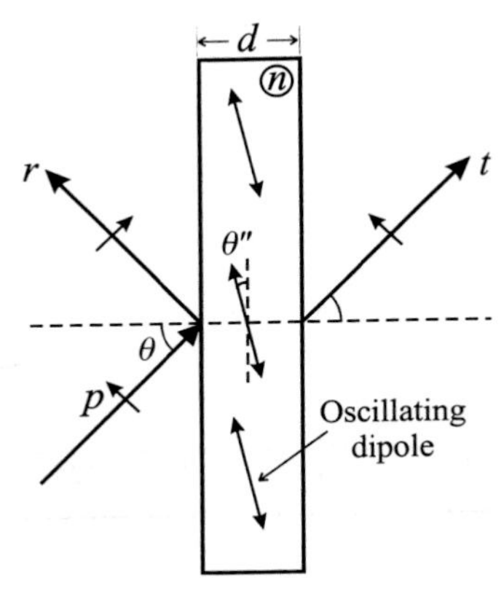

Figure 13.10 A p-polarized plane-wave is obliquely incident at an angle θ on a dielectric slab of thickness d and index n. The oscillating dipoles make an angle θ'' with the surface of the slab, radiating with different amplitudes in the forward and backward directions.

the function $1/W(\theta)$ exactly. Taking into account the ratio $W(\theta)$ between the forward and backward radiated beams, one can prove the Ewald–Oseen theorem as before.

Appendix

This chapter, when originally published in *Optics & Photonics News*, prompted the following criticism and reply.

"Editor:

While we are pleased that Masud Mansuripur has called attention in *OPN* to the rather basic Ewald–Oseen extinction theorem, we wish to take issue with certain parts of his article.[1]

"Mansuripur states that the goal of his article is 'to outline a general proof of the Ewald–Oseen theorem using arguments that are based primarily on thin-film optics.' We wish to note first that the proof he outlines, based on the field produced by a uniform sheet of dipole oscillators and the assumed form $\exp[2\pi i n z/\lambda_0]$ for the field inside the medium, is essentially the same approach used by Fearn, James, and Milonni.[2] Their proof is more general in that Fresnel coefficients (for normal incidence) are derived rather than assumed. Indeed, the derivation of the Fresnel coefficients assumes the extinction of the incident field inside the dielectric medium: Mansuripur's starting point implicitly assumes the very theorem

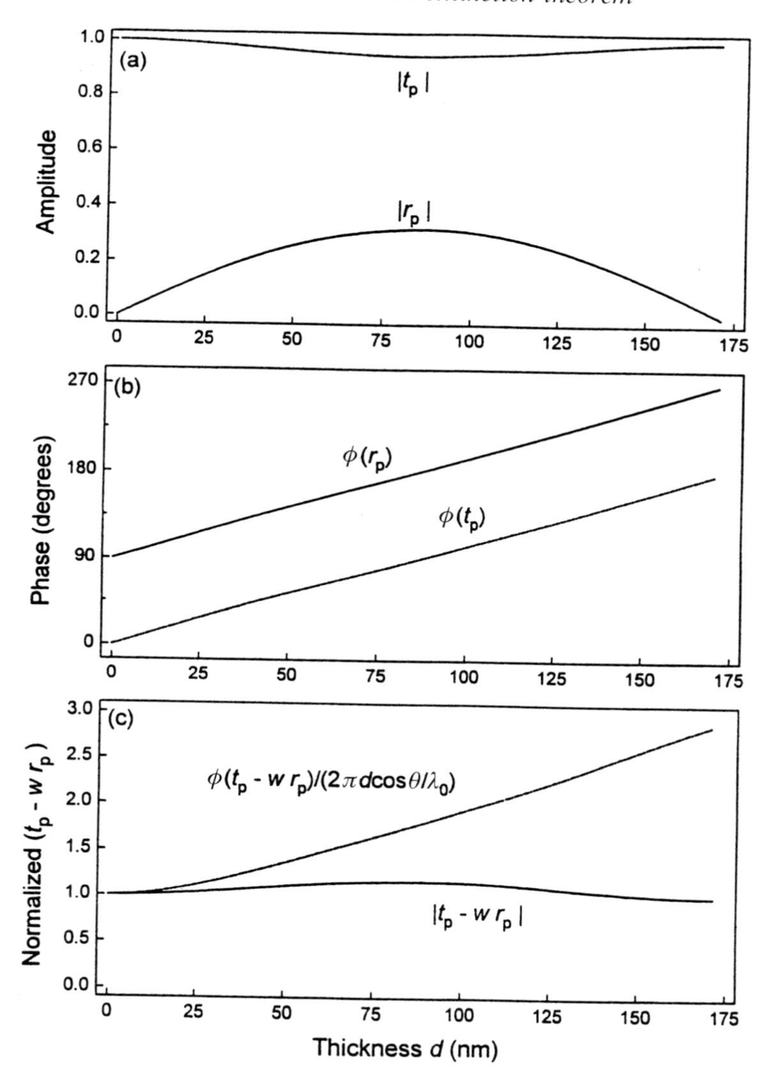

Figure 13.11 Computed plots of r and t for a slab of thickness d and index $n=2$, when a p-polarized plane wave with $\lambda_0 = 633$ nm illuminates the slab at $\theta = 50°$. The horizontal axis covers one cycle of variations in r and t, corresponding to a half-wave thickness of the slab at this particular angle of incidence.

he is trying to prove! In this connection we note that it was not claimed by Fearn *et al.* that they provided a 'general proof' of the extinction theorem. A general proof, valid for media bounded by surfaces of arbitrary shape, is given by Born and Wolf.[3]

"Mansuripur cites References 2 and 3 in support of his opinion that the proof of the extinction theorem is 'devoid of physical insight'. While it is true

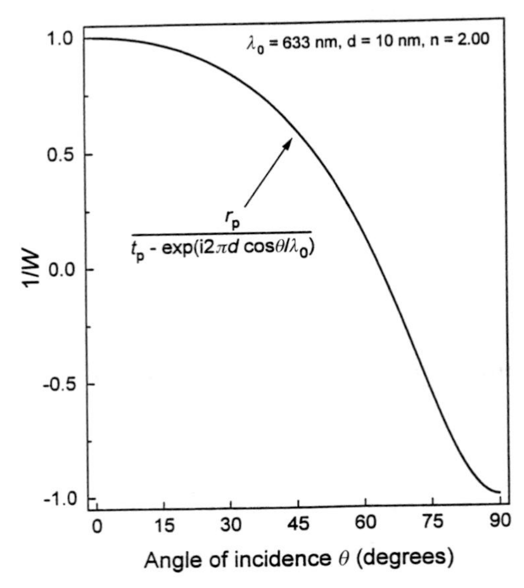

Figure 13.12 Computed ratio of the amplitudes of backward-propagating radiation and forward-propagating radiation for a dielectric slab 10 nm thick and with $n = 2$. A p-polarized plane wave with $\lambda_0 = 633$ nm is assumed to be obliquely incident on the slab at an angle θ.

that the proofs given in these references involve 'complicated integration over dipolar fields in three-dimensional space,' we do not think it is fair to say it [the proof] is devoid of physical insight. In Reference 3, page 101, the significance of the theorem is described in the following manner that could hardly be more physical: 'The incident wave may ... be regarded as extinguished at any point within the medium by interference with the dipole field and replaced by another wave with a different velocity (and generally also a different direction) of propagation.'

"Finally we note that various features of the extinction theorem have been interpreted differently by various authors: some of these differences have been discussed by Fearn *et al.*[2] It would be unfortunate if readers of Mansuripur's article were left with the impression that the theorem can somehow be based 'primarily on thin-film optics.'

1 M. Mansuripur, The Ewald–Oseen extinction theorem, *Opt. & Phot. News* **9** (8), 50–55 (1998).

2 H. Fearn *et al.*, Microscopic approach to reflection, transmission, and the Ewald–Oseen extinction theorem, *Am. J. Phy.* **64**, 986–995 (1996).

3 M. Born and E. Wolf, *Principles of Optics*, sixth edition, Cambridge University Press, Cambridge UK 1985, section 2.4.2.

Daniel James and Peter W. Milonni, Los Alamos National Laboratory,
Los Alamos NM
Heidi Fearn, California State University at Fullerton, Fullerton CA
Emil Wolf, University of Rochester, Rochester NY"

The author replied:

"It is puzzling that Fearn *et al.* consider starting from Fresnel's reflection coefficients a shortcoming of my method of proof. The Fresnel coefficients can be derived directly from Maxwell's equations without invoking the extinction theorem, they are available in many textbooks (including Born and Wolf, sixth edition, pp. 38–41), and their derivation from first principles does not in any way add to the value of a paper. I used Fresnel's coefficients to derive the radiation field for a sheet of dipoles (Equation 9 of my article), as this is a simple, accurate, and intuitive way of calculating the field, and also because its underlying principle is familiar to many practitioners of optics. Alternatively, one could derive the radiation field by integrating over individual dipoles within the sheet, as is done, for example, in *The Feynman Lectures on Physics* (my reference 3). After this step that establishes the radiation field from a dipolar sheet, the method of proof that I proposed (based on demonstrating self-consistency) is similar to that of Fearn *et al.*

"Although Fresnel's coefficients are derived from Maxwell's equations, nowhere in the standard derivation is it assumed that the incident beam is still present within the medium (albeit masked by the dipole radiations). Had the Ewald–Oseen theorem been somehow implicit in the standard derivation of Fresnel's coefficients, there would have been no need for the paper of Fearn *et al.* in the first place.

"I strongly disagree with the suggestion that the use of Fresnel's coefficients somehow renders my proof of the Ewald–Oseen theorem circular. I also dispute the assertion made by Fearn *et al.* that 'it would be unfortunate if readers ... were left with the impression that the theorem can somehow be based primarily on thin film optics.' Emphatically, the proof of the theorem *can* be based on thin film optics (this is exactly what I showed in the article), and it is far from 'unfortunate' indeed when a valid proof happens to be based on a simple physical picture.

"I erred in stating that I was going to 'outline a general proof of the ... theorem.' Mine was a general proof for the one-dimensional case, where the beam enters from free space through a plane boundary into an isotropic, homogeneous medium. My proof is more general than the proof of Fearn *et al.*, in that it covers both transparent and absorbing media, and also in that it considers the case of oblique incidence with p and s polarized light. The method described in Born and Wolf is obviously more general than both, because it applies to arbitrary boundaries. None of the above methods, however, is sufficiently general to embrace inhomogeneous, anisotropic, and optically active media, for which the theorem is presumably valid as well.

"Finally, my expressed opinion regarding the proof of the extinction theorem being 'devoid of physical insight' was meant as a commentary on the nature of the method, not as a reflection on the authors of the cited references. Ultimately, of course, such judgments are subjective and are best left to the readers."

References for chapter 13

1 P. P. Ewald, On the foundations of crystal optics, Air Force Cambridge Research Laboratories Report AFCRL-70-0580, Cambridge MA (1970). This is a translation by L. M. Hollingsworth of Ewald's 1912 dissertation at the University of Munich.
2 C. W. Oseen, Über die Wechselwirkung Zwischen Zwei elektischen Dipolen der Polarisationsebene in Kristallen und Flüssigkeiten, *Ann. Phys.* **48**, 1–56 (1915).
3 R. P. Feynman, R. B. Leighton, and M. Sands, *The Feynman Lectures on Physics*, chapters 30 and 31, Addison-Wesley, Reading, Massachusetts, 1963.
4 V. Weisskopf, How light interacts with matter, in *Lasers and Light, Readings from Scientific American*, W. H. Freeman, San Francisco, 1969.
5 M. Born and E. Wolf, *Principles of Optics*, sixth edition, Pergamon Press, Oxford, 1980.
6 H. Fearn, D. F. V. James, and P. W. Milloni, Microscopic approach to reflection, transmission, and the Ewald–Oseen extinction theorem, *Am. J. Phys.* **64**, 986–995 (1996).
7 H. A. Macleod, *Thin Film Optical Filters*, second edition, Macmillan, New York, 1986.

14

Reciprocity in classical linear optics

An informal survey of some colleagues and students revealed that the notion of reciprocity in optics is not widely appreciated. One colleague even justified the prevailing ignorance by drawing a parallel between reciprocity in optics and complementarity in quantum mechanics: "Both are true statements which have little, if any, practical value in their respective domains." This chapter is an attempt at explaining the concept of reciprocity, clarifying some associated misconceptions, and pointing out its practical applications.

Non-reciprocity of Faraday rotators

No one disputes that a Faraday rotator is a non-reciprocal element. The usual argument goes as follows. Let a linearly polarized beam of light be fully transmitted through a polarizing beam-splitter (PBS) before being directed through a 45° Faraday rotator, as shown in Figure 14.1. If the beam is reflected back (by an ordinary mirror, for example), it retraces its path through the rotator and emerges with its polarization vector rotated by a full 90°. At the PBS, therefore, the returning beam will be deflected away from its original path. (This, in fact, is a well-known method of isolating laser diodes from spurious reflections within a given system.) Since the reflected light does not return on its original path, and since the PBS is believed to be reciprocal, the argument is taken as proof of the non-reciprocity of the Faraday rotator.

Although it is true that Faraday rotators are non-reciprocal, there is a flaw in the above argument, which will become clear upon inspection of the system of Figure 14.2. In this system, which is similar to that of Figure 14.1, the Faraday rotator is replaced by a quarter-wave plate (QWP). The fast and slow axes of the plate are oriented at 45° to the direction of incident polarization, so that the light emerging from the plate in the forward path is

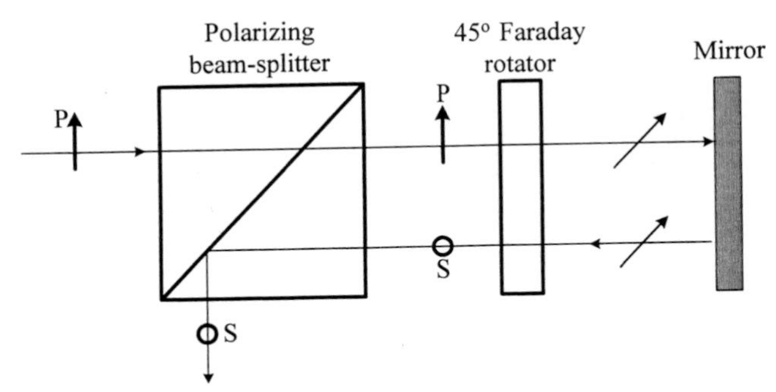

Figure 14.1 A Faraday rotator as used in an optical isolator. The incident p-polarized beam, having undergone two consecutive 45° rotations in its forward and backward paths through the rotator, becomes s-polarized, enabling the PBS to divert it away from its original direction.

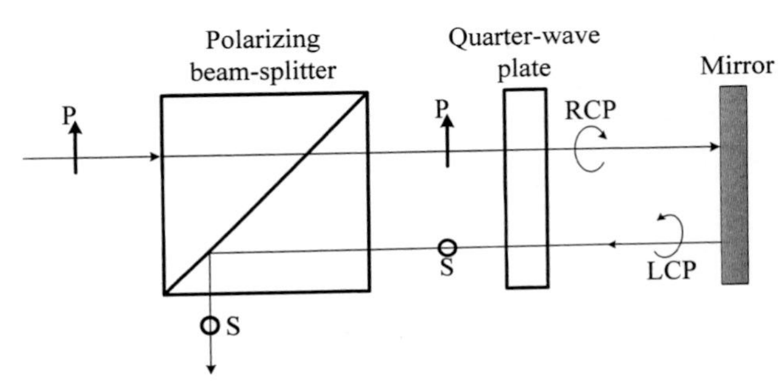

Figure 14.2 The quarter-wave plate as used in this system helps to separate the reflected beam from the incident beam. The key contribution is made by the (conventional) mirror, which converts the incident RCP beam into LCP upon reflection.

circularly polarized. (The system of Figure 14.2 is used in some optical disk drives with the optical disk acting as a mirror, the purpose being to separate the reflected beam from the incident beam efficiently, as well as to isolate the laser diode.) Although the system of Figure 14.2 behaves very much like that of Figure 14.1, no one claims that a QWP is non-reciprocal. This seeming paradox can be resolved after a careful examination of the concept of reciprocity, to which we now turn.

Is a polarizer reciprocal?

Consider the simple linear polarizer shown in Figure 14.3. A collimated beam of light entering from the left-hand side emerges from the polarizer linearly polarized along the transmission axis. The polarization state of the incident

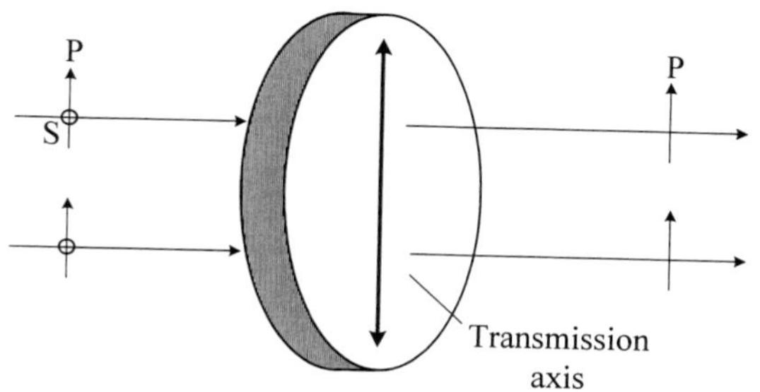

Figure 14.3 An ideal polarizer has a well-defined transmission axis (shown by the vertical double-ended arrow). The component of the incident beam that is polarized along the transmission axis goes through, while the component perpendicular to this axis is fully absorbed within the polarizer.

beam may be decomposed into two linear components, one parallel and the other perpendicular to the transmission axis. Assuming an ideal polarizer, the entire parallel component is transmitted while the entire perpendicular component is absorbed within the polarizer. If the direction of propagation of the transmitted beam is reversed, it will pass through the polarizer without any change. Since the original state of polarization of the incident beam is not recovered, the polarizer is a non-reciprocal element.

One might argue that in one sense the polarizer is reciprocal because, irrespective of whether the incident beam illuminates it from the left or from the right side, it behaves the same way. However, this turns out to be a poor way to define reciprocity, because it cannot be generalized to cover other optical elements. For example, consider the simple plano-convex lens shown in Figure 14.4. As will be shown below, lenses in general are reciprocal elements. However, a collimated beam of light shining on the convex surface of this lens comes to focus with less spherical aberration than a beam shining on its flat surface (see Figures 14.5 and 14.6). Therefore, if reciprocity required the identity of behavior from both sides of an element, one would end up with the undesirable result that a plano-convex lens, for example, is non-reciprocal. To avoid this outcome we return to our earlier definition that the beam transmitted through a reciprocal element, when "properly" reversed, must recreate the incident beam in the reverse direction. It is in this sense that the polarizer of Figure 14.3 is non-reciprocal.

Are lenses reciprocal in the above sense?

Consider an aberration-free lens that brings a collimated beam of light to focus as in Figure 14.7(a). A flat mirror placed in the focal plane of the lens

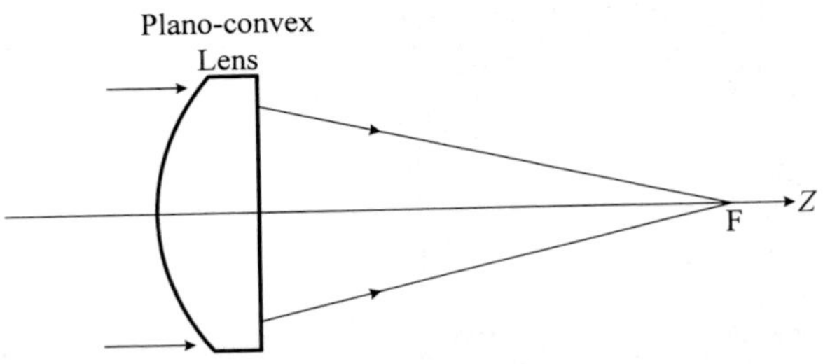

Figure 14.4 A simple plano-convex lens behaves differently depending on whether the light is incident from its plane side or from its convex side. The particular plano-convex lens used in the simulations had refractive index $n = 1.5$ at $\lambda_0 = 633$ nm, thickness $= 5$ mm, radius of curvature $= 10$ mm, and clear-aperture diameter $= 10$ mm. The 5 mm diameter incident beam was collimated and uniform. When the beam enters at the convex facet, the best focus (i.e., the circle of least confusion) appears at a distance of 16.49 mm in front of the plane facet. With the lens flipped and the beam entering the plane facet, the best focus occurs at 19.29 mm in front of the lens.

reflects the beam back towards the lens. Upon re-emerging from the lens the beam, now collimated once again, propagates in the reverse direction of the original incident beam. Is this sufficient proof that the lens is reciprocal? The answer is no, for the following reasons. What if the lens has aberrations? What if the incident beam is only illuminating one half of the lens's aperture, as in Figure 14.7(b)? What if the mirror is displaced from the focal plane of the lens, as in Figure 14.7(c)? In all these examples (and many more that can be conceived) the returning beam does not retrace the path of the incident beam. Does this mean that the lens is non-reciprocal? Again the answer is no. The culprit in all these examples is the mirror, which does not "properly" reverse the path of the beam.

What we need in place of the conventional mirror is a phase-conjugate mirror[1] (PCM) to reverse the wavefront properly. Suppose a PCM is placed perpendicular to the Z-axis at $z = z_0$. If the complex-amplitude distribution incident on the PCM is denoted by $A(x, y, z_0)$, then the reflected wavefront at the plane of the mirror will be $A^*(x, y, z_0)$, which propagates along the negative Z-axis and completely retraces the incidence path. Substituting the ordinary mirror by a PCM in Figure 14.7 ensures that the beam is properly reversed in each case, and proves beyond any doubt that lenses are reciprocal.

The quarter-wave plate

Returning now to the system of Figure 14.2, we re-examine the question of reciprocity of the QWP. Once again the mirror is recognized as the culprit:

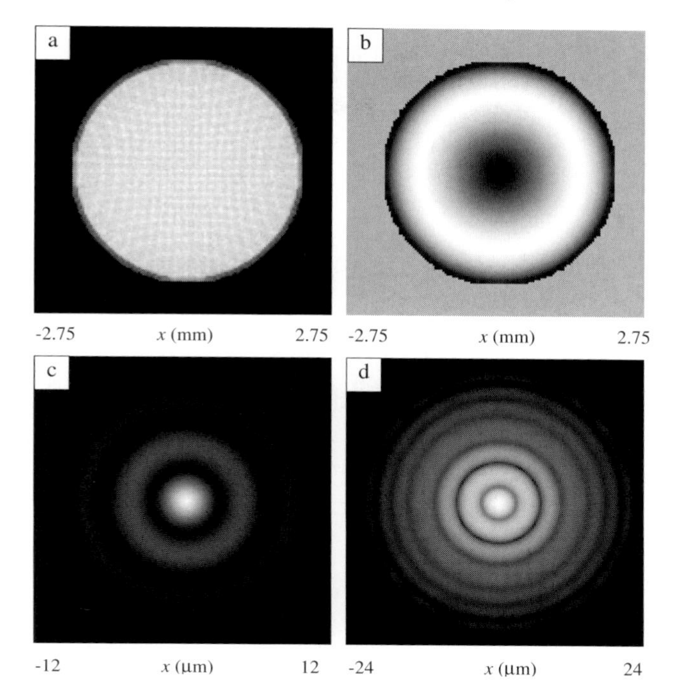

Figure 14.5 Plots of intensity and phase corresponding to the plano-convex lens of Figure 14.4, illuminated by a collimated and uniform beam from the convex side. (a) Intensity distribution immediately after the beam leaves the plane facet of the lens. (b) Residual phase distribution immediately after the beam leaves the plane facet. The curvature of the beam has been removed from the phase distribution, leaving only the residual spherical aberration balanced by a small amount of defocus. The r.m.s. value of these residual aberrations over the entire aperture is $0.17\lambda_0$. (c) Intensity distribution in the plane of best focus, i.e., at the circle of least confusion. (d) Same as (c) but on a logarithmic scale and over a larger area.

upon reflection from an ordinary mirror, a right circularly polarized (RCP) beam becomes left circularly polarized (LCP) and vice versa. The result is that the QWP in Figure 14.2 rotates the polarization of the beam by 90° in double pass, forcing it to change its propagation direction at the PBS. If the mirror is replaced by a PCM, the sense of circular polarization does not change upon reflection, and the beam emerges from the QWP with the same linear polarization as it had when it first entered the plate. The returning beam thus retraces its path, proving the reciprocity of the QWP.

The question arises as to what happens in the system of Figure 14.1 if the mirror is replaced by a PCM? Since the beam incident on the mirror is linearly polarized, it remains linear whether it is reflected from an ordinary mirror or from a PCM. Therefore, the path of the reflected light in Figure

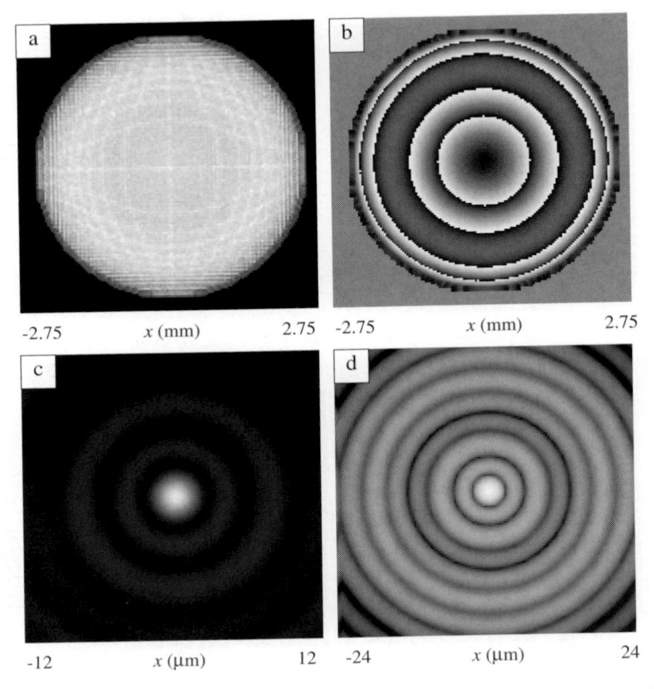

Figure 14.6 Same as Figure 14.5 for the case when the beam enters from the plane side of the lens. (a) Emergent intensity distribution at a plane tangent to the convex facet of the lens at its vertex. Note the larger diameter of the emergent beam compared with Figure 14.5(a). (b) Residual phase of the emergent beam within the tangent plane to the convex surface. The r.m.s. wavefront aberration over the entire aperture is $0.68\lambda_0$. (c) Distribution of intensity in the plane of best focus. (d) Same as (c) but on a logarithmic scale and over a larger area.

14.1 does not change as a result of changing the mirror, confirming our earlier conclusion that the Faraday rotator is non-reciprocal.

Reciprocity of conventional mirrors

In general, lossy elements are non-reciprocal by the above definition of reciprocity. The beam going through (or reflecting from) a lossy device becomes attenuated. Reversing the beam by a PCM reverses the propagation direction, but does not recover the losses incurred. A second path through the lossy element attenuates the beam even further. Thus the returning beam is twice attenuated, which means that it differs from the original incident beam, if not in its direction of propagation or phase or polarization state, at least in its amplitude. With the strict definition of reciprocity, which requires the

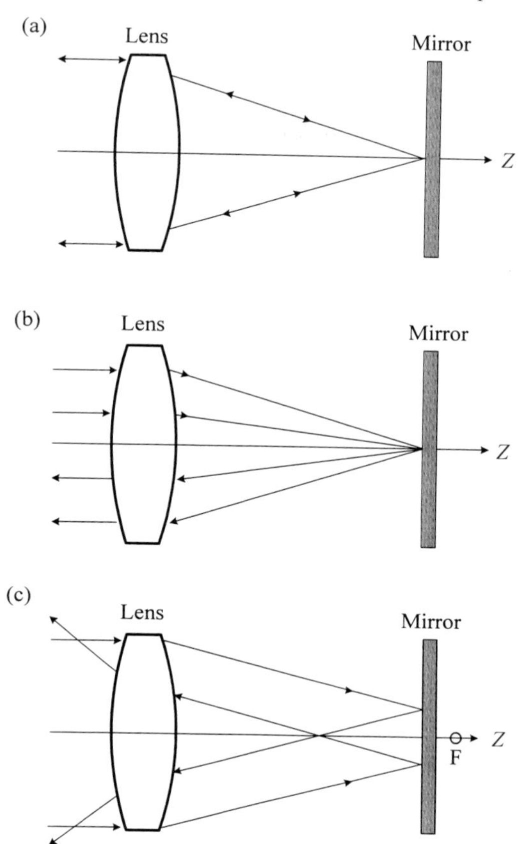

Figure 14.7 (a) A collimated and uniform beam, focused by an aberration-free lens and reflected by a plane mirror placed at the focal plane of the lens, retraces its path through the lens. (b) A collimated beam entering the upper half of the lens aperture and reflected from the mirror surface does not return on itself, but emerges from the lower half of the lens aperture. (c) When the mirror is displaced from the focal plane, the returning beam no longer retraces the incidence path.

beam to be fully recovered in the reverse path, attenuation is sufficient grounds for declaring lossy elements non-reciprocal.

A conventional mirror, such as a polished metallic surface, is lossy and therefore non-reciprocal. But consider a total internal reflection (TIR) device such as that shown in Figure 14.8. Here there are no losses and the only effect of the mirror on the incident beam is a change in its state of polarization. Suppose that the p and s components of the incident beam have complex amplitudes $a_p \exp(i\phi_p)$ and $a_s \exp(i\phi_s)$, respectively. Upon reflection from the TIR mirror these components retain their amplitudes but acquire different

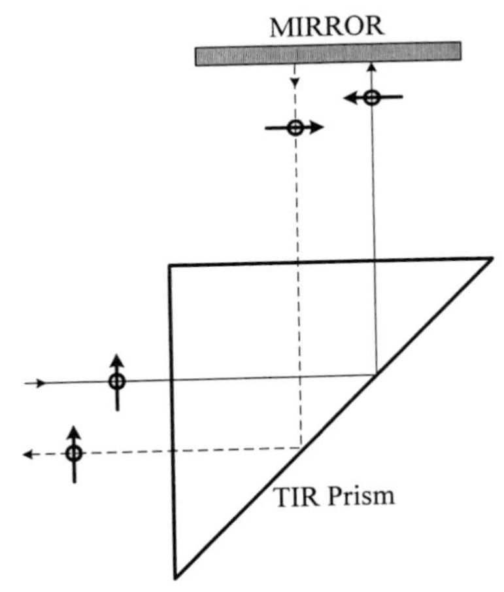

Figure 14.8 A collimated, uniform, and polarized beam of light is reflected from the rear facet of a TIR prism. If the beam is returned via a conventional mirror, the emergent beam will not, in general, have the same state of polarization as the incident beam. Use of a PCM mirror, however, ensures not only that the beam retraces its path but also that it will have the same state of polarization at any point along the path.

phases; the first one becomes $a_p \exp[i(\phi_p + \psi_p)]$, say, and the second becomes $a_s \exp[i(\phi_s + \psi_s)]$. If the direction of propagation of the beam is reversed by means of a conventional mirror, the phase angles ψ_p and ψ_s do not disappear from the returning beam; rather, they become twice as large. However, by now we have learned that a conventional mirror is not the proper device for reversing the beam. Instead, one must use a PCM to phase-conjugate the beam and launch it on its way back. When placed in the system of Figure 14.8, the PCM will return the two components of polarization as $a_p \exp[-i(\phi_p + \psi_p)]$ and $a_s \exp[-i(\phi_s + \psi_s)]$. The second reflection from the TIR mirror eliminates the acquired phases ψ_p and ψ_s and returns the conjugate of the original incident beam, which is exactly what is needed. A TIR mirror, therefore, is a reciprocal element.

A regular beam-splitter

There are many different ways of constructing a beam-splitter. For simplicity's sake, let us consider the specific beam-splitter shown in Figure 14.9. This flat piece of glass of thickness d and refractive index n has no coating layers and is used at a 45° angle of incidence. If the reflected and transmitted beams are returned by conventional mirrors, as shown in the figure, then, in general, a certain fraction of the light returns along the incidence path and the

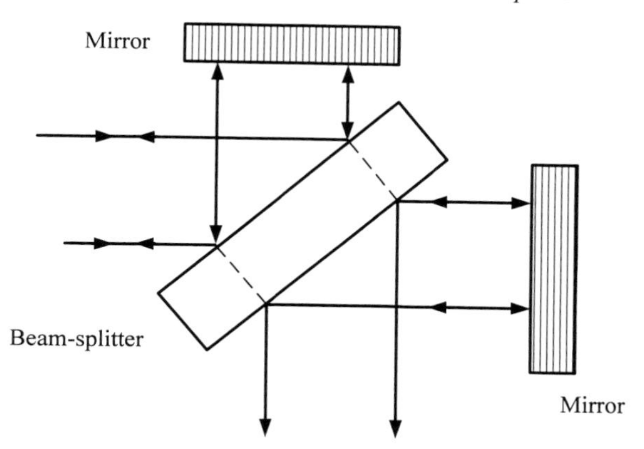

Figure 14.9 A parallel plate made of a glass slab of thickness d and refractive index n used as a beam-splitter. The collimated and uniform incident beam is partially reflected and partially transmitted at the slab. If conventional mirrors are used to return the reflected and transmitted beams back to the beam-splitter, in general a fraction of the beam will go back towards the source but the remainder will leave the beam-splitter in a fourth direction. However, if the mirrors are replaced by phase-conjugate mirrors, the entire beam will return along the incidence path.

remainder leaves the beam-splitter along a fourth direction. However, if the conventional mirrors in Figure 14.9 are replaced by PCMs, the entire beam will retrace its original path.

To see this we must first examine certain properties of the glass slab that forms the beam-splitter. Figure 14.10 shows computed plots of the reflection and transmission coefficients versus the thickness d of the slab. The assumed refractive index is $n = 2$, the angle of incidence is fixed at $\theta = 45^\circ$, and the incident beam is a coherent and monochromatic beam from a red HeNe laser ($\lambda_0 = 633\,\mathrm{nm}$). Only the range of thicknesses corresponding to one half-wavelength is shown in Figure 14.10, since the reflection and transmission coefficients are periodic with this period. The half-wave thickness of the slab is $d = \lambda_0/(2n\cos\theta') = 169.2\,\mathrm{nm}$. Here $\theta' = 20.7^\circ$, obtained from Snell's law, is the angle between the propagation direction within the slab and the slab's surface normal. The reflection and transmission coefficients for both p- and s-polarized light are shown in the figure.

Note in Figure 14.10 that, at any given thickness, $|r|^2 + |t|^2 = 1$ and $\phi_r - \phi_t = 90^\circ$. In fact, it may be shown that these two properties of the slab are quite general and hold not only for all thicknesses but also for all values of the refractive index n, angle of incidence θ, and wavelength λ_0. The first identity is a trivial statement of the principle of conservation of energy. The second, relating the phase angles of the reflected and transmitted beams,

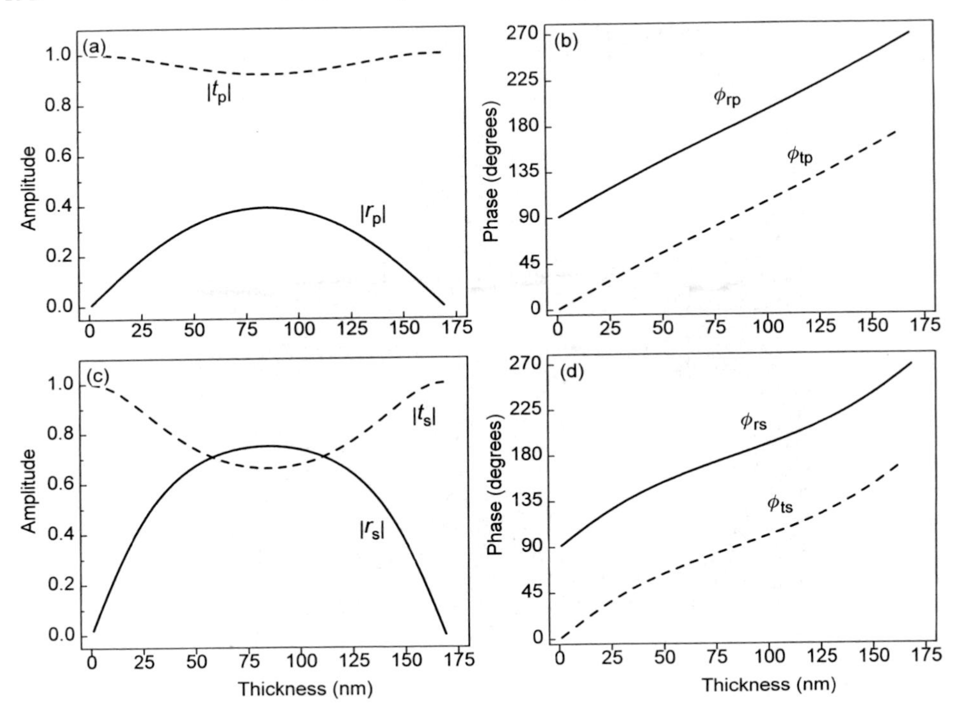

Figure 14.10 Computed plots of reflection and transmission coefficients versus slab thickness for the parallel-plate beam-splitter of Figure 14.9. The assumed refractive index of the glass material is $n = 2$, and the angle of incidence is fixed at $\theta = 45^\circ$. The incident beam is a coherent and monochromatic beam from a red HeNe laser ($\lambda_0 = 633\,\text{nm}$), and it is assumed to be linearly polarized either along the p- or the s-direction. The phase angles are evaluated at the front facet of the slab for the reflection coefficients and at the rear facet for the transmission coefficients. The reference phase angle is that of the incident beam at the front facet.

is more subtle, but its violation also results in non-conservation of energy, as we shall see shortly.

When the transmitted beam returns to the slab via a PCM it will have an amplitude t^*. Upon transmission (in the reverse direction) its amplitude becomes tt^*; it will then combine with the reversed reflected beam whose amplitude at this point is rr^*. The total returning amplitude is therefore $rr^* + tt^* = |r|^2 + |t|^2 = 1$. The remainder of the beam, leaving the beam-splitter in the fourth direction, will have a total amplitude $rt^* + r^*t = 2|rt|\cos(\phi_r - \phi_t)$, which is exactly zero because the phase difference between r and t is 90°. Thus the beams reversed by the two PCMs combine at the beam-splitter to yield the reverse propagating beam along the original path, leaving no other light to go in the fourth direction.

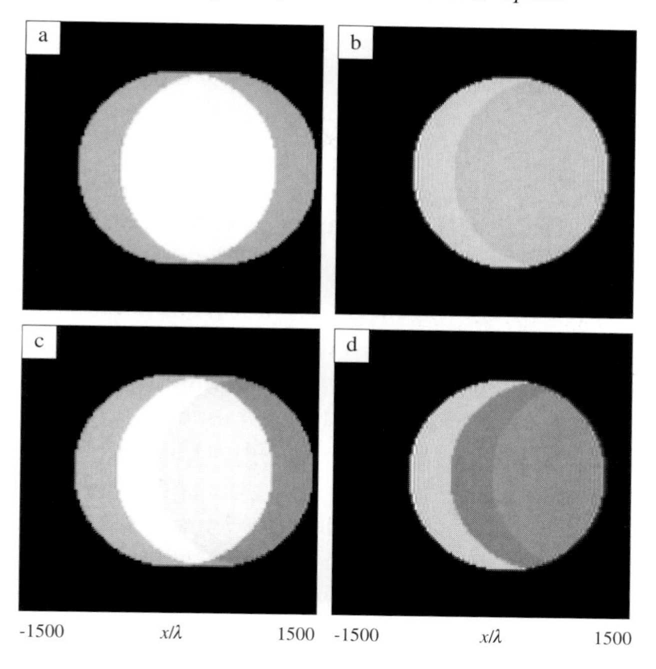

Figure 14.11 Plots of intensity distribution upon reflection or transmission of a collimated uniform beam from the beam-splitter of Figure 14.9. (Due to the limited range of the gray-scale, certain weak parts of the distributions are not visible.) The incident beam diameter is $2000\lambda_0$, where $\lambda_0 = 633$ nm. The beam-splitter, oriented at 45° to the propagation direction of the incident beam, has $n = 2$ and $d = 500\,\mu$m. (a) Logarithmic plot of the reflected intensity distribution for p-polarized incident beam. Since reflection from each surface is weak, only the first- and second-order reflected beams are observed. (b) Transmitted intensity distribution for p-polarized incident beam. (c) Logarithmic plot of the reflected intensity distribution for s-polarized incident beam. Since reflection from each surface is strong, the effect of the third-order reflection can also be seen in this figure. (d) Transmitted intensity distribution for s-polarized incident beam.

Although the above proof for reciprocity of the glass slab was given for plane waves, one can show its validity in the general case of a finite-size incident beam as well. To appreciate the effects of finite size, consider the plots of intensity distribution in Figure 14.11, computed for a HeNe beam of diameter $2000\lambda_0$ upon reflection from and transmission through a slab 500 μm thick of $n = 2$ glass. Near the edges of the beam the various reflected (or transmitted) orders do not overlap and, consequently, give rise to varying degrees of brightness in these regions. Instead of considering the edges separately, however, the appropriate proof of reciprocity for a finite-size beam involves the consideration of such beams as a superposition of a large number of plane waves traveling in different directions (i.e., angular spectrum decomposition). Since the reciprocity applies to each such plane wave, it must, of necessity, also apply to their linear superposition.

Reciprocity and Maxwell's equations

The principle of reciprocity in classical linear optics is rooted in the fact that electromagnetic waves obey Maxwell's equations and that these equations admit reciprocal solutions. Consider a distribution of electromagnetic waves in a region of space occupied by matter represented by the dielectric tensor $\varepsilon(x, y, z)$. Assume that the fields oscillate harmonically at a given frequency ω, and that the time-dependence factor $\exp(-i\omega t)$ has been eliminated from Maxwell's equations.[2] Suppose now that the propagation direction is reversed everywhere, so that any plane-wave component of the field that was propagating along a given k-vector is now propagating along the negative direction of that same k-vector. If we replace the E-fields by E^* and the H-fields by $-\mathrm{H}^*$ everywhere, Maxwell's equations remain satisfied so long as the dielectric tensor of the material environment obeys the relation $\varepsilon = \varepsilon^*$ at all points of space. This latter relation holds, for example, if the medium is isotropic and lossless (i.e., ε is a real-valued scalar), or if the medium is birefringent but non-absorptive (i.e., ε is a real-valued symmetric matrix), or if the medium has optical activity of the type observed in sugar crystals. If, however, the medium is absorptive, or if it has magneto-optical activity such as that exhibited by a Faraday rotator, then $\varepsilon \neq \varepsilon^*$, in which case the reverse-propagating beam(s) violate Maxwell's equations and, consequently, reciprocity breaks down.

Multilayer dielectric stack

The power of the reciprocity principle may be demonstrated by the following analysis of a multilayer dielectric stack. Adopting the approach pioneered by Sir George Gabriel Stokes (1819–1903)[3] we prove that any stack consisting of an arbitrary number of dielectric (i.e., non-absorbing) layers exhibits symmetric behavior between its front facet and rear facet reflectivity (or transmissivity). To prove this statement consider returning both the reflected beam and the transmitted beam back to the stack via two PCMs, as shown in Figure 14.12. Denoting the front facet reflection and transmission coefficients by r and t, and the corresponding rear facet coefficients by r' and t' we must have, by reciprocity, the following identities:

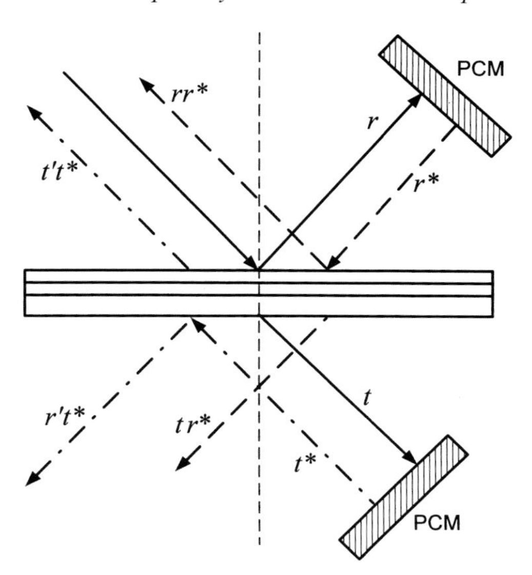

Figure 14.12 Multilayer stack consisting of an arbitrary number of dielectric layers. A unit-amplitude beam is partially reflected and partially transmitted at the top facet of the stack. If the reflected and transmitted beams are returned to the stack via phase-conjugate mirrors (PCMs), the principle of reciprocity requires that the beam must retrace its path. Thus the total amplitude along the reverse incidence direction must be unity and the total amplitude emerging from the bottom facet of the stack must be zero.

$$rr^* + t't^* = 1, \tag{14.1}$$

$$tr^* + r't^* = 0. \tag{14.2}$$

Equation 14.1, in conjunction with the principle of conservation of energy, yields $t = t'$, proving that the complex transmission coefficient is the same from both facets of the stack. From Eq. (14.2) one obtains $r' = -tr^*/t^*$, which proves that the amplitude of the reflection coefficient is the same from the two facets, that is, $|r| = |r'|$. As for the phase angles we have:

$$\tfrac{1}{2}(\phi_\mathrm{r} + \phi'_\mathrm{r}) = \phi_\mathrm{t} \pm 90^\circ. \tag{14.3}$$

These relations are readily verified for the specific quadrilayer stack whose performance characteristics are depicted in Figure 14.13. Needless to say, the symmetry of reflection and transmission from the two facets of a multilayer stack applies quite generally unless one or more layers are absorptive or magneto-optically active. In fact, the media of incidence and emergence on the two sides of the stack do not have to be identical either. Using the method of proof outlined above, one can readily show that the behavior of dielectric

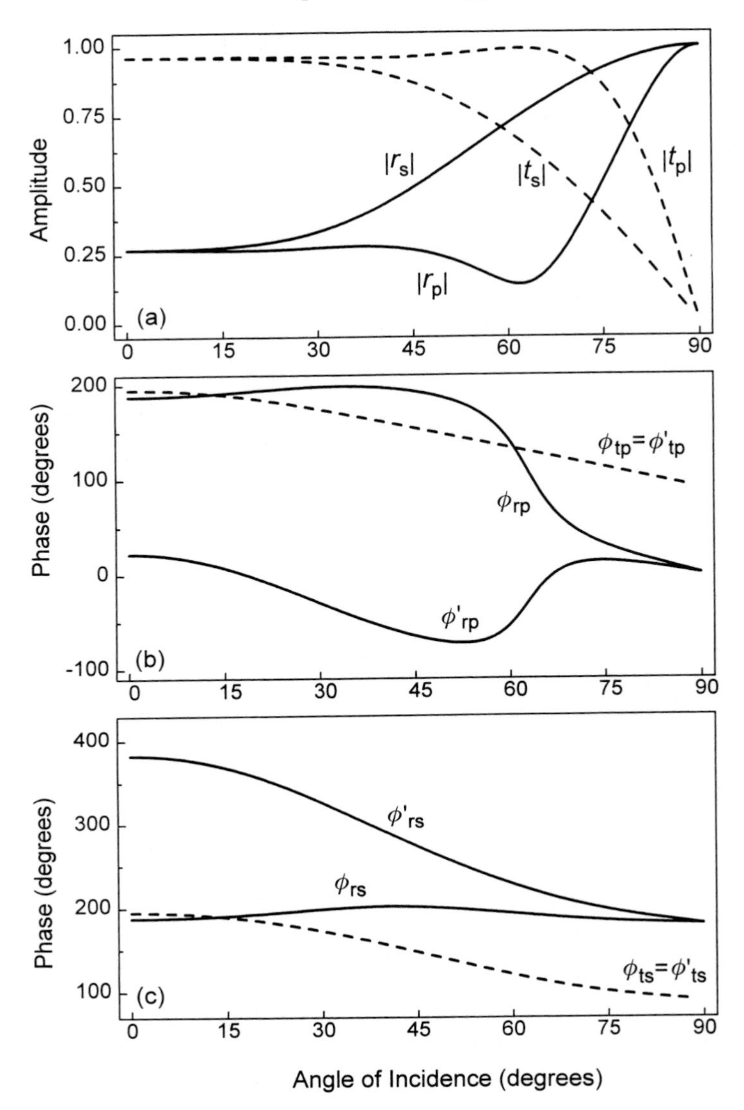

Figure 14.13 Computed plots of reflection and transmission coefficients versus the angle of incidence for a quadrilayer dielectric stack surrounded by free space. The layer thickness d and refractive index n for consecutive layers starting at the top of the stack are as follows: 140 nm, 2.2; 200 nm, 1.8; 80 nm, 2.0; 100 nm, 1.5. The magnitudes of the various reflection and transmission coefficients shown in (a) are the same whether the beam is incident from the top side or from the bottom side of the stack. The phase angles of the transmission coefficients, ϕ_{tp} and ϕ_{ts}, are also the same for top and bottom incidence. The phase angles of the reflection coefficients, however, depend on the side of the stack at which the beam is directed. In (b) and (c) ϕ_{rp} and ϕ_{rs} are the phase angles for p- and s-reflectivities when the beam is incident from the top of the stack. The corresponding primed quantities refer to incidence from the bottom.

stacks remains symmetrical even when the media above and below the stack have arbitrary refractive indices n_1 and n_2, provided that proper account is made of the difference in beam cross-section and the dependence of power on the refractive index.

Another interesting property of multilayer stacks arises when one or more of the layers happen to be absorptive. Since reciprocity no longer applies to this case, it should come as no surprise that the reflectivities of the two sides of the stack are, in general, different. What is surprising is that, even in the presence of absorption, the transmissivity continues to be the same from both sides. This property can be proven using standard methods of thin-film-stack calculation[4] and has been verified numerically in several situations. A simple proof for the symmetric behavior of the transmissivity under quite general conditions is given in the following appendix.

Appendix

We prove that the Fresnel transmission coefficient t for a multilayer stack consisting of metal and dielectric layers does not depend on whether the light is incident from the top or the bottom of the stack. For stacks consisting solely of dielectric layers this property has been proved in the present chapter, using reciprocity. Reciprocity, however, breaks down in the presence of absorptive layers, and one needs to resort to an alternative method of proof, such as that outlined below.

A general stack consists of an arbitrary number of layers, each having thickness d_j and complex refractive index $(n + \mathrm{i}k)_j$, the subscript j referring to the layer number. For an incident plane wave of wavelength λ, arriving at the top of the stack at angle θ, the Fresnel reflection and transmission coefficients of the stack are denoted by r and t, respectively. Similarly, when the beam is incident from the bottom side on the stack (again at angle θ), the Fresnel coefficients are denoted r' and t'. Our goal is to demonstrate the equality of t and t', even though, in general, r and r' may differ from each other.

Consider the hypothetical situation shown in Figure A14.1, where the stack is split along an interfacial plane into two smaller stacks separated by an air gap d. The upper stack, identified as stack 1, has reflection and transmission coefficients from top and bottom denoted by r_1, t_1, r'_1, t'_1. Similarly, the corresponding parameters of the lower stack, stack 2, are r_2, t_2, r'_2, t'_2. The transmissivity t of the entire stack (in the presence of the air gap) can be obtained by adding an infinite number of terms corresponding to the beams bouncing back and forth in the gap, namely,

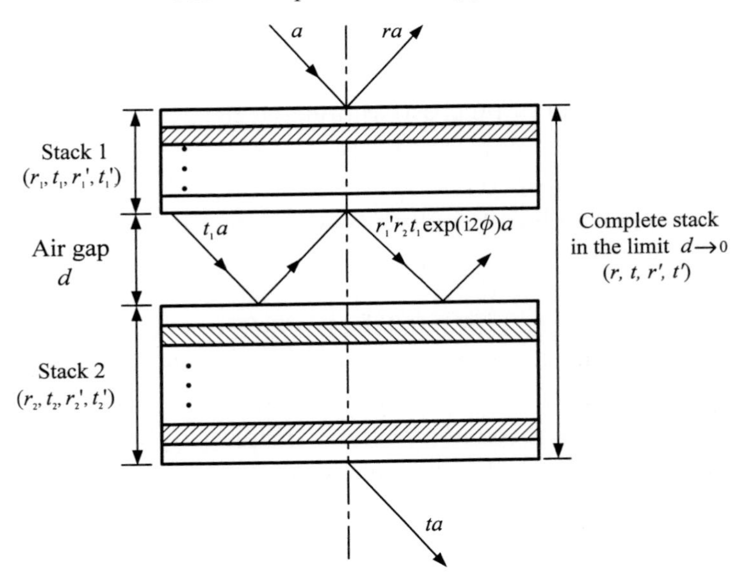

Figure A14.1 A multilayer stack consisting of metal and dielectric layers is split into two sub-stacks along an arbitrary interfacial plane. The upper stack has Fresnel reflection and transmission coefficients r_1, t_1 when the beam of light is incident from the top. The corresponding coefficients when the light is incident from the bottom are r_1', t_1'. Similarly, the lower stack has reflection and transmission coefficients r_2, t_2, r_2', t_2'. The width of the air gap separating the two sub-stacks is d. The overall transmission coefficient t of the entire stack can be obtained by adding the contributions of the infinite number of beams that bounce back and forth in the air-gap region.

$$\begin{aligned} t &= t_1 t_2 \exp(\mathrm{i}\phi) + t_1 r_2 r_1' t_2 \exp(\mathrm{i}3\phi) + t_1 r_2^2 r_1'^2 t_2 \exp(\mathrm{i}5\phi) + \ldots \\ &= t_1 t_2 \exp(\mathrm{i}\phi)/[1 - r_1' r_2 \exp(\mathrm{i}2\phi)]. \end{aligned} \tag{A14.1}$$

Here $\phi = 2\pi d \cos\theta/\lambda$ is the phase delay due to one passage of the beam through the gap. In the limit of a vanishing gap (i.e., $d \to 0$) we find a simple expression for t in terms of the parameters of stacks 1 and 2:

$$t = t_1 t_2/(1 - r_1' r_2). \tag{A14.2}$$

In similar fashion, the reverse-direction transmissivity t' of the stack (bottom illumination) is found to be

$$t' = t_1' t_2'/(1 - r_1' r_2). \tag{A14.3}$$

The argument for the equality of t and t' flows readily from Eqs. (A14.2) and (A14.3), using proof by induction as follows . It is clear that if the individual sub-stacks are such that $t_1 = t_1'$ and $t_2 = t_2'$, then $t = t'$ is guaranteed. For each sub-stack the reduction to a pair of smaller stacks can be repeated until each sub-stack is a single-layer, in which case $t_1 = t_1'$ and $t_2 = t_2'$ obviously hold. The proof is thus complete.

References for chapter 14

1 A. Yariv and D. M. Pepper, Amplified reflection, phase conjugation, and oscillation in degenerate four-wave mixing, *Opt. Lett.* **1**, 16–18 (1977).
2 M. Born and E. Wolf, *Principles of Optics*, sixth edition, Pergamon Press, Oxford, 1980.
3 E. Hecht, *Optics*, third edition, Addison-Wesley, Reading, Massachusetts, 1998.
4 H. A. Macleod, *Thin Film Optical Filters*, second edition, Macmillan, New York, 1986.

15

Linear optical vortices[†]

An optical vortex is a phase singularity nested within the cross-sectional profile of a coherent beam of light.[1–3] Such vortices occur naturally in the electromagnetic mode structure of certain optical cavities.[4] They may also be created artificially by computer-generated holograms designed to impart to an incident beam of light the desired phase and amplitude characteristics of a vortex.[5] In recent years the study of vortices has become the focus of several research groups around the world, as potential applications have emerged. Noteworthy examples of such applications are the manipulation of small objects by optical tweezers[6] and the control of atomic or molecular beams via the exchange of angular momentum with optical vortices.[7]

Mathematical description

The complex-amplitude distribution of a simple vortex of order m centered at (x_0, y_0) in the cross-sectional plane of a Gaussian beam may be written as

$$A(x, y, z = 0) = [(x - x_0) + \mathrm{i}\ \mathrm{sign}(m)\,(y - y_0)]^{|m|} \exp[-(x^2 + y^2)/r_0^2]. \quad (15.1)$$

The sign of the integer m determines whether the vorticity is clockwise or counterclockwise, and the magnitude of m is the number of 2π phase shifts in one cycle around the singularity. For the amplitude distribution to be single-valued it must go to zero at the center, as is indeed the case in Eq. (15.1). The host is here a circular Gaussian beam having 1/e radius r_0 at the waist, and the beam's propagation direction is along the Z-axis.

Figure 15.1 shows an $m = +1$ vortex centered at $(x_0, y_0) = (0, 0)$ within a Gaussian beam of radius $r_0 = 10\lambda$, where λ is the wavelength of the light. The

[†]This chapter was coauthored with Ewan M. Wright of the Optical Sciences Center, University of Arizona.

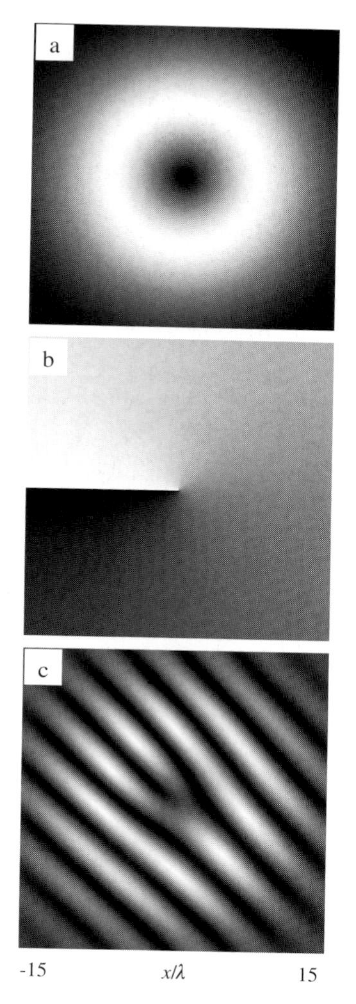

Figure 15.1 A Gaussian beam, having 1/e radius $r_0 = 10\lambda$ at the waist, hosts an $m = +1$ vortex at its center. (a) Intensity and (b) phase distribution at the beam waist. (c) Interferogram with a tilted plane wave.

intensity distribution in Figure 15.1(a) has a hole at the center, and the phase distribution in Figure 15.1(b) displays a continuous variation from 0 to 2π around the vortex. If this vortex is made to interfere with a tilted plane wave, the resulting fringe pattern would resemble that in Figure 15.1(c). The fork at the center of the fringe pattern created by the splitting of a single fringe is characteristic of all first-order vortices.

An important feature of vortices is that they maintain their identity as they propagate through space. Figure 15.2 shows the intensity and phase distribu-

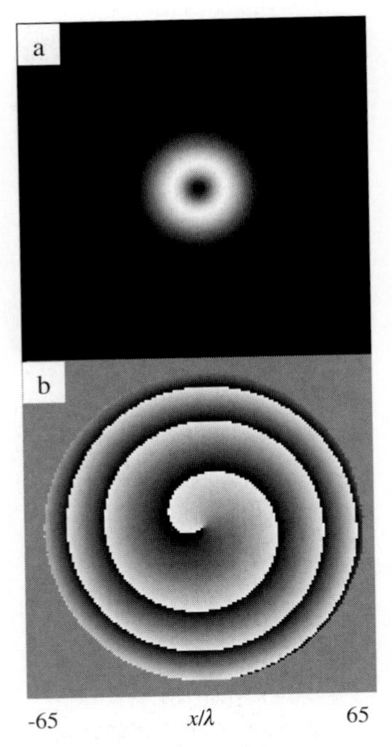

Figure 15.2 The beam of Figure 15.1 propagates to its Rayleigh range at $z = 314\lambda$. (a) Intensity, (b) phase. The phase singularity is now mixed with the wavefront curvature.

tions of the vortex of Figure 15.1 after propagating to the Rayleigh range of the host Gaussian beam at $z = \pi r_0^2/\lambda = 314\lambda$. Clearly the central hole in the intensity pattern and the 2π phase variation around the singularity are preserved, even though the phase is mixed with the curvature acquired during propagation.

The flow of energy

Figure 15.3 shows another example of an optical vortex, this one of order $m = -3$. The intensity distribution in Figure 15.3(a) is similar to that of the first-order vortex in Figure 15.1(a). However, the phase distribution depicted in Figure 15.3(b) shows a continuous 6π variation around the center. The fringe pattern in Figure 15.3(c), produced by allowing the vortex to interfere with a tilted plane wave, has the characteristic fork; here one fringe splits into four.

Figure 15.4 shows the distribution of the Poynting vector $\boldsymbol{S}$ for the above vortex. Shown from top to bottom are the x-, y-, and z- components of $\boldsymbol{S}$ encoded in gray-scale (black corresponds to a minimum, white to a max-

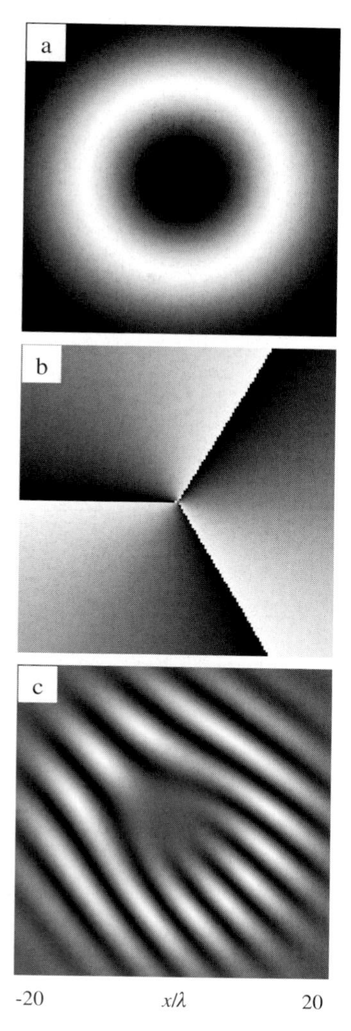

Figure 15.3 Same as Figure 15.1 for an $m = -3$ vortex.

imum). The normalized ranges of values are: $-0.04 \leq S_x \leq 0.04$, $-0.04 \leq S_y \leq 0.04$, and $0 \leq S_z \leq 1$. So, for example, in Figure 15.4(a) the bright regions indicate that S_x is directed along $+X$, while in the dark regions S_x is directed along $-X$. Similarly, S_y on the left-hand side of Figure 15.4(b) is directed along $+Y$, while on the right-hand side it is directed along $-Y$. In Figure 15.4(c), where $S_z \geq 0$, large positive values appear bright whereas those in the vicinity of zero are dark. Overall, $\boldsymbol{S}$ spirals around the Z-axis. The electromagnetic energy, therefore, does not flow straightforwardly along the optical axis but twists and turns as the beam moves forward.

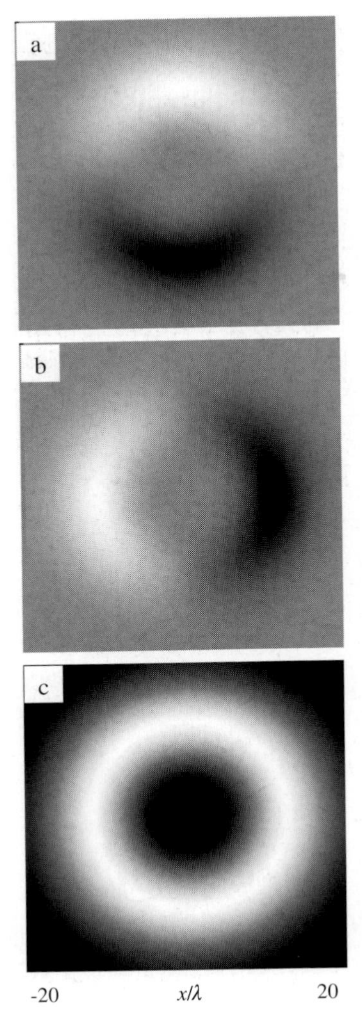

Figure 15.4 From top to bottom: x-, y-, and z- components of the Poynting vector $\boldsymbol{S}$ for the vortex of Figure 15.3. The minimum value of each function is shown as black and its maximum value as white, the intermediate values being covered by the gray-scale. The depicted ranges of values (in normalized units) are: $-0.04 \leq S_x \leq 0.04$, $-0.04 \leq S_y \leq 0.04, 0 \leq S_z \leq 1$. The Poynting vector here has clockwise circulation around the optical axis.

When the vortex of Figure 15.3 propagates to the Rayleigh range at $z = 314\lambda$, the intensity and phase patterns of Figure 15.5 are obtained. As before, the central hole in the intensity pattern and the singularity of the phase pattern are preserved, but the phase is now mixed with the curvature of the diverging wavefront.

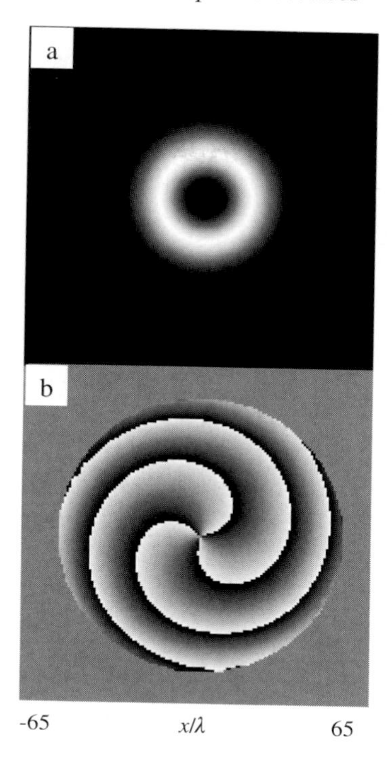

Figure 15.5 The beam of Figure 15.3 is propagated to its Rayleigh range at $z = 314\lambda$. (a) Intensity, (b) phase. The phase singularity is mixed with the wavefront curvature.

Vortex pair

The complex amplitude of a beam containing multiple vortices may be written as the product of terms similar to those appearing in Eq. (15.1), namely,

$$A(x, y, z = 0) = \left\{ \prod_{n=1}^{N} [(x - x_n) + \mathrm{i}\,\mathrm{sign}(m_n)\,(y - y_n)]^{|m_n|} \right\} \exp[-(x^2 + y^2)/r_0^2]. \tag{15.2}$$

This equation represents N vortices nested in a Gaussian beam of radius r_0; the nth vortex whose order is m_n is centered at the point (x_n, y_n) within the XY-plane.

Figure 15.6 shows two identical $m = +1$ vortices, separated by a distance $d = 11\lambda$ in a Gaussian beam with $r_0 = 10\lambda$. From left to right, the distributions represent the beam at its waist ($z = 0$), at the Rayleigh range ($z = 314\lambda$), and in the far field ($z = 2000\lambda$). Note that the three cross-sections in this figure are plotted on different scales. The beam expands along the propaga-

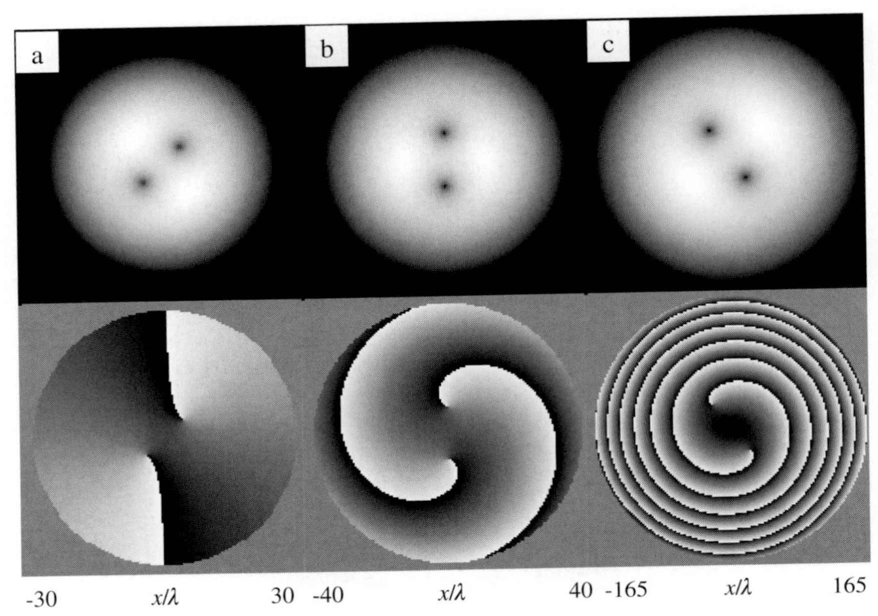

Figure 15.6 A pair of identical $m = +1$ vortices separated by $d = 11\lambda$, nested within a Gaussian beam having $r_0 = 10\lambda$ at the waist. Shown are the logarithmic plots of intensity (top row) and phase (bottom row) at several distances from the waist. (a) At the waist of the beam. (b) At the Rayleigh range, $z = 314\lambda$. (c) At $z = 2000\lambda$. Note that the vortex pair survives into the far field while rotating by 90°.

tion path, of course, but the vortices maintain their relative shape and position while undergoing a collective 90° rotation around the optical axis between the waist and the far field.[8]

The case of two vortices of opposite helicity is shown in Figure 15.7. Here $m = -1$ for one vortex and $m = +1$ for the other. The initial separation between the vortices at the beam waist is $d = 11\lambda$. As the beam propagates through free space, the vortices appear to spread out and combine with each other. Eventually, they carve out a circular niche for themselves, but the phase discontinuity near the beam center survives all the way to the far field.

A somewhat different behavior will be observed when two vortices of opposite polarity are separated at the beam waist by $d < r_0$. The corresponding intensity and phase patterns remain more or less the same as those in Figure 15.7 (which are representative of the case $d > r_0$) but, at some distance z from the waist, the phase discontinuity near the beam center disappears.[4] This behavior is reminiscent of fluid vortices of opposite chirality, which collide and annihilate when they happen to be within each other's basin of attraction.

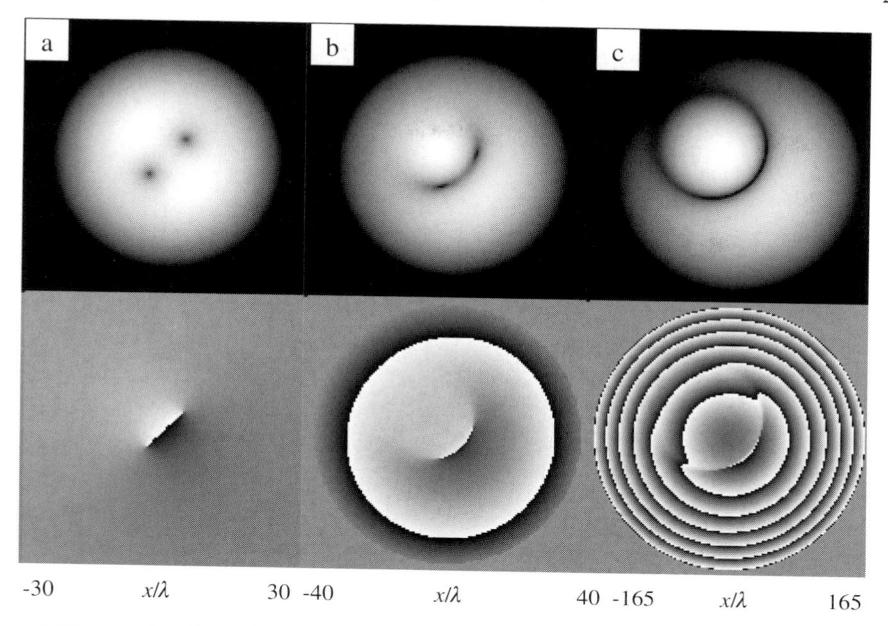

Figure 15.7 A pair of vortices of opposite helicity, $m = +1$ and $m = -1$, separated by $d = 11\lambda$, nested within a Gaussian beam having $r_0 = 10\lambda$ at the waist. Shown are the logarithmic plots of intensity (top row) and phase (bottom row) at several distances from the waist. (a) At the waist of the beam. (b) At the Rayleigh range, $z = 314\lambda$. (c) At $z = 2000\lambda$. Note that the phase singularity survives into the far field.

Relation to Gauss–Hermite (or Laguerre) polynomials

It can be shown that N vortices of the type described by Eq. (15.2) can be written as a superposition of Gauss–Hermite (or Gauss–Laguerre) polynomials of order $\leq N$. These polynomials, which describe the eigenmodes of certain waveguides, are also the eigenmodes of free-space propagation in the paraxial regime. The vortices formed by the superposition of such modes propagate in free space but, while individual modes maintain their identities, different modes accrue different phase shifts. As a result of these differing phase shifts the pattern of vortices might change along the optical axis, but the main topological features of their singularities are usually preserved.[4,9]

Resolving adjacent vortices

An interesting question concerning vortices is whether one can densely pack them on a given waveform (or on a given surface) and use the resulting pattern for data communication (or for information storage). Figure 15.8

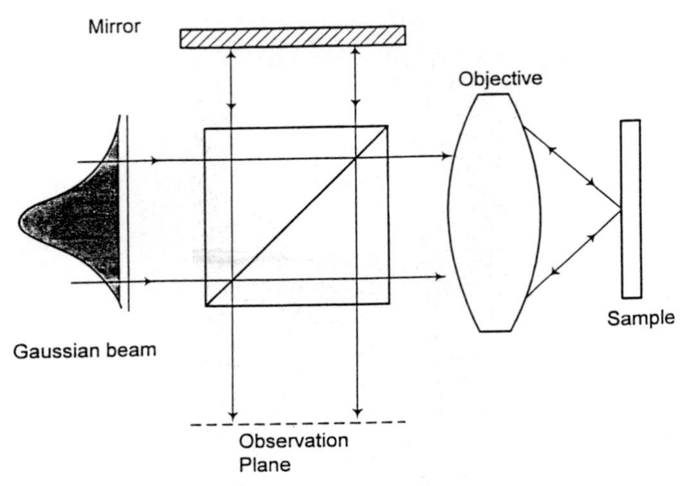

Figure 15.8 Densely packed vortices imprinted upon a sample's flat surface may be observed through a coherent-light microscope. The incident Gaussian beam has 1/e radius $r_0 = 900\lambda$. The entrance pupil of the $0.95NA$ objective lens, having a radius of 3000λ, allows the Gaussian beam through with negligible truncation. The beam reflected from the sample picks up the amplitude and phase patterns of the vortices and returns through the objective lens. The phase structure may be extracted by interference with the original Gaussian beam reflected from the reference mirror.

is the schematic of a coherent-light microscope that might be used to retrieve a dense pattern of vortices recorded on a flat surface. The Gaussian beam entering the system is narrow enough that truncation at the objective's aperture may be considered negligible. Upon focusing through the $0.95NA$ lens, the FWHM diameter of the focused spot becomes $\approx 1.33\lambda$. The focused spot is modulated by the amplitude and phase reflectivity of the sample before returning to the objective lens. At the beam-splitter the returning beam is diverted towards the observation plane, where the intensity pattern may be examined directly, and the phase pattern may be obtained by interference with a reference beam (supplied by the mirror).

Figure 15.9 shows the patterns of intensity and phase at the focal plane of the objective lens immediately after the beam is reflected from the sample. There appear here a total of seven vortices within the focused beam area, all with the same helicity, $m = +1$. The pair in the middle, having a center-to-center spacing of $\lambda/2$, is at the resolution limit of conventional optical microscopy. When the reflected beam reaches the observation plane, the patterns shown in Figure 15.10 are obtained. Note that both the intensity and the phase distribution at the observation plane are magnified versions of the

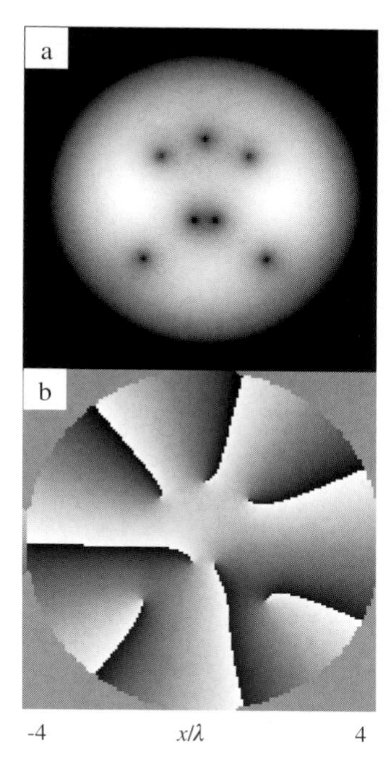

Figure 15.9 (a) Intensity and (b) phase distribution imparted to the focused Gaussian beam in Figure 15.8 immediately after reflection from the sample's surface. There is a total of seven $m = +1$ vortices; the distance between the closest pair, near the center, is 0.5λ.

original distributions at the sample, albeit with an inconsequential 90° rotation around the optical axis.

As was the case for two vortices of the same helicity (see Figure 15.6), the beam in the present case has also preserved the seven equal-helicity vortices all the way to the far field. To observe the phase structure, however, it is necessary to interfere the beam returning from the sample with a reference beam. The resulting interference pattern is shown in Figure 15.10(c). The split fringes corresponding to five of the vortices are clearly distinguishable in this pattern, even for those vortices that are far from the beam center. However, the split fringes for the two adjacent vortices near the center are hard to recognize and, in practice, where the signal-to-noise ratio is limited, it is unlikely that these vortices can be resolved. It appears, therefore, that the Rayleigh criterion for resolution in image-forming systems applies to these vortices as well, even though the image quality here is extremely good and no information seems to have been lost between the sample and the observation plane.

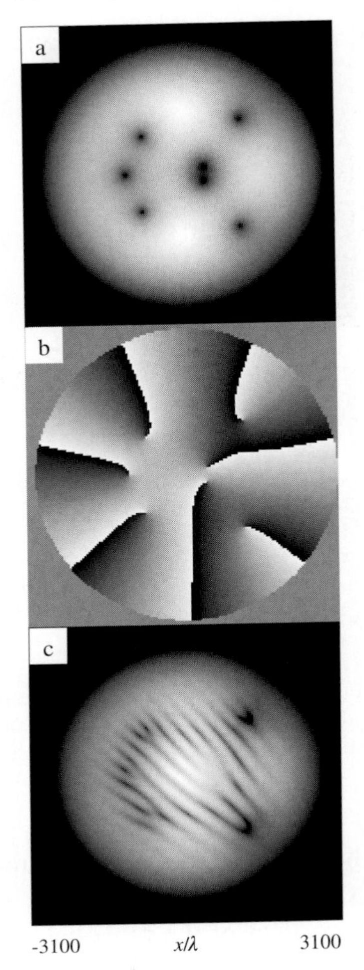

Figure 15.10 Distributions of (a) intensity and (b) phase at the observation plane of Figure 15.8 corresponding to the seven vortices of Fig. 15.9. In (a) and (b) the reference beam is blocked. In (c) the reference beam interferes with the beam returning from the sample, thus creating fringes. The vortices may be identified by the forks within these fringes.

References for chapter 15

1 J. F. Nye and M. V. Berry, Dislocations in wave trains, *Proc. Roy. Soc. London A* **336**, 165–190 (1974).
2 N. B. Baranova *et al.*, Wavefront dislocations: topological limitations for adaptive systems with phase conjugation, *J. Opt. Soc. Am.* **73**, 525–528 (1983).
3 J. M. Vaughan and D. V. Willetts, Temporal and interference fringe analysis of TEM^*_{01} laser modes, *J. Opt. Soc. Am.* **73**, 1018–1021 (1983).

4 G. Indebetouw, Optical vortices and their propagation, *J. Mod. Opt.* **40**, 73–87 (1993).

5 N. R. Heckenberg *et al.*, Laser beams with phase singularities, *Opt. Quant. Electronics* **24**, S951–S962 (1992).

6 K. T. Gahagan and G. A. Swartzlander, Optical vortex trapping of particles, *Opt. Lett.* **21**, 827–829 (1996).

7 H. He *et al.*, Direct observation of transfer of angular momentum to absorptive particles from a laser beam with a phase singularity, *Phys. Rev. Lett.* **75**, 826–829 (1995).

8 D. Rozas, Z. S. Sacks and G. A. Swartzlander, Experimental observation of fluidlike motion of optical vortices, *Phys. Rev. Lett.* **79**, 3399–3402 (1997).

9 M. W. Beijersbergen *et al.*, Astigmatic laser mode converters and transfer of orbital angular momentum, *Opt. Comm.* **96**, 123–132 (1993).

16

Geometric-optical rays, Poynting's vector, and the field momenta

In isotropic media the rays of geometrical optics are usually obtained from the surfaces of constant phase (i.e., wavefronts) by drawing normals to these surfaces at various points of interest.[1] It is also possible to find the rays from the eikonal equation, which is derived from Maxwell's equations in the limit when the wavelength λ of the light is vanishingly small.[2] Both methods provide a fairly accurate picture of beam-propagation and electromagnetic-energy transport in situations where the concepts of geometrical optics and ray-tracing are applicable. The artifact of rays, however, breaks down near caustics and focal points and in the vicinity of sharp boundaries, where diffraction effects and the vectorial nature of the field can no longer be ignored.

It is possible, however, to define the rays in a rigorous manner (consistent with Maxwell's electromagnetic theory) such that they remain meaningful even in those regimes where the notions of geometrical optics break down. Admittedly, in such regimes the rays are no longer useful for ray-tracing; for instance, the light rays no longer propagate along straight lines even in free space. However, the rays continue to be useful as they convey information about the magnitude and direction of the energy flow, the linear momentum of the field (which is the source of radiation pressure), and the angular momentum of the field. Such properties of light are currently of great practical interest, for example, in developing optical tweezers, where focused laser beams control the movements of small objects.[3–6] Similarly, the manipulation of atoms and molecules with laser beams is presently an active area of research that has tremendous potential for future applications.[7]

Computing the Poynting vector

For a coherent, monochromatic beam of light the time-averaged Poynting vector $\boldsymbol{S}$ at any point (x, y, z) in space can represent the direction and magnitude of the corresponding ray. Computing $\boldsymbol{S}$ is fairly straightforward and involves only a few fast-Fourier transformations (FFTs). Outlined below is a method of calculating $\boldsymbol{S}$ for a beam in free space, but the method can readily be generalized to material environments as well. Throughout this chapter the adopted system of units is MKS, c is the speed of light in vacuum, ε_0 is the permittivity of free space ($\varepsilon_0 c^2 = 10^7/4\pi$), and h is Planck's constant.

With the beam's propagation direction fixed along the Z-axis, consider the distribution of the $\boldsymbol{E}$-field in the beam's cross-sectional plane, XY. The only components of $\boldsymbol{E}$ needed for calculating $\boldsymbol{S}$ are E_x and E_y. To compute the Poynting vector, decompose the beam into its plane-wave spectrum. This requires one FFT for $E_x(x, y)$ and another for $E_y(x, y)$. For each plane wave thus obtained compute the Z-component $E_z(x, y)$ of the $\boldsymbol{E}$-field using the requirement $\boldsymbol{k} \cdot \boldsymbol{E} = 0$. Here $\boldsymbol{k} = 2\pi\boldsymbol{\sigma}/\lambda$ is the wave-vector for the plane wave propagating along the unit vector $\boldsymbol{\sigma}$. The knowledge of $\boldsymbol{E}$ and $\boldsymbol{k}$ for each plane wave leads directly to the corresponding magnetic field $\boldsymbol{B} = \boldsymbol{\sigma} \times \boldsymbol{E}/c$. At this point all six components of $\boldsymbol{E}$ and $\boldsymbol{B}$ for each plane wave are determined; therefore, an inverse FFT on each such component would yield the complete $\boldsymbol{E}$- and $\boldsymbol{B}$- fields within the cross-sectional XY-plane. Finally, the time-averaged Poynting vector may be obtained from $\boldsymbol{S} = \frac{1}{2}\varepsilon_0 c^2 \mathrm{Real}(\boldsymbol{E} \times \boldsymbol{B}^*)$.[8]

Rays of a linearly polarized Gaussian beam

As an example, consider a Gaussian beam of wavelength λ at its waist, having 1/e (amplitude) radii $R_x = 15\lambda$ along X, $R_y = 10\lambda$ along Y. Assuming that the beam is linearly polarized in the X-direction, its Poynting vector may be computed by the aforementioned method. Figure 16.1(a) shows the distribution of intensity for the x-component of polarization, $I_x = |E_x|^2$. By definition, this beam is linearly polarized along X and has no E_y, but E_z, albeit very small, is not zero, as can be seen in Figure 16.1(b). The phase of E_z (not shown) is $+90°$ on the left-hand side and $-90°$ on the right-hand side of the beam's cross-section. For this beam it turns out that $S_x = S_y = 0$, and only $S_z \neq 0$; a plot of S_z at the waist of the beam is shown in Figure 16.1(c). Clearly S_z is strongest at the beam center and decays with increasing distance from the center, behaving very much like I_x does. The fact that $\boldsymbol{S}$ in this case is everywhere parallel to the Z-axis is consistent with one's intuitive expectation that, at its waist, the Gaussian beam should be collimated.

When the beam propagates away from the waist, it acquires curvature and the rays exhibit behavior characteristic of a divergent beam. The Cartesian components of the Poynting vector (S_x, S_y, S_z) all turn out to be nonzero in

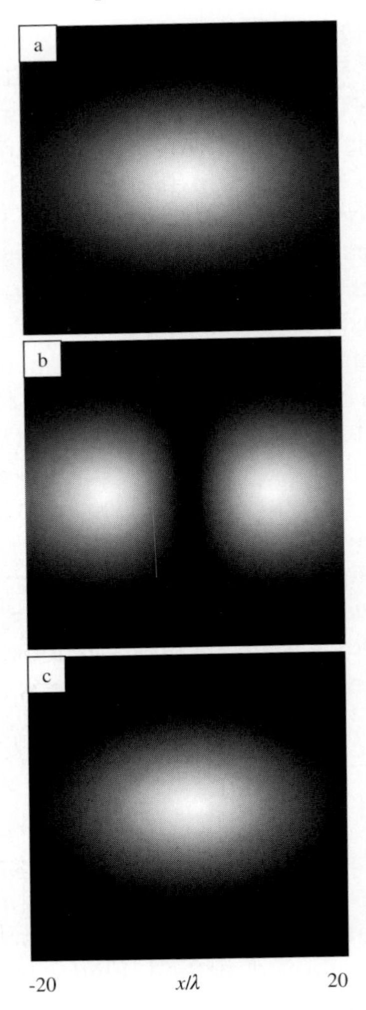

Figure 16.1 Various distributions at the waist of a Gaussian beam having 1/e (amplitude) radii $R_x = 15\lambda$, $R_y = 10\lambda$; the beam is linearly polarized along the X-axis. (a) Intensity of the x-component of polarization, $I_x = |E_x|^2$. (b) Intensity of the z-component of polarization, $I_z = |E_z|^2$. In (a) and (b) the peak intensities are in the ratio $I_x : I_z = 1.0 : 0.83 \times 10^{-4}$. (c) A plot of S_z, the projection of the Poynting vector $\boldsymbol{S}$ along the optical axis. $S_z(x, y) \geq 0$ is encoded in gray-scale (black, minimum; white, maximum). The other components of $\boldsymbol{S}$, namely, S_x and S_y, are exactly zero at this cross-section.

this case. Figure 16.2 shows various distributions for the above Gaussian beam at a distance of $z = 800\lambda$ from the waist. Shown in the left-hand column are the intensity profiles for the three Cartesian components of $\boldsymbol{E}$. The peak intensities in these figures are in the ratios $I_x : I_y : I_z = 1.0 : 0.39 \times 10^{-8}: 0.83 \times 10^{-4}$. Whereas the beam at the waist is elongated

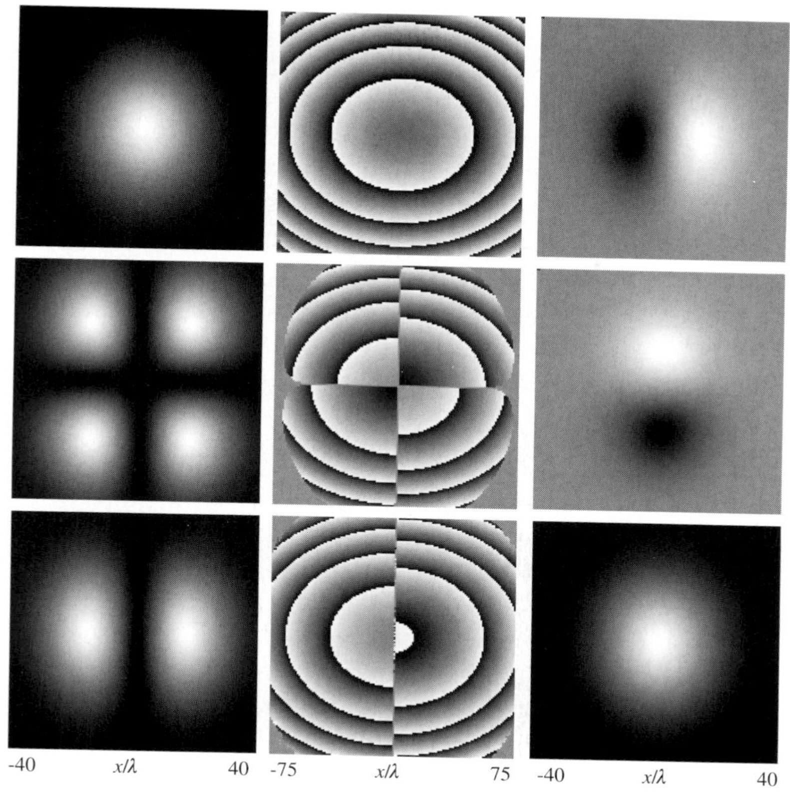

Figure 16.2 The Gaussian beam of Figure 16.1 after propagating a distance of $z = 800\lambda$ in free space. The left-hand column shows, from top to bottom, the distributions of intensity for the x-, y-, and z- components of polarization; the peak intensities are in the ratios $I_x : I_y : I_z = 1.0 : 0.39 \times 10^{-8} : 0.83 \times 10^{-4}$. The middle column shows the corresponding phase plots for E_x, E_y, E_z; the gray-scale covers the range $-180°$ (black) to $+180°$ (white). The third column shows the Cartesian components of the Poynting vector, S_x, S_y, S_z, in gray-scale (black, minimum; white, maximum). Here the normalized ranges of values are: $-0.48 \le S_x \le 0.48$, $-0.9 \le S_y \le 0.9$, $0 \le S_z \le 100$. Symmetry with respect to the optical axis ensures that the angular momentum of the field around this axis is zero. Note that the dimensions are not the same in the three columns.

along X, at $z = 800\lambda$ it is elongated along Y; this is a natural consequence of diffractive propagation. The phase plots in Figure 16.2, middle column, reveal the acquired curvature of the beam, as well as a π phase difference between the adjacent quadrants of E_y and the two halves of E_z. The general structure of the intensity and phase patterns depicted here may be readily understood in terms of the symmetries of the Gaussian beam and the basic properties of electromagnetic radiation.

Shown in the right-hand column of Figure 16.2 are, from top to bottom, the x-, y-, and z-components of $\boldsymbol{S}$ encoded in gray-scale (black corresponds to a minimum, white to a maximum). The normalized ranges of values are: $-0.48 \le S_x \le 0.48$, $-0.9 \le S_y \le 0.9$, $0 \le S_z \le 100$. So, for example, in the top frame the bright regions indicate that S_x is directed along $+X$, while in the dark regions S_x points toward $-X$. Similarly, S_y in the upper half of the middle frame points along $+Y$, while it is directed along $-Y$ in the lower half. In the bottom frame where $S_z \ge 0$, the large positive values appear bright and those in the vicinity of zero appear dark. As expected, these plots of S_x, S_y, S_z represent a divergent beam.

The case of circular polarization

Let us consider once again the Gaussian beam for which the computed Poynting vector at the waist was shown in Figure 16.1(c). This time, however, we assume that the polarization state of the beam is circular rather than linear. The Cartesian components of $\boldsymbol{S}$ for this circularly polarized Gaussian beam at the waist are shown in Figure 16.3. The normalized ranges of values are: $-0.96 \le S_x \le 0.96$, $-0.64 \le S_y \le 0.64$, $0 \le S_z \le 100$. Although S_x and S_y are nonzero at the waist, they exhibit neither a convergent nor a divergent behavior. Indeed the projection of $\boldsymbol{S}$ in the XY-plane, $S_x\boldsymbol{x} + S_y\boldsymbol{y}$, shows only a counterclockwise circulation. (Reversing the sense of circular polarization of the beam would reverse the circulation of $\boldsymbol{S}$ as well.) From the standpoint of geometrical optics this behavior of the rays is totally unexpected, since the state of polarization should affect neither the magnitude nor the direction of the rays. Nonetheless, taking into account the full distribution of the fields (especially the components E_z and B_z) yields for circularly polarized light a non-zero projection of $\boldsymbol{S}$ in the XY-plane, in sharp contrast to the case of linear polarization where $S_x = S_y = 0$.

Linear and angular momenta of the field

It is well known that the momentum density of the field is directly proportional to $\boldsymbol{S}$. (Feynman *et al.*[8] give a beautiful exposition of the concept of field momentum density and its relation to the Poynting vector, $\boldsymbol{p} = \boldsymbol{S}/c^2$.) The field's angular momentum is then computed by integrating $\boldsymbol{r} \times \boldsymbol{p}$ over the volume of interest. Here $\boldsymbol{r}$ is the position vector and $\boldsymbol{p}$ is the field's momentum density at location $\boldsymbol{r}$. Since the momentum distribution of the circularly polarized Gaussian beam depicted in Figure 16.3 has a net circulation in the XY-plane, it follows that the beam carries a net angular momentum

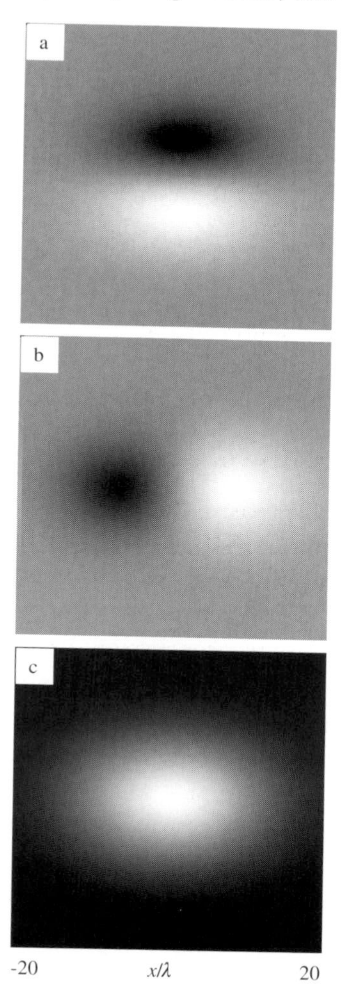

Figure 16.3 From top to bottom: plots of S_x, S_y, S_z at the waist of a circularly polarized Gaussian beam having 1/e radii $(R_x, R_y) = (15\lambda, 10\lambda)$. The normalized ranges of values are: $-0.96 \le S_x \le 0.96$, $-0.64 \le S_y \le 0.64$, and $0 \le S_z \le 100$. The counterclockwise circulation of $\boldsymbol{S}$ around the optical axis gives rise to the beam's angular momentum around this axis.

around the optical axis Z. If, for instance, such a beam is absorbed by a particle, it will exert a torque on the particle due to the transferred angular momentum.[9]

If one expands the Gaussian beam by enlarging its cross-sectional area (while maintaining its total optical power), S_x and S_y decrease faster than S_z and, in the limit of an infinitely large beam (i.e., a plane wave), S_x and S_y vanish. Does this mean that a circularly polarized plane wave does not carry angular momentum? The answer is no, because while S_x and S_y diminish with the expansion of the beam they also spread over a larger area, yielding the same final value for the integrated $\boldsymbol{r} \times \boldsymbol{p}$ over the beam's cross-section.[10] This is also in agreement with the quantum picture of light, where a circularly polarized photon of frequency ν carries energy $h\nu$ and a unit of angular momentum $h/2\pi$.

Spin versus orbital angular momentum

In chapter 15 we discussed optical vortices and showed that their Poynting-vector distribution over the beam's cross-section exhibits a circulation similar to that seen here in Figure 16.3. In the case of these vortices the state of polarization was linear and the circulation of $\boldsymbol{S}$ arose from the particular phase structure of the beam, whereas in the present case the phase is uniform but the polarization is circular. These two cases have often been compared to the orbital and spin angular momenta of bound electrons but, in reality (unless the beam is treated in the paraxial approximation), the two contributions to angular momentum are intermixed, making it difficult to distinguish one from the other. All one can say in the general case is that the field has a net angular momentum, which is obtained by integrating $\boldsymbol{r} \times \boldsymbol{p}$ over the beam's cross-section.[10,11]

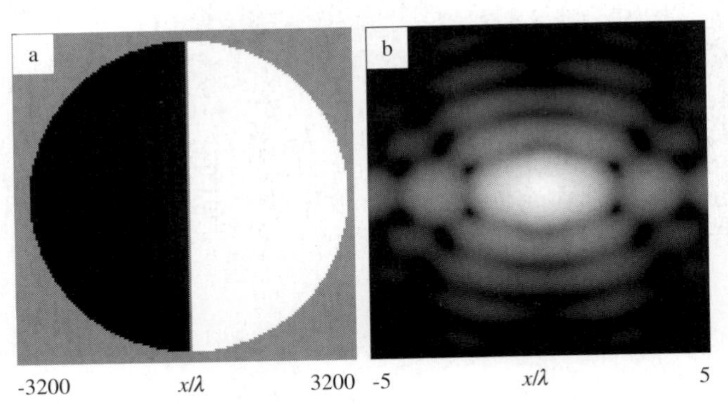

Figure 16.4 A coherent, monochromatic, and collimated beam having constant amplitude and phase but nonuniform polarization enters the pupil of an aberration-free, $0.5NA$ lens. (a) Distribution of the beam's polarization angle at the entrance pupil. Both halves are linearly polarized, the right at $+45°$ and the left at $-45°$ with respect to the X-axis. (b) Logarithmic plot of total E-field intensity at the focal plane of the lens.

Rays at the focal plane of a lens

As a final example, we show the complex pattern of ray distribution that can be obtained by focusing a relatively simple beam through a diffraction-limited microscope objective lens. Consider a beam of constant amplitude and phase, but non-uniform polarization, in which one side is linearly polarized at +45° and the other side at −45° relative to the X-axis. The distribution of polarization angle over the beam's cross-section is shown in Figure 16.4(a). Let this beam be brought to focus by an aberration-free $0.5NA$ lens. The distribution of total E-field intensity (i.e., $|E_x|^2 + |E_y|^2 + |E_z|^2$) at the focal plane is shown in Figure 16.4(b). Note the elongation of the focused spot

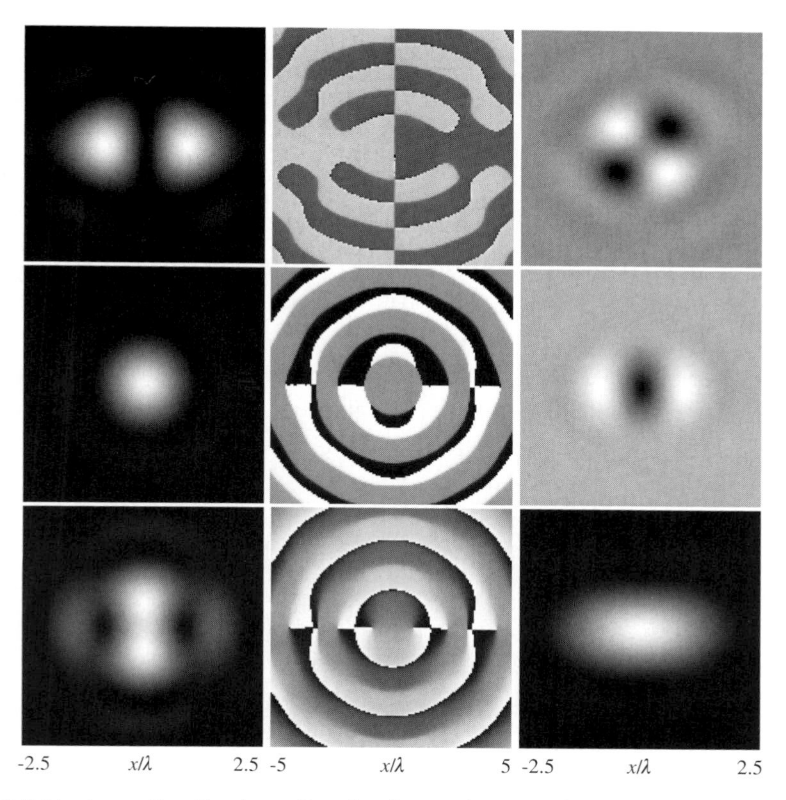

Figure 16.5 Various distributions for the focused spot of Figure 16.4(b). The left-hand column shows, from top to bottom, plots of intensity for the x-, y-, and z-components of polarization. The peak intensities are in the ratios $I_x : I_y : I_z = 0.49 : 1.0 : 0.06$. The corresponding phase plots appear in the middle column, where the gray-scale covers the range −180° (black) to +180° (white). The right-hand column shows plots of S_x, S_y, S_z in gray-scale (black, minimum; white, maximum). The normalized ranges of values are: $-9.5 \le S_x \le 9.5$, $-22.6 \le S_y \le 12.9$, $0 \le S_z \le 100$. Note that the dimensions are not the same in the three columns.

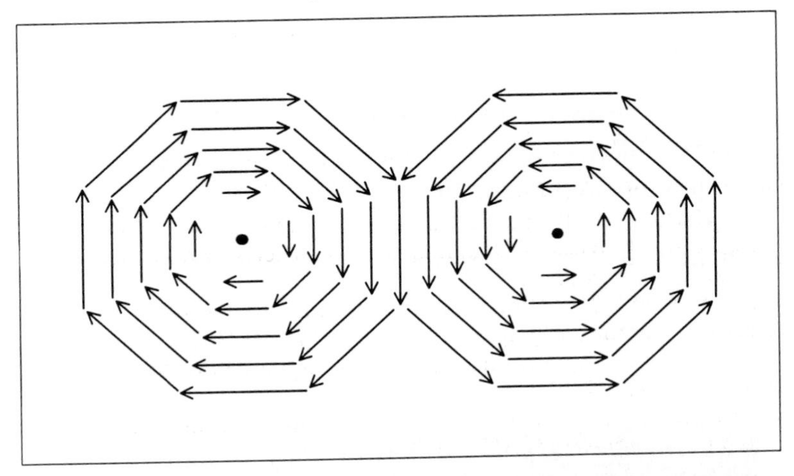

Figure 16.6 A schematic diagram showing the vortex structure of the Poynting vector for the focused spot depicted in Figures 16.4(b) and 16.5. The arrows represent the projection of $\boldsymbol{S}$ in the focal plane, namely, $S_x\boldsymbol{x} + S_y\boldsymbol{y}$.

along the X-axis, which is a consequence of the particular polarization pattern of the incident beam. Figure 16.5, left column, shows the computed intensity distributions for the x-, y- and z- components of polarization at the focal-plane. The corresponding phase patterns are shown in the middle column. Of particular interest here are the focal-plane distributions of S_x, S_y, S_z, shown in the right-hand column. There are two equal but opposite vortices in this picture, which may be discerned by considering the combined effects of S_x and S_y. A schematic diagram of the projection of $\boldsymbol{S}$ in the focal plane, namely, $S_x\boldsymbol{x} + S_y\boldsymbol{y}$, is given in Figure 16.6. These, as well as more complex momentum distributions, can now be routinely created in the laboratory and used to trap and manipulate small objects within the confines of the focal region of a microscope.

References for chapter 16

1 M. V. Klein, *Optics*, Wiley, New York (1970).
2 M. Born and E. Wolf, *Principles of Optics*, sixth edition, Pergamon Press, Oxford, 1980.
3 A. Ashkin, J. M. Dziedzic, J. E. Bjorkholm, and S. Chu, Observation of a single-beam gradient force optical trap for dielectric particles, *Opt. Lett.* **11**, 288–290 (1986).
4 K. T. Gahagan and G. A. Swartzlander, Optical vortex trapping of particles, *Opt. Lett.* **21**, 827–829 (1996).

5 H. He *et al.*, Direct observation of transfer of angular momentum to absorptive particles from a laser beam with a phase singularity, *Phys. Rev. Lett.* **75**, 826–829 (1995).
6 M. W. Berns, Laser scissors and tweezers, *Scientific American* **278**, 62–67 (April 1998).
7 E. A. Cornell and C. E. Wieman, The Bose–Einstein condensate, *Scientific American* **278**, 40–45 (March 1998).
8 R. P. Feynman, R. B. Leighton, and M. Sands, *The Feynman Lectures on Physics*, Addison-Wesley, Reading, Massachusetts (1964). See Vol. I, section 34–9, and Vol. II, chapter 27.
9 M. Kristensen and J. P. Woerdman, Is photon angular momentum conserved in a dielectric medium?, *Phys. Rev. Lett.* **72**, 2171–2174 (1994).
10 H. A. Haus and J. L. Pan, Photon spin and the paraxial wave equation, *Am. J. Phys.* **61**, 818–821 (1993).
11 S. M. Barnett and L. Allen, Orbital angular momentum and nonparaxial light beams, *Opt. Commun.* **110**, 670–678 (1994).

17

Diffraction gratings†

John William Strutt, Lord Rayleigh (1842–1919), graduated from Trinity College Cambridge in 1864. From 1879 to 1884 he was the Cavendish professor of experimental physics at Cambridge, succeeding James Clerk Maxwell. His theory of scattering (1871) provided the first correct explanation of the blue color of the sky. Rayleigh's discovery of the inert gas argon (1895) earned him the 1904 Nobel Prize for Physics. The Rayleigh–Sommerfeld theory of diffraction is one of the pillars of the classical theory. (Photo: courtesy of AIP Emilio Segré Visual Archives, Physics Today Collection.)

Diffraction gratings have been used in spectroscopy and other studies of electromagnetic phenomena for nearly two centuries.[1–4] Josef Fraunhofer (1787–1826), the discoverer of the dark lines in the solar spectrum, built the first gratings in 1819 by winding fine wires around two parallel screws.[5] Henry Rowland made significant contributions to the fabrication of precise, large-area, high-frequency ruled gratings in the 1880s.[6] Robert Wood, who succeeded Rowland in the chair of experimental physics at Johns Hopkins

†This chapter's coauthors are Lifeng Li and Wei-Hung Yeh.

University in 1901, used these ruled gratings extensively in his researches and discovered, among other things, the "anomalous" behavior of metallic gratings, which he first published in 1902.[7] John William Strutt (Lord Rayleigh) developed a theoretical model of these gratings around 1907 and was successful in explaining certain features of Wood's anomalies.[8] However, it is only during the past thirty years or so that a thorough understanding of nearly all aspects of the behavior of diffraction gratings has been achieved through the consistent application of Maxwell's equations with the help of advanced analytical and numerical techniques.[2,9,10]

Modern gratings having a few thousand lines per millimeter with near-perfect periodicity are fabricated over fairly large areas (grating diameters of around one meter or so are possible). The groove shapes can be controlled to be sinusoidal, rectangular, triangular, or trapezoidal, and one can obtain shallow or deep grooves (relative to the groove width) by current manufacturing techniques. These gratings can be made on various metal, plastic, and glass substrates and, when necessary, they can be coated with thin-film metal and/or dielectric stacks. The primary applications of diffraction gratings are still in spectroscopy, where they are used for analyzing the frequency content of electromagnetic radiation (visible light, ultraviolet, X-rays, infrared, microwave), but they are also used as wavelength selectors in tunable lasers, beam-sampling mirrors in high-power lasers, band-pass filters, pulse compressors, and polarization-sensitive optics, among other applications.

The goal of the present chapter is to describe some of the basic properties of gratings and to point out through several examples the complex behavior of these devices. These examples are by no means comprehensive, but they should make it amply clear that there is no simple way to predict a grating's diffraction efficiency. Although the number and the propagation direction of diffracted orders can be readily obtained from simple principles, the computation of diffraction efficiencies requires the complete solution of Maxwell's equations in conjunction with the appropriate boundary conditions. The results of these calculations are often non-intuitive and depend strongly on a number of factors such as the period of the grating, the geometry of the grooves, the (complex) refractive index of the material(s) comprising the grating, and the wavelength as well as the polarization state and the propagation direction of the incident beam of light. Fortunately, powerful computer programs now exist that take all the relevant factors into account and provide a reliable solution to the electromagnetic equations that govern the behavior of diffraction gratings.

Grating theories

The simplest theory of gratings treats them as corrugated structures that modulate the amplitude and/or phase of the incident beam in proportion to the local reflectivity and the height or depth of the surface relief features. The modulated reflected (or transmitted) wavefront is then decomposed into its Fourier spectrum to yield the various diffracted orders. Known as the scalar theory of gratings, this elementary treatment yields the correct number and direction of propagation for the diffracted orders, but it does not provide an accurate estimate of the amplitude, phase, and polarization state of each order. Rayleigh made a substantial contribution to the understanding of gratings by representing the diffracted field as the superposition of a number of homogeneous (i.e., propagating) and inhomogeneous (i.e., evanescent) plane waves.[8] He then determined the complex amplitudes of the various plane waves by imposing the electromagnetic boundary conditions at the grating surface.

Although Rayleigh's method was far superior to the scalar theory – it could account for some of the observed anomalies and, in fact, provided exact solutions to the electromagnetic field equations in certain cases of practical interest – it failed to provide a comprehensive solution that would be applicable under general conditions. A satisfactory analysis of the diffraction from gratings requires a numerically stable solution to Maxwell's equations constrained by the relevant boundary conditions. Several such methods have been discovered and elaborated over the past 30 years by a number of researchers from around the world.[2,9–11] The results presented in this chapter are based on the differential method of Chandezon, which uses the so-called coordinate transformation technique.[11]

Diffraction orders

Figure 17.1 shows the cross-section of a metallized grating with a trapezoidal groove geometry. The grating period is denoted by p, the groove depth by d, and the duty cycle, which is the ratio of the land width to the grating period, by c. In this symmetric grating both side walls make the same angle α with the horizontal plane. The metal layer, specified by its complex refractive index (n, k), is assumed to be thick enough to render the grating opaque.

Referring to Figure 17.2, the plane of the grating is XY, and its surface normal is the Z-axis. The plane of incidence is XZ, θ being the angle of incidence. When the incident E-field is in the plane of incidence, the beam is p-polarized, and when the E-field is along the Y-axis it is s-polarized. In an

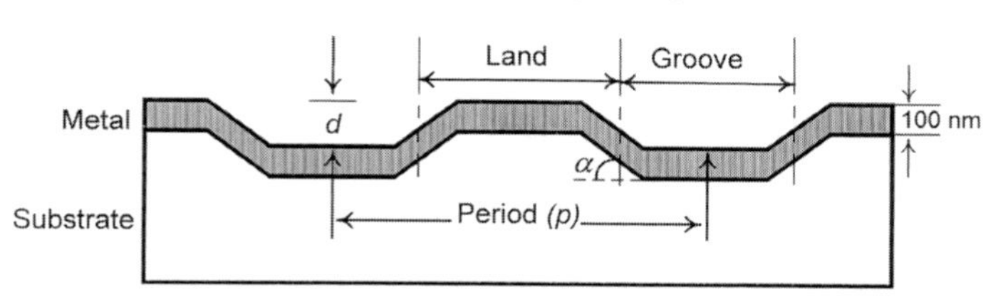

Figure 17.1 Cross-section of a metallized grating. Throughout this chapter the side-wall angle $\alpha = 60^\circ$ and the duty cycle c, which is the ratio of the land width to the grating period, is 60%. At $\lambda_0 = 0.633\,\mu\text{m}$ the substrate's refractive index $n = 1.5$ and the metal layer's complex index is given by $(n, k) = (2, 7)$.

alternative nomenclature, in Figure 17.2(a), the polarization is transverse electric (TE) when the incident E-field is parallel to the grooves and transverse magnetic (TM) when it is perpendicular to the grooves. Although the grating may be mounted with its grooves in an arbitrary direction within the XY-plane, we shall consider only two situations. In the first case, depicted in Figure 17.2(a) and referred to as "classical mount", the grooves are perpendicular to the plane of incidence. In this case all diffracted orders remain in the XZ-plane, their propagation vectors k given by

$$\boldsymbol{k}^{(m)} = (2\pi/\lambda_0)(\sigma_x^{(m)} + \sigma_z^{(m)}\boldsymbol{z}) = (2\pi/\lambda_0)\{[\sin\theta + (m\lambda_0/p)]\boldsymbol{x} + \sigma_z^{(m)}\boldsymbol{z}\}. \quad (17.1)$$

Here λ_0 is the vacuum wavelength of the light, the integer m specifies the diffraction order, the unit vector $\boldsymbol{\sigma} = (\sigma_x, \sigma_y, \sigma_z)$ is along the propagation direction, and the medium of incidence is implicitly assumed to be air. With $\sigma_y = 0$, it is necessary that $\sigma_x^2 + \sigma_z^2 = 1$, from which σ_z can be determined once σ_x is known. To keep σ_z real, $\sigma_x^{(m)} = \sin\theta + m\lambda_0/p$ must be in the range $(-1, +1)$, a constraint that determines the number of propagating orders.

In the second case, depicted in Figure 17.2(b) and referred to as "conical mount", the grooves are parallel to the plane of incidence. Here all diffracted orders (other than the zeroth) are outside the XZ-plane and their propagation vectors are given by

$$\boldsymbol{k}^{(m)} = (2\pi/\lambda_0)(\sigma_x^{(m)}\boldsymbol{x} + \sigma_y^{(m)}\boldsymbol{y} + \sigma_z^{(m)}\boldsymbol{z}) = (2\pi/\lambda_0)[(\sin\theta)\boldsymbol{x} + (m\lambda_0/p)\boldsymbol{y} + \sigma_z^{(m)}\boldsymbol{z}]. \quad (17.2)$$

Again, the integer m is the diffraction order, the implicitly assumed medium of incidence is air, and the constraint $\sigma_x^2 + \sigma_y^2 + \sigma_z^2 = 1$ specifies σ_z once σ_x and σ_y are identified. The inequality $\sigma_x^2 + \sigma_y^2 = \sin^2\theta + (m\lambda_0/p)^2 \leq 1$ determines the number of propagating orders. This mounting is called conical because the various diffracted orders reside on the surface of a cone.

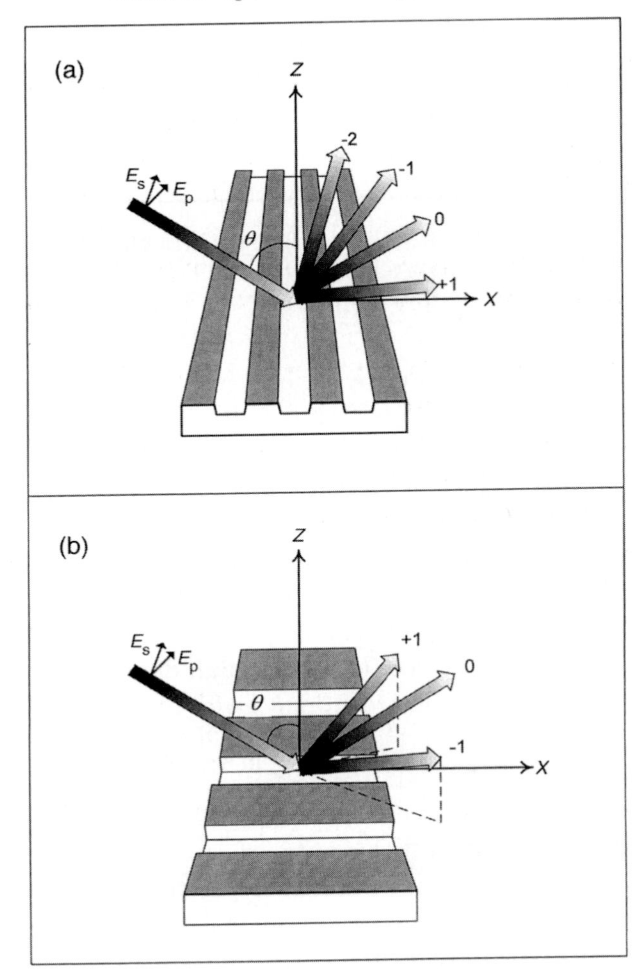

Figure 17.2 A monochromatic beam of light breaks up into multiple diffraction orders upon reflection from a grating. The incidence angle θ is measured from the Z-axis. When the incident E-field is in the XZ-plane of incidence, the beam is p-polarized, and when $\boldsymbol{E}$ is perpendicular to XZ, the beam is s-polarized. In (a) the plane of incidence is perpendicular to the direction Y of the grooves. In this so-called classical mount all diffracted orders remain within the XZ-plane. In (b), where the grooves are parallel to the plane of incidence (conical mount), diffracted orders appear on both sides of the XZ-plane.

Technically speaking, the mount is conical whenever the grooves deviate from the normal to the plane of incidence.[2] In this chapter, however, whenever the mount is said to be conical, the grooves will be strictly parallel to the plane of incidence.

Location of diffracted beams

A simple experimental setup for observing the beams diffracted from a grating appears in Figure 17.3. The coherent beam of a red HeNe laser ($\lambda_0 = 0.633\,\mu\text{m}$) is focused at oblique incidence θ onto the grating through a long-focal-length lens ($NA = 0.065$). The diffraction-limited spot diameter is $1.22\lambda_0/NA \approx 12\,\mu\text{m}$, which, if the grating period p is sufficiently small, will cover several land-groove pairs. The various diffracted beams are then collected and collimated by a microscope objective lens ($NA = 0.8$, $f = 2000\lambda_0$). In the following examples the system is arranged in such a way that the zeroth-order beam always appears at the center of the collimating lens. This lens being aplanatic, if we denote the angle between a diffracted beam and the zeroth order by χ, then the diffracted beam's distance from the center of the exit pupil will be $f \sin \chi$ rather than $f \tan \chi$.

Figure 17.4 shows computed patterns of intensity distribution at the exit pupil of the collimating lens for the grating of Figure 17.1 in the case of classical mount ($\lambda_0 = 0.633\,\mu\text{m}$, $p = 4\lambda_0$, $d = \lambda_0/8$; $\theta = 0$ in Figures 17.4(a), (b), $\theta = 40^\circ$ in Figures 17.4(c), (d)).[12] The incident beam is p-polarized or TM (i.e., E-field parallel to the XZ-plane), but the beams appearing in the exit pupil have both the p- and s- components of polarization. In Figure 17.4 the intensity patterns on the left represent the component of polarization that stays within the XZ-plane, while those on the right correspond to the component along Y. In all cases $E_\perp$ is much weaker than $E_{||}$, the ratio of the peak intensities $|E_\perp|^2 : |E_{||}|^2$ being 0.65×10^{-5} in the case of normal incidence and 0.009 in the case $\theta = 40^\circ$.

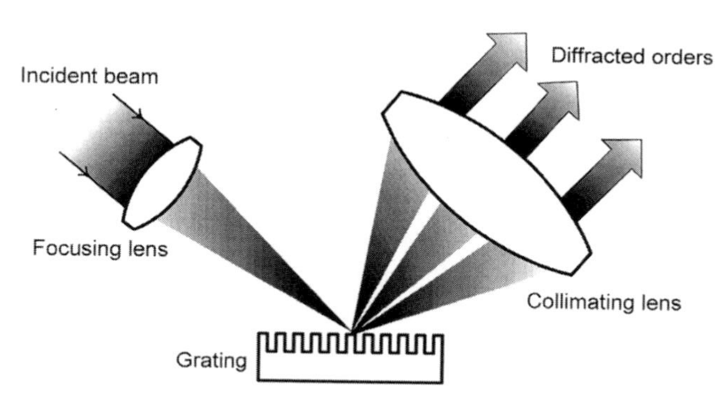

Figure 17.3 A monochromatic beam of light is focused by a low-NA lens onto a grating. Compared with the grating period, the focused spot is large, covering several land-groove pairs at the grating's surface. The diffracted orders, collected and collimated by a high-NA aplanatic lens, may be observed at the exit pupil.

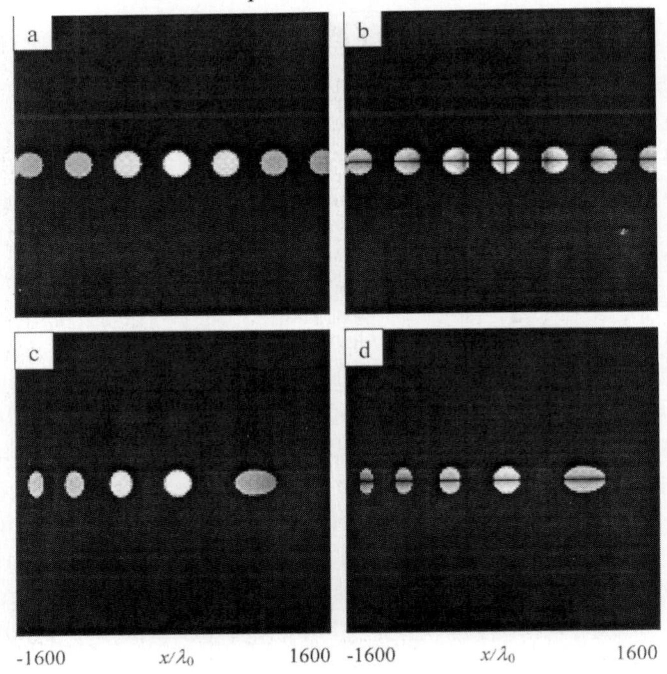

Figure 17.4 Computed plots of intensity distribution at the exit pupil of the collimating lens in Figure 17.3, when the beam is diffracted from the grating of Figure 17.1 ($\lambda_0 = 0.633\,\mu\text{m}, p = 4\lambda_0,\ d = \lambda_0/8$). The grooves are perpendicular to the plane of incidence, as in Figure 17.2(a), and the incident beam is p-polarized. The frames on the left correspond to the component of polarization parallel to the XZ-plane ($E_{||}$), while those on the right correspond to the component along the Y-axis ($E_{\perp}$). In (a) and (b) the incidence is normal, whereas in (c) and (d) $\theta = 40^{\circ}$. The ratio of the peak intensity in (b) to that in (a) is 0.65×10^{-5}. Similarly, the peak intensity ratio of (d) to (c) is 0.009. These results are based on full vector-diffraction calculations.

Figure 17.5 is similar to Figure 17.4, except that the grating is rotated by 90° in the XY-plane to bring the grooves parallel to the plane of incidence (i.e., conical mount). In Figures 17.5(a), (b) the incidence is normal, whereas in (c), (d) it is oblique at $\theta = 30^{\circ}$. In both cases the incident beam is p-polarized, but the diffracted beams contain a certain amount of s-polarization as well.[12] At the exit pupil of the lens, the ratio of the peak intensities perpendicular and parallel to the XZ-plane is fairly small, $|E_{\perp}|^2 : |E_{||}|^2$ being 0.97×10^{-4} at normal incidence and 0.025 at $\theta = 30^{\circ}$.

In both the above cases if the scalar theory of diffraction is used (instead of the full vector theory), the picture that emerges will show the diffracted orders in their correct locations but the amplitude, phase, and polarization state of the various orders will be substantially incorrect.

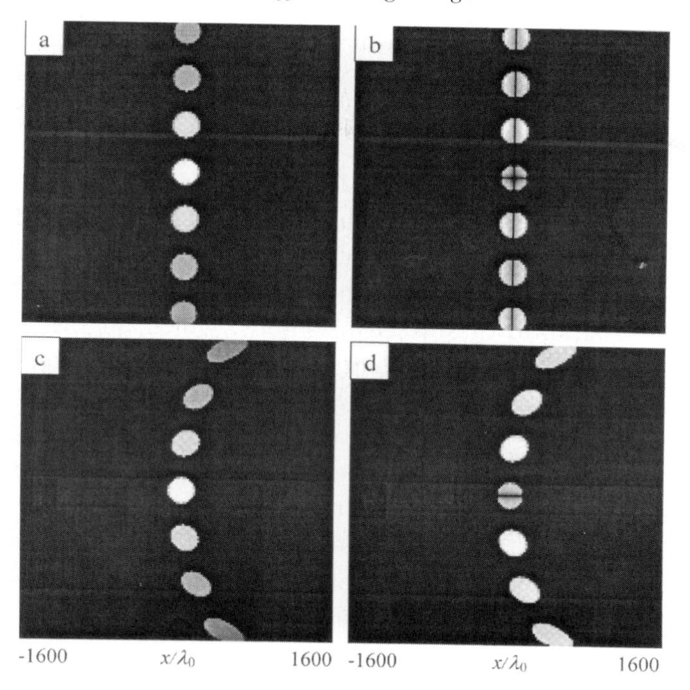

Figure 17.5 Computed plots of intensity distribution at the exit pupil of the collimating lens of Figure 17.3, when the beam is diffracted from the grating of Figure 17.1 ($\lambda_0 = 0.633\,\mu\text{m}, p = 4\lambda_0,\ d = \lambda_0/8$). The grooves are parallel to the plane of incidence, as in Figure 17.2(b), and the incident beam is p-polarized. The frames on the left correspond to the component of polarization parallel to the XZ-plane ($E_{\|}$), while those on the right correspond to the component along the Y-axis ($E_{\perp}$). In (a) and (b) the incidence is normal, whereas in (c) and (d) $\theta = 30^\circ$. The ratio of the peak intensity in (b) to that in (a) is 0.97×10^{-4}. Similarly, the peak intensity ratio of (d) to (c) is 0.025. These results are based on full vector-diffraction calculations.

Diffraction efficiency

We denote by E the amplitude of the incident beam at angle θ and by $E^{(m)}$ the amplitude of the mth-order reflected (or transmitted) beam emerging at $\theta^{(m)}$. It is further assumed that the incidence medium is air and, in the case of a transmission grating, that the transparent medium into which the diffracted orders emerge has refractive index n_0. For the mth-order reflected (transmitted) beam the diffraction efficiency $\rho^{(m)}$ ($\tau^{(m)}$) can be written as

$$\rho^{(m)} = |E^{(m)}|^2 \cos\theta^{(m)}/(|E|^2 \cos\theta), \tag{17.3a}$$

$$\tau^{(m)} = n_0|E^{(m)}|^2 \cos\theta^{(m)}/(|E|^2 \cos\theta). \tag{17.3b}$$

Here the squared amplitude is the beam's intensity, and the cosine factor keeps track of the change in the beam's cross-sectional area upon diffraction.

Figure 17.6 shows computed plots of diffraction efficiency versus θ for the zeroth- and $\pm$ first-order beams for the grating of Figure 17.1 ($\lambda_0 = 0.633\,\mu\text{m}$, $p = 3\lambda_0$, $d = \lambda_0/8$).[12] In each frame there are four curves, representing the diffraction efficiency of the corresponding order when the incident beam is either p- or s-polarized and when the mount is either classical (ρ_p, ρ_s) or conical (ρ'_p, ρ'_s). The sharp peaks and valleys appearing in these plots are caused by the excitation of surface plasmons, which, in the case of metal gratings, exist only when the incident beam has an E-field component perpendicular to the grooves (see chapter 9, "What in the world are surface plasmons?"). The arrows at the bottom of each figure point to the angles of incidence associated with the Rayleigh anomalies; these are points at which a particular diffraction order appears or disappears. In Figure 17.6(b), for example, ρ_p and ρ_s terminate at $\theta = 41.81°$, which is where the + first-order beam becomes parallel to the surface and subsequently vanishes. In the case of ρ'_p and ρ'_s (conical mount) the cutoff of both the $\pm$ first orders occurs at $\theta = 70.53°$. When the metallic grating has a large conductivity, the surface plasmon features and Rayleigh anomalies are usually located pairwise, close to each other.

Dependence of diffraction efficiency on the grating period

The efficiency curves become somewhat erratic as the period p of the grating decreases, but they approach a limiting behavior with increasing p. Figure 17.7 shows computed plots of the zeroth-order efficiency versus θ for the grating of Figure 17.1 with (a) $p = \lambda_0$ and (b) $p = 5\lambda_0$ (in both cases $\lambda_0 = 0.633\,\mu\text{m}$, $d = \lambda_0/8$).[12] These plots should be compared with those of Figure 17.6(a) for which $p = 3\lambda_0$. Notice the substantial departure of the curves in Figure 17.7(a) from those in Figure 17.6(a). However, there are similarities between Figures 17.6(a) and 17.7(b), stemming from the fact that in both cases the grating period is fairly large and the grooves are rather shallow.

Figure 17.8 shows plots of $\rho^{(0)}$ versus p at a fixed angle of incidence ($\theta = 30°$, $\lambda_0 = 0.633\,\mu\text{m}$, $d = \lambda_0/8$). The solid (broken) arrows at the bottom (top) of the figure indicate the locations of Rayleigh anomalies for the classical (conical) mount. It appears that as the period increases the various zeroth-order efficiencies approach a limiting value in the vicinity of 55%. The remainder of the incident energy in this case is partly absorbed by the metal layer and partly distributed among other diffracted orders. As $p \to \infty$

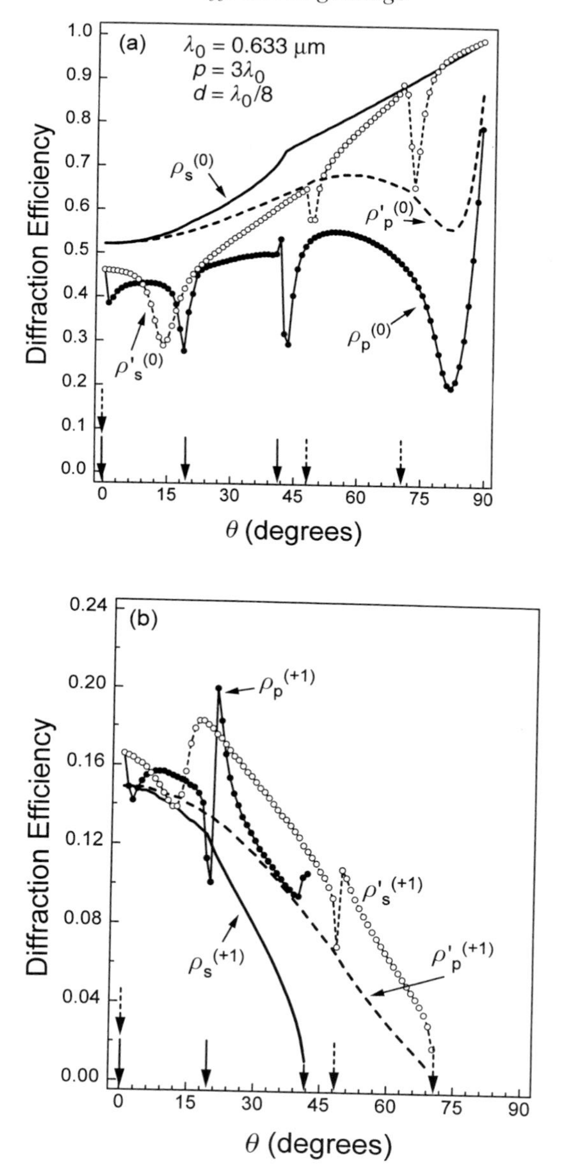

Figure 17.6 Computed plots of diffraction efficiency versus the angle of incidence θ for ρ_p, ρ_s (classical mount) and ρ_p', ρ_s' (conical mount, i.e., grooves parallel to the incidence plane). The solid (broken) arrows indicate the locations of Rayleigh anomalies for the classical (conical) mount. (a) zeroth-order, (b) +first-order, and (c) −first-order diffracted beams upon reflection from the grating of Figure 17.1 ($\lambda_0 = 0.633\,\mu\text{m}$, $p = 3\lambda_0$, $d = \lambda_0/8$).

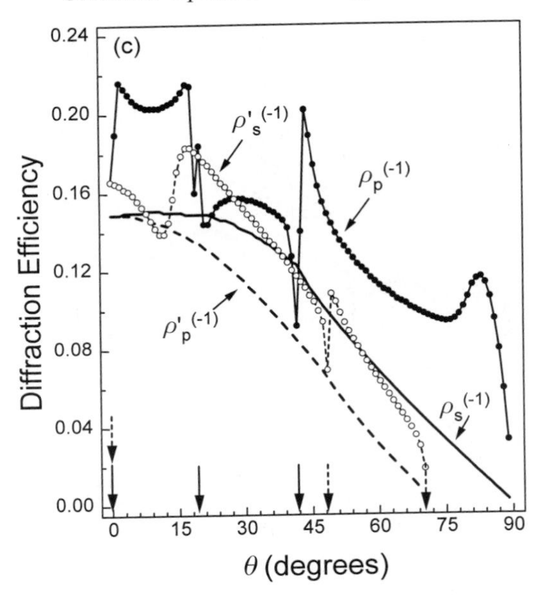

Figure 17.6 (*continued*)

the orders that carry the bulk of the reflected energy converge towards the zeroth order line. At the same time, the overall reflectance, which becomes more and more concentrated around the direction of specular reflection, approaches the specular reflectivity of the flat metal layer at 30° incidence (i.e., 84% for p-light, 88% for s-light). In the opposite extreme, $p \to 0$, the reflectivity curves once again show a limiting behavior. Although there are no other diffracted orders in this case, the limiting value of $\rho^{(0)}$ is not necessarily the same as the specular reflectance of the flat metal layer but should be calculated from an "effective medium" theory.

Effect of the groove depth

Another factor that complicates the behavior of a grating is the dependence of its efficiency on the groove depth d. Figure 17.9 shows plots of $\rho^{(0)}$ versus θ for reflection from the grating of Figure 17.1 when the groove depth $d = 0.2\,\mu\text{m}$ ($\lambda_0 = 0.633\,\mu\text{m}$, $p = 3\lambda_0$). These curves are quite different from those of Figure 17.6(a), which correspond to a similar grating with shallower grooves. The lower values of $\rho^{(0)}$ in the case of a deep-groove grating indicate that more light is being channeled into other diffracted orders.

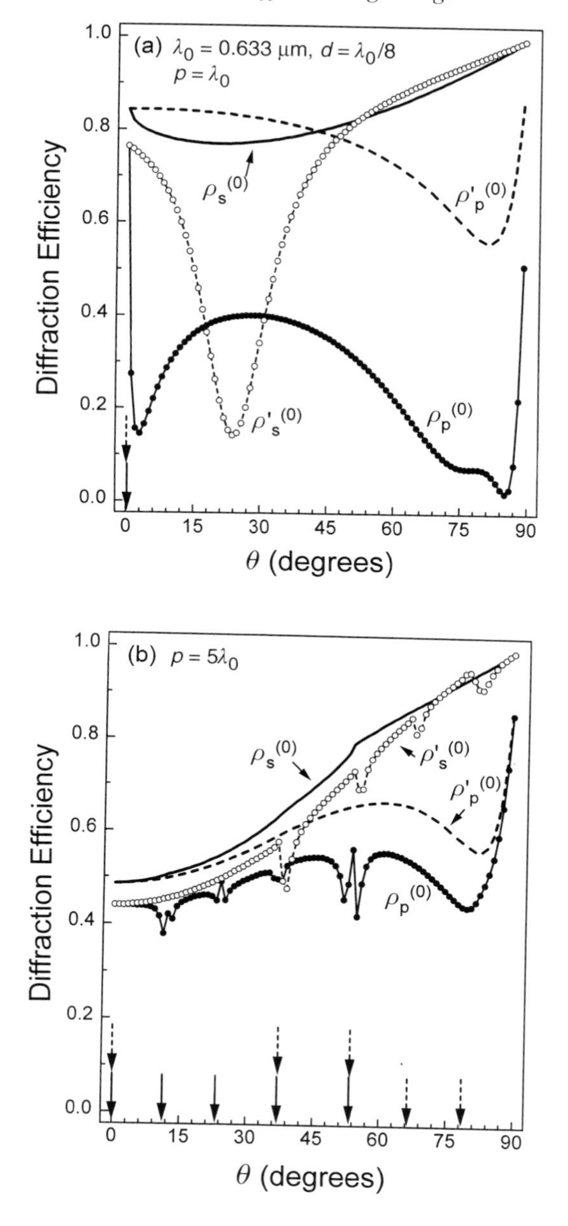

Figure 17.7 Computed plots of diffraction efficiency versus θ for the zeroth-order diffracted beam upon reflection from the grating of Figure 17.1 ($\lambda_0 = 0.633\,\mu$m, $d = \lambda_0/8$). In (a) the grating period $p = \lambda_0$ while in (b) $p = 5\lambda_0$. The solid (broken) arrows indicate the locations of Rayleigh anomalies for the classical (conical) mount.

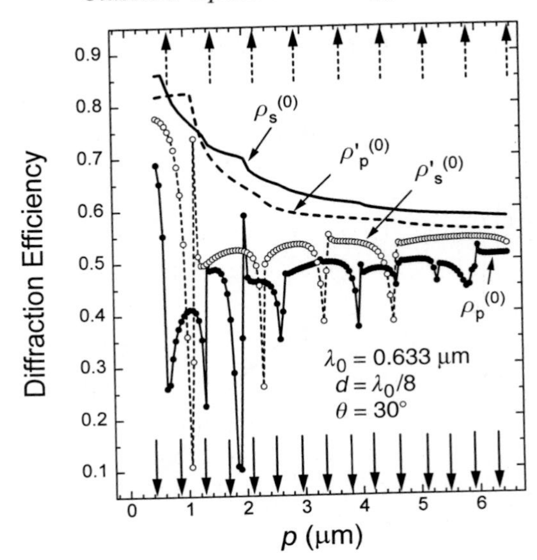

Figure 17.8 Computed plots of the zeroth-order efficiency versus the grating period p for the grating of Figure 17.1 ($\lambda_0 = 0.633\,\mu\text{m}$, $d = \lambda_0/8$, $\theta = 30°$). The solid (broken) arrows indicate the locations of Rayleigh anomalies for the classical (conical) mount.

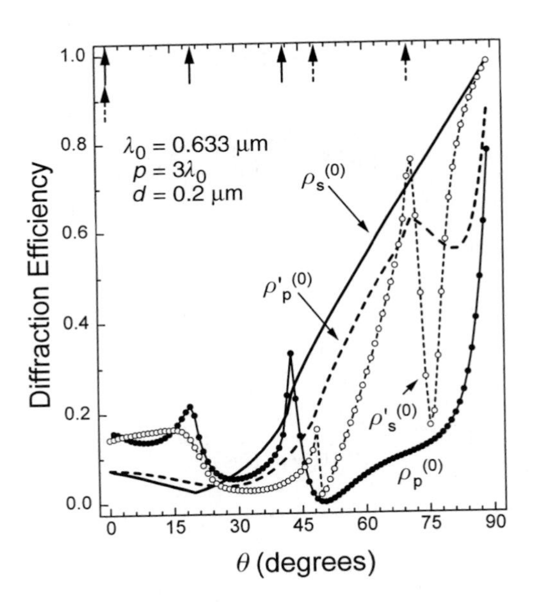

Figure 17.9 Computed plots of the zeroth-order diffraction efficiency versus the angle of incidence for the grating of Figure 17.1 ($\lambda_0 = 0.633\,\mu\text{m}$, $p = 3\lambda_0$, $d = 0.2\,\mu\text{m}$). The solid (broken) arrows indicate the locations of Rayleigh anomalies for the classical (conical) mount.

Reciprocity theorem

There exists a powerful and quite unexpected reciprocity relation between the beam incident on a grating and any of the resulting diffracted orders. Suppose the incident beam arrives at the grating at an angle θ and the mth diffracted order emerges at an angle $\theta^{(m)}$, having diffraction efficiency $\rho^{(m)}$ or, in the case of a transmitted order, $\tau^{(m)}$. If the direction of incidence is now changed in such a way that the incident beam is along the path of the mth-order beam (in the reverse direction, of course), there emerges a $-m$th diffracted order along the path of the original incident beam (again in the reverse direction). The reciprocity theorem states that the efficiency of this particular diffracted order will be exactly equal to $\rho^{(m)}$ (or $\tau^{(m)}$). This theorem can be rigorously proved under general conditions.[2] In Figure 17.6 the $\pm$ first-order efficiency curves in the classical mount, i.e., $\rho_s^{(\pm 1)}$ and $\rho_p^{(\pm 1)}$, show several manifestations of the reciprocity theorem. A few more consequences of reciprocity will be pointed out in the examples that follow.

Resolving power

Consider a grating of period p having a total of N grooves. The width of the mth-order diffracted beam that covers the entire grating is $Np\cos\theta^{(m)}$. If this beam is brought to diffraction-limited focus by a lens of focal length f, the focused spot diameter D will be[1]

$$D \approx \lambda_0 f/(Np\cos\theta^{(m)}). \tag{17.4}$$

Spectroscopists are interested in the focused spots formed by two nearby wavelengths, λ_0 and $\lambda_0 + \Delta\lambda$. According to Eq. (17.1) the diffraction angle $\theta^{(m)}$ in the classical mount is given by $\sin\theta^{(m)} = \sin\theta + m\lambda_0/p$, in which case for a small change of wavelength $\Delta\lambda$ we have

$$\cos\theta^{(m)}\Delta\theta^{(m)} \approx (m/p)\Delta\lambda. \tag{17.5}$$

Therefore, in the focal plane of the lens, a shift of the wavelength from λ_0 to $\lambda_0 + \Delta\lambda$ causes a shift of the focused spot by the following amount:

$$f\Delta\theta^{(m)} \approx mf\Delta\lambda/(p\cos\theta^{(m)}). \tag{17.6}$$

The two wavelengths are just resolved when the above shift equals the spot diameter D in Eq. (17.4), that is, when $f\Delta\theta^{(m)} \approx D$. This leads to the following expression for the resolving power:

$$\lambda_0/\Delta\lambda \approx mN. \tag{17.7}$$

It is thus seen that the resolving power of a grating is directly proportional to N, its total number of illuminated grooves, and to m, the order of diffraction. The resolving power is completely independent of such seemingly relevant factors as the groove period, the groove geometry, and the incidence angle.

Littrow mount and blazed gratings

To build compact spectrometers, it is desirable that one of the diffracted orders should return along (or almost along) the direction of incidence. In the so-called Littrow mount, the nth-order beam, where n is negative, returns along the direction of incidence. For instance, in the −first-order Littrow mount, we find from Eq. (17.1)

$$2 \sin \theta = \lambda_0 / p. \tag{17.8}$$

Under this condition, if $p < 1.5\lambda_0$, then the only possible diffracted orders are the zeroth and the −first. Furthermore, if the efficiency for the zeroth order can be reduced to zero, all the available power that is not absorbed by the grating will return along the −first reflected order, thus maximizing the sensitivity of the spectrometer. Gratings that direct all or most of the incident optical power into a single diffracted order are known as blazed gratings. Although in the early days ruled gratings having a triangular groove profile satisfied the blaze condition, a triangular cross-section is no longer a prerequisite to the blazing property. Gratings with triangular cross-section and a 90° apex angle are now more appropriately referred to as "echelette" gratings.

Figure 17.10 shows a metallic prism with an inclination angle α. When a plane wave is normally incident on the inclined facet of this prism, the specularly reflected light returns along the direction of incidence. Let the lengths of the equidistant lines drawn on the prism parallel to the direction of incidence be integer multiples of $m\lambda_0/2$, where m is an arbitrary (but fixed) integer. If the metal prism is cut along these lines and its segments rearranged, one obtains an echelette grating with period $p = m\lambda_0/(2 \sin \alpha)$, as shown in the lower part of the figure. With an incidence angle $\theta = \alpha$ on this grating, Littrow's condition for the negative mth diffracted order will be satisfied. In the geometric-optical approximation, this grating should be equivalent to the original prism, because the various reflected rays from its individual facets suffer phase delays in multiples of 2π only, making the grating's reflected wavefront indistinguishable from that of the prism. In reality, however, the electromagnetic field "feels" the groove structure, and the actual diffraction efficiency of the beam returning along the direction of

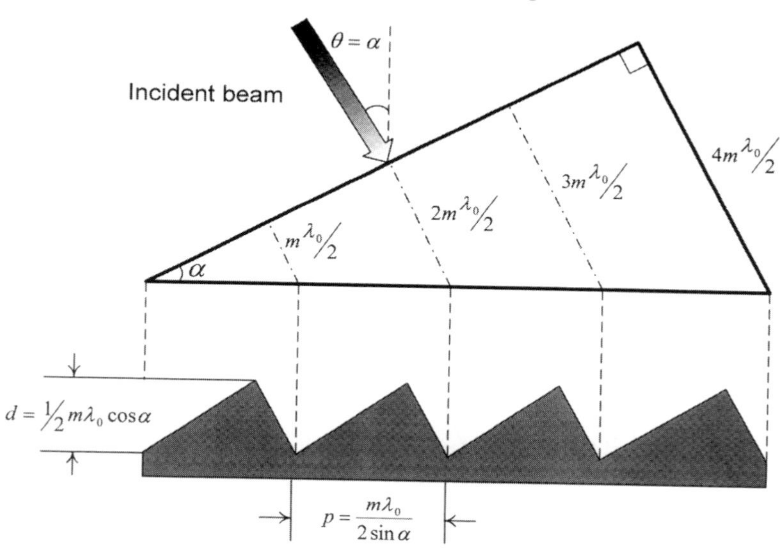

Figure 17.10 A normally incident beam of light is specularly reflected from the inclined facet of a metallic prism (inclination angle α). For a given integer m, imagine cutting the prism along the broken and dotted lines, which are parallel to the direction of incidence and have lengths that are multiples of $m\lambda_0/2$. The various sections are then rearranged to form the echelette grating shown in the lower part of the figure. If the grating is similarly illuminated at $\theta = \alpha$, the diffracted order that retraces the incidence path in the reverse direction will be quite strong, which is why this kind of grating has come to be known as a blazed grating.

incidence will not always be the same as the specular reflectance of the polished metal prism, although they are usually close.

Figure 17.11 shows computed efficiency curves in the classical mount for the echelette grating of Figure 17.10 having $\alpha = 30^\circ$, $p = 2\lambda_0$, and $(n, k) = (2, 7)$ at $\lambda_0 = 0.633\,\mu\text{m}$.[12] The horizontal axis depicts $\sin\theta$, the incidence angle θ being positive (negative) when incidence is from the side of the large (small) facet of the triangular grooves. The arrows at the top of each frame indicate the locations of Rayleigh anomalies, in the neighborhood of which resonance features and slope discontinuities are seen to occur. The zeroth-order efficiency curves for p- and s-polarized light are shown in Figure 17.11(a). Despite the asymmetrical groove geometry, the plots of $\rho_p^{(0)}$ and $\rho_s^{(0)}$ are perfectly symmetric around $\theta = 0$, which is a manifestation of the reciprocity theorem mentioned earlier. The +first-order efficiency curves in Figure 17.11(b) show the same kind of symmetry around $\theta = -14.48^\circ$ (i.e., $\sin\theta = -0.25$), which is the angle of incidence for the +first-order Littrow mount. Similarly, the −first-order curves in Figure 17.11(c) show the reciprocity theorem at work around $\theta = 14.48^\circ$, the angle of incidence for the

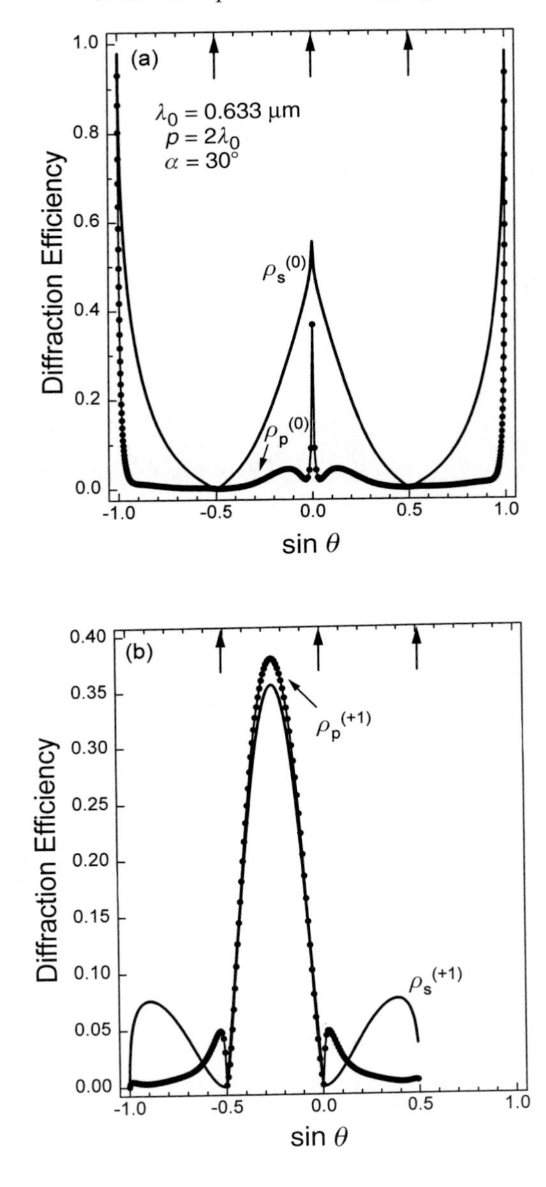

Figure 17.11 Computed plots of diffraction efficiency versus sin θ, where θ is the angle of incidence on the echelette grating of Figure 17.10 ($\lambda_0 = 0.633\,\mu$m, $\alpha = 30°$, $p = 2\lambda_0$, $(n, k) = (2, 7)$). When $\theta > 0$, incidence is from the large-facet side of the triangular grooves while when $\theta < 0$ incidence is from the small-facet side. The displayed efficiencies are for p- and s-polarized incident light in the classical mount. (a) Zeroth order, (b) +first order, (c) −first order, (d) ± second order, (e) ± third order. The arrows at the top of each frame indicate the locations of Rayleigh anomalies.

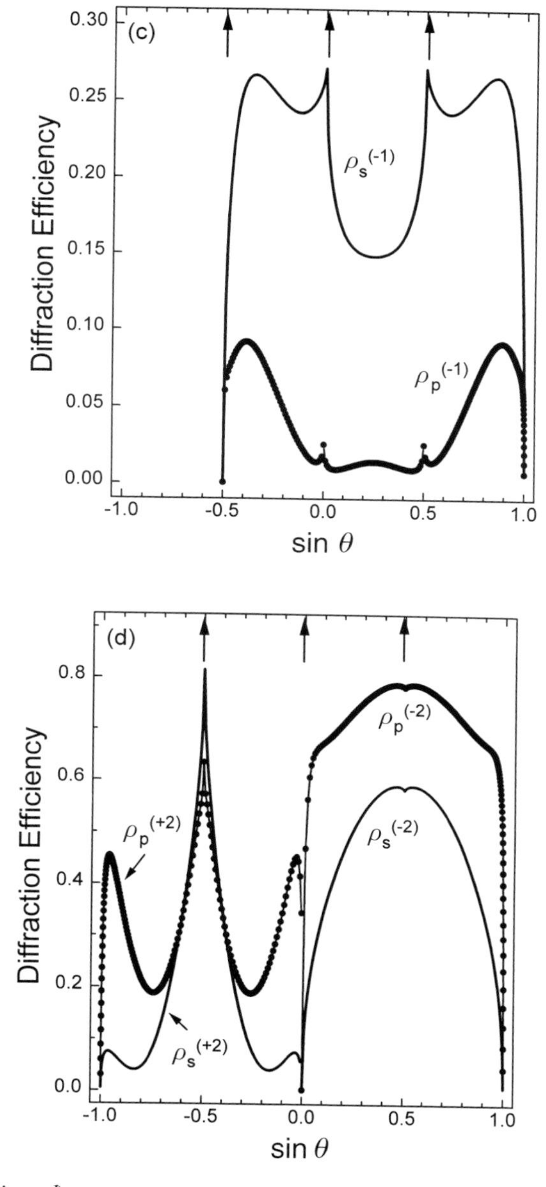

Figure 17.11 (*continued*)

−first-order Littrow mount. The Rayleigh anomalies at $\theta = \pm 30^\circ$ (i.e., $\sin\theta = \pm 0.5$) mark the disappearance of the ±first-order beams beyond these angles, as may be seen clearly in Figures 17.11(b) and 17.11(c).

The ±second-order efficiency curves are shown in Figure 17.11(d). These curves peak at, and are symmetrical around, $\theta = \pm 30^\circ$, where the Littrow

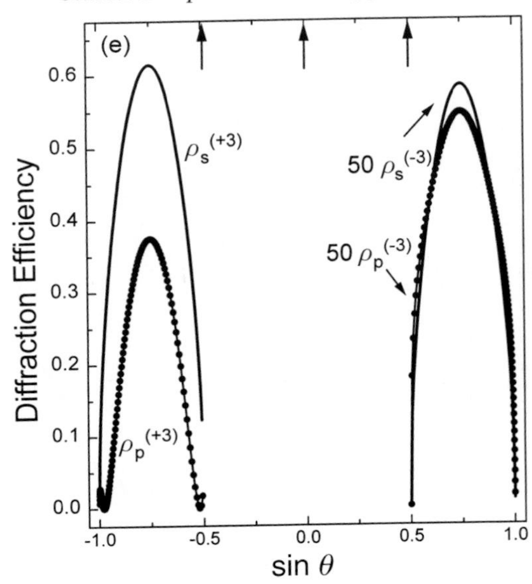

Figure 17.11 (*continued*)

condition for the ±second-order beams is satisfied. Reciprocity between the incident beam and the ±second-order reflected beams is evident in the symmetrical values of efficiency around $\theta = \pm 30°$. Note in the case of the p-polarized beam incident at $\theta = 30°$, where the −second-order efficiency reaches 80% while that of all other orders essentially vanishes, that the remaining 20% of the incident power must have been absorbed by the grating. A similar consideration applies to both $\rho_p^{(+2)}$ and $\rho_s^{(+2)}$ at $\theta = -30°$. The ±third-order beams exist only at large angles of incidence, as may be inferred from Figure 17.11(e). Again note the symmetry of these curves (due to reciprocity) around $\sin\theta = \pm 0.75$; these values of θ correspond to the Littrow mount in the ±third-order.

For the sake of completeness we present in Figure 17.12 computed efficiency curves in the case of conical mount for the same echelette grating as discussed above.[12] Here the grooves are parallel to the plane of incidence, and symmetry with respect to $\theta = 0$ obviates the need for displaying the results for negative values of θ. In this conical mount only the zeroth and ±first diffracted orders are allowed; even then, the ±first-order beams disappear beyond $\theta = 60°$. Note that, because of the asymmetrical groove shape, the +first-order efficiency curves are quite different from those of the −first-order. Also note that, beyond $\theta = 60°$, where the zeroth-order beam is the only beam reflected from the grating, the relatively small values of $\rho_p'^{(0)}$ and $\rho_s'^{(0)}$ indicate substantial absorption within the grating medium.

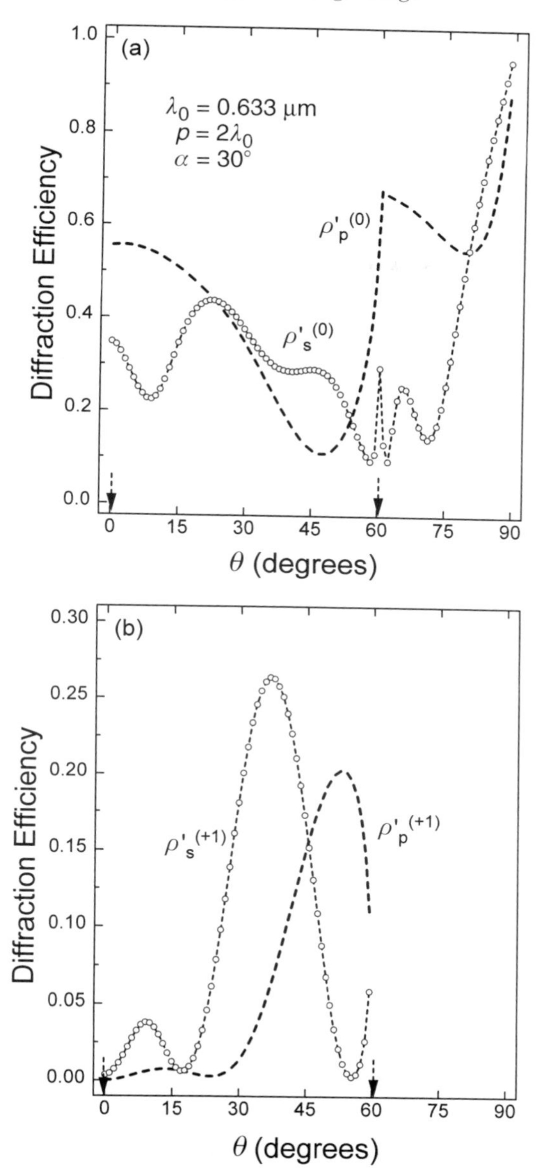

Figure 17.12 Computed plots of diffraction efficiency versus the angle of incidence on the echelette grating of Figure 17.10 ($\lambda_0 = 0.633\,\mu\text{m}$, $\alpha = 30°$, $p = 2\lambda_0$, $(n, k) = (2, 7)$). The displayed efficiencies are for p- and s-polarized incident light in the conical mount. (a) zeroth order, (b) +first order, (c) −first order. The arrows at the bottom of each frame indicate the locations of Rayleigh anomalies.

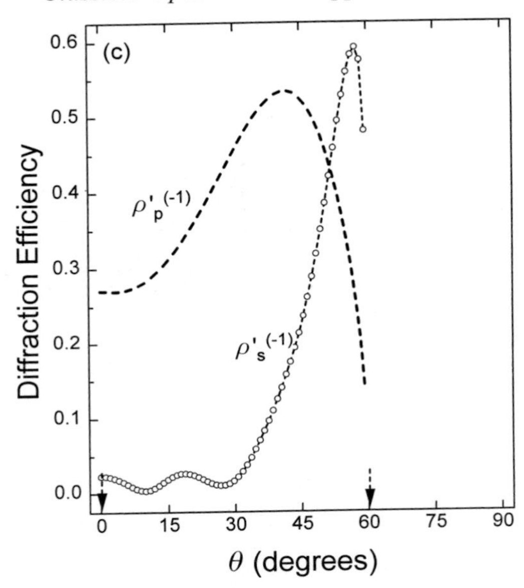

Figure 17.12 (*continued*)

Transmission grating

Consider a grooved glass plate such as that depicted in Figure 17.13(a). When a plane wave is incident at θ on this grating, the directions of the reflected orders may be found from Eqs. (17.1) and (17.2), but the transmitted orders inside the glass plate obey different equations. In the classical mount the transmitted orders emerge at angles $\theta^{(m)}$, where

$$n_0 \sin\theta^{(m)} = \sin\theta + m\lambda_0/p. \tag{17.9}$$

Here n_0 is the refractive index of the substrate. The number of diffracted orders in the substrate could, therefore, be greater than the number reflected into the air. However, when the transmitted orders attempt to exit the bottom of the substrate, those incident at an angle higher than the critical angle for total internal reflection will be fully reflected. The beams that do exit the substrate will emerge at angles greater than $\theta^{(m)}$, in accordance with Snell's law; the coefficient n_0 on the left-hand side of Eq. (17.9) is effectively canceled. Consequently, the beams emerging from the bottom of the substrate have exactly the same number and (aside from being mirror images) the same directions as those reflected from the top of the grating. Nonetheless, the transmitted diffracted orders may be observed in their native form by using a hemispherical substrate, as shown in Figure 17.13(b).

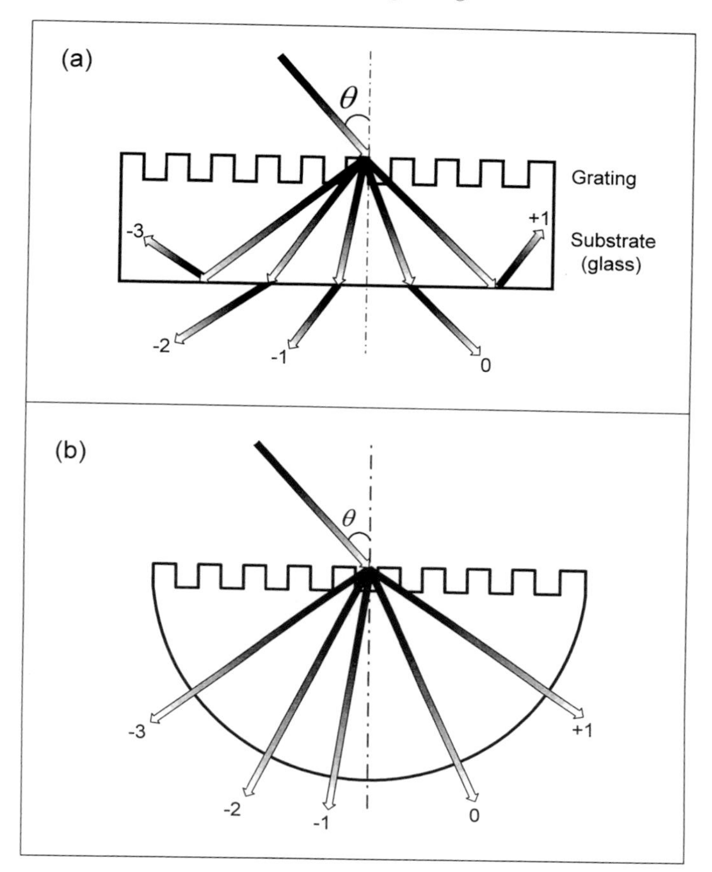

Figure 17.13 A simple transmission grating may be obtained by ruling or etching a glass substrate, or by a holographic method. The substrate's refractive index being greater than unity, the diffraction angles inside the substrate are smaller than those observed upon reflection from the same grating into the air. (a) When the substrate bottom is flat, Snell's law of refraction reorients the beams as they emerge into the air, making the diffraction angles equal to those observed in reflection. However, one or more diffracted orders may be missing, owing to total internal reflection at the substrate bottom. (b) If the grating is made on the flat surface of a glass hemisphere, the transmitted orders emerge into the air undisturbed.

In the case of conical mount similar arguments apply, so that the mth-order beam inside the substrate will have a propagation direction given by the unit vector $\sigma^{(m)}$, where

$$\sigma^{(m)} = \sigma_x^{(m)}\boldsymbol{x} + \sigma_y^{(m)}\boldsymbol{y} + \sigma_z^{(m)}\boldsymbol{z} = (1/n_0)[(\sin\theta)\boldsymbol{x} + (m\lambda_0/p)\boldsymbol{y}] + \sigma_z^{(m)}\boldsymbol{z}. \quad (17.10)$$

Again, σ_z is determined from the relation $\sigma_x^2 + \sigma_y^2 + \sigma_z^2 = 1$. As above, when this beam emerges into air from the bottom of a flat substrate, Snell's law multiplies σ_x and σ_y by the refractive index n_0, ensuring that the emergent beams (aside from being mirror images) have the same propagation directions as the corresponding beams reflected from the top of the grating.

Figure 17.14 shows the location of the transmitted diffracted orders from a glass grating.[12] The assumed grating in this case is similar to that of Figure 17.1, except that the metal layer is absent. The observation system is also similar to that in Figure 17.3, except for the position of the collimating lens, which is moved to the opposite side of the grating to collect the transmitted orders. The incident beam, arriving at $\theta = 30^\circ$ in the conical mount, is p-polarized. The pictures on the left-hand side of Figure 17.14 represent the

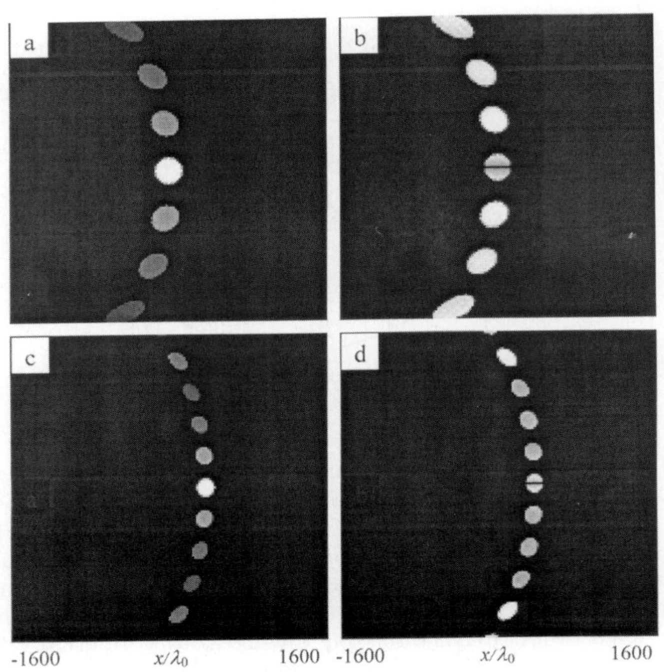

Figure 17.14 Computed plots of intensity distribution at the exit pupil of the collimating lens of Figure 17.3, when the system is rearranged to allow observation of transmitted orders from the grating of Figure 17.1, from which the metal layer has been removed ($\lambda_0 = 0.633\,\mu\text{m}$, $p = 4\lambda_0$, $d = \lambda_0/8$). In this case of conical mount at 30° incidence the grooves are parallel to the plane of incidence, as in Figure 17.2(b), and the incident beam is p-polarized. The pictures on the left correspond to the component of polarization in the XZ-plane, while those on the right represent the polarization component along the Y-axis. In (a) and (b) the substrate bottom is flat, as in Figure 17.13(a), whereas in (c) and (d) it is hemispherical, as in Figure 17.13(b). The ratio of the peak intensity in (b) to that in (a) is 0.21×10^{-4}. Similarly, the peak-intensity ratio of (d) to (c) is 0.89×10^{-4}. These results are based on full vector-diffraction calculations.

component of polarization in the XZ-plane ($E_{||}$), while those on the right correspond to polarization along the Y-axis ($E_{\perp}$). The top row shows the intensity distribution at the exit pupil of the collimating lens when the substrate bottom is flat; the bottom row corresponds to the case of a hemispherical substrate. As expected, in the latter case there are more diffracted orders, the orders are more closely spaced, and the individual beam diameters are smaller. For the flat substrate the peak-intensity ratio $|E_{\perp}|^2 : |E_{||}|^2 = 0.21 \times 10^{-4}$, while for the hemispherical substrate $|E_{\perp}|^2 : |E_{||}|^2 = 0.89 \times 10^{-4}$.

Dielectric-coated grating

Figure 17.15 is a diagram of a dielectric-coated transmission grating on a hemispherical glass substrate. In the example that follows it is assumed that $\lambda_0 = 0.633\,\mu\text{m}$, the grating period $p = \lambda_0$, the groove depth $d = \lambda_0/8$, the side-wall inclination angle $\alpha = 60°$, and the duty cycle $c = 60\%$. The coatings are conformal to the grating surface, both dielectric layers are 100 nm thick, and their refractive indices are 2.1 and 1.5, as indicated. Because there are no metallic layers in this case there will be no surface plasmon excitations, but there is the possibility of guided-mode coupling to the dielectric waveguide formed by the coating layers. The hemispherical substrate allows all trans-

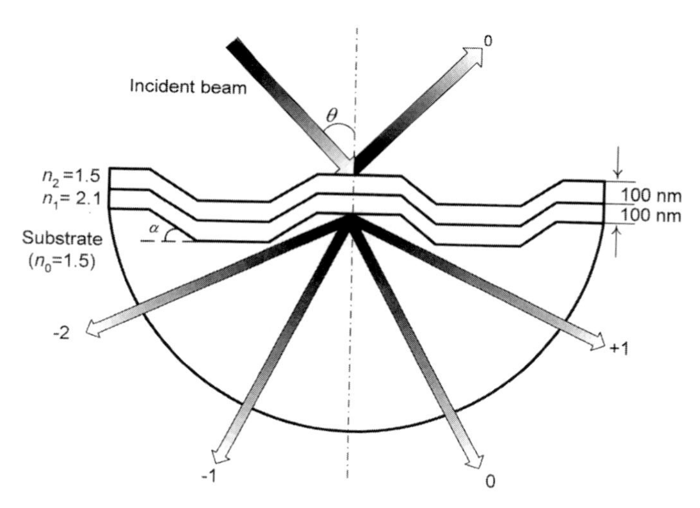

Figure 17.15 Cross-section of a dielectric-coated diffraction grating. The side-wall angle $\alpha = 60°$, and the duty cycle c, which is the ratio of the land width to the grating period, is 60%. Both coating layers are 100 nm thick and (at $\lambda_0 = 0.633\,\mu\text{m}$) their refractive indices are $n_1 = 2.1$ and $n_2 = 1.5$. For the substrate, which is also transparent, $n_0 = 1.5$.

mitted orders to exit and be measured in air. The bottom of the hemisphere is antireflection coated, to avoid losses as the beams exit the substrate.

Figure 17.16 shows computed plots of diffraction efficiency versus θ for the grating of Figure 17.15.[12] The case of conical mount does not show interesting phenomena, as evidenced by the featureless plots of ρ' and τ' for the various orders. This is not surprising, considering that no guided modes can be launched in the dielectric layers in this case. However, for the classical mount ρ_p, ρ_s, τ_p and τ_s show peaks and valleys that are indicative of resonant behavior. Figure 17.16(b) shows plots of ρ_p and ρ_s for the $-$first-order reflected beam, which carries as much as 8% of the incident beam into this particular direction at several angles of incidence. Reciprocity between the incident beam and the $-$first-order reflected beam is evident in Figure 17.16(b), in the symmetrical values of efficiency before and after $\theta = 30°$. Note that, unlike surface plasmon excitations in metals, which occur in p-polarization only, the waveguide modes of dielectric layers can be excited by both p- and s-polarized light. For the classical mount, Figure 17.16(d) shows

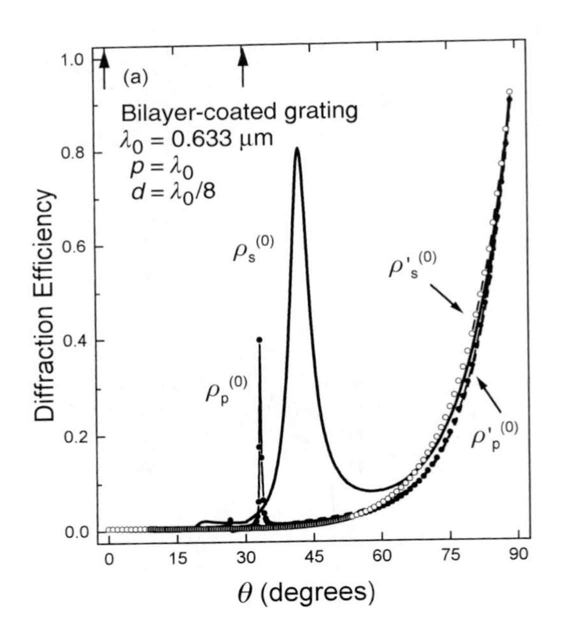

Figure 17.16 Computed diffraction efficiencies versus θ for the dielectric-coated grating of Figure 17.15 ($\lambda_0 = 0.633$ µm, $p = \lambda_0$, $d = \lambda_0/8$). *Reflected beams*: (a) zeroth order, (b) $-$first order. *Transmitted beams*: (c) zeroth order, (d) +first order, (e) $-$first order, (f) $-$second order (classical mount only). The arrows at the top or the bottom of each frame indicate the locations of Rayleigh anomalies in the classical mount.

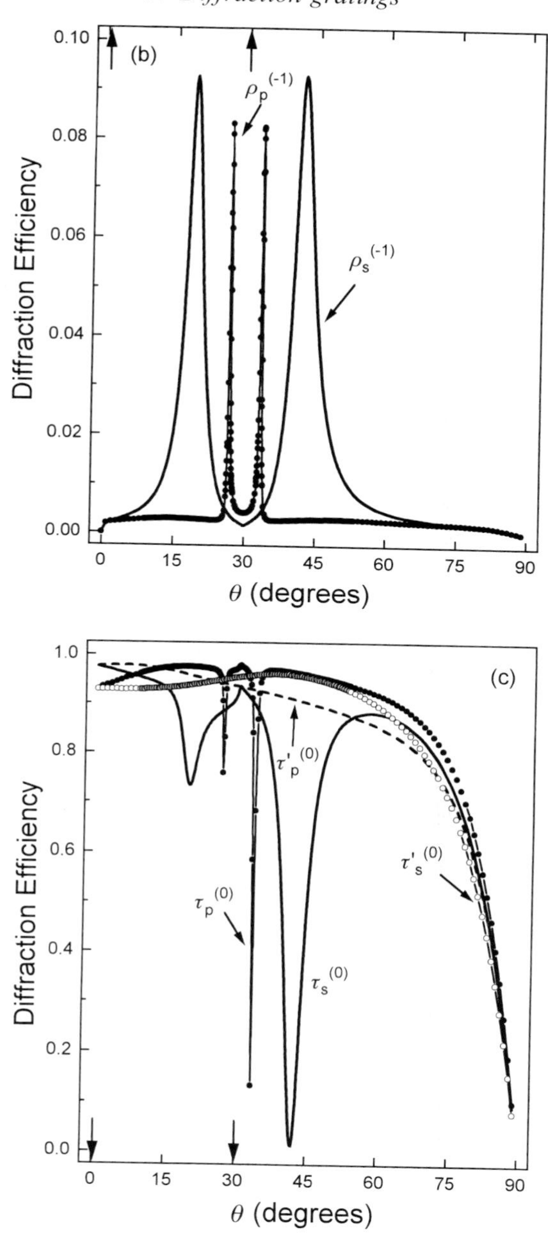

Figure 17.16 (*continued*)

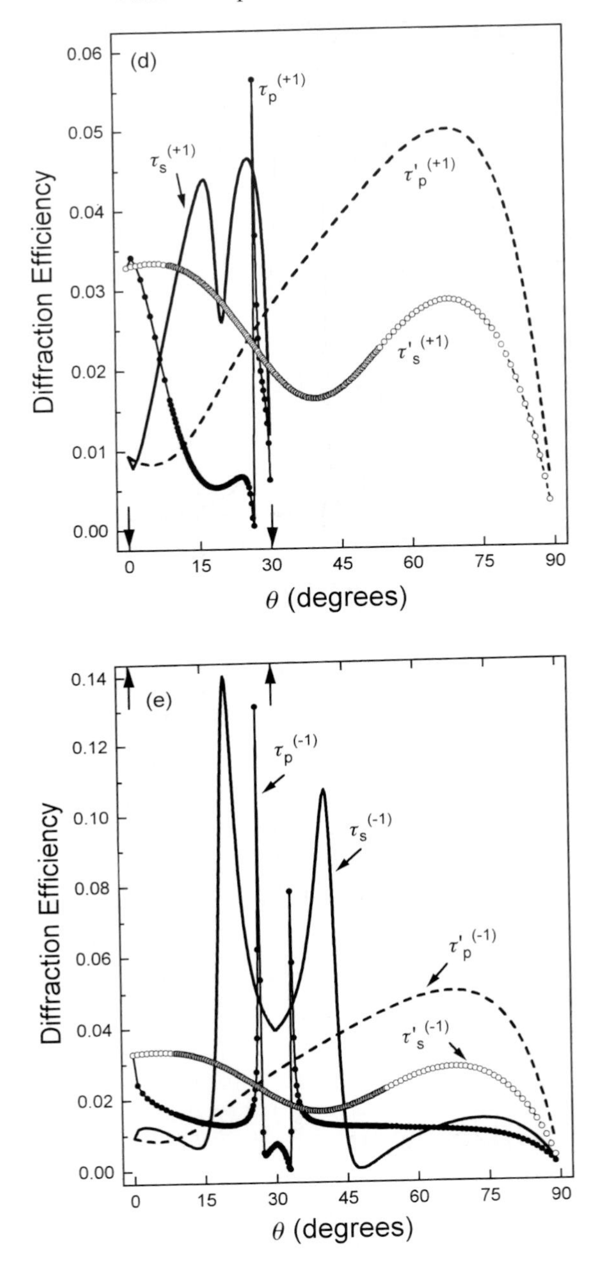

Figure 17.16 (*continued*)

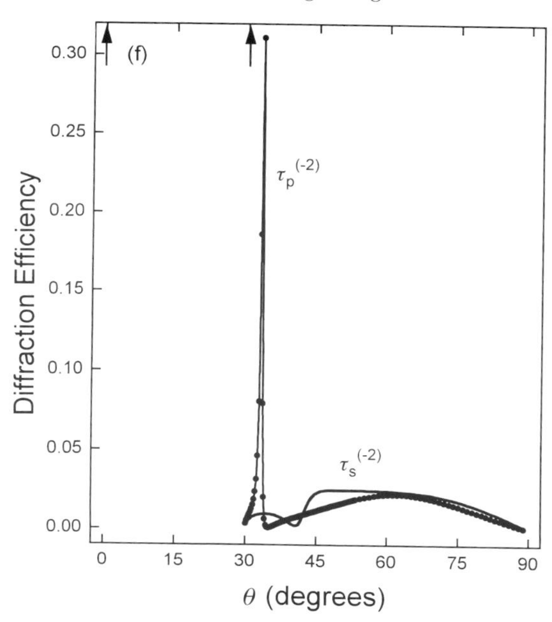

Figure 17.16 (*continued*)

that the +first-order transmitted beam is cut off beyond $\theta = 30^\circ$. In its place the −second-order transmitted beam shown in Figure 17.16(f) appears and shows fairly high efficiency for p-polarized light in a narrow range of angles around $\theta = 33^\circ$.

It is impossible to describe in a brief survey the entire range of physical phenomena that occur in diffraction gratings and their potential applications. We hope, however, to have brought to the reader's attention the richness and complexity of the physics of gratings, and to have encouraged further exploration of this fascinating subject.

References for chapter 17

1 M. Born and E. Wolf, *Principles of Optics*, 6th edition, Pergamon Press, Oxford, 1980.
2 R. Petit, editor, *Electromagnetic Theory of Gratings*, Vol. 22 of *Topics in Current Physics*, Springer Verlag, Berlin, 1980.
3 M. C. Hutley, *Diffraction Gratings*, Academic Press, New York, 1982.
4 E. G. Loewen and E. Popov, *Diffraction Gratings and Applications*, Marcel Dekker, New York, 1997.
5 J. Fraunhofer, *Ann. d. Physik* **74**, 337 (1823), reprinted in his collected works, 117 (Munich, 1888).
6 H. A. Rowland, *Phil. Mag.* (5), **13**, 469 (1882).

7 R. W. Wood, On a remarkable case of uneven distribution of light in a diffraction grating spectrum, *Phil. Mag.* **4**, 396–402 (1902).
8 J. W. S. Rayleigh, *Proc. Roy. Soc. London A* **79**, 399 (1907).
9 D. Maystre, Rigorous vector theories of diffraction gratings, in *Progress in Optics*, Vol. **21**, 1–67, ed. E. Wolf, Elsevier, Amsterdam, 1984.
10 D. Maystre, ed., *Selected Papers on Diffraction Gratings*, SPIE Milestone series, Vol. **MS 83**, SPIE, Bellingham, 1993.
11 Lifeng Li, Multilayer-coated diffraction gratings: differential method of Chandezon *et al.* revisited, *J. Opt. Soc. Am. A* **11**, 2816–2828 (1994).
12 The simulations in this chapter were performed by DELTA, a program developed by Lifeng Li for grating calculations, and by DIFFRACT™, a product of MM Research Inc., Tucson, Arizona.

18
The Talbot effect

The Talbot effect, also referred to as self-imaging or lensless imaging, was originally discovered in the 1830s by H. F. Talbot.[1] Over the years, investigators have come to understand different aspects of this phenomenon, and a theory of the Talbot effect based on classical diffraction theory has emerged which is capable of explaining the various observations.[2–4] For a detailed description of the Talbot effect and related phenomena, as well as a historical perspective on the subject, the reader may consult references 3 and 4 and further references cited therein. Since many of the standard optics textbooks do not even mention the Talbot effect, it is worthwhile to bring to the reader's attention the essential features of this phenomenon.

Lensless imaging of a periodic pattern

The Talbot effect is observed when, under appropriate conditions, a beam of light is reflected from (or transmitted through) a periodic pattern. The pattern may have one-dimensional periodicity (as in traditional gratings), or it may exhibit periodicity in two dimensions (e.g., a surface relief structure or a photographic plate imprinted with identical features on a regular lattice).

In what follows we shall present the diffraction patterns obtained from a periodic array of cross-shaped apertures in an otherwise opaque screen. Because the diffraction pattern of a single aperture differs markedly from that of a periodic array of such apertures, we begin by examining the behavior of an individual aperture under coherent illumination. Consider the cross-shaped opening in an opaque screen shown in Figure 18.1(a). A collimated beam of coherent light, wavelength λ, illuminates the screen at normal incidence; the assumed length and height of the aperture are each 20λ. Logarithmic plots of intensity distribution at distances $z = 100\lambda$, 200λ, and 600λ beyond the screen are computed and shown in Figures 18.1(b)–(d),

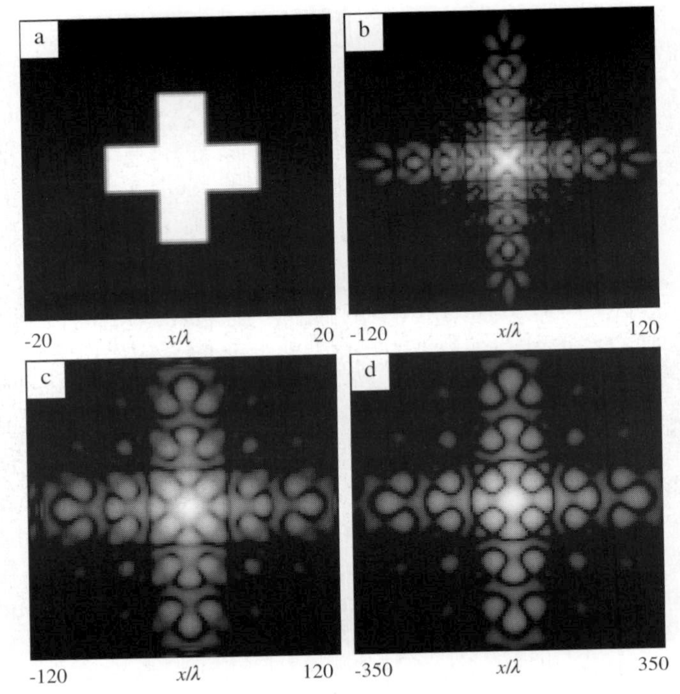

Figure 18.1 (a) A cross-shaped aperture in an opaque screen, illuminated by a normally incident plane wave of wavelength λ. The length and the height of the cross are each 20λ. Also shown are the computed plots of intensity distribution (logarithmic) at various distances z from the aperture: (b) $z = 100\lambda$, (c) $z = 200\lambda$, (d) $z = 600\lambda$. Note that the scale varies.

respectively (note the different scales of these figures). For $z > 600\lambda$ the intensity distribution will have the far field pattern of Figure 18.1(d), although its size will scale with distance from the screen. Under no circumstances do we obtain an intensity pattern that closely resembles the cross shape of the aperture itself.

Now consider the periodic array of cross-shaped apertures shown in Figure 18.2(a); each aperture is identical to that in Figure 18.1(a). The center-to-center spacing between adjacent apertures along the X- and Y- directions is $p = 60\lambda$. (For simplicity we have assumed the periodic pattern to extend to infinity, although, for practical purposes, a finite number of apertures in a periodic arrangement will suffice.) When the pattern in Figure 18.2(a) is illuminated by a normally incident, coherent beam of light, the cross shape of the apertures is abundantly reproduced in the intensity patterns obtained at certain distances from the screen. Figures 18.2(b)–(f) show the computed patterns of intensity distribution at distances $z = 600\lambda$, 1200λ, 1800λ, 2700λ,

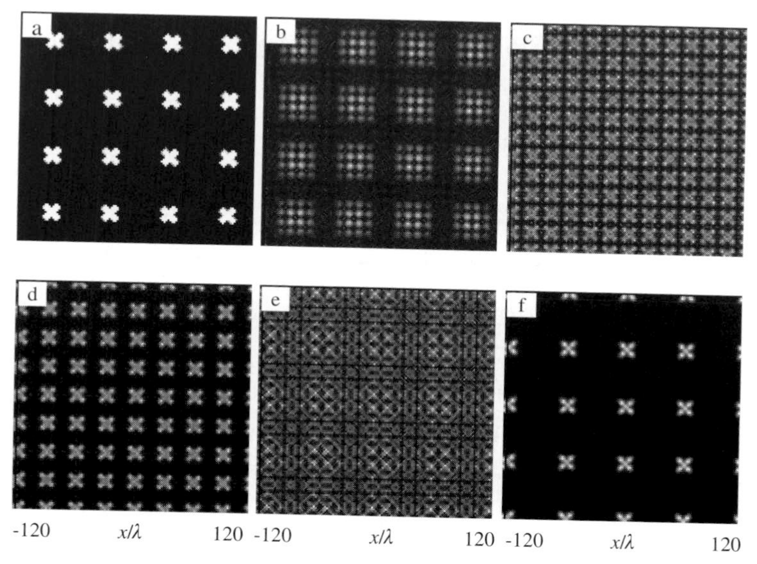

Figure 18.2 (a) A periodic array of cross-shaped apertures in an opaque screen, illuminated by a normally incident plane wave of wavelength λ. As in Figure 18.1, the crosses are 20λ wide on each side. Also shown are the computed plots of intensity distribution at various distances z from the aperture: (b) $z = 600\lambda$, (c) $z = 1200\lambda$, (d) $z = 1800\lambda$, (e) $z = 2700\lambda$, (f) $z = 3600\lambda$. Note that the scale is the same for all the various pictures.

and 3600λ, respectively. (Note that all pictures in Figure 18.2 have the same scale.) When the distance from object to image $z = p^2/\lambda$, as is the case in Figure 18.2(f), the original pattern of the apertures is reproduced, albeit with a half-period shift in both the X- and the Y-direction. In Figure 18.2(d), the distance to the image is $p^2/(2\lambda)$, and not only is the original pattern replicated but also its frequency (along both X and Y) has doubled. In Figure 18.2(c), where the distance to the image is $z = p^2/(3\lambda)$, the pattern is repeated with three times the original frequency along both X- and Y- axes. By showing the intensity distribution at other distances from the object, Figures 2(b), 2(e) emphasize that perfect reproduction of the shapes in the original pattern does not occur everywhere but only at certain special planes.

A hint as to why these periodic patterns are reproduced at certain intervals may be gleaned from the following argument. A plane wave normally incident on a periodic structure creates a discrete spectrum of plane waves propagating along the directions

$$\boldsymbol{k} = (k_x, k_y, k_z) = 2\pi\left(m/p, n/p, \sqrt{(1/\lambda)^2 - (m/p)^2 - (n/p)^2}\right). \tag{18.1}$$

The z-component of this vector may be approximated as follows:

$$k_z \approx (2\pi/\lambda)\left[1 - \tfrac{1}{2}(m\lambda/p)^2 - \tfrac{1}{2}(n\lambda/p)^2\right] \tag{18.2}$$

provided that p/λ is large enough that, for all m, n values of interest, the above Taylor-series expansion to first order suffices. The acquired phase after a propagation distance of z will then be

$$k_z z \approx (2\pi z/\lambda) - \pi z(m^2 + n^2)\lambda/p^2. \tag{18.3}$$

Now, since m, n are integers, if z happens to be an even-integer multiple of p^2/λ then the above phase will differ from the constant value $2\pi z/\lambda$ by a multiple of 2π only. Since all plane waves emanating from the object will thus arrive at the image plane with the same phase factor, their superposition will recreate the original pattern.

It turns out that z does not need to be an even-integer multiple of p^2/λ for self-imaging to occur. At odd-integer multiples of p^2/λ, for instance, a replica of the original pattern will also emerge, but with a half-period shift. Multiple images of the pattern will appear at certain non-integer multiples of p^2/λ as well. These aspects of the Talbot effect will be further clarified below, when we present a more rigorous analysis.

Although the mathematical argument supporting the Talbot effect depends on periodicity of the object in the XY-plane, certain patterns that are not globally periodic, but appear to be so locally, will also produce self-images. For example, the concentric ring pattern shown in Figure 18.3(a), when illuminated by a normally incident coherent beam, will yield the patterns of Figures 18.3(b)–(d) at distances $z = 18\lambda$, 27λ, and 36λ, respectively. The period p of the rings is 6λ and the width of the bright rings is 2λ. Clearly, the self-images break down near the center and near the outer edge, because (local) periodicity is no longer valid in these regions. But a near self-image at $z = p^2/\lambda$ and a frequency-doubled image at $z = p^2/(2\lambda)$ are clearly observed. Another example is shown in Figure 18.4, where a spiral pattern with period $p = 9\lambda$ is propagated to distances $z = p^2/(2\lambda)$, $3p^2/(4\lambda)$, and p^2/λ. Again in Figures 18.4(b), (d) the center and the outer rings are not well reproduced, but nearly everything else is.

The Talbot effect is much more general than the above limited exposition may indicate. The pattern periodicities may be in one or two dimensions; the object may modulate both the amplitude and the phase of the light beam; certain applications rely on the use of incoherent light sources; in the case of two-dimensional periodic patterns, the underlying lattice may be square, rectangular, hexagonal, etc.; the incident beam may be a plane wave or a spherical wavefront originating at a point source; applications are not limited

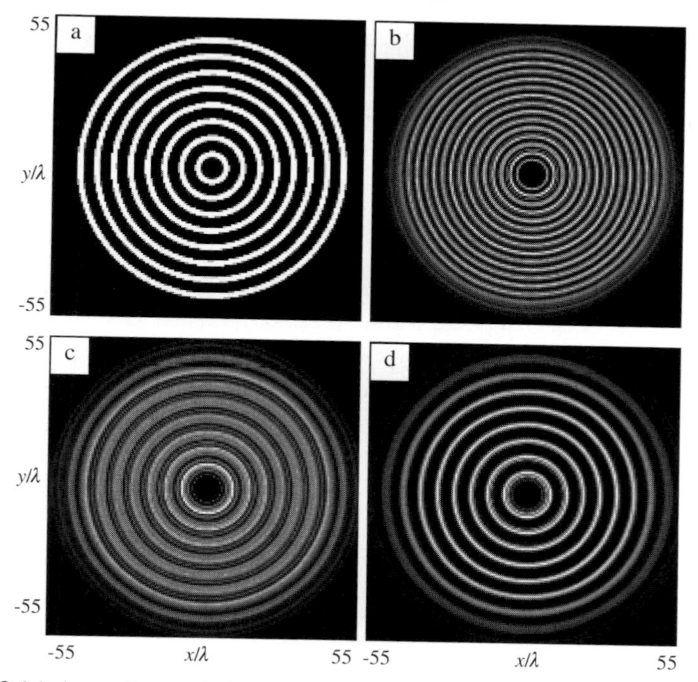

Figure 18.3 (a) A mask consisting of eight concentric rings (width$= 2\lambda$, spacing$= 6\lambda$) is illuminated by a normally incident plane wave of wavelength λ. The computed intensity distributions shown here are at distances of (b) $z = 18\lambda$, (c) $z = 27\lambda$, and (d) $z = 36\lambda$ from the mask. A bright spike appearing in the central region of each image has been blocked off in order to improve the image contrast.

to visible light but extend to X-rays and microwaves, as well as to electron and atom optics. To appreciate the variety of arrangements that lead to useful and interesting images the reader is encouraged to consult the published literature.

A simple analysis

Consider the point source shown in Figure 18.5, located at $(x, y, z) = (x_0, y_0, 0)$ and radiating a spherical wavefront into the region $z > 0$ of space. In this analysis we assume that all spatial dimensions are normalized by the vacuum wavelength λ of the light; as a result, λ will not appear explicitly in any of the following equations. In the $z = z_0$ plane, the complex-amplitude distribution may be written

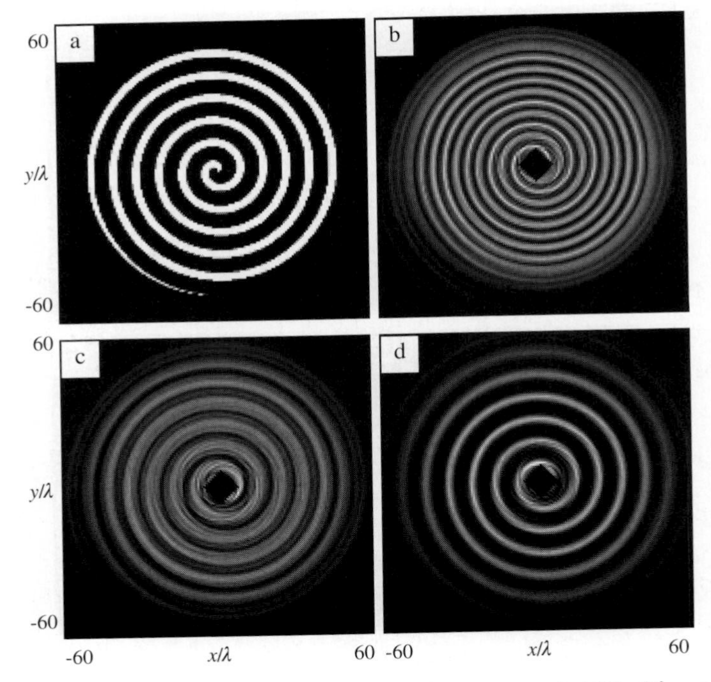

Figure 18.4 (a) A mask consisting of a spiral aperture (width 3λ, spacing 9λ) is illuminated by a normally incident plane wave of wavelength λ. The computed intensity distributions shown here are at distances of (b) $z = 40.5\lambda$, (c) $z = 60.75\lambda$, and (d) $z = 81\lambda$ from the aperture. As in the previous figure, a bright spike appearing in the central region of each image has been blocked off in order to improve the image contrast.

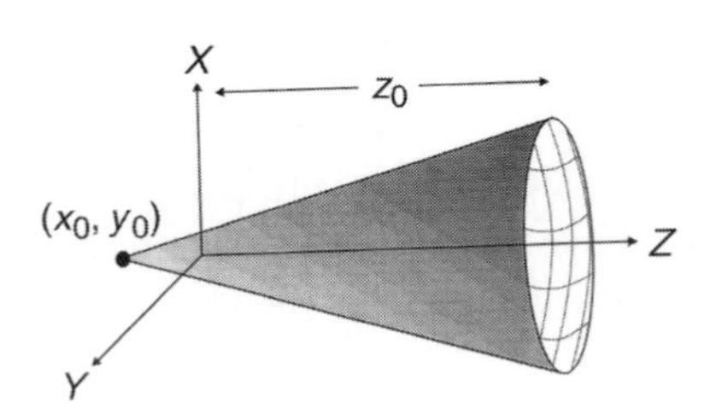

Figure 18.5 A quasi-monochromatic point source located at $(x, y, z) = (x_0, y_0, 0)$ radiates a cone of light into the half-space $z > 0$.

$$
\begin{aligned}
A(x, y, z = z_0) &= (1/r)\exp(\mathrm{i}2\pi r) \\
&= \left\{1/\sqrt{(x - x_0)^2 + (y - y_0)^2 + z_0^2}\right\} \\
&\quad \times \exp\left\{\mathrm{i}2\pi\sqrt{(x - x_0)^2 + (y - y_0)^2 + z_0^2}\right\} \\
&\approx (1/z_0)\exp(\mathrm{i}2\pi z_0) \times \exp\left[\mathrm{i}\pi\left(x^2 + y^2\right)/z_0\right] \\
&\quad \times \exp\left[\mathrm{i}\pi\left(x_0^2 + y_0^2\right)/z_0\right] \times \exp[-\mathrm{i}2\pi(xx_0 + yy_0)/z_0].
\end{aligned}
\tag{18.4}
$$

In deriving the above approximate expression we have used, for the exponent, the first term in the Taylor series expansion

$$\sqrt{1 + x^2} = 1 + \tfrac{1}{2}x^2 + \ldots \tag{18.5}$$

Now, the first two terms on the right-hand side of Eq. (18.4) are the approximate form of the spherical wavefront emanating from a point source at the origin of the plane $z = 0$. The next term is a constant phase factor that depends on the position (x_0, y_0) of the point source within the XY-plane and the last term is a linear phase factor in x and y.

Next, let us assume that a periodic mask, having periods a_x and a_y along the X- and Y- axes, is placed at $z = z_0$ (see Figure 18.6). In the general case, where the mask modulates the phase and/or the amplitude of the light beam, its complex-amplitude transmission function may be written

$$t(x, y) = \sum\sum C_{mn}\exp[\mathrm{i}2\pi(mx/a_x + ny/a_y)]. \tag{18.6}$$

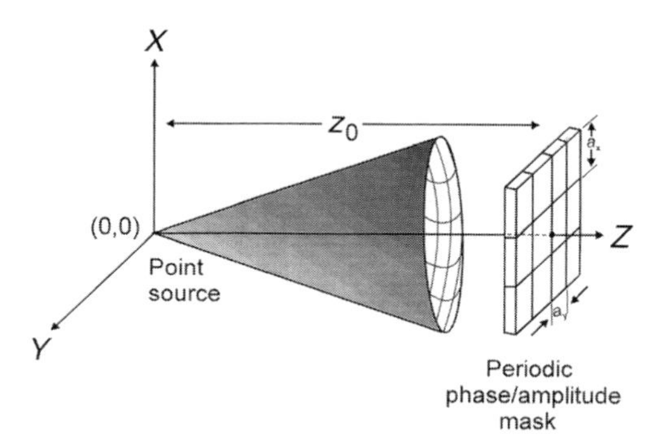

Figure 18.6 A quasi-monochromatic point source located at the origin of the coordinate system illuminates a periodic phase and/or amplitude mask placed parallel to the XY-plane at $z = z_0$. The periods of the mask's pattern are a_x along the X-axis and a_y along the Y-axis.

When the incident spherical wavefront is multiplied by $t(x, y)$, each Fourier component of $t(x, y)$ will create a different spherical wavefront which, according to Eq. (18.4), appears to originate at a different point $(x_0, y_0) = (-mz_0/a_x, -nz_0/a_y)$ within the XY-plane. In addition, each such point source appears to have the following phase factor:

$$\exp(\mathrm{i}\phi_{mn}) = \exp[-\mathrm{i}\pi(x_0^2 + y_0^2)/z_0] = \exp[-\mathrm{i}\pi z_0(m^2/a_x^2 + n^2/a_y^2)]. \quad (18.7)$$

The net effect of the mask, therefore, is to replace the single point source with a periodic array of point sources, as shown in Figure 18.7, where the magnitude of each point source is $C_{mn} \exp(\mathrm{i}\phi_{mn})$. At the observation plane, each point source will give rise to a spherical wavefront that will obey Eq. (18.4), except that the distance z_0 is replaced by $z_0 + z_1$. We thus have

$$\begin{aligned} A(x, y, z = z_0 + z_1) \approx\ & [1/(z_0 + z_1)] \exp[\mathrm{i}2\pi(z_0 + z_1)] \\ & \times \exp[i\pi(x^2 + y^2)/(z_0 + z_1)] \\ & \times \sum\sum C_{mn} \exp[-\mathrm{i}\pi z_0(m^2/a_x^2 + n^2/a_y^2)] \\ & \quad \times \exp[\mathrm{i}\pi(m^2/a_x^2 + n^2/a_y^2)z_0^2/(z_0 + z_1)] \\ & \quad \times \exp\{\mathrm{i}2\pi[x(mz_0/a_x) + y(nz_0/a_y)]/(z_0 + z_1)\}. \end{aligned} \quad (18.8)$$

The first two factors in the above equation correspond to a spherical wavefront with radius of curvature $z_0 + z_1$; we need not keep track of them any longer. The last factor can be simplified if we define a magnification factor $M = (z_0 + z_1)/\ z_0$, in which case it is written as

$$\exp\{\mathrm{i}2\pi[mx/(Ma_x) + ny/(Ma_y)]\}. \quad (18.9)$$

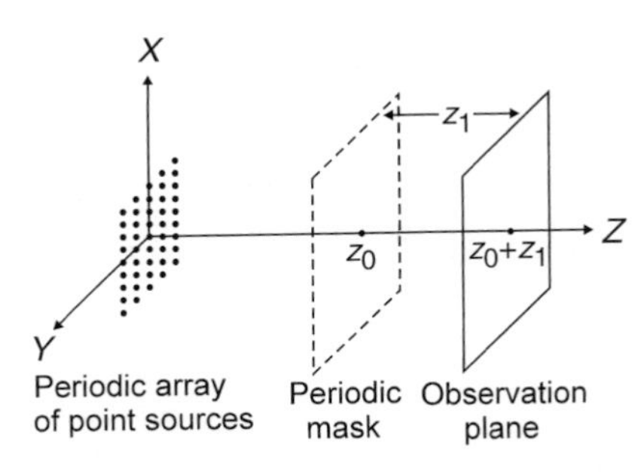

Figure 18.7 Interaction between the periodic mask and the cone of light shown in Figure 18.6 gives rise to an array of (virtual) point sources, each having a certain phase and amplitude depending on the structure of the mask and its location z_0 along the Z-axis. To determine the light distribution at the observation plane one may replace the mask by this "equivalent" array of point sources.

This is just the (m, n)th plane-wave component of the spectrum, whose periods a_x, a_y are magnified by a factor M. Except for this scale factor, the Fourier basis functions have not changed in going from the plane of the mask ($z = z_0$) to the observation plane ($z = z_0 + z_1$). The main factors in Eq. (18.8), therefore, are the first two factors in the double sum; these can be written as follows:

$$\exp\left[-\mathrm{i}\pi\left(m^2/a_x^2 + n^2/a_y^2\right)z_0 z_1/(z_0 + z_1)\right] = \exp\left[-\mathrm{i}\pi\left(z_1/M\right)\left(m^2/a_x^2 + n^2/a_y^2\right)\right]. \tag{18.10}$$

Let us now assume that a_x^2 and a_y^2 have a least common multiple in the following sense:

$$\mu a_x^2 = \nu a_y^2 = a^2, \tag{18.11}$$

where both μ and ν are integers. Then the phase factor in Eq. (18.10) may be written

$$\exp\{-\mathrm{i}\pi[z_1/(Ma^2)](\mu m^2 + \nu n^2)\}. \tag{18.12}$$

Since $\mu m^2 + \nu n^2$ is an integer, if z_1 is chosen to be $2\kappa M a^2$ with κ integer, then the phase factor in Eq. (18.12) will become unity for all values of m and n and can therefore be ignored. Under such circumstances Eq. (18.8) will yield a magnified image of the mask at the observation plane. This is the essence of the Talbot effect.

By allowing z_0 to approach infinity, the above results can be readily extended to the case of plane-wave illumination. The magnification factor M will become unity in this case, but no other change will be necessary in the preceding equations.

Image multiplicity

The appearance of multiple images at the observation plane may be readily explained in the special case where the periodicity is one dimensional and the frequency of the image is twice that of the object. The explanation, nonetheless, captures the essence of the phenomenon and can be easily extended to periodicity in two dimensions and to higher multiplicities. Consider the periodic function $f(x)$ shown in Figure 18.8(a). Note that the period a_x is much larger than the width of the individual "features" of the function, so that there is plenty of space to insert additional features. Let the Fourier-series representation of this function be

$$f(x) = \sum C_m \exp(\mathrm{i}2\pi\, mx/a_x). \tag{18.13}$$

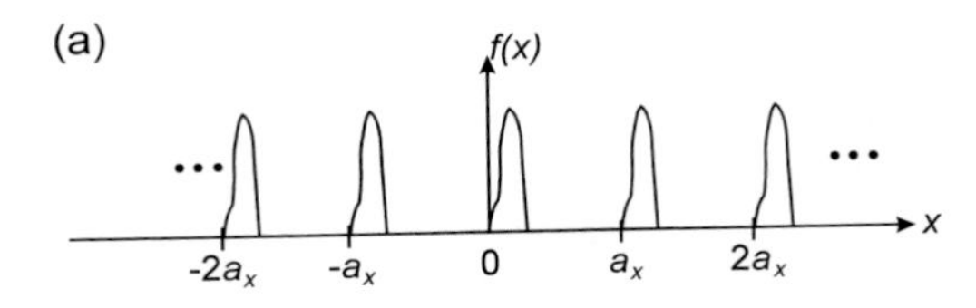

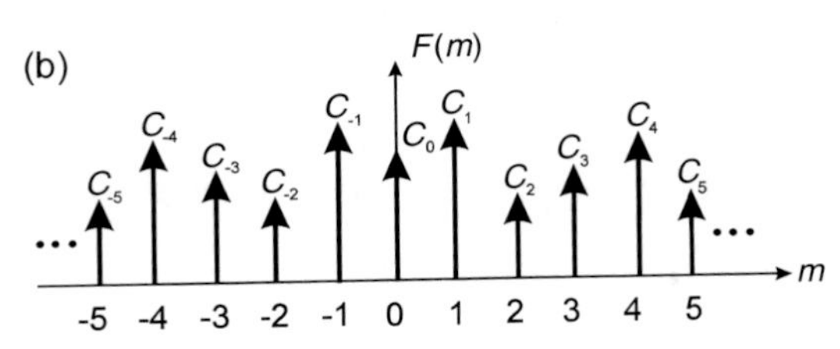

Figure 18.8 (a) A periodic function $f(x)$ in one-dimensional space; the individual "features" of the function are much narrower than its period a_x. (b) The Fourier transform of $f(x)$ consists of a sequence of delta functions located at integer multiples of $1/a_x$ in the Fourier domain.

In the Fourier domain, the Fourier transform $F(m)$ of $f(x)$ is a "comb" function with period $1/a_x$, where the delta function at position m is multiplied by the corresponding Fourier coefficient C_m, as shown in Figure 18.8(b).

Now, let us assume that the odd coefficients of $F(m)$ are multiplied by a complex constant β. (This would happen in Eq. (18.12), for instance, if $\mu = 1$, $\nu = 0$, and $z_1 = \frac{1}{2}Ma^2$, in which case $\beta = -\mathrm{i}$.) We can then separate the Fourier coefficients of $f(x)$ into even and odd terms, as shown in Figure 18.9. Both the resulting comb functions in the Fourier domain will have twice the period of the original comb function; therefore, their inverse transforms in the x-domain will have twice the frequency. The second comb function in Figure 18.9 is also shifted by a half-period, which means that its inverse transform must be multiplied by $\exp(\mathrm{i}2\pi x/a_x)$. The resulting comb functions in the x-domain are shown in Figure 18.10. The net result is that when we add the two comb functions of Figure 18.10 and convolve the resultant with the unit-period function $f_0(x)$, we will find the function shown in Figure 18.11. Because the width of $f_0(x)$ is less than half the period a_x, the new features added to the function will not overlap with the old ones, yielding a function with an apparently increased frequency. However, the periodicity is only in the amplitude of the function, since the phase of each feature differs from the phase of its neighbors. In any event, this description explains why the apparent periodicity of the pattern in Figure 18.2 increases at certain distances between the object and the image.

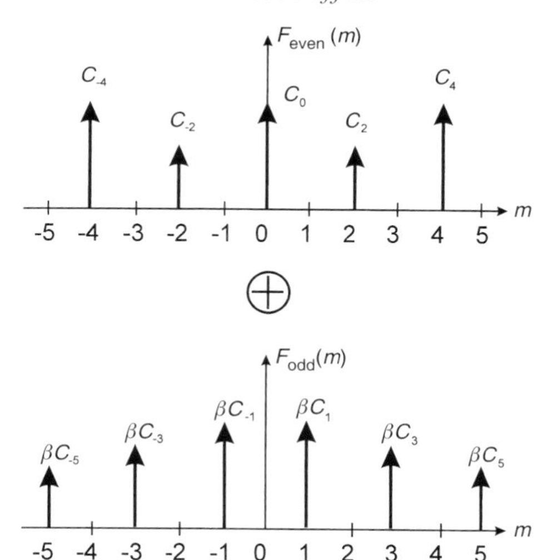

Figure 18.9 In Figure 18.8(b), when the odd components of the Fourier-transformed function $F(m)$ are multiplied by a constant β, the function may be resolved into two "comb" functions, $F_{\text{even}}(m)$ and $F_{\text{odd}}(m)$. In these new functions the spacing between adjacent delta functions is $2/a_x$ and, in the case of $F_{\text{odd}}(m)$, the function is shifted by a half-period.

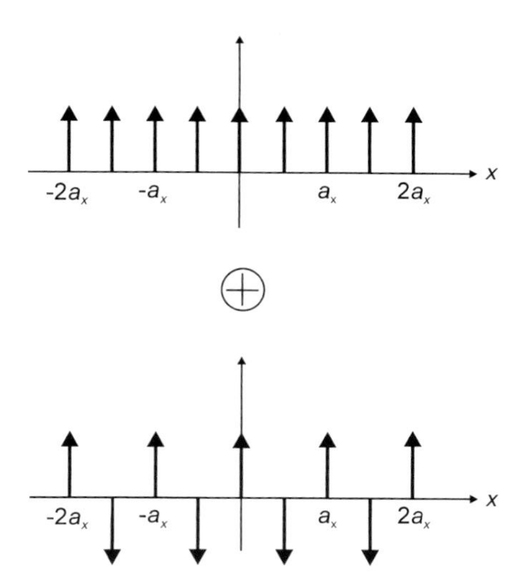

Figure 18.10 The comb function corresponding to $F_{\text{even}}(m)$, when inverse-transformed to the x-domain, will yield a comb function that has twice the frequency of the original function $f(x)$. Likewise, the inverse transform of the comb function corresponding to $F_{\text{odd}}(m)$ will have a spacing of $\frac{1}{2}a_x$ between its adjacent delta functions but, because of the half-period shift in the Fourier domain, every other delta function is flipped over.

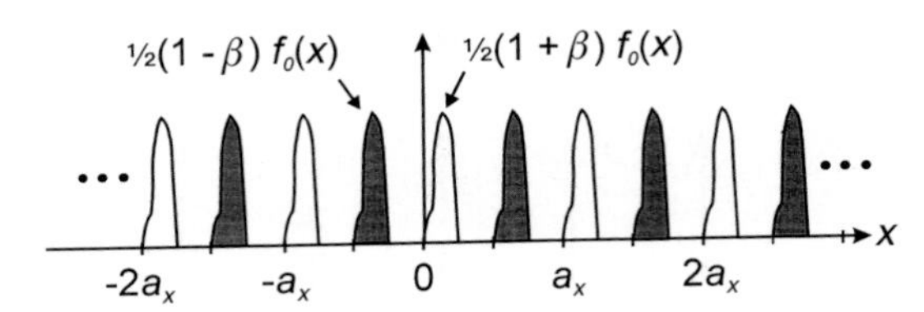

Figure 18.11 When the sum of the two comb functions in Figure 18.10 is convolved with the individual features $f_0(x)$ of $f(x)$, the resulting function appears to have twice the frequency of the original $f(x)$. Note, however, that the "features" of the new function are alternately multiplied by $\frac{1}{2}(1+\beta)$ and $\frac{1}{2}(1-\beta)$.

References for chapter 18

1 H. F. Talbot, *Phil. Mag.* **9**, 401 (1836).
2 Lord Rayleigh, *Phil. Mag.* **11**, 196 (1881).
3 O. Bryngdahl, Image formation using self-imaging techniques, *J. Opt. Soc. Am.* **63**, 416–419 (1973).
4 J. F. Clauser and M. W. Reinsch, New theoretical and experimental results in Fresnel optics with applications to matter-wave and *X*-ray interferometry, *Appl. Phys. B* **54**, 380–395 (1992).

19

Some quirks of total internal reflection

Readers are undoubtedly familiar with the phenomenon of total internal reflection (TIR), which occurs when a beam of light within a high-index medium arrives with a sufficiently great angle of incidence at an interface with a lower-index medium. What is generally not appreciated is the complexity of phenomena that accompany TIR. For instance, consider the simple optical setup shown in Figure 19.1, where a uniform beam of light is brought to focus by a positive lens, being reflected, somewhere along the way, at the rear facet of a glass prism. Assuming a refractive index $n = 1.65$ for the prism material, the critical angle of incidence is readily found to be $\theta_{\text{crit}} = \sin^{-1}(1/n) = 37.3^\circ$. Let the lens have numerical aperture $NA = 0.2$ (i.e., f-number $= 2.5$). Then the range of angles of incidence on the prism's rear facet will be $(33.5^\circ, 56.5^\circ)$. The majority of the rays thus suffer total internal reflection and converge, as depicted in Figure 19.1, towards a common focus in the observation plane.

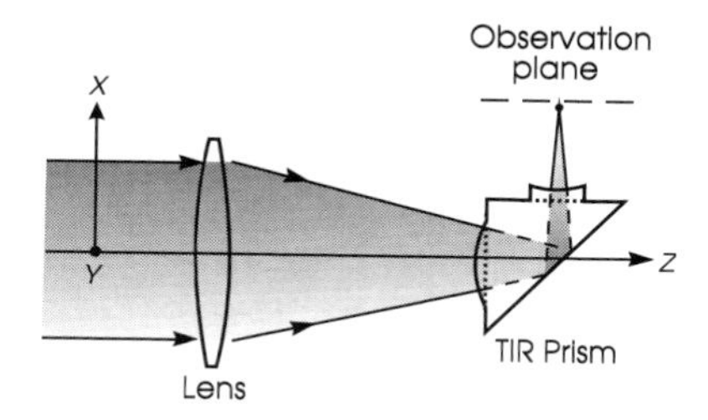

Figure 19.1 Focusing of a uniform beam through a TIR prism. The incident beam is linearly polarized along the X-axis, the numerical aperture of the lens is 0.2, and the refractive index of the prism material is 1.65. The entrance and exit facets of the prism are assumed to be spherical so that ray-bending by Snell's law at these surfaces is avoided, thus eliminating the corresponding spherical aberrations.

Figure 19.2 shows computed plots of intensity and phase at the observation plane, indicating that the focused spot essentially has the Airy pattern, albeit with minor deviations from the ideal. The diameter of the first dark ring, for example, is approximately 6λ, which is close to the theoretical value of $1.22\lambda/NA$ for the Airy disk.[1] The coma-like tail appearing on the right-hand side of the focused spot is caused by those rays that strike the prism in the neighborhood of the critical TIR angle, θ_{crit}, thus introducing apodization and aberration. (Apodization is due to a reduction of the reflectivity of the prism below the critical angle, and aberration is caused by deviations from linearity of phase as a function of angle of incidence.) One noteworthy feature of the focused spot of Figure 19.2 is that it is not centered on the optical axis, but is shifted to the right by about one wavelength. This shift is known as the Goos–Hänchen effect,[2–4] and its cause will become clear in the course of the following discussion.

For the prism of Figure 19.1 the computed amplitude and phase of Fresnel's reflection coefficients at the glass-to-air interface are presented in Figure 19.3.[1] The curves for both p- and s-components of polarization are shown, even though in our example we are primarily concerned with p-polarized light. Note that beyond the critical angle the phase of the reflected p-light has a very large slope. To the extent that this phase may be approximated by a straight line (within the range of incidence angles of interest) it imparts a linear phase shift to the beam upon reflection from the prism's rear facet. This linear phase shift is nothing other than a wavefront tilt, which causes a displacement of the focused spot; in other words, it gives rise to the Goos–Hänchen effect. One might phrase the same explanation in the lan-

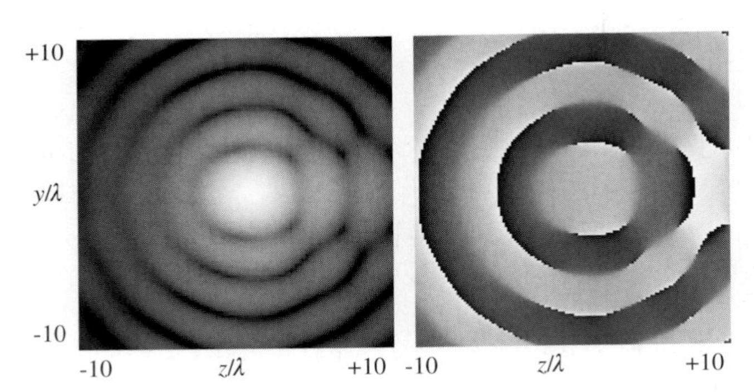

Figure 19.2 Plots of (a) logarithmic intensity distribution and (b) phase, at the focal plane of the lens. The center of the bright spot is shifted to the right by about one wavelength in consequence of the Goos–Hänchen effect. The light and dark rings in the phase plot correspond to regions of 0° and 180° phase, respectively.

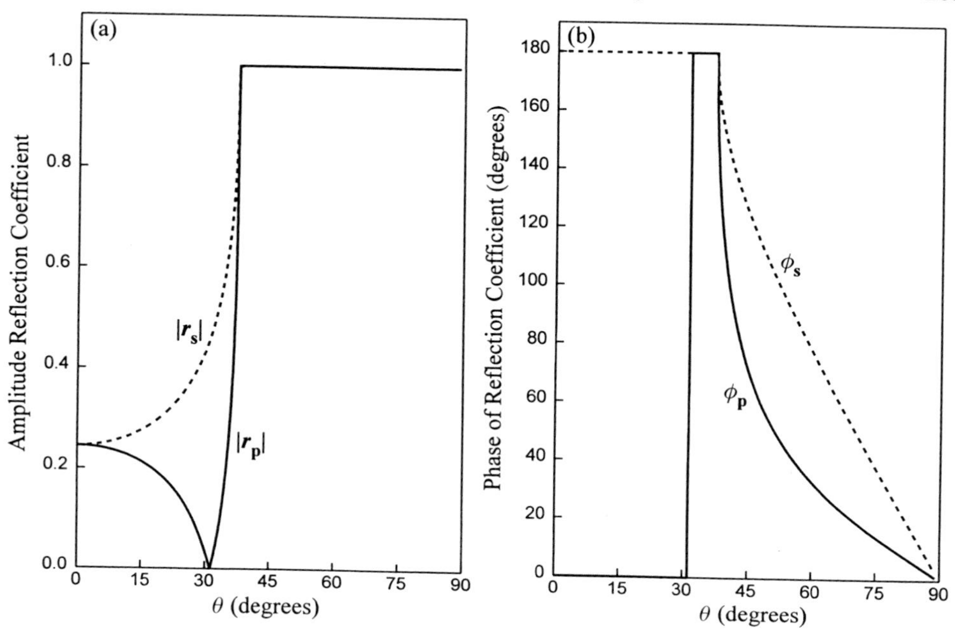

Figure 19.3 Plots of amplitude and phase for the reflection coefficients of the p- and s-components of polarization at a glass–air interface. The assumed index of the glass is $n = 1.65$. The critical angle for TIR is $\theta_{\text{crit}} = \sin^{-1}(1/n) = 37.3°$, and the Brewster angle is $\theta_B = \tan^{-1}(1/n) = 31.2°$.

guage of Fourier-transform theory by stating that when a function is multiplied by a linear phase factor, its Fourier transform is displaced by an amount proportional to the slope of that phase factor.

Note that the largest slopes of the phase plots in Figure 19.3(b) occur immediately after the critical angle; therefore, the greatest effects would be observed when the incident beam's angular spectrum is confined to the vicinity of θ_{crit}. In our example, of course, the range of incident angles is fairly large (33.5° to 56.5°), and deviations from linearity of the phase function show up as higher-order aberrations (e.g., coma, astigmatism, spherical aberration, defocus). It is this deviation from linearity that is mainly responsible for the aberration of the focused spot seen in Figure 19.2.

A question frequently asked about TIR concerns the balance of energy among the incident beam, the reflected beam, and the evanescent waves that exist in the medium beyond the prism. If all the light is reflected at the glass–air interface, then how can there be any energy in the form of electromagnetic fields in the region immediately beyond the interface? To answer this question one must distinguish between the steady state of the system, which prevails once the waves have established themselves throughout space, and the transient state, which exists in the earlier stage immediately after the light source has been turned on. In the transient state, some of the incident energy goes into developing the evanescent waves, which are established early on and remain for as long as the system remains undisturbed. If one calculated for the evanescent field the component of the Poynting vector perpendicular to the interface, one would find that the electric and magnetic components of this field are exactly 90° out of phase and, therefore, that the perpendicular component of the Poynting vector is zero. In other words, no energy is

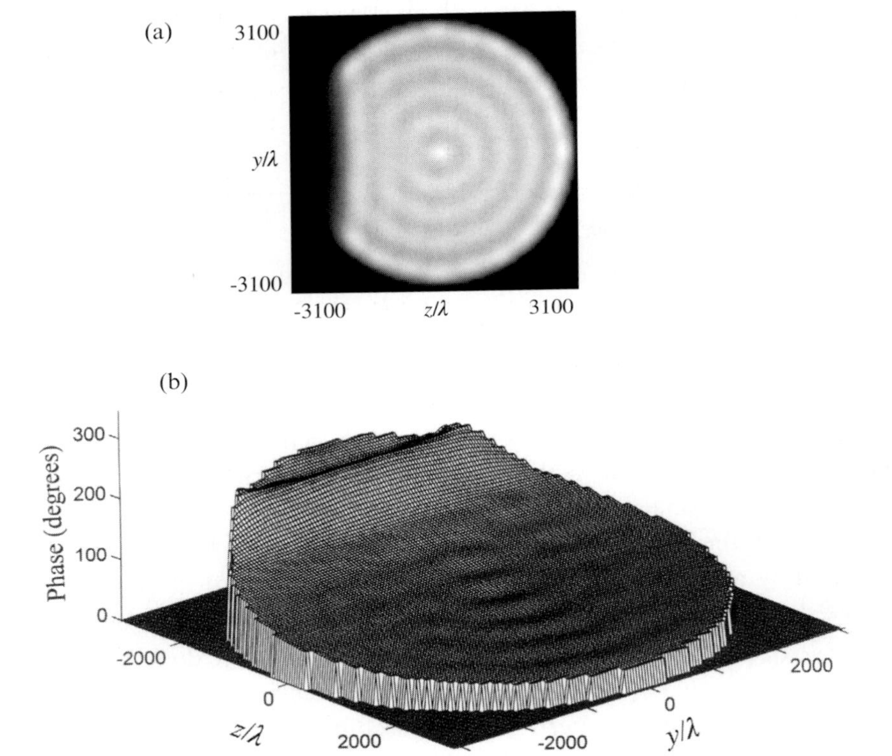

Figure 19.4 (a) Plot of intensity distribution at the exit pupil of the collimating lens. The low-contrast rings are caused by diffraction effects during propagation and by loss of the high-spatial-frequency content of the spectrum. The rays on the left side of the beam, having been below the critical angle for TIR, have been partially transmitted through the prism. (b) Distribution of phase at the exit pupil of the collimating lens. The small linear slope is responsible for the Goos–Hänchen displacement of the focused spot. The plateau on the left-hand side is caused by the (partially reflected) rays that fall below the critical angle. The sharp rise immediately before reaching the plateau is due to the rapidly decreasing phase of the reflected rays just above the critical angle.

carried away from the interface by these evanescent waves. Consequently, all the incident optical energy in the steady state is carried away by the reflected beam.

Next, we consider the effect of a collimating lens (identical to the original focusing lens), placed so as to capture the radiation emanating from the focused spot. (In the system of Figure 19.1, this lens would be placed one focal length above the observation plane and parallel to it.) The resulting collimated beam is depicted in Figure 19.4, which shows computed plots of intensity and phase at the exit pupil of the collimator. Note, in particular, the strong attenuation of the left edge of the beam (owing to a loss of rays below θ_{crit}), and also the near-linearity of the phase plot in regions far from the critical angle. As the light rays approach θ_{crit} from above, the phase pattern in Figure 19.4(b) rises rather sharply and then flattens. This is precisely what one would expect based on the behavior of ϕ_{p} in the interval (33.5°, 56.5°), shown in Figure 19.3(b).

One cannot leave the subject of TIR without at least mentioning the fascinating phenomena associated with frustrated TIR, which occur when a second prism is brought to the vicinity of the interface at which TIR occurs. Consider a pair of identical glass hemispheres separated by an air gap of width Δ, as shown in Figure 19.5. Displayed in Figure 19.6 are computed plots of amplitude reflection coefficients $|r_{\text{p}}|$ and $|r_{\text{s}}|$ versus the angle of incidence θ for three different values of Δ. In Figure 19.6(a), where $\Delta = 100\,\text{nm}$, one can see close similarities to Figure 19.3(a), albeit with TIR completely suppressed: the Brewster angle at $\theta_{\text{B}} = 31.2°$ is still there,

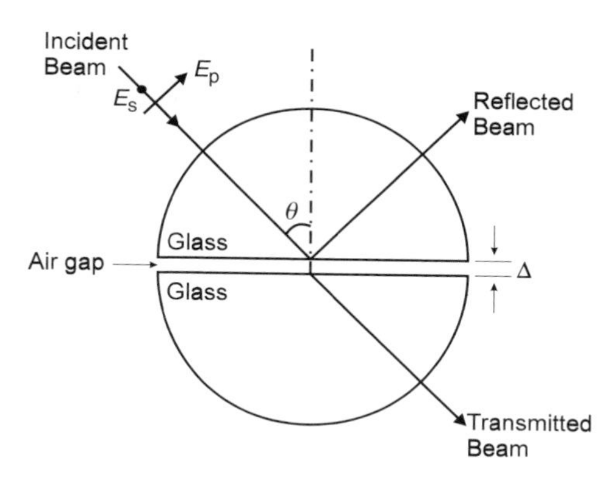

Figure 19.5 A pair of glass hemispheres separated by an air gap may be used to demonstrate the phenomenon of frustrated TIR. The coherent beam of light is directed at the center of the upper hemisphere at incidence angle θ. The width Δ of the air gap is adjustable.

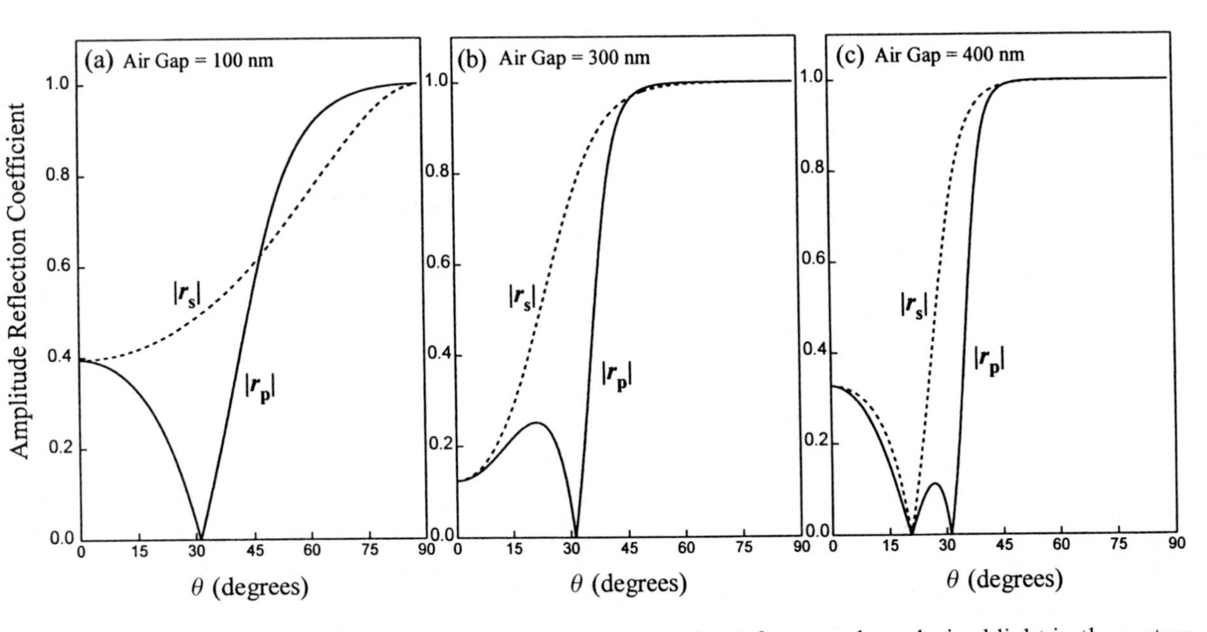

Figure 19.6 Computed amplitude reflection coefficients, $|r_p|$ and $|r_s|$, for p- and s-polarized light in the system of Figure 19.5. The refractive index of the glass hemispheres is $n = 1.65$, the wavelength of the incident beam is $\lambda = 650\,\text{nm}$, and the width Δ of the air gap is (a) 100 nm, (b) 300 nm, (c) 400 nm.

but there are no sharp transitions to 100% reflectivity. In Figure 19.6(b), Δ is set to 300 nm and the curves are beginning to look more like those in Figure 19.3(a); it appears that by increasing Δ one can make a rather smooth transition to TIR. But wait! In Figure 19.6(c), where $\Delta = 400$ nm, there is a radical departure from the presumed "smooth transition". Specifically, at $\theta = 20.7^\circ$ both r_p and r_s vanish identically. What is going on here? What will happen if the gap width Δ keeps increasing? These questions are not difficult to answer but require some thought. Essentially, at a certain gap width Δ and at some angle of incidence θ both r_p and r_s vanish. The gap width is such that, at this angle, $\Delta \cos\theta = \lambda/2$ exactly. Now, whenever a non-absorbing layer's thickness becomes an integer multiple of a half-wavelength, that layer will have no effect on the multiple-beam interferences and can, therefore, be eliminated from consideration. Removing the air gap would bring the two hemispheres into contact, in which case all the incident light will naturally pass from one hemisphere to the other, leaving no reflected light whatsoever for either type of polarization.

If the flat surface of the bottom hemisphere in Figure 19.5 is coated with a metallic layer, one would observe the phenomenon of attenuated TIR.[5] Figure 19.7(a) shows plots of $|r_p|$ and $|r_s|$ versus θ for the case of an aluminum-coated surface separated from the top hemisphere by a 875 nm air gap. The s-polarized light does not exhibit any interesting effects, but the drop in p-light reflectivity around $\theta = 37.4^\circ$ (just 0.1° above θ_{crit}) is quite impressive. In fact, when the angle of incidence θ is properly selected, it is possible to modulate the reflectivity of the p-light from essentially 0% to 100% by adjusting the gap width, without ever bringing the two surfaces into contact (or near contact). This has provided the mechanism for a novel light-intensity modulator, which was patented some years ago.[6] Figure 19.7(b) shows the component of the Poynting vector perpendicular to the gap as a function of the vertical distance from the top surface into the gap; the assumed angle of incidence within the top hemisphere is 37.4°. The optical energy is seen to propagate unattenuated through the 875 nm gap, before being fully absorbed within the top 30 nm of the aluminum layer.

The physics of attenuated TIR involves the excitation of surface plasmons in the metallic layer by p-polarized evanescent waves in the air gap. Surface plasmons are fairly easy to describe and to understand; in fact, they are just inhomogeneous plane-wave solutions to Maxwell's equations in absorptive media. See chapter 9, "What in the world are surface plasmons?", for a more comprehensive discussion of this subject.

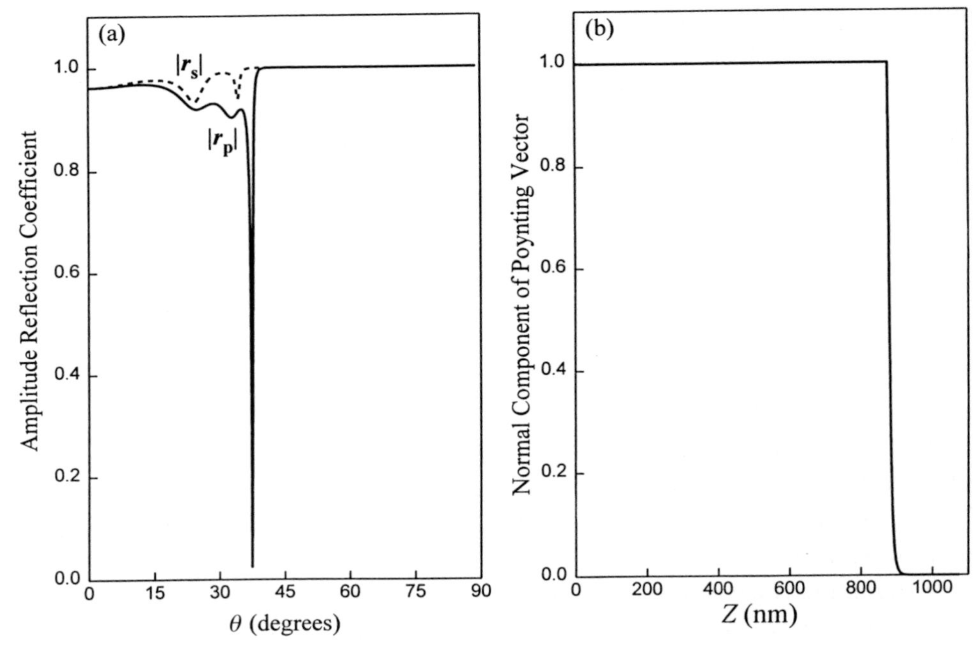

Figure 19.7 (a) Computed amplitude reflection coefficients, $|r_p|$ and $|r_s|$, for p- and s-polarized light in the system of Figure 19.5, when the flat surface of the bottom hemisphere is coated with a thick layer of aluminum: $(n, k) = (1.47, 7.8)$, thickness $= 200$ nm. The top glass hemisphere is assumed to have refractive index $n = 1.65$, the wavelength of the incident beam is $\lambda = 650$ nm, and the width Δ of the air gap is 875 nm. This particular gap width was chosen because it brought the minimum in r_p close to zero. At other gap widths the behavior is qualitatively the same but the minimum of reflectivity is higher. (b) Component of the Poynting vector perpendicular to the gap, computed at $\theta = 37.4°$. The horizontal axis is the distance measured from the top of the air gap towards the aluminized surface at the bottom. The optical energy flows unattenuated through the air before being fully absorbed in the top 30 nm of the aluminum layer.

References for chapter 19

1 M. Born and E. Wolf, *Principles of Optics*, sixth edition, Pergamon Press, New York, 1983.
2 F. Goos and H. Hänchen, *Ann. Phys. Lpz.* (6) **1**, 333 (1947).
3 F. Goos and H. Lindberg-Hänchen, *Ann. Phys. Lpz.* (6) **5**, 251 (1949).
4 H. K. V. Lotsch, Beam displacement at total reflection: the Goos–Hänchen effect, *Optik* **32**: part I, 116–137, part II, 189–204 (1970); part III, 299–319, part IV, 553–569 (1971).
5 A. Otto, *Zeit für Physik* **216**, 398 (1968).
6 G. T. Sincerbox and J. G. Gordon, *Appl. Opt.* **20**, 1491–1494 (1981).

20

Evanescent coupling

Evanescent electromagnetic waves abound in the vicinity of luminous objects. These waves, which consist of oscillating electric and magnetic fields in regions of space immediately surrounding an object, do not transfer their stored energy to other regions and, therefore, remain localized in space. Like all electromagnetic waves, the behavior of evanescent waves is governed by Maxwell's equations, and their presence in the vicinity of an object helps to satisfy the requirements of field continuity at the object's boundaries. Evanescent fields decay exponentially with distance away from the object's surface, making them exceedingly difficult to detect at distances much greater than a wavelength.[1]

When a beam of light shines on a diffraction grating, for example, various diffracted orders partake of the energy of the incident beam and carry it away in different directions. At the same time, evanescent waves are created around the grating, which ensure the continuity of the field at the grating's corrugated surface. Similarly, a beam of light shining on an aperture or on a small particle sets up evanescent fields around the boundaries of these objects. Perhaps the best-known example of evanescence, however, is provided by total internal reflection (TIR) from an internal facet of a prism (see Figure 20.1). Here the evanescent field is formed in the free-space region behind the prism, and remains distinct and isolated from the propagating (i.e., incident and reflected) beams; this phenomenon was discussed briefly in chapter 19.

Bringing an object to the vicinity of another object that has an established evanescent field in its neighborhood could change the distribution of the electromagnetic field throughout the entire space. For example, if a material object is placed behind the prism of Figure 20.1, close enough to sense the evanescent field but not close enough for the two to make physical contact, photons will tunnel through the small gap thus created, diverting a fraction of the incident beam across the gap and into the latter object. This is the

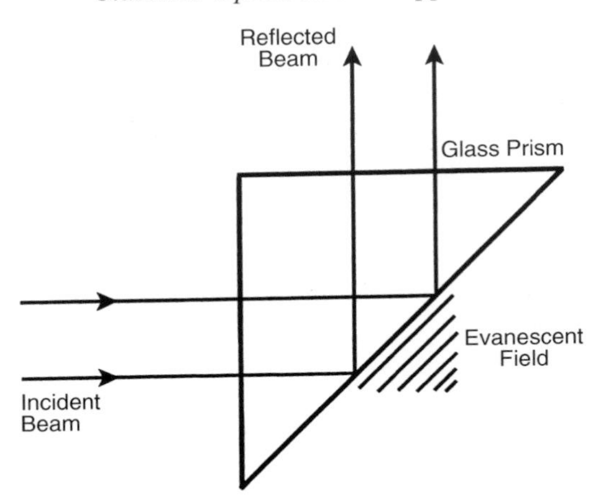

Figure 20.1 A beam of light is totally internally reflected from the rear facet of a glass prism. The electromagnetic field lurking in the free space region behind the prism is evanescent; both its electric and magnetic components decay exponentially with distance from the interface, and the projection of its Poynting vector perpendicular to the interface is zero. The energy stored in the evanescent field is deposited there at the time when the light source is first turned on. In the steady state, energy is neither added to nor removed from the evanescent field; all the incoming optical energy is reflected at the rear facet of the prism.

essence of evanescent coupling, of which we present several examples in this chapter. The well-known phenomena of frustrated TIR and attenuated TIR, which are of relevance here, were discussed in previous chapters (see chapter 9, "What in the world are surface plasmons?", and chapter 19, "Some quirks of total internal reflection").

Focusing through a glass hemisphere

We begin by considering the system of Figure 20.2, in which a uniform, collimated, linearly polarized beam of light (vacuum wavelength $\lambda_0 = 633\,\text{nm}$) is brought to focus by an aberration-free $0.8NA$ objective lens. A glass hemisphere of refractive index $n = 2$, also referred to as a solid immersion lens (SIL), is placed over the focal plane so that the focused spot rests at its flat facet.[2] For simplicity, we assume that the objective lens and the spherical surface of the SIL are antireflection coated; thus the only reflected light originates at the flat facet of the SIL. For rays that arrive at this flat facet at an angle below the critical TIR angle ($\theta_c = \arcsin(1/n) = 30°$) the reflectance is fairly small (about 11% at normal incidence). For $\theta > \theta_c$, however, reflectivity is 100%, so that the cone of light covering the range of ray angles from critical to marginal is fully reflected. The computed intensity

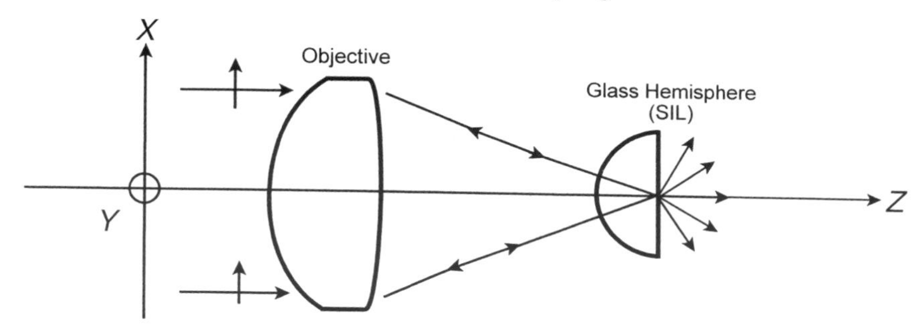

Figure 20.2 A collimated beam of light, uniform, monochromatic ($\lambda_0 = 633$ nm), and linearly polarized along X, enters an aplanatic $0.8NA$ objective lens ($f = 3750\lambda_0$). A glass hemisphere – also known as a SIL – of refractive index $n = 2$ is placed so that its flat facet coincides with the objective's focal plane. The surfaces of the objective as well as the spherical surface of the SIL are antireflection coated, but the flat facet of the SIL is bare. The light reflected from this flat facet returns to the objective, is collimated by it, and appears at the exit pupil.

distribution for the reflected light at the exit pupil of the objective is shown in Figure 20.3(a). The bright ring resulting from TIR is clearly visible in this plot. The central region of the aperture is not totally dark either, but to discern it requires a picture with better contrast. Figure 20.3(b), a logarithmic plot of the same distribution as in Figure 20.3(a), shows the structure of the central region. The two dark spots inside the ring along the horizontal axis arise from low reflectance at and around the Brewster angle. The overall reflectivity at the flat facet of the hemisphere (as measured at the objective's exit pupil) is 66%.

Because Fresnel's reflection coefficients at the flat facet differ for the p- and s-polarized light, the reflected beam appearing at the exit pupil is no longer linearly polarized. Figures 20.3(c), (d) show distributions of the x- and y-components of polarization, respectively. E_x contains about two-thirds of the reflected optical power, while E_y contains the remaining one-third. What is more, the relative phase of E_x and E_y varies over the aperture, thus creating a non-uniform state of polarization. The computed distribution of the polarization ellipticity η is shown in Figure 20.3(e); here the gray-scale encodes angles from $-37°$ (black) to $+37°$ (white). The distribution of the polarization rotation angle ρ is shown in Figure 20.3(f), where the gray-scale represents angles from $-90°$ (black) to $+90°$ (white). Clearly the state of polarization in the TIR region is quite complex.

Suppose now that an identical hemisphere is placed in front of the SIL and separated from it by a narrow air gap (see Figure 20.4). Under these circumstances, evanescent coupling causes a good fraction of the beam to be transmitted through to the second hemisphere. Figure 20.5 shows the computed distributions of the reflected light at the exit pupil of the objective lens for a 100 nm air gap. These distributions should be compared directly with those in Figure 20.3. The overall reflectance is now 43%, of which two-thirds is again in the x-component of polarization (Figure 20.5(c)) and one-third in the y-component (Figure 20.5(d)). Where there was a bright ring of light at the exit

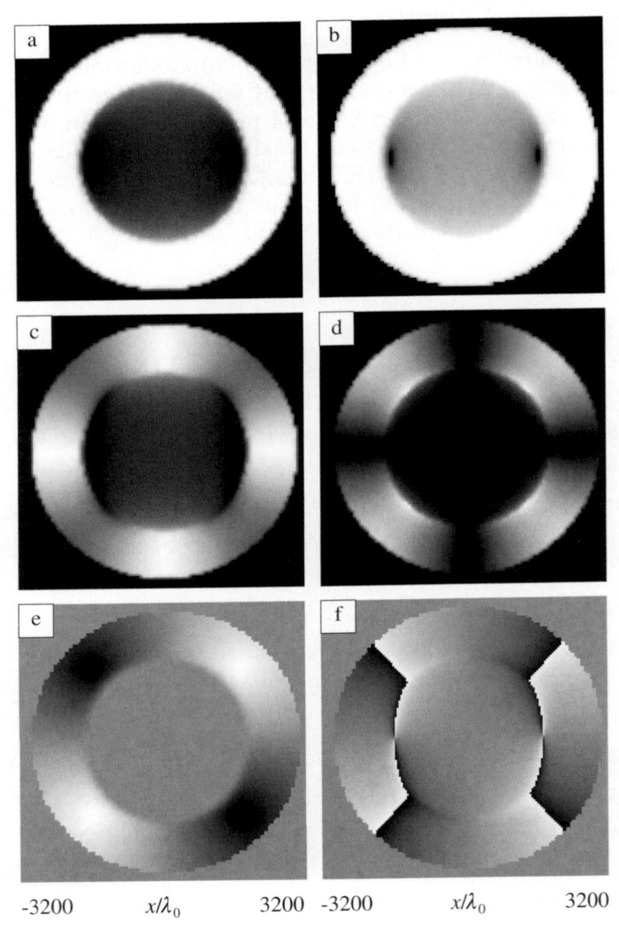

Figure 20.3 Various distributions of the reflected light at the exit pupil of the objective lens of Figure 20.2. (a) Plot of reflected intensity corresponding to a 66% overall reflectivity at the flat facet of the hemisphere. (b) Logarithmic plot of the reflected intensity. (c) Intensity distribution for the x-component of polarization, E_x. (d) Intensity distribution for the y-component of polarization, E_y. (e) The polarization ellipticity η encoded by gray-scale, covering a range from -37° (black) to $+37^\circ$ (white). (f) The polarization rotation angle ρ encoded by gray-scale, covering a range from -90° (black) to $+90^\circ$ (white).

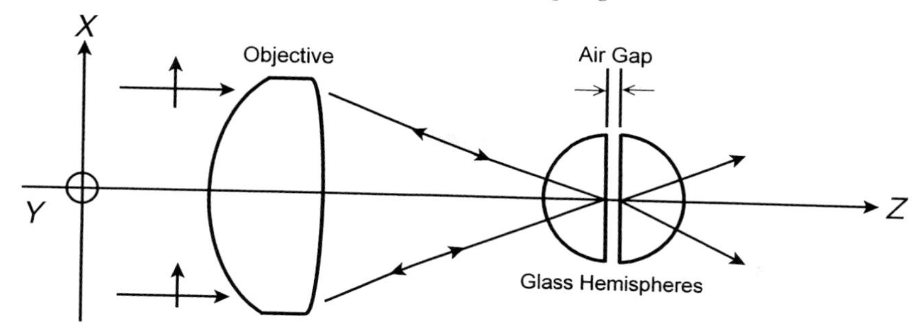

Figure 20.4 A collimated beam of light, uniform, monochromatic ($\lambda_0 = 633$ nm), and linearly-polarized along X, enters an aplanatic $0.8NA$ objective lens ($f = 3750\lambda_0$). Two glass hemispheres of refractive index $n = 2$, separated by an air gap, are arranged in such a way that the focal plane of the objective coincides with the mid-plane of the air gap. Both hemispheres are antireflection coated on their spherical surfaces, but are left bare on their flat surfaces. The light reflected at the air gap returns to the objective, is collimated by it, and appears at the exit pupil.

pupil in Figure 20.3(a), now there is a gradual brightening toward the margins in Figure 20.5(a), indicating the gradual decrease in evanescent coupling with increasing angle of incidence. The two dark spots in the vicinity of the Brewster angle are clearly visible in the logarithmic plot of Figure 20.5(b). The ellipticity η shown in Figure 20.5(e) varies over the aperture in the range $\pm 29.5^\circ$, while the polarization rotation angle ρ has the distribution shown in Figure 20.5(f).

Up to this point we have considered the effects of evanescent coupling upon a full cone of light, which contains rays both below and above the critical TIR angle, θ_c. Let us now place a circular mask in the path of the incident beam in Figure 20.4 in order to block those rays that arrive at the air gap at angles below θ_c. The semi-hollow cone of light thus formed by the objective lens contains only rays with $\theta > \theta_c$. Figure 20.6 is the computed plot of reflectance versus gap width for the system of Figure 20.4 augmented with a mask that completely blocks the central part of the beam (i.e., the region with no contribution to evanescent coupling). The reflectance curve is seen to start at zero, when the hemispheres are in contact and the light is fully transmitted. With a widening gap, however, the reflectance increases rapidly and saturates at 100% before the gap width reaches even one wavelength of the light.

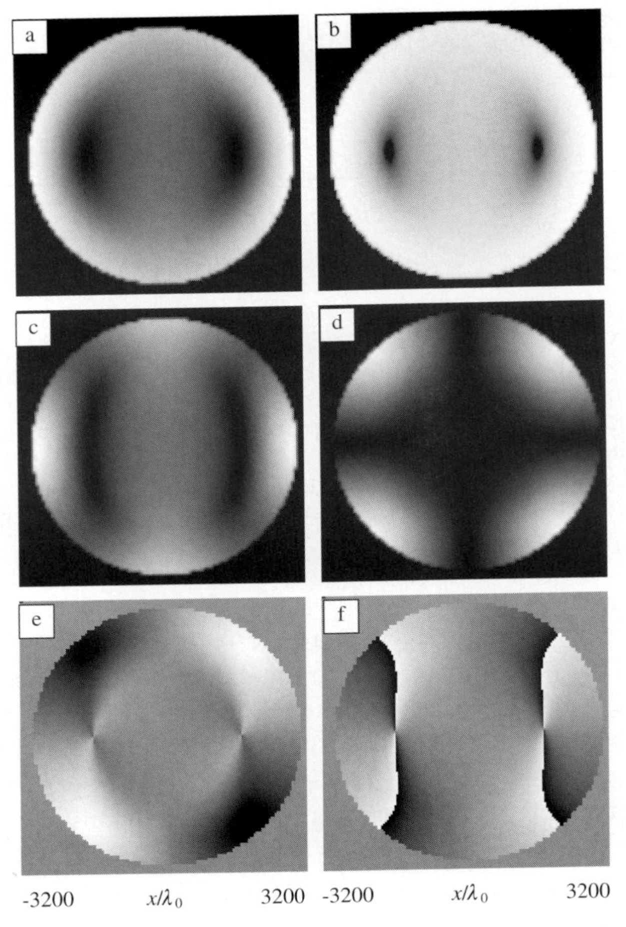

Figure 20.5 Various distributions of the reflected light at the exit pupil of the objective lens of Figure 20.4; the gap width is fixed at 100 nm. (a) Plot of the reflected intensity corresponding to a 43% overall reflectivity at the air gap. (b) Logarithmic plot of the reflected intensity. (c) Intensity distribution for E_x. (d) Intensity distribution for E_y. (e) The polarization ellipticity η encoded by gray-scale, covering a range from -29.5° (black) to $+29.5^\circ$ (white). (f) The polarization rotation angle ρ encoded by gray-scale, covering a range from -90° (black) to $+90^\circ$ (white).

Evanescent coupling to a metallic film

Suppose now that the flat facet of the second hemisphere of Figure 20.4 is coated with a metallic layer, say, a layer of aluminum 50 nm thick ($n = 1.4, k = 7.6$). For a 100 nm air gap, Figure 20.7 shows the various distributions of the reflected light at the exit pupil of the objective lens. The plot of reflected intensity in Figure 20.7(a) shows high reflectance everywhere

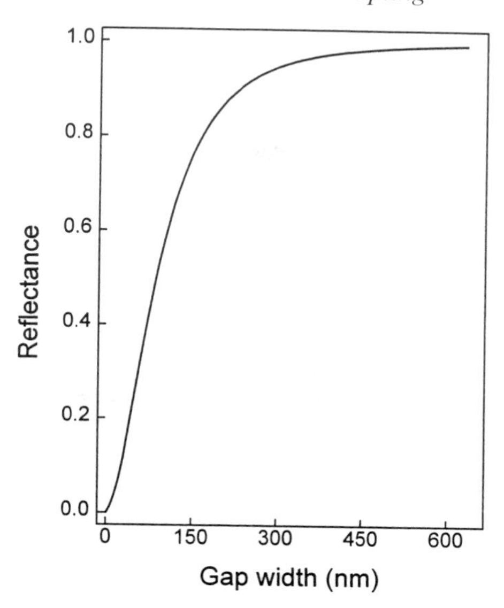

Figure 20.6 Computed plot of reflectance versus gap width in the system of Figure 20.4, when a circular mask is placed in the incident beam's path to block the rays that arrive at the interface between the hemispheres at or below the critical TIR angle.

except in two crescent-shaped areas, which correspond to a dip in the Fresnel reflection coefficient for p-polarized light. The overall reflectance is 92%, of which 66% has x-polarization and 26% has y-polarization. The remaining 8% of the light has been absorbed by the aluminum layer. (At 50 nm thickness, the aluminum film is opaque; the incident light is partly absorbed and partly reflected from the film's surface, practically no light being transmitted through the film.) The absorbed light comes partly from the central region of the incident beam, which is transported through the gap by ordinary (i.e., propagating) waves, and partly from the remaining annular region, which is transported by evanescent coupling. As before, the reflected polarization state is quite complex, regions near the critical angle being RCP in two quadrants and LCP in the other two quadrants. (The coarseness of the mesh used in these calculations does not reveal the resonant absorption by surface-plasmon excitation. This type of absorption occurs within a narrow range of angles just above the critical angle for p-polarized light. Because the angular range of resonant absorption is extremely narrow, however, its contribution to the overall absorption within the aluminum film may be neglected.)

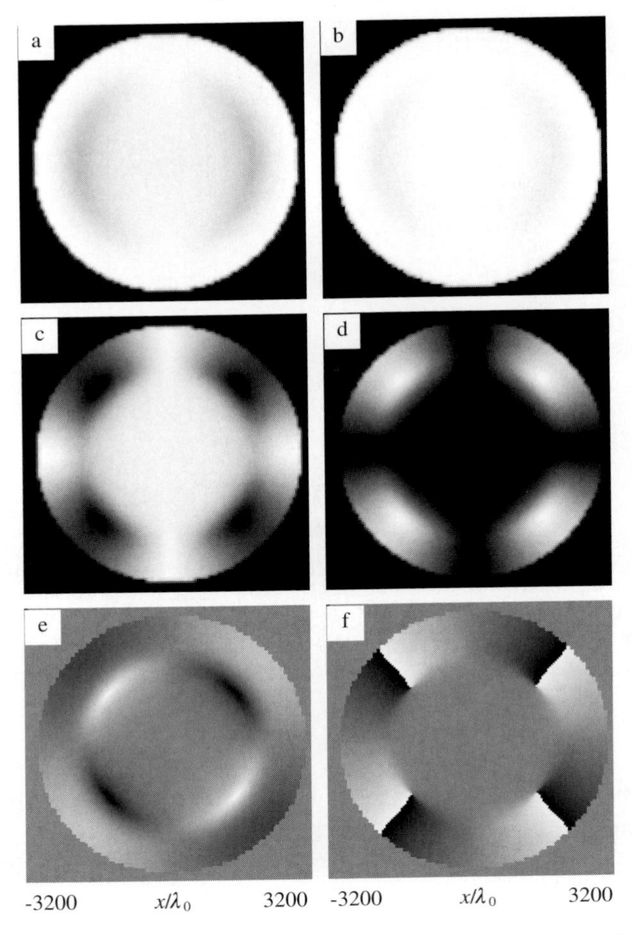

Figure 20.7 Various distributions of the reflected light at the exit pupil of the objective lens of Figure 20.4, when the flat facet of the second hemisphere is coated with a layer of aluminum 50 nm thick. The gap width is fixed at 100 nm. (a) Plot of the reflected intensity, showing a 92% overall reflectance. (If the SIL is removed, the reflectivity drops to 90%.) (b) Logarithmic plot of the reflected intensity. (c) Intensity distribution for E_x. (d) Intensity distribution for E_y. (e) The polarization ellipticity η ranging from $-43.4°$ (black) to $+43.4°$ (white). (f) The polarization rotation angle ρ, ranging from $-90°$ (black) to $+90°$ (white).

To compute the fraction of light absorbed by the aluminum film through evanescent coupling, we place once again a circular mask in the central region of the incident beam, blocking all the rays that would arrive at the gap below the critical angle. The results of calculations in this case are shown in Figure 20.8 for a bare aluminum film (solid curve) as well as a coated film (broken curve). With increasing gap width the reflectance of the bare film drops

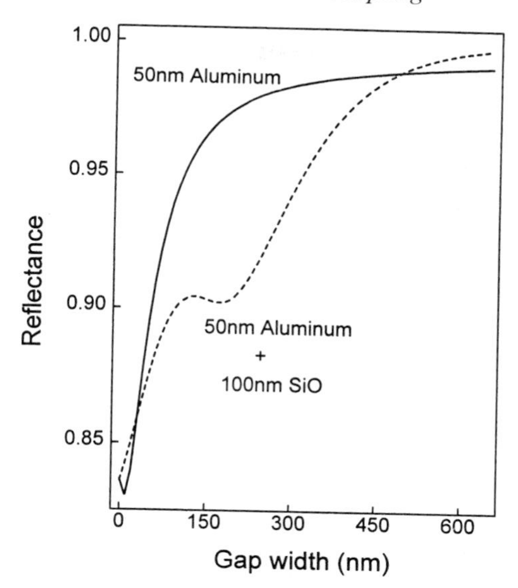

Figure 20.8 Computed plots of reflectivity versus gap width for two different samples. The solid curve shows the reflectance when an aluminum layer 50 nm thick coats the flat facet of the second hemisphere of Figure 20.4. The broken curve corresponds to the case where a layer of SiO 100 nm thick and with $n = 2$ coats the 50 nm thick aluminum film. A mask blocks the central region of the beam in both cases.

slightly at first, then rises rapidly to saturate at 100%. When the aluminum film is in contact with the SIL (i.e., zero gap width) it absorbs about 16% of the light, but by the time the gap widens to 150 nm the absorption is down to a mere 3%. One can improve upon this situation by applying a dielectric coating over the aluminum layer. The broken curve in Figure 20.8 is a plot of reflectance versus gap width for an aluminum film 50 nm thick coated with a layer of SiO 100 nm thick. It is clear that evanescent coupling now takes place over a wider range of gaps; in particular, between gap widths of 150 nm and 200 nm there is a plateau of about 10% absorption. (In the absence of the mask blocking the center of the beam, the dielectric-coated aluminum film absorbs a total of 19% of the incident power at a gap width of 100 nm.)

The above aluminum–SiO bilayer is discussed for illustration purposes only; it does not represent the most efficient multilayering scheme for coupling the light to the aluminum layer. One can do somewhat better by adding additional layers on top of the aluminum film and/or beneath the flat facet of the SIL and by optimizing the thickness and refractive index of each such layer.

Magneto-optical disk

A major application area for evanescent coupling is the field of magneto-optical (MO) disk data storage.[2–4] Here a disk, which is typically a multilayer stack of metallic and dielectric layers on a glass or plastic substrate, is placed under a solid immersion lens (SIL). As the disk spins, the SIL rides on an air cushion, which separates the two by a fixed gap width. Two of the most important questions in this area are: (1) how much of the focused optical energy is absorbed within the optical disk?; (2) how does the reflected MO signal depend on the air gap?

Before answering these questions, however, we must give a brief overview of the physical mechanisms involved in MO recording and readout. The disk consists of a thin magnetic layer sandwiched between two dielectric layers coated atop a reflector such as an aluminum-coated substrate (see Figure 20.9). The layer thicknesses and refractive indices shown in the figure are typical but in fact can vary somewhat, depending on the configuration of the drive for which the disk is intended. The disk is used in reflection, and the multilayer stack is designed to take advantage of optical interference in order

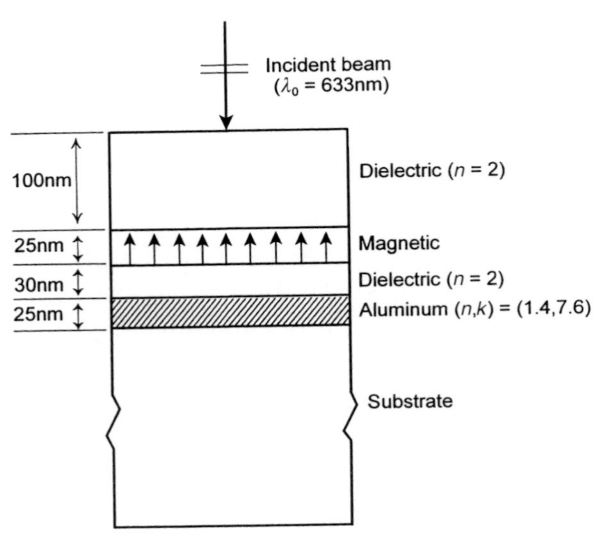

Figure 20.9 Quadrilayer structure of a typical MO disk used in conjunction with the SIL. Within the drive, the disk rotates at several thousand r.p.m., and the SIL flies over the top dielectric layer of the disk, supported on a cushion of air several tens of nanometers thick. The local state of magnetization (up or down) represents the state of the recorded bit (0 or 1). The size of the focused spot directed through the SIL at the magnetic layer determines the minimum mark size that can be recorded and read out.

to maximize the coupling of the laser beam to the magnetic layer.[3] The aluminum reflector is an important component of this optical interference device, but it also serves as a heat sink to remove from the magnetic layer the thermal energy deposited there by the focused laser beam. The dielectric layers protect the magnetic film from the environment and, through their thickness and refractive index, provide the necessary degrees of freedom for adjusting the optical characteristics of the stack. Also, the dielectric layer between the magnetic film and the aluminum layer controls the flow of heat between these two metallic layers.

The optical properties of the magnetic film are fully specified by its dielectric tensor, namely,

$$\boldsymbol{\varepsilon} = \begin{pmatrix} \varepsilon & \varepsilon' & 0 \\ -\varepsilon' & \varepsilon & 0 \\ 0 & 0 & \varepsilon \end{pmatrix}. \tag{20.1}$$

For conventional MO media, typical values of the diagonal and off-diagonal elements of this tensor at $\lambda_0 = 633\,\text{nm}$ are $\varepsilon = -8 + 27\text{i}$ and $\varepsilon' = 0.6 - 0.2\text{i}$. The off-diagonal elements are responsible for cross-coupling the x- and y-components of polarization, this being the origin of optical activity in these media. If ε' is set to zero, MO activity vanishes, as if the magnetization of the material had disappeared. Reversing the direction of magnetization causes a sign reversal of ε'.

Suppose a plane wave, linearly polarized along the X-axis, is directed at normal incidence onto the MO stack of Figure 20.9. Upon reflection from the stack the beam will have two components of polarization: E_x along the original X-axis and E_y along the Y-axis. The y-component of the reflected polarization is created by the optical activity of the magnetic layer. If the magnetization of the sample is reversed, E_x does not change at all, and E_y undergoes a sign change only. If E_x and E_y happen to be in phase, the reflected polarization appears to have been rotated from its original direction; this is referred to as MO Kerr rotation. If, however, E_y is 90° out of phase with respect to E_x, the returning beam will have pure Kerr ellipticity. In general, the polarization components have an arbitrary relative phase and, therefore, the reflected beam exhibits both Kerr rotation and ellipticity. Both Kerr angles convey information about the state of magnetization of the disk: when the magnetization reverses, the sign of E_y is switched, in which case both angles (i.e., rotation and ellipticity) change sign.[3]

In practice a disk is both recorded and read out using a focused laser beam. The writing involves local heating of the magnetic film in the presence of an external magnetic field. At high enough temperature the external magnetic field succeeds in reversing the direction of local magnetization. Obviously, a small focused spot yields a small recorded mark (i.e., a small magnetic domain). The SIL is valued in optical recording precisely because it does produce a small focused spot[4] (see chapter 26, "Scanning optical microscopy"). During readout, the same focused spot is used, albeit at a lower power to avoid heating the media. The sign of the Kerr rotation (or ellipticity) then provides the read signal for the detection system. Once again, the usefulness of the SIL for this application becomes apparent when one recognizes that by producing a small focused spot the SIL helps to resolve small recorded marks.[2,4]

Differential detection

The standard method of detecting the MO signal in conventional optical disk drives is shown in Figure 20.10.[3] The beam reflected from the disk and collected by the objective lens is sent through a Wollaston prism and thus divided between two identical detectors. Typically the detection module is oriented at 45° with respect to the original direction of polarization of the laser, so that E_x and E_y are equally split between the two detectors. Whereas both the magnitude and phase of E_x at the two detectors are identical, E_y arrives at each detector with a different sign. The total light amplitude arriving at the detectors is thus $(E_x \pm E_y)/\sqrt{2}$, the plus sign corresponding to one detector and the minus sign corresponding to the other. If the phase differ-

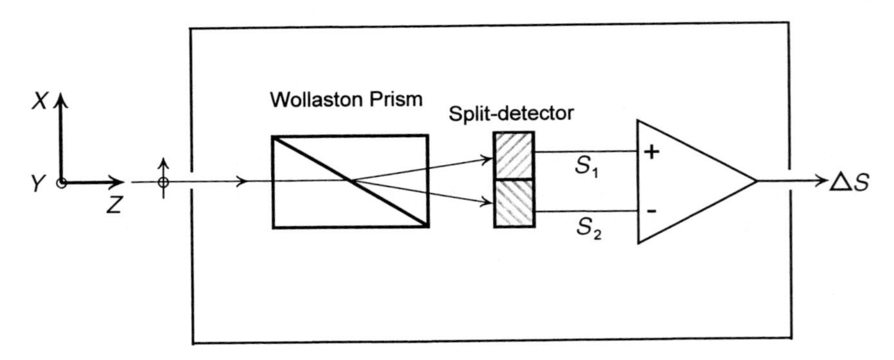

Figure 20.10 Schematic diagram of a differential detection module consisting of a Wollaston prism and two identical photodetectors. Since it can rotate around the optical axis Z, the module may be oriented arbitrarily relative to the ellipse of polarization of the incident beam. In particular, if the original linear polarization of the laser beam is along the X-axis and the magneto-optically generated component of polarization is along the Y-axis, then the module may be aligned in such a way that the transmission axes of the Wollaston prism make 45° angles with X and Y.

ence between E_x and E_y is denoted $\phi_x - \phi_y$ then the net differential signal may be written as

$$\begin{aligned}\Delta S = S_1 - S_2 &= \gamma \int (\tfrac{1}{2}|E_x + E_y|^2 - \tfrac{1}{2}|E_x - E_y|^2)\mathrm{d}x\,\mathrm{d}y \\ &= 2\gamma \int |E_x E_y| \cos(\phi_x - \phi_y)\mathrm{d}x\,\mathrm{d}y.\end{aligned} \tag{20.2}$$

In this equation, γ is the responsivity of the detectors (in volts per watt of optical energy) and the integrals are over the individual detector areas. Note that when the magnetization direction at the disk is reversed the sign of E_y reverses, resulting in a sign reversal for ΔS. Also note that any phase difference between E_x and E_y reduces the output signal by the cosine factor in the above equation. In principle, this phase difference may be eliminated by a properly patterned phase plate placed immediately before the Wollaston prism. In practice, however, unless $\phi_x - \phi_y$ is fairly uniform over the aperture, it is difficult to correct the effects of this relative phase.

Evanescent coupling to an optical disk

Consider the typical MO disk structure of Figure 20.9 placed under the SIL of Figure 20.2. When in contact with the SIL and at normal incidence, the disk has reflectance 36%, Kerr rotation angle 0.66°, and Kerr ellipticity 0.05°. Focusing the beam by the 0.8*NA* objective through the SIL changes these parameters only slightly, as long as the disk and the SIL remain in contact. However, a small air gap between the disk and the SIL can change the disk's performance drastically.

Figure 20.11 shows computed distributions at the exit pupil of the objective lens for a 100 nm gap width. Figure 20.11(a) is the intensity distribution for the reflected E_x, containing 36% of the incident power. The dark oval-shaped region in the middle indicates an area of strong absorption by the disk. The phase of E_x, shown in Figure 20.11(b), is non-uniform over the aperture, ranging in value from −180° (black) to +180° (white). At the center the phase is about +100°, and drops continuously along the *X*-axis to −150° at the edge.

Figure 20.11(c) is a plot of intensity distribution for the reflected E_y, which contains 11.5% of the incident power. This *y*-component is due mainly to the Fresnel reflection coefficients at the interface between the SIL and the multilayer stack. The fraction of E_y created by MO activity is relatively small and, although embedded in Figure 20.11(c), is difficult to recognize at this point. The phase of E_y depicted in Figure 20.11(d) shows the well-known π shift

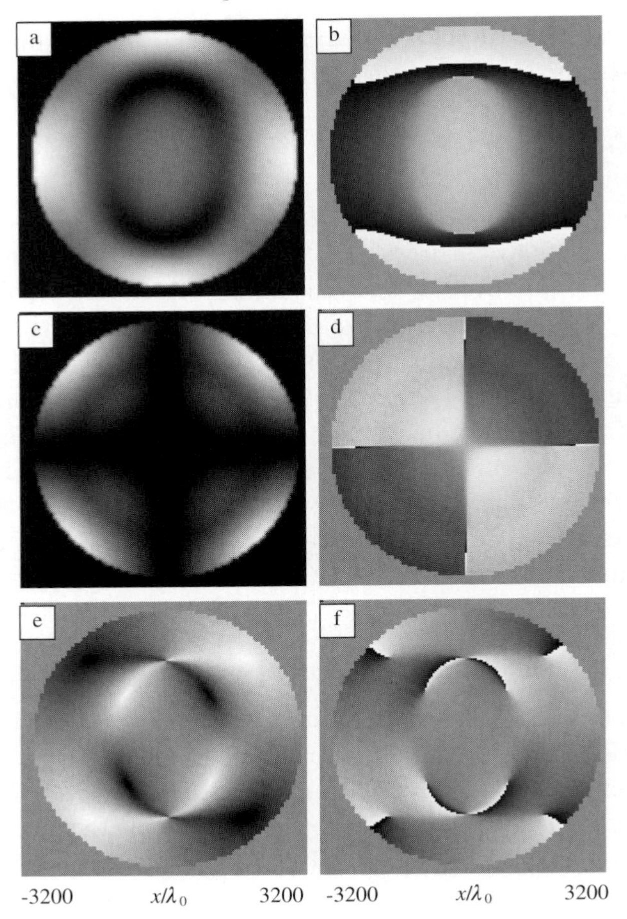

Figure 20.11 Various distributions of the reflected light at the exit pupil of the objective lens of Figure 20.2, when the MO stack of Figure 20.9 is placed in front of the SIL with a 100 nm air gap. (a) Intensity distribution for E_x, containing 36% of the incident optical power. (b) Phase distribution for E_x; the gray-scale covers the range -180° (black) to $+180^\circ$ (white). (c) Intensity distribution for E_y, containing 11.5% of the incident power. (d) Phase distribution for E_y; the phase difference between adjacent quadrants is nearly 180°. (e) The polarization ellipticity η ranging from -45° (black) to $+45^\circ$ (white). (f) The polarization rotation angle ρ ranging from -90° (black) to $+90^\circ$ (white).

between adjacent quadrants. The polarization distribution over the exit pupil (see Figures 20.11(e), (f)) is highly non-uniform and contains all possible states of polarization, i.e., linear, elliptical, and circular.

To observe the contribution to E_y by MO activity, we eliminate the magnetization of the disk by setting to zero the off-diagonal element ε' of the tensor, then subtracting the complex-amplitude distributions at the exit pupil with and without the magnetization. In doing so the x-component cancels out exactly, showing that there are no magnetic contributions to the reflected E_x. However, the y-component shows a residual distribution ΔE_y. Figure 20.12 shows the intensity and phase plots for ΔE_y at the exit pupil of the objective for a 100 nm gap width. Notice that this MO contribution to E_y has circular symmetry; moreover, it is large in the region where absorption by the disk is strong (compare the position of the bright ring in Figure 20.12(a) with that of the dark oval-shaped region in Figure 20.11(a)).

The total optical power contained in the distribution of Figure 20.12 is 0.37×10^{-4} of the incident power; the corresponding value for the case of zero gap width is 0.44×10^{-4}. Despite the fact that interference effects at the air gap have boosted ΔE_y in the central region of the aperture, a substantial reduction in evanescent coupling has caused an overall reduction of ΔE_y. Also notice the phase non-uniformity of ΔE_y in Figure 20.12(b), ranging from 246° at the center to −70° at the edge of the lens. Variations over the aperture of the relative phase between E_x and ΔE_y (see Figures 20.11(b) and 20.12(b)) have negative implications for the readout signal from the disk, as will be discussed shortly.

To isolate the contribution to the MO signal made by evanescent coupling, we place a mask in the central region of the beam, blocking all the rays below the critical angle. Figure 20.13 shows computed plots of the total reflectivity (i.e., $|E_x|^2 + |E_y|^2$ integrated over the aperture), versus the gap width, and the total contribution to E_y by the MO activity (i.e., the integrated value of $|\Delta E_y|^2$) versus the gap width. With increasing gap width the reflectance increases, leaving less light to be coupled to the magnetic film. In conse-

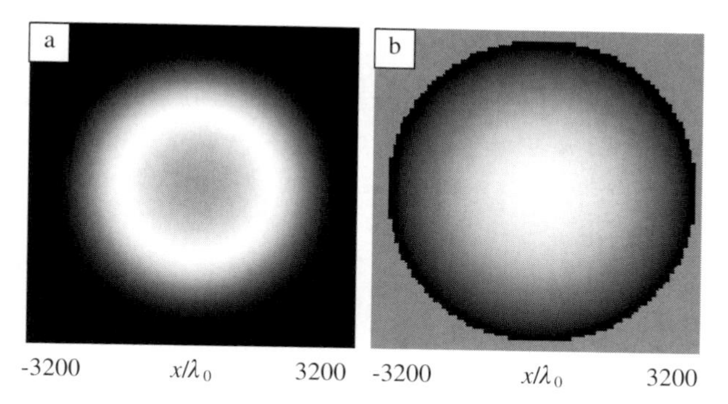

Figure 20.12 Plots of intensity and phase for the MO contribution to the reflected light, ΔE_y, at the exit pupil of the objective lens of Figure 20.2. The multilayer stack of Figure 20.9 is assumed to be in front of the SIL with a 100 nm gap. (a) Intensity distribution, containing a fraction 0.37×10^{-4} of the incident optical power. (b) Phase distribution, ranging from −70° (black) to +246° (white).

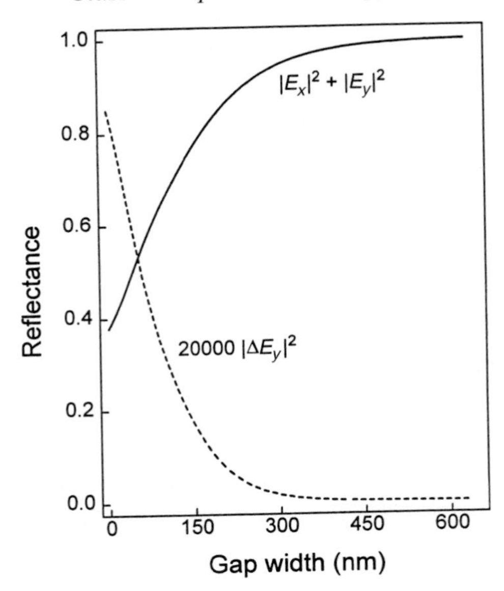

Figure 20.13 Total reflectivity (solid line) and the integrated intensity of the MO signal (broken line) at the exit pupil as functions of the gap width. These calculations correspond to the system of Figure 20.2 in conjunction with the quadrilayer MO stack of Figure 20.9, when a mask blocks the central region of the beam.

quence of this reduced coupling, the MO content of E_y progressively decreases; by the time the gap width reaches 200 nm, there is hardly any E_y left from the evanescently coupled MO interaction.

A similar trend may be seen in the normalized differential signal, $(S_1 - S_2)/(S_1 + S_2)$, which is plotted versus the gap width in Figure 20.14. (See Figure 20.10 and Eq. (20.2) for the definition of S_1, S_2.) Again we have blocked the central region of the incident beam in order to concentrate on the effects of evanescent coupling. With the SIL and the disk in contact, the normalized differential signal is close to its ideal value, which is twice the tangent of the Kerr rotation angle, namely, $2 \tan 0.66^\circ = 0.023$. As the gap widens, the differential signal drops sharply: at 100 nm gap width, for instance, the signal is down by a factor of four. Roughly one-half of this drop may be attributed to the reduction in ΔE_y and the corresponding rise in reflectivity (see Figure 20.13). The remaining half, however, is due to variations over the beam's cross-section of the relative phase $\phi_x - \phi_y$ of E_x and ΔE_y.

It must be emphasized that the quadrilayer stack of Figure 20.9 is not specifically optimized for operation with the system of Figure 20.2. By changing the thicknesses and the refractive indices of the various layers and/or by

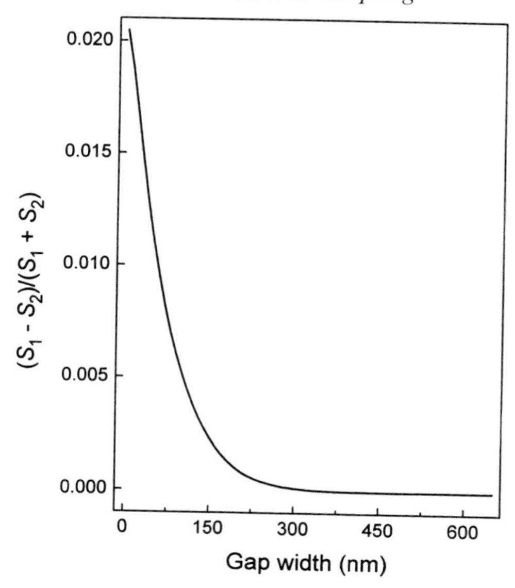

Figure 20.14 A computed plot of the normalized differential signal, $(S_1 - S_2)/(S_1 + S_2)$, versus the gap width. This result corresponds to the system of Figure 20.2 in conjunction with the quadrilayer MO stack of Figure 20.9 and the differential detector of Figure 20.10. It is assumed that a mask blocks the central region of the incoming laser beam, thus eliminating all the rays below the critical angle.

introducing dielectric coatings at the bottom of the SIL, it might be possible to improve upon the aforementioned performance figures. It is highly unlikely, however, that one can achieve significant gains in terms of the coupling efficiency and the magnitude of the MO Kerr signal over what we have already reported.

References for chapter 20

1 M. Born and E. Wolf, *Principles of Optics*, 6th edition, Pergamon Press, Oxford, 1980.
2 S. M. Mansfield, W. R. Studenmund, G. S. Kino, and K. Osato, High numerical aperture lens system for optical storage, *Opt. Lett.* **18**, 305–307 (1993).
3 T. W. McDaniel and R. H. Victora, eds., *Handbook of Magneto-optical Recording*, Noyes Publications, Westwood, New Jersey, 1997.
4 B. D. Terris, H. J. Mamin, and D. Rugar, Near-field optical data storage using a solid immersion lens, *Appl. Phys. Lett.* **65**, 388–390 (1994).

21

Internal and external conical refraction

Sir William Rowan Hamilton (1805–1865). Irish mathematician and astronomer who put forward the theory of quaternions, a landmark in the development of algebra, and discovered the phenomenon of conical refraction. His unification of dynamics and optics has had a lasting influence on mathematical physics, even though the significance of this work was not fully appreciated until after the rise of quantum mechanics. Hamilton had learned Latin, Greek, and Hebrew by the time he was five years old and learned many more languages afterwards. In 1827, while still an undergraduate, he was appointed Professor of Astronomy at Dublin's Trinity College. Hamilton published his third supplement to *Theory of Systems of Rays* in 1832. Near the end of this work he applied the characteristic function to study Fresnel's wave surface. From this he predicted conical refraction and asked Humphrey Lloyd, a professor of physics at Trinity College, to try to verify his prediction experimentally. Lloyd's confirmation two months later of conical refraction brought great fame to Hamilton. (Photo: courtesy of AIP Emilio Segrè Visual Archives.)

The phenomenon of conical refraction was predicted by Sir William Rowan Hamilton in 1832 and its existence was confirmed experimentally two months later by Humphrey Lloyd.[1,2] (James Clerk Maxwell was only a toddler at the time.) The success of this experiment contributed greatly to the general acceptance of Fresnel's wave theory of light.

Conical refraction has been known for nearly 170 years now,[1,2] and a complete explanation based on Maxwell's electromagnetic theory has emerged, which is accessible through the published literature.[3,4] The complexity of the physics involved, however, is such that it prevents us from attempting to give a simple explanation. We shall, therefore, confine our efforts to presenting a descriptive picture of internal and external conical refraction by way of computer simulations based on Maxwell's equations.

Overview

To observe internal conical refraction one must obtain a slab of biaxial birefringent crystal, such as aragonite, that has been cut with one of its optic axes perpendicular to the polished parallel surfaces of the slab (see Figure 21.1). When a collimated beam of light (say, from a HeNe laser) is directed at normal incidence towards the front facet of the slab, the beam enters the crystal and spreads out in the form of a hollow cone of light. Upon reaching the opposite facet, the beam emerges as two concentric hollow cylinders, propagating in the same direction as the original, incident beam.

External conical refraction is, in a way, the above phenomenon in reverse. Specifically, a hollow cone of light, converging towards a point on the surface

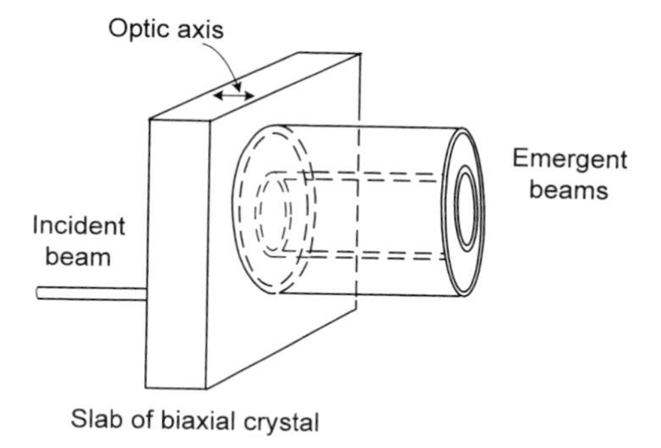

Figure 21.1 Internal conical refraction. A normally incident coherent beam arriving at the front facet of a slab of biaxial birefringent crystal propagates inside the slab in the form of a cone of light, and emerges from the rear facet as two hollow concentric cylinders. The crystal is cut so that one of its optic axes is perpendicular to the polished parallel surfaces of the slab.

of a biaxial crystal slab, becomes collimated along the optic-ray axis of the crystal and continues to propagate along that axis for as long as the beam remains within the crystal slab (see Figure 21.2). When the beam reaches the opposite facet of the slab, it emerges as an expanding cone of light. The focused cone thus "remains in focus" in its entire path through the crystal and diverges only after exiting the slab.

There are certain subtle differences between internal and external conical refraction; for instance, the *optic axis of wave normals* along which the beam propagates in the former case is not the same as the *optic-ray* axis in the latter. This and other differences will become clear in the course of the following discussions.

Biaxial birefringent crystals and their optic axes

In general, a birefringent crystal has three different refractive indices along the directions of its three principal axes. Assuming that the principal axes are the X-, Y-, and Z- axes of a Cartesian coordinate system, the principal indices may be denoted n_x, n_y, n_z. The index ellipsoid of this crystal has semi-axis lengths n_x, n_y, n_z along the coordinate axes, as shown in Figure 21.3. For a plane wave propagating along a given wave-vector $\boldsymbol{k}$, the plane passing through the center of the ellipsoid and perpendicular to $\boldsymbol{k}$ will, in

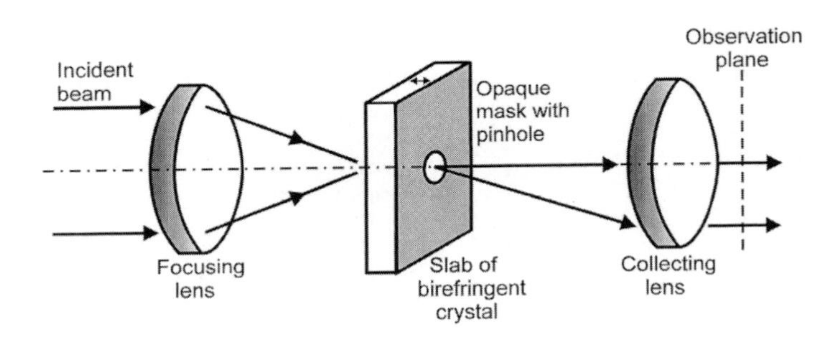

Figure 21.2 External conical refraction. A coherent monochromatic beam of light (wavelength λ_0) is focused by a lens at a biaxial birefringent crystal slab, which is cut with its polished surfaces perpendicular to one of its optic-ray axes. The exit facet of the crystal is painted black, except for a small aperture in the middle that is left clear to allow rays that propagate near the optic-ray axis to exit the crystal. The exiting rays propagate to a second lens where they are collected and recollimated. In our simulations the incident beam is uniform over the entrance pupil of the focusing lens, both lenses have $NA = 0.075$ and $f = 46667\lambda_0$, the crystal slab has thickness $= 5000\lambda_0$ and principal refractive indices $n_x = 1.533$, $n_y = 1.500$, $n_z = 1.565$, and the pinhole diameter $d = 100\lambda_0$.

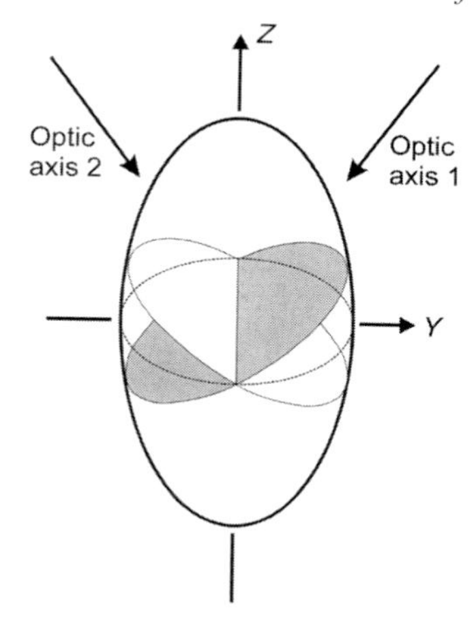

Figure 21.3 The index ellipsoid has semi-axes of length n_x, n_y, n_z along the principal axes X, Y, Z of the crystal. For a plane wave propagating in a given direction, a plane through the center of the ellipsoid and perpendicular to the wave normal will have an elliptical cross-section with the index ellipsoid. A propagation direction for which the cross-sectional ellipse becomes a circle is known as an *optic axis*. Similarly, the ray ellipsoid has semi-axes of length $1/n_x$, $1/n_y$, $1/n_z$ along the principal axes. For a given ray direction, a plane through the center of the ray ellipsoid and perpendicular to the ray will have an elliptical cross-section with the ray ellipsoid. A propagation direction for which the cross-sectional ellipse becomes a circle is known as an *optic-ray* axis. In general, biaxial crystals have two optic axes and two optic-ray axes.

general, have an elliptical cross-section with the index ellipsoid. The semi-axes of this cross-sectional ellipse yield the refractive indices associated with the two orthogonal polarizations of the beam. If the wave-vector $\boldsymbol{k}$ happens to be in such a direction that its corresponding cross-sectional ellipse becomes a circle, then the beam will "see" a single refractive index, irrespective of its state of polarization. The propagation direction corresponding to this circular cross-section is known as the optic axis. Crystals in which the three principal indices of refraction n_x, n_y, n_z are all different exhibit two such optic axes and are, therefore, referred to as biaxial. A crystal in which one index of refraction differs from the other two exhibits one optic axis and is known as a uniaxial birefringent crystal. Conical refraction occurs only in biaxial birefringent crystals.

Birefringent crystals also have a ray ellipsoid with semi-axis lengths $1/n_x$, $1/n_y$, $1/n_z$ along the principal axes. The ray ellipsoid, therefore, is different from the index ellipsoid, whose semi-axis lengths are the refractive indices themselves. While the index ellipsoid is relevant to the discussion of internal conical refraction, it is the ray ellipsoid that plays the central role in the case of external conical refraction. For a ray propagating along a given direction, the plane passing through the center of the ray ellipsoid and perpendicular to the ray will, in general, have an elliptical cross-section with the ellipsoid. If a ray happens to be in such a direction that its corresponding cross-sectional ellipse becomes a circle, then the direction of that ray defines an *optic-ray axis*. In general, the optic-ray axis is different from the optic axis, which is obtained in a similar fashion from the index ellipsoid. Biaxial birefringent crystals thus possess two optic-ray axes in addition to their two optic axes. Assuming $n_y < n_x < n_z$, it is not difficult to show that the optic-ray axis is in the YZ-plane and makes an angle θ with the Z-axis, where

$$\tan\theta = \sqrt{(n_x^2 - n_y^2)/(n_z^2 - n_x^2)}.$$

Internal conical refraction

To give a specific example, let us consider a slab of crystal having three principal refractive indices $n_x = 1.533$, $n_y = 1.500$, $n_z = 1.565$, and thickness $= 25000\lambda_0$, where λ_0 is the vacuum wavelength of the incident beam. (For the red HeNe wavelength of 633 nm, for example, the assumed thickness of the slab would be about 1.6 cm.) It is not difficult to show that the optic axes of this crystal are in the YZ-plane, located symmetrically with respect to the Z-axis at angles of $\pm 46.35°$ from it. We assume the slab is cut with one of its optic axes perpendicular to its polished surfaces.

Next, we assume that a collimated beam of coherent monochromatic light is normally incident on this slab; the beam has a Gaussian profile and its $1/e$ diameter is $150\lambda_0$. The intensity distribution for this beam is shown as a small bright spot in Figure 21.4(a). (The coordinate system is now redefined in such a way that the incident beam propagates along the Z-axis, and the polished surfaces of the crystal are parallel to the XY-plane.) The incident beam, upon entering the crystal, breaks up into a multitude of rays that propagate as a cone of light through the crystal, and emerge from the opposite facet of the slab in the form of two concentric hollow cylinders; the plot in Figure 21.4(b) shows the computed intensity distribution immediately after the beam exits the crystal. The scale of Figure 21.4(b) is the same as that of Figure 21.4(a), so one can compare the size and position of the bright rings with those of the incident beam. Note, in particular, that the incident beam is *not* at the center of the emerging cylinder, but at its bottom; see Figure 21.1. (Had the crystal been cut in such a way that its other optic axis was perpendicular to the facets, the incident beam would have been at the top of the cylinder.)

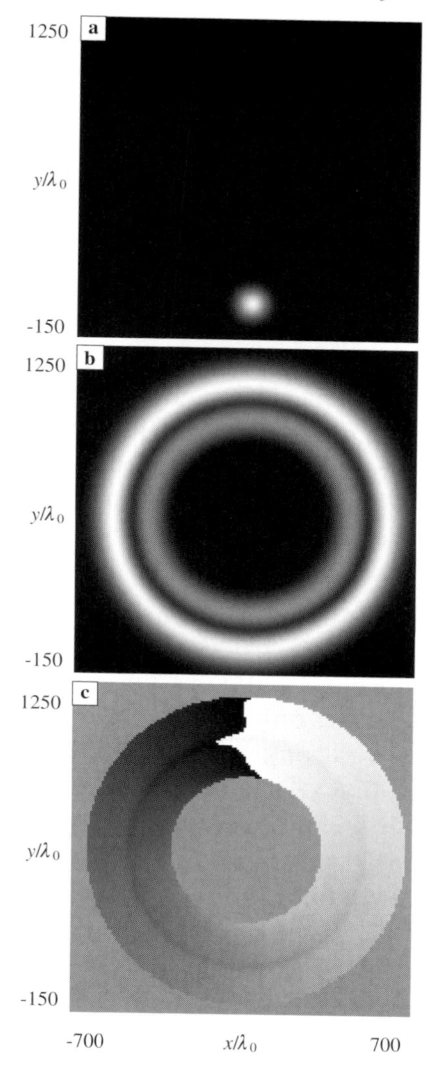

Figure 21.4 (a) The incident intensity distribution at the front facet of the crystal slab in Figure 21.1. (b) Emergent intensity distribution at the rear facet of the slab, corresponding to the circularly polarized incident beam shown in (a). (c) Distribution of the angle of the emergent polarization vector to the X-axis. The gray-scale is such that a white pixel represents a $+90°$ angle while a black pixel corresponds to a $-90°$ angle. The emergent polarization is linear at any given point on the beam's cross-section, but its direction varies from point to point. At the top of the rings there is an apparent 180° discontinuity in the direction of polarization. The jaggedness of the discontinuity is caused by small numerical errors that are inevitable when computing the state of polarization in the dark regions around the rings.

To obtain the full rings seen in the present example we have assumed the state of polarization of the incident beam to be circular; states of both right and left circular polarization (RCP and LCP) yield the same results. Alternatively, the incident beam may be assumed to be unpolarized for the full rings to emerge. As we shall see later, with linearly polarized light a certain part of the rings will be missing.

Polarization and phase patterns of the refracted beam

The polarization and phase distributions emerging in internal conical refraction are quite interesting. At any given point on the beam's cross-section, the state of polarization is linear, but the direction of the *E*-field varies as one scans the cross-section of the beam. The gray-scale plot in Figure 21.4(c) shows the distribution of the angle between the polarization vector and the *X*-axis at the exit facet of the crystal. In this picture a black pixel represents a -90° angle, a white pixel corresponds to $+90^\circ$, and the gray pixels represent angles in between. At the bottom of the emergent rings of light the polarization angle is 0°, that is, the *E*-field is parallel to the *X*-axis. The angle increases continuously from 0° to 90° as one moves from the bottom to the top on the right side of the rings. Similarly, on the left side, the orientation angle of the *E*-field varies continuously from 0° at the bottom to -90° at the top. Thus the polarization vector rotates by 180° as the point of observation moves a full circle around the beam's cross-section. This seems to imply a discontinuity in the *E*-field distribution at the top of Figure 21.4(c). In reality, however, this discontinuity does not occur, because the phase of the *E*-field (not shown here) also undergoes a 180° change in a full circle around the rings. Thus the polarization vector rotates by 180° and, at the same time, its phase changes by 180° in a round trip of the circumference of the rings, so that the *E*-field distribution is continuous at all locations.

We emphasize once again that the incident beam in the case shown in Figure 21.4 is circularly polarized. Whether this beam is RCP or LCP, however, is immaterial because the chirality of the incident beam affects neither the intensity distribution nor the polarization state of the emergent beam. In other words, if an observer facing the beam scans the rings in a clockwise sense, the polarization vector also appears to rotate clockwise, whether the incident beam is RCP or LCP. The only way to determine the state of incident polarization is by examining the phase distribution of the emergent beam, which increases clockwise in one case and counterclockwise in the other. It is also interesting to note that unpolarized light (i.e., light containing equal amounts of RCP and LCP that have randomly varying amplitudes and phases) gives exactly the same distributions as in Figure 21.4. In the case of unpolarized light, however, the phase distribution is meaningless, because it varies randomly with time and with location on the beam's cross-section.

To gain an appreciation for the phase distribution over the beam's cross-section, we show in Figure 21.5 two computed interferograms corresponding to the superposition of the beam emerging from the exit facet of the crystal and a uniform reference beam. The beam entering the crystal is assumed to be RCP in all cases, but the reference beam is RCP in Figure 21.5(a) and LCP in Figure 21.5(b). We notice that in Figure 21.5(a) the outer ring has interfered constructively with the reference beam, whereas the inner ring shows destructive interference. As a general rule, there is a 180° phase shift between the inner and outer rings at radially adjacent locations, irrespective of the state of incident polarization. This phase difference aside, the two rings are identical in their polarization and phase distributions. The interferogram of Figure 21.5(b) is more complicated than that of Figure 21.5(a); nevertheless, it can be fully explained in terms of the states of polarization and the distribution of phase over the rings, which we have already described.

Effect of linear incident polarization

Figure 21.6 shows the computed intensity distributions at the exit facet of the crystal for three cases of linear incident polarization: (a) parallel to the X-axis; (b) at 45° to X; and (c) parallel to Y. In all three cases the emergent state of polarization (not shown) is similar to that in Figure 21.4(c). A segment

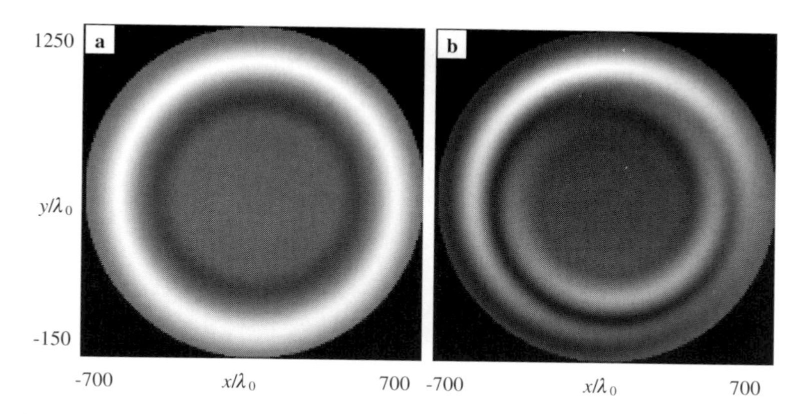

Figure 21.5 Interferograms showing the intensity distribution resulting from the superposition of the emergent beam (at the rear facet of the crystal slab) with a uniform reference beam. The beam incident on the crystal is RCP in both cases, but the reference beam is RCP in (a) and LCP in (b).

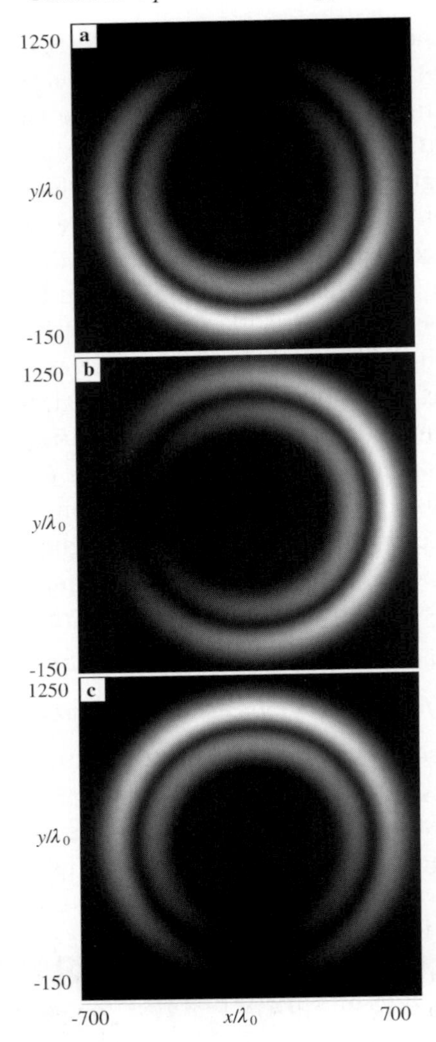

Figure 21.6 When the incident beam is linearly polarized, the emergent rings of light will be incomplete. This figure shows the intensity distribution at the rear facet of the slab in the cases where the incident E-field is (a) parallel to the X-axis, (b) at 45° to X, and (c) parallel to the Y-axis.

from the top of the rings is missing in Figure 21.6(a); this is the region that would have had polarization parallel to Y had the incident beam contained the corresponding E-field component. Similarly, the bottom of the rings is missing in Figure 21.6(c); this region would have had polarization along the X-axis. Unlike the distribution of polarization over the rings, which is independent of the state of incident polarization, the phase of the rings is very much a function of the polarization of the incident beam. When the incident beam is linearly polarized, as in Figure 21.6, the emergent phase (not shown) will have a constant value over the entire area of each ring. (As before, the two rings will have a 180° phase difference.) One may verify the above statements by considering the various linearly polarized incident beams as superpositions of RCP and LCP beams and by analyzing the corresponding superpositions at the exit facet of the crystal.

External conical refraction

Consider the system of Figure 21.2, which consists of a focusing lens, a slab of biaxial birefringent crystal, a pinhole, and a collimating lens. The incident beam is uniform, coherent, and monochromatic with a vacuum wavelength of λ_0. The crystal slab has refractive indices $n_x = 1.533$, $n_y = 1.500$, $n_z = 1.565$, and its thickness is $5000\lambda_0$. (For the red HeNe wavelength of 633 nm, for example, this slab would be 3.165 mm thick.) The optic-ray axes of this crystal are located symmetrically in the YZ-plane at angles $\theta = \pm 45.14°$ from Z. The slab is cut with its polished flat surfaces perpendicular to one of its optic-ray axes. (The coordinate system is now redefined to be such that the incident beam propagates along the Z-axis and the polished surfaces of the crystal are parallel to the XY-plane.)

When the incident rays enter the crystal slab they will propagate, in general, in various directions, but the rays that happen to be on a special cone, namely, the cone of external conical refraction, propagate strictly along the optic-ray axis and will emerge from a point opposite the point of entry into the crystal. A small pinhole (of diameter $100\lambda_0$ in the present example) on the exit facet of the slab allows only these axial rays to emerge. The emergent rays diverge in a hollow cone as they propagate towards a collecting lens, where they are recollimated and directed towards the observation plane.

Figure 21.7 shows computed plots of the intensity distribution, polarization ellipticity, and polarization rotation angle at the observation plane, corresponding to a circularly polarized incident beam. Note that the emergent rings of light in Figure 21.7(a) are in the bottom half of the exit pupil. (Had the crystal been cut with its other optic-ray axis perpendicular to the polished surfaces, the rings would have appeared in the top half of the exit pupil instead.) The ellipticity plot in Figure 21.7(b) is coded in gray-scale, black corresponding to −45° (i.e., LCP) and white to +45° (i.e., RCP). The relevant part of the plot, which is the region in the bottom half of the pupil where the emergent beam's intensity is non-vanishing, shows zero ellipticity. The emergent rings of light, therefore, are linearly polarized. The direction of

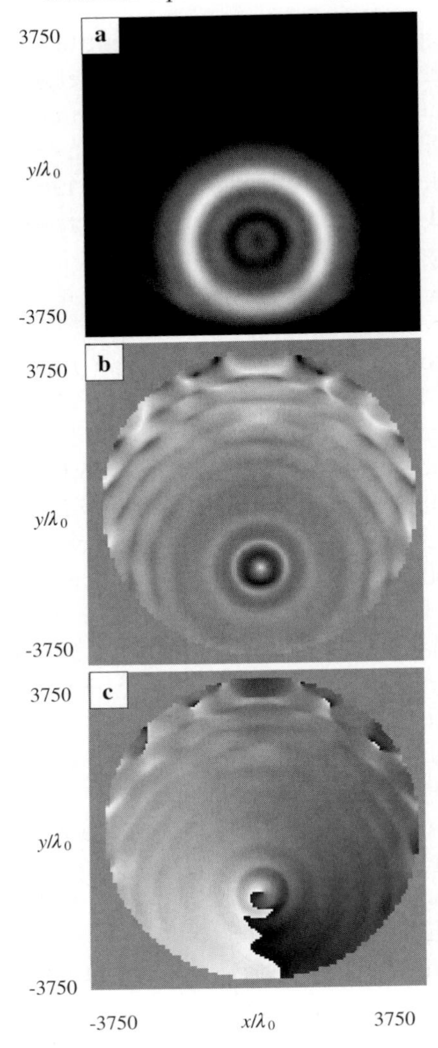

Figure 21.7 The distributions of (a) intensity, (b) polarization ellipticity, and (c) polarization rotation angle at the observation plane. The incident beam at the entrance pupil of the focusing lens is assumed to be circularly polarized. The ellipticity plot in (b) is coded in gray-scale, black corresponding to -45° (i.e., LCP) and white to $+45^\circ$ (i.e., RCP). The distribution of polarization rotation angle depicted in (c) is also coded in gray-scale, but the black pixels in this case represent -90° rotation from the X-axis and the white pixels represent $+90^\circ$ rotation. As before, the jaggedness of the transition from black to white in the lower part of (c) is caused by small numerical errors; since the discontinuity represented by this transition is not a physical discontinuity, this jaggedness has no physical significance.

this linear polarization varies over the rings, however, as the plot of polarization rotation angle in Figure 21.7(c) indicates. (The gray-scale used here assigns black to $-90°$ and white to $+90°$.)

According to Figure 21.7(c), over the circumference of the rings the polarization vector rotates from $-90°$ at the bottom (i.e., E-field antiparallel to Y-axis) to $0°$ at the top (E-field parallel to X) and back to $+90°$ at the bottom (E-field parallel to Y). The apparent discontinuity of polarization direction at the bottom of the rings does not signify a physical discontinuity, as before, because the phase of the rings (not shown here) also exhibits a $180°$ change during one full cycle around the rings. The overall E-field distribution turns out to be continuous after all.

Character of the emergent beam at the pinhole and the effect of incident polarization

The beam emerging from the pinhole in the system of Figure 21.2 possesses certain interesting features. Figure 21.8 shows computed plots of (a) intensity distribution, (b) polarization ellipticity, and (c) polarization rotation angle at the pinhole, for a circularly polarized incident beam. The intensity plot in Figure 21.8(a) shows a bright spot at the center of the pinhole, surrounded by a diffuse, more or less uniform background distribution. The origin of the diffuse light may be traced back to those incident rays that were outside the cone of external refraction and, therefore, once inside the crystal, did not become aligned with the optic-ray axis. The plot of polarization ellipticity in Figure 21.8(b) shows that the state of polarization varies from RCP in the bright white rings to LCP in the dark rings, covering the full gamut of elliptical polarization in the intervening regions. The plot of polarization rotation angle in Figure 21.8(c) indicates that the orientation of the ellipse of polarization is not uniform over the aperture but rotates through $180°$ around certain circular bands. All in all, this is a complex and fascinating state of affairs compared to the dull, uniform polarization state of the focused spot that first entered the crystal.

As was the case with internal conical refraction, the full cone of external refraction appears only when the incident beam contains all possible polarization directions. This is the case with RCP or LCP as well as with unpolarized light. When the incident beam happens to have linear polarization, however, certain parts of the emergent cone of light will be missing. This is shown in Figure 21.9 for an incident beam that is linearly polarized along the X-axis. The distributions of intensity, polarization ellipticity, and polarization rotation angle at the observation plane shown in Figure 21.9 are analo-

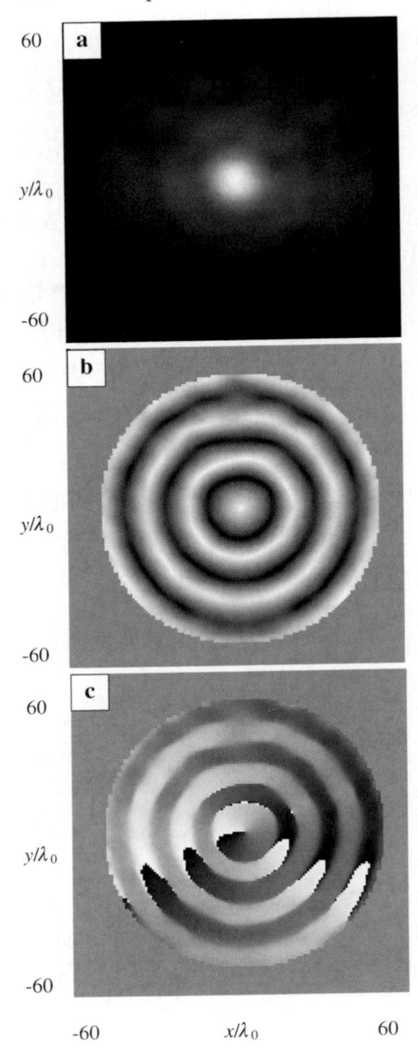

Figure 21.8 Distributions of (a) intensity, (b) polarization ellipticity, and (c) polarization rotation angle within the pinhole at the exit facet of the crystal slab. The incident beam at the entrance pupil of the focusing lens is assumed to be circularly polarized. The ellipticity plot in (b) is coded in gray-scale, black corresponding to -45° (i.e., LCP) and white to $+45^\circ$ (i.e., RCP). The distribution of polarization rotation angle depicted in (c) is also coded in gray-scale, but the black pixels in this case represent -90° rotation from the X-axis and the white pixels represent $+90^\circ$ rotation.

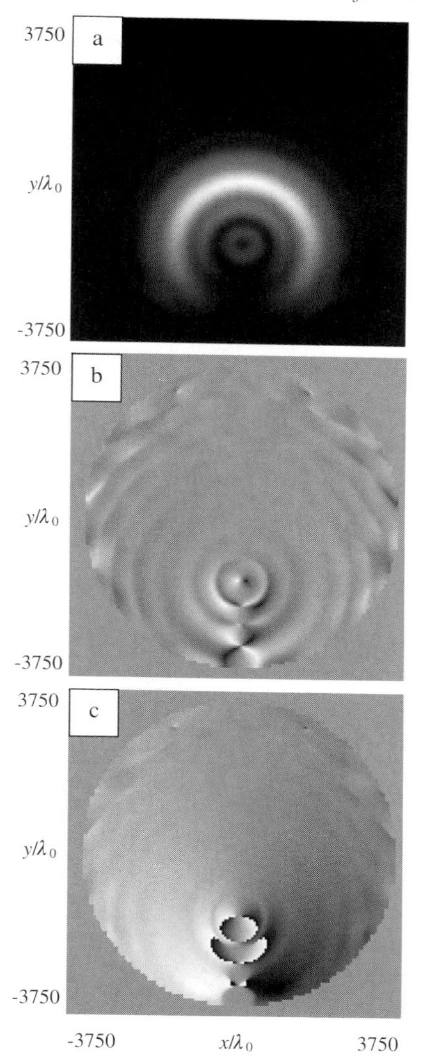

Figure 21.9 Same as Figure 21.7 except for the state of polarization of the incident beam, which is linear along X in the present case.

gous to those in Figure 21.7, where the incident beam is circularly polarized. The lower part of the rings in Figure 21.9(a), however, is missing simply because the corresponding polarization, linear along Y, is not present in the incident beam. Aside from this missing segment, other features of the emergent beam shown in Figure 21.9 are quite similar to those in Figure 21.7.

References for chapter 21

1 W. R. Hamilton, *Trans. Roy. Irish Acad.* **17**, 1 (1833).
2 H. Lloyd, *Trans. Roy. Irish Acad.* **17**, 145 (1833).
3 M. Born and E. Wolf, *Principles of Optics*, 6th edition, chapter 14, Pergamon Press, Oxford, 1980.
4 M. V. Klein, *Optics*, Wiley, New York, 1970.

22

The method of Fox and Li

The electromagnetic fields within a waveguide or a resonator cannot have arbitrary distributions. The requirements of satisfying Maxwell's equations as well as the boundary conditions specific to the waveguide (or the resonator) confine the distribution to certain shapes and forms. The electromagnetic field distributions that can be sustained within a device are known as its stable modes of oscillation.[1,2]

When the device and its geometry are simple, the stable modes can be determined analytically. For complex systems and complicated geometries, however, numerical methods must be used to solve Maxwell's equations in the presence of the relevant boundary conditions. The method of Fox and Li is an elegant numerical technique that can be applied to certain waveguides and resonators in order to obtain the operating mode of the device. Instead of solving Maxwell's equations explicitly, the method of Fox and Li uses the Fresnel–Kirchhoff diffraction integral to mimic the physical process of wavefront propagation within the device, thus arriving at its stable mode of operation after several iterations.[3,4]

To illustrate the method of Fox and Li we focus our attention on the confocal resonator shown in Figure 22.1(a). Let us assume that the two mirrors are aberration-free parabolas with an effective numerical aperture $NA = 0.01$ and focal length $f = 62\,500\lambda_0$ (λ_0 is the vacuum wavelength of the light confined within the cavity). The clear aperture of each mirror will therefore have a diameter of $1250\lambda_0$. (For the HeNe wavelength of $\lambda_0 = 0.633\,\mu\text{m}$, for example, this resonator will be a filament 8 cm long and 0.8 mm wide.) The resonator of Figure 22.1(a) may be modeled as the periodic-lens waveguide depicted in Figure 22.1(b). The beam starts at the focal plane of the first lens, becomes collimated, reaches the second lens, is focused by the second lens, and the process repeats itself over and over again. The essence of the method of Fox and Li for the computation

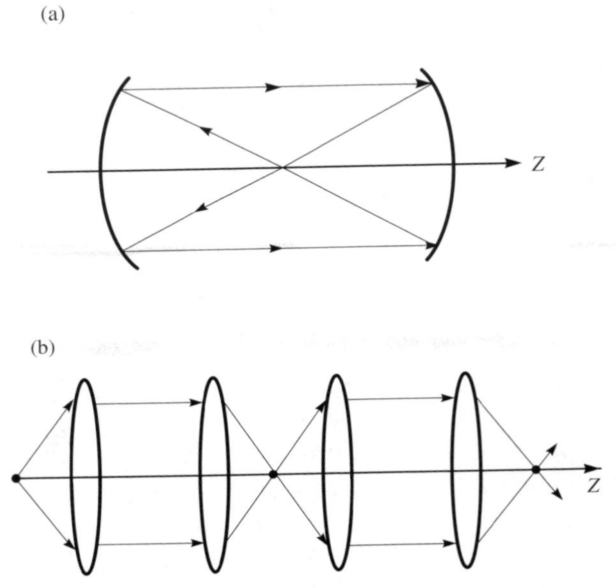

Figure 22.1 (a) Schematic diagram of a confocal optical resonator consisting of two parabolic mirrors. The mirrors are identical, each having numerical aperture NA and focal length f. (b) A periodic-lens waveguide that can be used to simulate the behavior of the resonator.

of the stable mode within this cavity may now be described as follows. An initial distribution is propagated through the periodic-lens waveguide until a steady-state distribution is reached. In the steady state the shape of the complex-amplitude distribution within the cavity will no longer change with successive iterations, but its power content will decline at a constant rate due to losses in the cavity. These ideas may best be explained by several examples.

The lowest-order mode

The mode of the cavity that is easiest to obtain by the method of Fox and Li is the lowest-order mode. Typically, just about any arbitrary initial distribution that one picks will converge to the lowest-order mode. From a practical standpoint this is very useful, because the lowest-order mode is also the mode in which the resonator operates, under most practical conditions. In Figure 22.2(a) we show a uniform initial distribution within a fairly large circular aperture. After going through about 80 iterations, this distribution settles into the mode known as the 0,0 mode of the cavity and shown in Figures 22.2(b)–(d). The 0,0 mode is essentially Gaussian in character, although, as

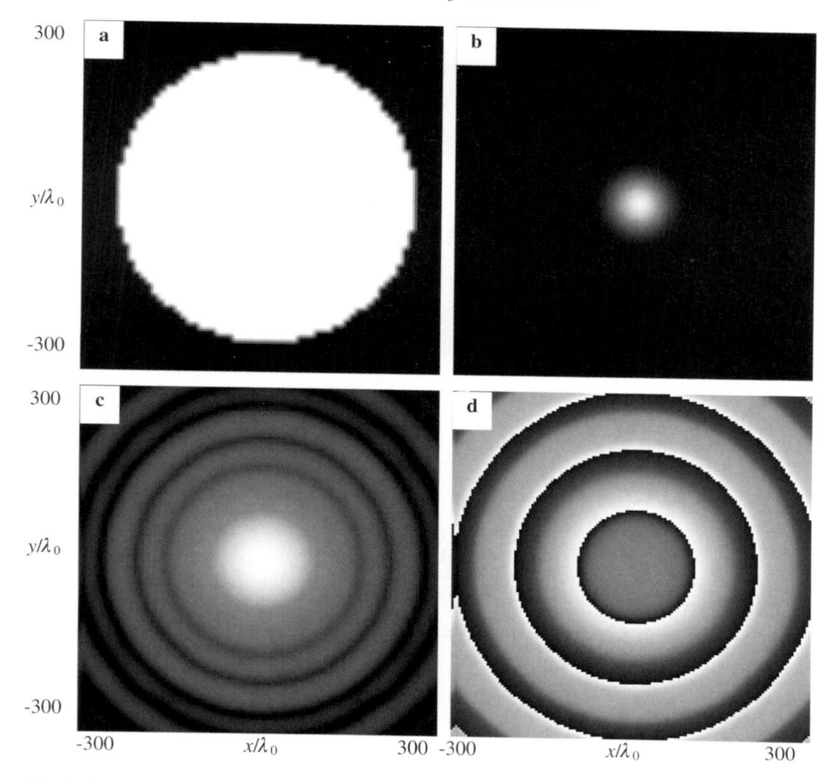

Figure 22.2 Computing the 0,0 mode of the resonator shown in Figure 22.1 using the method of Fox and Li. (a) The assumed initial distribution, having uniform amplitude and constant phase across a wide circular aperture. (b) Computed intensity distribution at the mid-plane of the cavity, obtained after 80 iterations. (c) Same as (b) but showing the logarithm of intensity on a scale of 1 (white) to 10^{-5} (black). (d) Distribution of the phase in the mid-plane of the cavity corresponding to the steady-state intensity distribution shown in (b) and (c). In this picture a white pixel represents a $+180^\circ$ phase angle, a black pixel represents a 180° phase angle, and the gray pixels represent the continuum of values in between.

the logarithmic plot of intensity in Figure 22.2(c) and the phase plot of Figure 22.2(d) show, it has an oscillating tail. The oscillation is caused by the finite apertures of the mirrors, which truncate the ideal, Gaussian mode.

For the above simulation a plot of the power attenuation coefficient γ versus iteration number is shown in Figure 22.3. γ is the ratio of the optical power contained in the beam after a given iteration to the same quantity before the iteration. It thus represents, for the particular mode under consideration, the fractional losses of the cavity during one round trip of the beam. The steady-state value of γ is also related to the eigenvalue of the

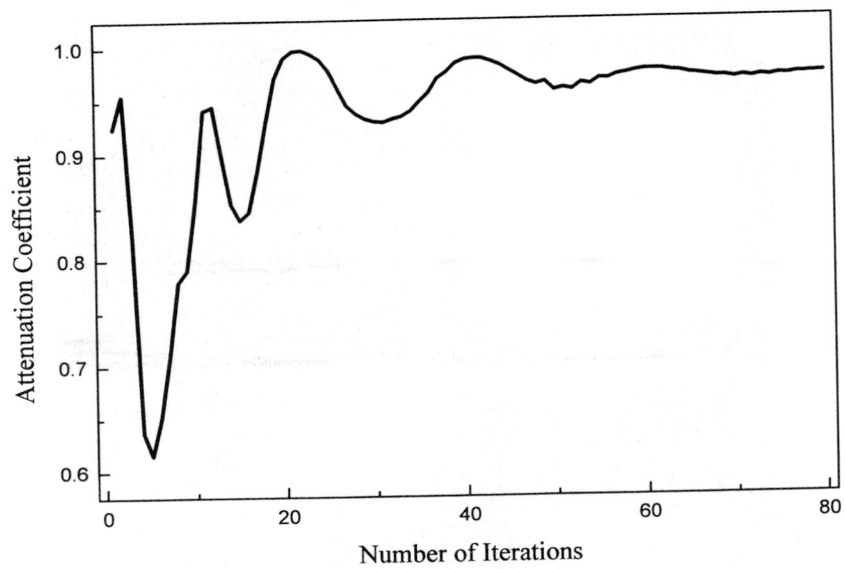

Figure 22.3 Evolution of the power attenuation coefficient γ during the simulation that led to the 0,0 mode shown in Figure 22.2. The computation stabilizes after about 80 iterations, and the steady-state value of γ is close to 0.97.

mode under consideration; the mode itself is an eigenfunction of the cavity. In the present example, where the steady-state value of γ is 0.97, the losses for the lowest-order mode are indeed very small.

Higher-order modes

Although the method of Fox and Li is ideally suited for computation of the lowest-order mode of the cavity, under special circumstances (and sometimes with the aid of special tricks) it is possible to compute some of the higher-order modes as well. As an example, consider the initial distribution shown in Figure 22.4(a), which consists of four identical lobes, each having the same uniform intensity distribution. Although not shown, it is also assumed that the phase is 0° for the pair of lobes along one diagonal and 180° for the opposite pair. The stage is thus set for excitation of the so-called 1,1 mode of the cavity.

Figures 22.4(b)–(d) show the computed 1,1 mode obtained from the initial distribution of Figure 22.4(a) after 64 iterations. The plot of attenuation coefficient γ versus iteration number shown in Figure 22.5 reveals that the steady-state is reached after only about 40 iterations and that the final value

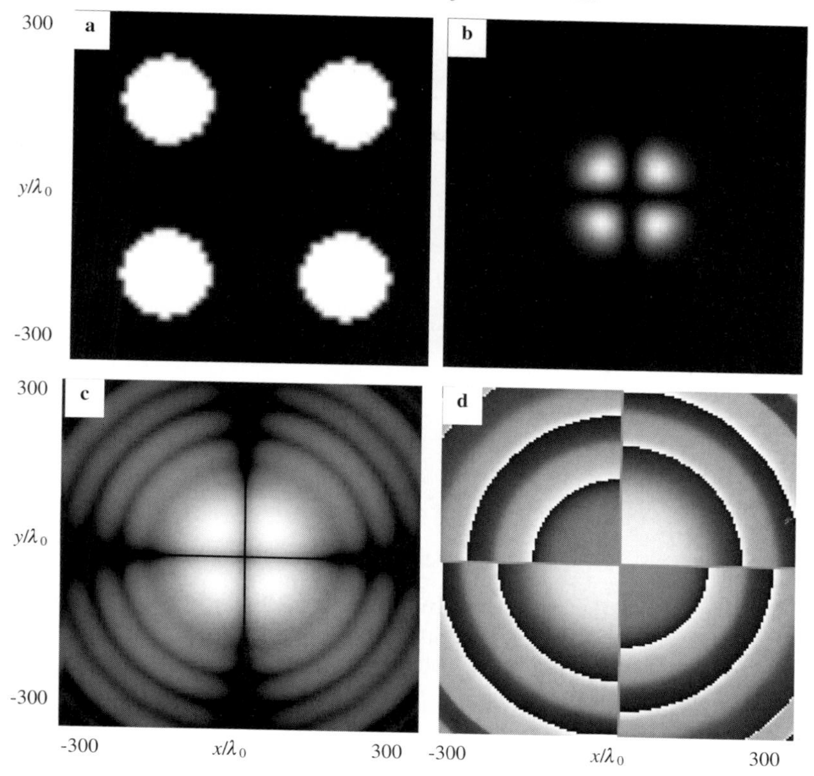

Figure 22.4 Computation of the 1,1 mode of the confocal resonator in Figure 22.1 begins with the initial distribution shown in (a). Here the four lobes of the initial pattern have uniform and equal intensities, but the phase of each lobe (not shown) differs from that of its adjacent lobes by 180°. The steady-state distribution in the mid-plane of the cavity is obtained after 64 iterations. (b) Plot of the intensity distribution in the steady state. (c) The same as (b) but showing the logarithm of intensity on a scale of 1 (white) to 10^{-4} (black). (d) The distribution of phase in the steady state. (For a description of the gray-scale see the caption to Figure 22.2(d).)

of γ is 0.87. That this value of γ is less than that for the 0,0 mode is consistent with the observation that the 1,1 mode is more spread out and, therefore, must suffer higher truncation losses at the apertures of the mirrors. For comparison with the steady-state distribution, two of the intermediate distributions obtained in this simulation are shown in Figure 22.6. In this figure the intensity plots appear on the left-hand side and the corresponding log(intensity) plots appear on the right-hand side. The patterns in Figures 22.6(a), (b) are obtained after 6 and 17 iterations, respectively.

Another example of a high-order mode is shown in Figure 22.7. Here the starting distribution of Figure 22.7(a) has eight lobes, each having the same

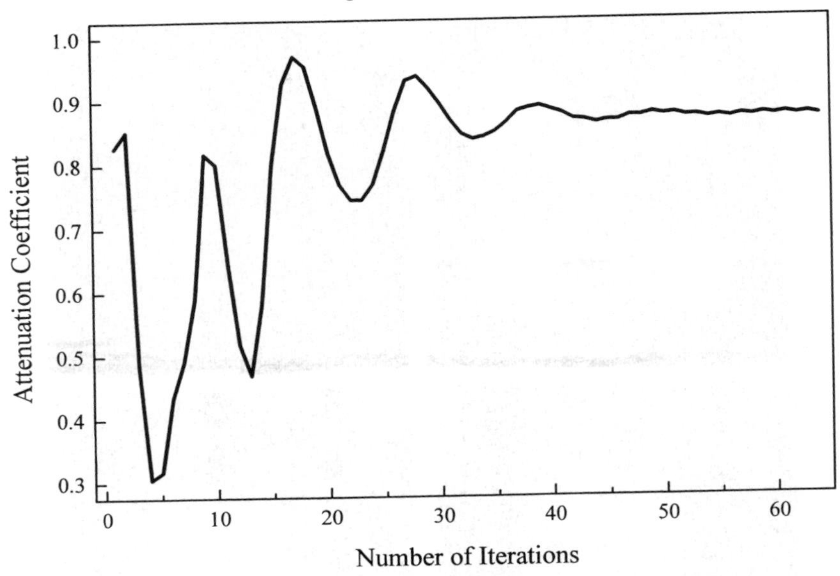

Figure 22.5 Evolution of the power attenuation coefficient γ during the simulation that led to the 1,1 mode shown in Figure 22.4. The computation stabilizes after about 40 iterations, and the final value of γ is approximately 0.87.

uniform amplitude; the phase of the adjacent lobes alternates between 0° and 180°. After 16 iterations the distribution of Figures 22.7(b)–(d) is obtained. Although this is very close to one of the high-order modes of the cavity, the simulation does not converge at this point, but continues to evolve towards the 0,0 mode. Figure 22.8, which is the corresponding plot of attenuation coefficient γ versus iteration number, clearly demonstrates the situation. Although after about 20 iterations the simulation appears to be stabilizing, small numerical errors disturb the system and push it away from the high-order mode. We confirmed that the steady-state distribution in this case was the same as the 0,0 mode of Figure 22.2; also notice that the steady-state value of γ in Figure 22.8 is 0.97, in agreement with our previous estimate of γ for the lowest-order mode.

To get an idea of how the pattern in Figure 22.7 reconfigures itself to resemble that of the 0,0 mode, we show in Figure 22.9 an intermediate state obtained from the initial state of Figure 22.7(a) after 33 iterations. Notice that some of the lobes have moved towards the center and have begun to merge, giving rise to a central bright spot which, thanks to its lower losses, will eventually overtake the higher-order mode.

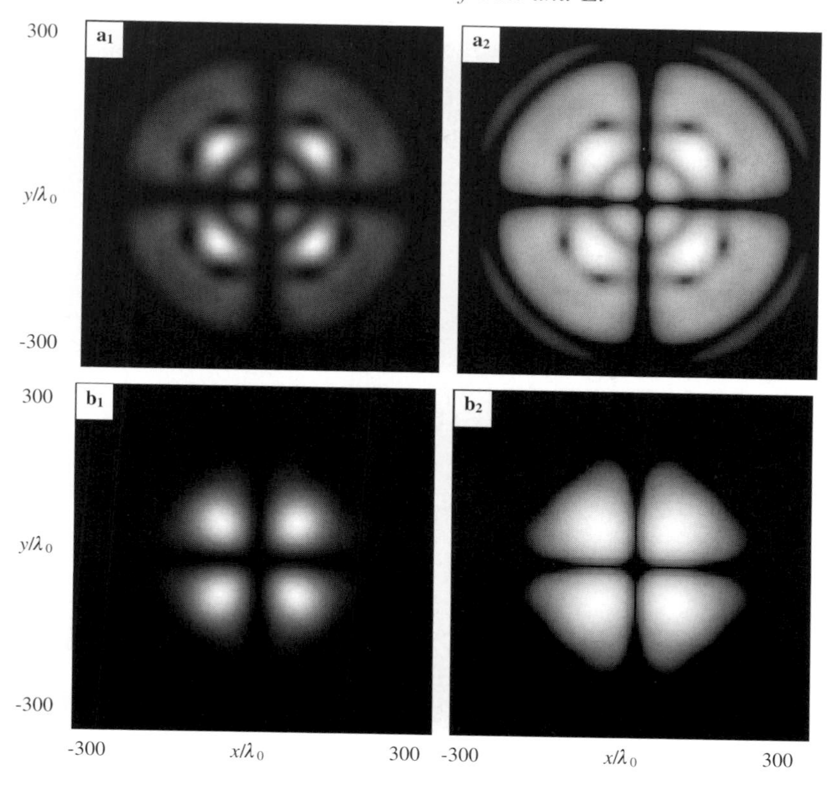

Figure 22.6 Intermediate patterns of intensity distribution in the cavity's mid-plane during the computation of the 1,1 mode shown in Figure 22.4. For each intensity plot on the left-hand side the corresponding logarithmic plot is shown on the right-hand side. The scale of the logarithmic plots is from 1 (white) to 10^{-2} (black). (a) After six iterations; (b) after 17 iterations.

Effect of misalignments and aberrations

One of the great advantages of the method of Fox and Li is that in the presence of misalignments and other imperfections, when analytical methods become intractable, this numerical scheme continues to be effective in calculating the stable mode of the resonator. As an example consider the same resonant cavity as that of Figure 22.1(a), but now suppose that one of the mirrors has two waves of primary coma. The results of computer simulations pertaining to this case are shown in Figures 22.10 and 22.11. Note that the stable mode in this case is a somewhat elongated version of the 0,0 mode, exhibiting a comatic tail. Note also that the steady-state attenuation coefficient γ is slightly reduced from its value in the unaberrated case.

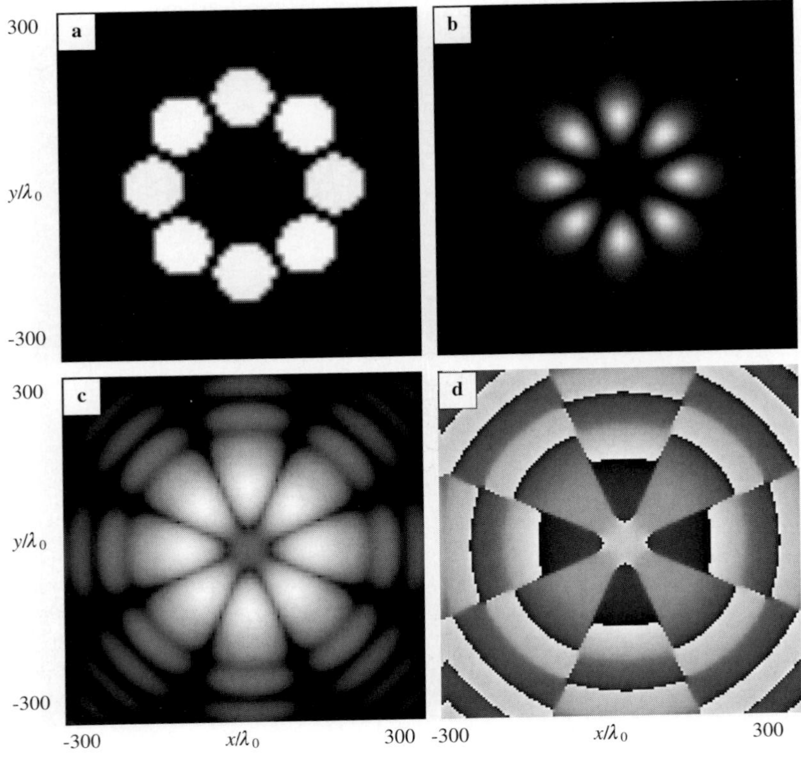

Figure 22.7 Computed results for a high-order mode of the confocal resonator shown in Figure 22.1. The assumed initial distribution in the cavity's mid-plane has eight lobes of uniform amplitude, as shown in (a), but its phase distribution (not shown) alternates between 0 and 180° from lobe to adjacent lobe. (b) Intensity distribution in the cavity's mid-plane after 16 iterations. (c) Same as (b), but showing the logarithm of intensity on a scale of 1 (white) to 10^{-3} (black). (d) Distribution of phase in the mid-plane of the cavity after 16 iterations, corresponding to the intensity patterns in (b) and (c). (For a description of the gray-scale see the caption to Figure 22.2(d).)

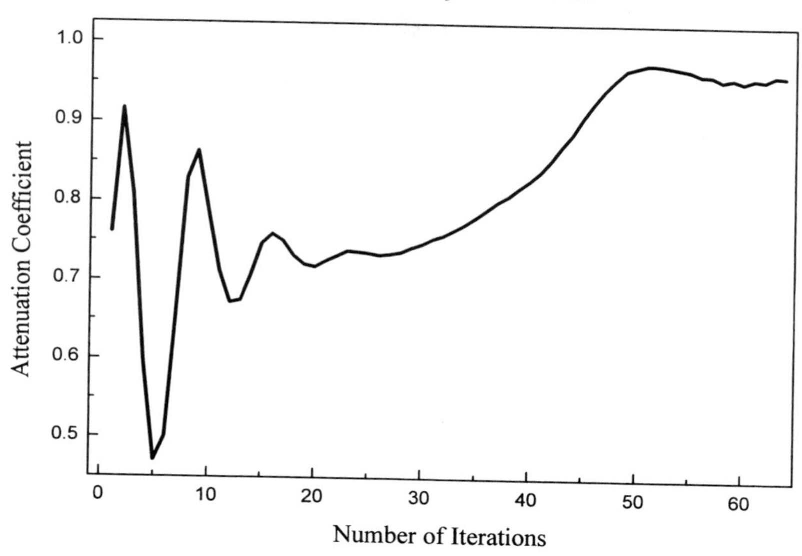

Figure 22.8 Evolution of the power attenuation coefficient γ during the simulation that started with the distribution of Figure 22.7(a) and went through the state shown in Figures 22.7(b)–(d). At first the simulation appears to stabilize with γ around 0.73, but instability sets in after about 25 iterations, forcing the system towards the 0,0 mode and a value of $\gamma \approx 0.97$.

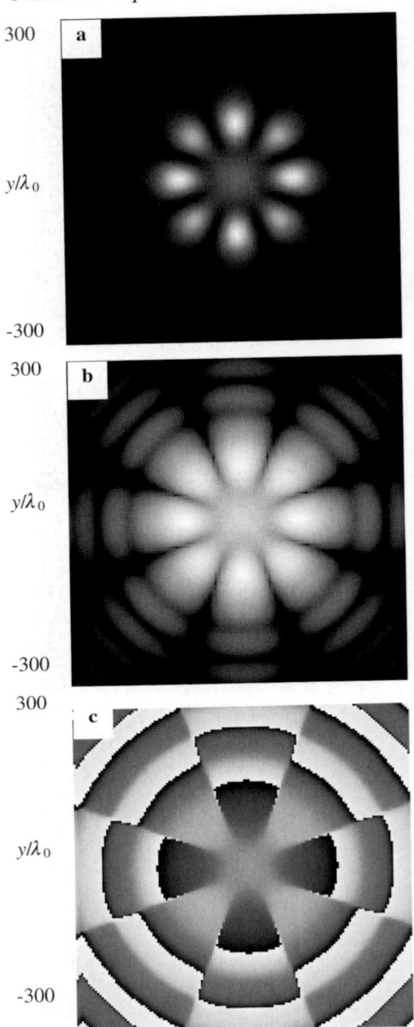

Figure 22.9 Distributions of (a) intensity, (b) log (intensity), and (c) phase at the cavity's mid-plane after a total of 33 iterations, starting in the initial state of Figure 22.7(a). This is a snap-shot from an intermediate state in the simulation whose other results are depicted in Figures 22.7 and 22.8. Note that four of the lobes have moved towards the center and started to merge into a bright central spot. This is the spot that will eventually become the dominant 0,0 mode.

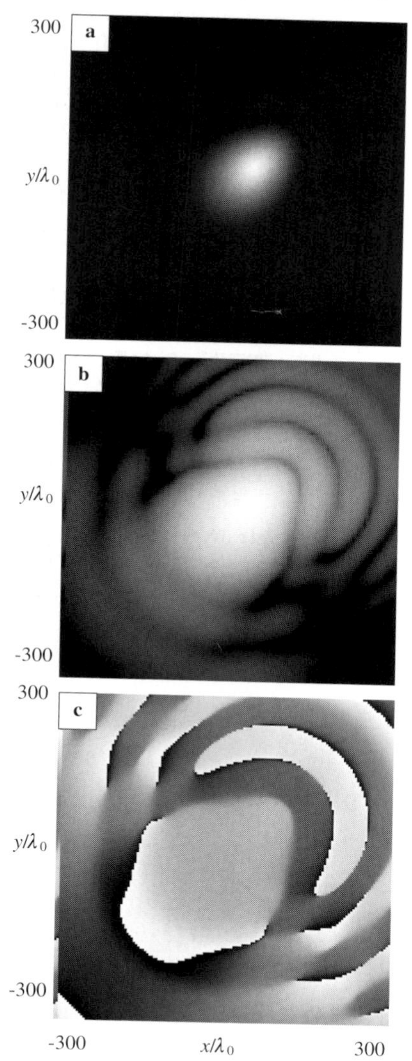

Figure 22.10 Computing the lowest-order mode of the confocal resonator of Figure 22.1 when one of the mirrors has two waves of primary coma. The assumed initial distribution has uniform amplitude and constant phase across a wide, circular aperture, as shown in Figure 22.2(a). (a) Computed intensity distribution at the mid-plane of the cavity, obtained after 64 iterations. (b) Same as (a) but showing the logarithm of intensity on a scale of 1 (white) to 10^{-5} (black). (c) Distribution of phase in the mid-plane of the cavity corresponding to the steady-state intensity distribution shown in (a) and (b). (For a description of the gray-scale see the caption to Figure 22.2(d).)

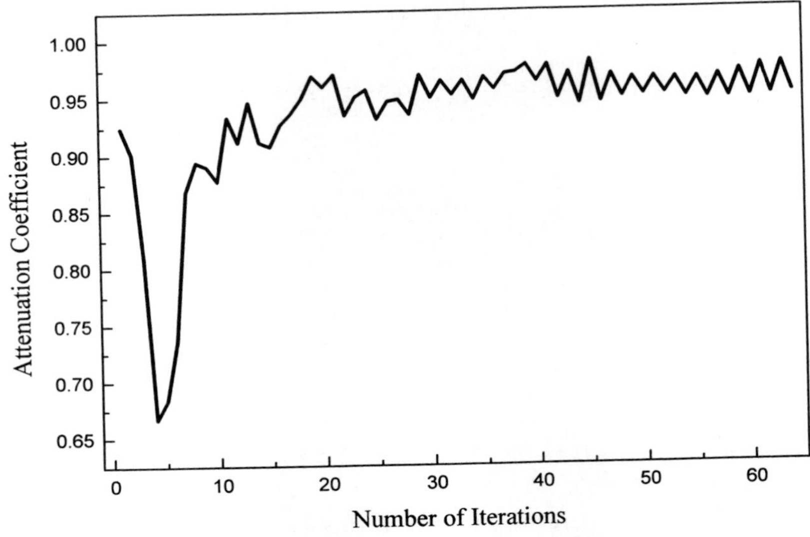

Figure 22.11 Evolution of the power attenuation coefficient γ during the simulation that led to the stable mode shown in Figure 22.10. The computation stabilizes after about 30 iterations, and the steady-state value of γ is close to 0.95.

References to chapter 22

1 A. E. Siegman, *An Introduction to Lasers and Masers*, McGraw-Hill, New York (1971).
2 H. Kogelnik and T. Li, Laser beams and resonators, *Proc. IEEE* **54**, 1312 (1966).
3 A. G. Fox and T. Li, Resonant modes in a maser interferometer, *Bell Syst. Tech. J.* **40**, 453 (1961).
4 A. G. Fox and T. Li, Modes in a maser interferometer with curved and tilted mirrors, *Proc. IEEE* **51**, 80 (1964).

23

The beam propagation method†

The beam propagation method (BPM) is a simple numerical algorithm for simulating the propagation of a coherent beam of light through a dielectric waveguide (or other structure).[1] Figure 23.1 shows the split-step technique used in the BPM, in which the diffraction of the beam and the phase-shifting action of the guide are separated from each other in repeated sequential steps, of separation Δz. One starts a BPM simulation by defining an initial cross-sectional beam profile in the XY-plane. The beam is then propagated (using classical diffraction formulas) a short distance Δz along the Z-axis before being sent through a phase/amplitude mask. The properties of the mask are derived from the cross-sectional profile of the waveguide (or other structure) in which the beam resides. The above steps of diffraction followed by transmission through a mask are repeated until the beam reaches its destination or until one or more excited modes of the guide become stabilized.[2,3]

Instead of propagating continuously along the length of the guide, the beam in BPM travels for a short distance in a homogeneous isotropic medium, which has the average refractive index of the guide but lacks the guide's features (e.g., core, cladding, etc.). After this diffraction step, a phase/amplitude mask is introduced in the beam path. To account for the refractive index profile of the guide, the mask must phase-shift certain regions of the beam relative to others. The mask must also adjust the beam's amplitude distribution to simulate the effects of regions that absorb or amplify, when the guide happens to contain such regions.

A good approximation to the real physical situation is obtained in the limit when $\Delta z \rightarrow 0$ and the phase/amplitude modulation imparted by the mask is

†The coauthors of this chapter are Ewan M. Wright and Mahmoud Fallahi of the Optical Sciences Center, University of Arizona.

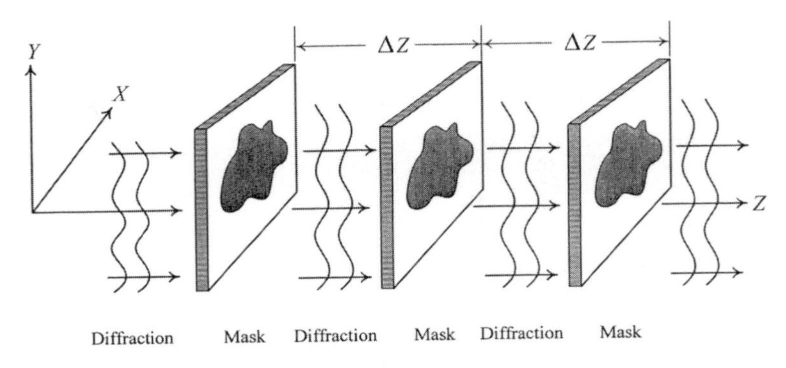

Figure 23.1 The split-step technique used in the BPM. Instead of continuously propagating the beam in an inhomogeneous environment, the method alternates between diffracting the beam a short distance through a homogeneous medium and then modulating its phase/amplitude through a mask. The mask imparts to the incident beam the cumulative effect of phase shifts and amplitude attenuations (or amplifications) during each propagation step.

scaled in proportion to Δz. In practice, the BPM works quite well without the need to make Δz excessively small. The various examples presented in this chapter should make the capabilities of the BPM abundantly clear.

Single-mode step-index fiber

Figure 23.2 shows a basic setup for injecting a laser beam into an optical fiber. The Gaussian beam of the laser diode is captured and truncated by the lens, then focused onto the entrance facet of the fiber. The focused beam typically loses about 4% of its power to reflection at the cleaved facet, but the remaining 96% enters the fiber. A fraction of this optical energy couples into the fiber's propagating mode and travels along the axis in the Z-direction; the rest radiates away from the axis and disappears in the region beyond the cladding.

For a $0.2NA$ diffraction-limited lens, Figure 23.3(a) shows the computed intensity profile of the focused spot in the XY-plane immediately after entering the fiber. The average refractive index n of the silica glass fiber is assumed to be 1.5, and so the wavelength λ of the light in this medium is λ_0/n, where λ_0 is the vacuum wavelength. The light amplitude distribution of Figure 23.3(a) serves in this example as the initial distribution for the BPM.

To simulate a single-mode step-index fiber, we use the phase mask of Figure 23.3(b) and choose $\Delta z = 2.5\lambda$. The assumed core and cladding diameters are 5λ and 30λ, respectively, and the index difference of

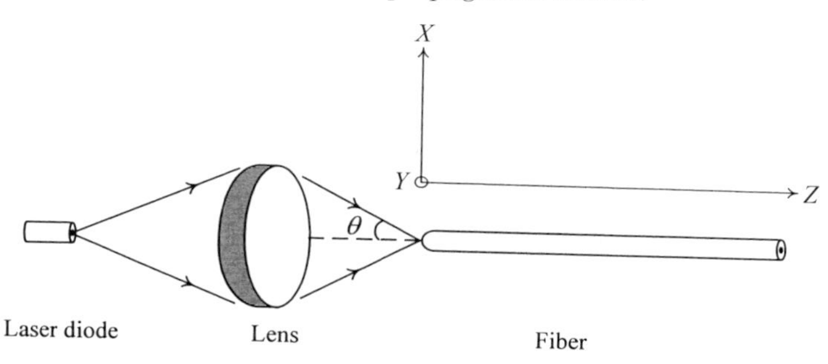

Figure 23.2 The emergent beam from a semiconductor laser diode is captured by a lens and focused onto the cleaved facet of a fiber. The numerical aperture of the focused cone of light is $NA = \sin\theta$, where θ is the half-angle of the cone of light arriving at the fiber. A small fraction of the incident beam is typically lost by reflection from the facet, while the remaining light penetrates the fiber, entering the core and cladding. Depending on the modal structure of the fiber and the cross-sectional profile of the injected beam, a certain fraction of the input optical power is coupled into the guided mode, which will then propagate along the fiber's axis. The remaining (uncoupled) light radiates away from the core and is lost in the surrounding regions.

$\Delta n = n_{\text{core}} - n_{\text{clad}} = 0.0125$ results in a 3° phase shift per distance λ of propagation. The mask depicted in Figure 23.3(b) is therefore required to advance the phase by 7.5° in its core region (relative to the cladding) during each BPM step. The light amplitude distribution inside the fiber reaches a steady state after a few hundred iterations; the intensity profile shown in Figure 23.3(c) is obtained at $z = 1500\lambda$. The mesh used in this simulation had 512×512 pixels, and the entire computation on a modern personal computer took less than one hour.

The fraction of the beam that radiates away from the core during a BPM simulation should not be allowed to reach the mesh boundary. The reason is that the periodic boundary condition imposed on the mesh by the fast Fourier transform (FFT) algorithm used in diffraction calculations tends to return the radiation modes into the computational region, via aliasing. In the present simulation we solved this problem by our choice of the mask, which, in addition to a core and cladding, contains a strongly absorbing region beyond the cladding (transmission coefficient zero for $r > 15\lambda$). The transition from cladding to absorber is tapered to minimize back-reflections into the core and cladding.

Figure 23.4(a) is a cross-sectional plot of the stabilized light amplitude distribution in the fiber; the vertical broken lines mark the core–cladding boundary. This guided mode is essentially trapped in the core but has eva-

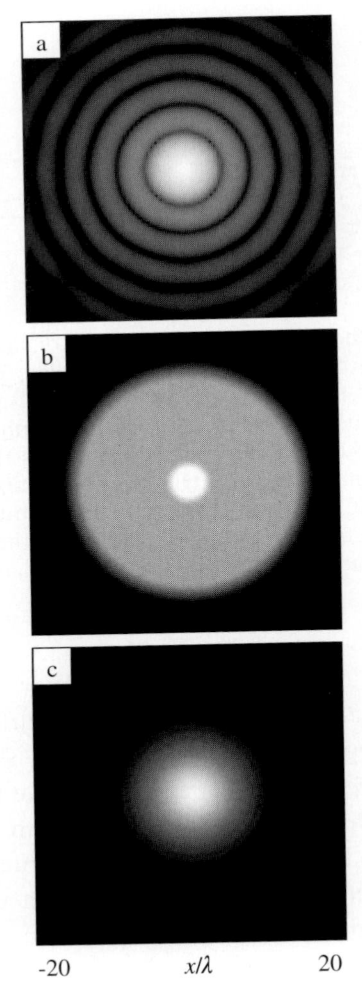

Figure 23.3 (a) Logarithmic plot of intensity distribution immediately after the front facet of a silica glass fiber, produced by a laser beam of wavelength λ_0 focused through a 0.2*NA* lens. (b) Phase mask used in the BPM simulation of a single-mode step-index fiber. The core and cladding diameters are 5λ and 30λ, respectively, and the phase shift imparted by the core relative to that imparted by the cladding is 3° per λ; here λ is the wavelength inside the fiber. The region beyond the cladding is absorptive. (c) Logarithmic plot of intensity distribution in the cross-section of the fiber at a distance $z = 1500\lambda$ from the fiber's front facet.

nescent tails in the cladding. The significant penetration of the evanescent waves into the cladding is a direct consequence of the small index-contrast Δn chosen for this particular example.[2,3]

Figure 23.4(b) shows the power content of the beam versus z throughout the simulation. (The power at any given point along the Z-axis is obtained by

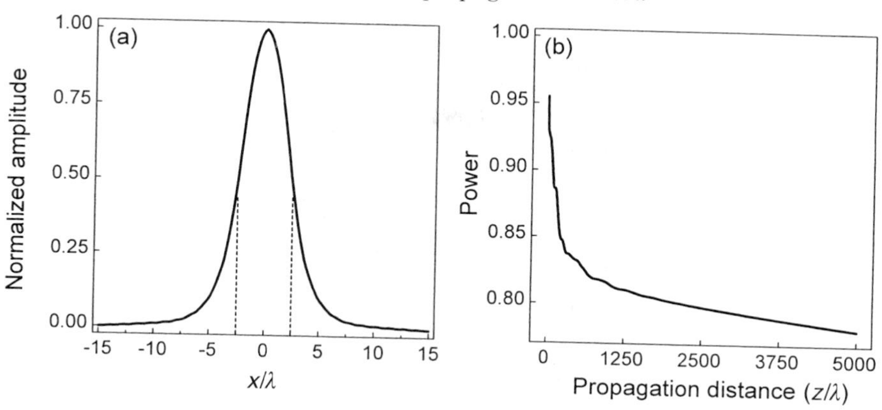

Figure 23.4 (a) Computed amplitude of the stable mode along the X-axis for the single-mode fiber of Figures 23.2 and 23.3 (the vertical lines denote the boundary between core and cladding). The mode stabilizes at about $z = 1500\lambda$ and does not change afterwards. (b) Computed optical power along the length of the fiber when the incident power at the front facet is set to unity.

integrating the beam's intensity over the entire cross-sectional area in the XY-plane.) The power of the beam, which is set to unity at the entrance pupil of the lens (see Figure 23.2), drops to ~ 0.96 immediately after the beam enters the fiber. It then drops rapidly until $z \sim 1000\lambda$ while the beam adjusts itself to the fiber, shedding some of its energy by radiating into the absorption region. The slight decline of optical power for $z > 1000\lambda$ is partly due to the slow decay of the radiative modes. Nonetheless, the curve in Figure 23.4(b) continues to exhibit a small negative slope even after a long propagation distance. This behavior is indicative of the presence of a small loss factor despite the fact that the simulated guide is lossless. So long as the diffraction step is treated non-paraxially, this small loss factor, which is a consequence of the discrete approximation to an inherently continuous problem, will remain an unavoidable feature of the BPM.[3]

Fiber with a complex core structure

Figure 23.5 shows the cross-section of a special fiber with a core that contains 19 low-index filaments symmetrically arranged around its axis. Such structures, known as photonic crystals or photonic bandgap structures, currently command worldwide attention because of their unique optical properties.

For our BPM simulation of the fiber depicted in Figure 23.5 we chose core and cladding diameters of 50λ and 100λ, respectively, and an index contrast of $\Delta n = 0.0125$, while placing a tapered absorber outside the cladding region.

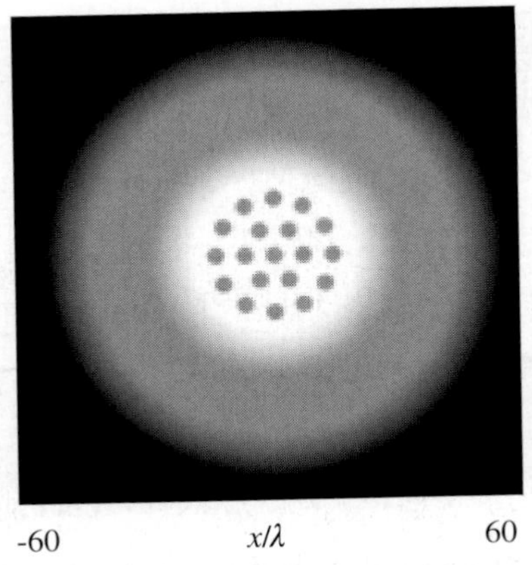

Figure 23.5 Phase mask used in simulating a special fiber containing 19 low-index filaments within its core. The core and cladding radii are 25λ and 50λ, respectively, and the region beyond the cladding is absorptive. The filaments each have a diameter of 4λ and the same refractive index as the cladding. All transition regions are tapered. The phase shift imparted to the beam by the high-index regions of the core (relative to the low-index cladding and the filaments) is 3° per λ.

The core filaments each had diameter 4λ and the same index of refraction as the cladding. A uniform beam, having a circular cross-section of diameter 40λ and unit optical power, was used as the initial distribution. Figure 23.6 shows the various cross-sectional patterns of intensity obtained along a propagation path 5000λ long. It is clear that several modes of the fiber have been excited and that interference among these modes gives rise to the observed patterns. Note also that the light tends to avoid the low-index filaments at all times. The beam's power content, plotted versus z in Figure 23.7, is seen to decline slowly with the propagation distance. The power stabilizes when the radiative modes leave the core and cladding to disappear into the surrounding absorber; however, a small negative slope similar to the one mentioned in connection with Figure 23.4(b) remains even after stabilization.

Y-branch beam-splitter

Figure 23.8 is a diagram of a Y-branch channel waveguide. This structure is typically embedded in a lower-index medium that plays the role of cladding.[2] The beam, injected on the left-hand side, establishes in the initial section a guided mode that propagates along the Z-axis; in Figure 23.8 the length of this initial section of the guide is z_1. The waveguide then slowly opens up over a distance z_2, and the beam follows this expansion adiabatically (i.e., without significant loss of power and without exciting higher-order modes). Once the

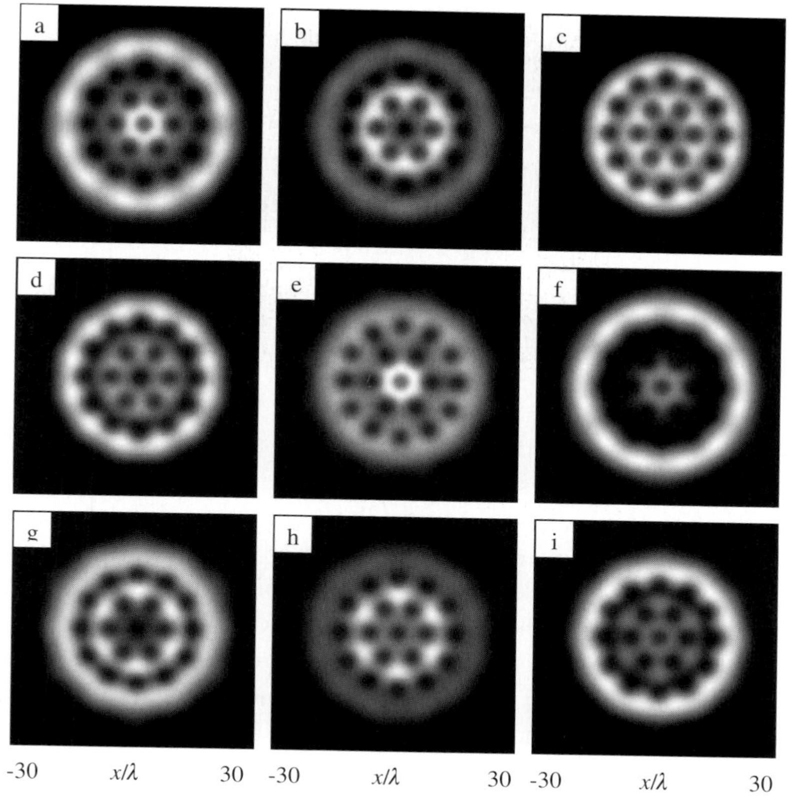

Figure 23.6 Computed intensity profiles in the core region of the fiber of Figure 23.5 obtained in a BPM simulation. The assumed split steps are of length $\Delta z = 2.5\lambda$. The initial distribution is a uniform circularly symmetric beam of diameter 40λ, which enters the fiber at $z = 0$. The distributions in (a) to (i) correspond to propagation distances $z/\lambda = 500, 1250, 1750, 2500, 3250, 3750, 4000, 4500, 5000$.

guide has been sufficiently broadened, it splits into two channels that slowly recede from each other over a distance z_3 until they are optically isolated. Afterwards the two channels may remain parallel for a distance z_4. Thus a beam injected into the initial section of the guide will split in two, each of which may be extracted from a separate channel.

A set of phase masks for simulating a symmetric Y-splitter is shown in Figure 23.9(a). From top to bottom these masks represent the initial section of the guide ($5\lambda \times 5\lambda$ square), the end of the expanded region in which the width of the mask increases to 10λ, and the length along the split section where the center-to-center separation of the two channels slowly increases from zero to 45λ (branching angle = 1.15°). The assumed lengths of the various sections are $z_1 = z_2 = 1000\lambda$ and $z_3 = z_4 = 2000\lambda$ (see Figure 23.8).

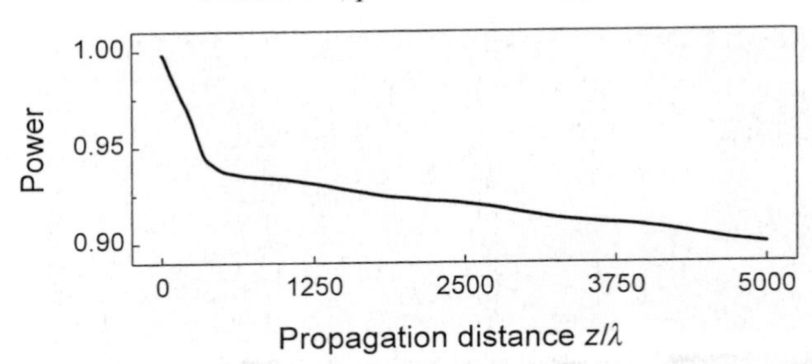

Figure 23.7 Total power of the beam versus propagation distance along the length of the fiber for the BPM simulation depicted in Figure 23.6. The incident power at $z = 0$ is set to unity.

Each mask imparts a 15° phase shift to the incident beam after a propagation step of $\Delta z = 5\lambda$; this corresponds to a 3° phase shift per λ. For the initial distribution at $z = 0$ we chose a uniform beam having a circular cross-section of diameter 14λ. The output of the device at $z = 6000\lambda$ is shown in Figure 23.9(b). The intensity profile in the broad section of the guide at $z = 2000\lambda$ (just before branching) is shown in Figure 23.9(c), while a plot of the phase distribution at the same location appears in Figure 23.9(d). The phase plot indicates the propagation of the radiative modes away from the core and into the absorbing region beyond the cladding.

Figure 23.10 shows the computed amplitude distributions at several cross-sections of the Y-splitter of Figure 23.8. Evident in this picture is the evolution of the guided mode from a narrow beam in the initial section of the guide

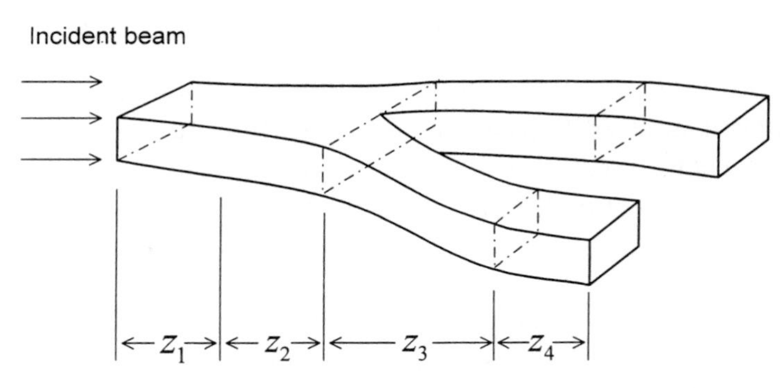

Figure 23.8 Core region of a Y-branch channel waveguide, which is typically embedded in a lower-index cladding. For adiabatic operation the initial section of the guide (length z_1) is slowly broadened over a distance z_2 before splitting into two branches. The branches then move apart at a small angle (typically less than 1°) over a distance z_3 until they are optically isolated from each other. Afterwards the two channels remain parallel for a distance z_4.

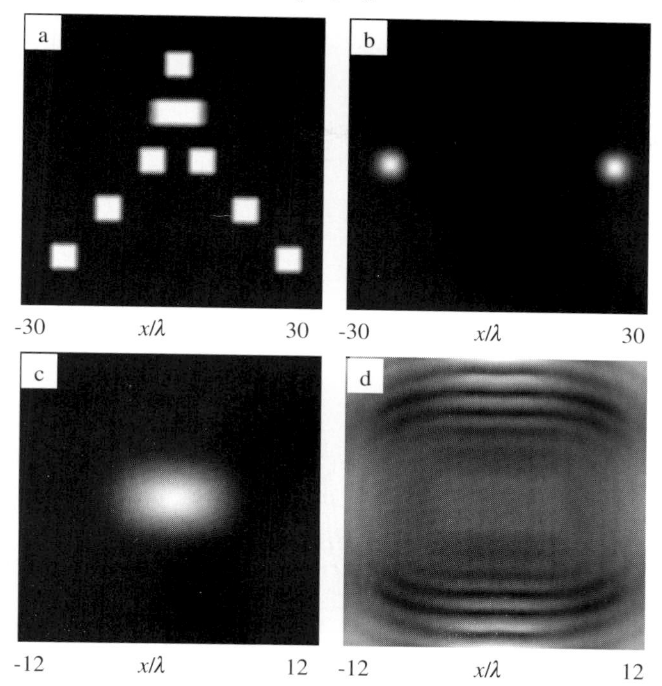

Figure 23.9 (a) A set of phase masks used to simulate the Y-splitter of Figure 23.8. From top to bottom: at the start of the guide; at the end of the expanded region; at three locations in the split section. (b) The computed intensity pattern at the end of the guide, $z = 6000\lambda$, showing two output beams confined to their respective channels. (c), (d) Intensity and phase distributions in the broad section of the guide, just before branching. In the phase plot the remnants of the incident beam, which are not coupled into a guided mode, are seen to be radiating away from the core.

to a wider beam in the broadened section, and onwards to a pair of well-confined beams in the divided channel. The power content of the entire beam is plotted versus z in Figure 23.11, indicating the losses in various sections of the guide and confirming the stabilization of power in the output channels once they are sufficiently separated from each other.

Directional coupler

Figure 23.12 shows a channel waveguide known as a directional coupler, which has applications such as switching in optical communication systems.[4] A beam of light injected into channel 1 propagates along that channel until it reaches a point where channel 2 is close enough to sense the evanescent tail of the guided mode. At this point the beam leaks into channel 2 and, after a certain distance, moves entirely into the second channel. If the parallel sec-

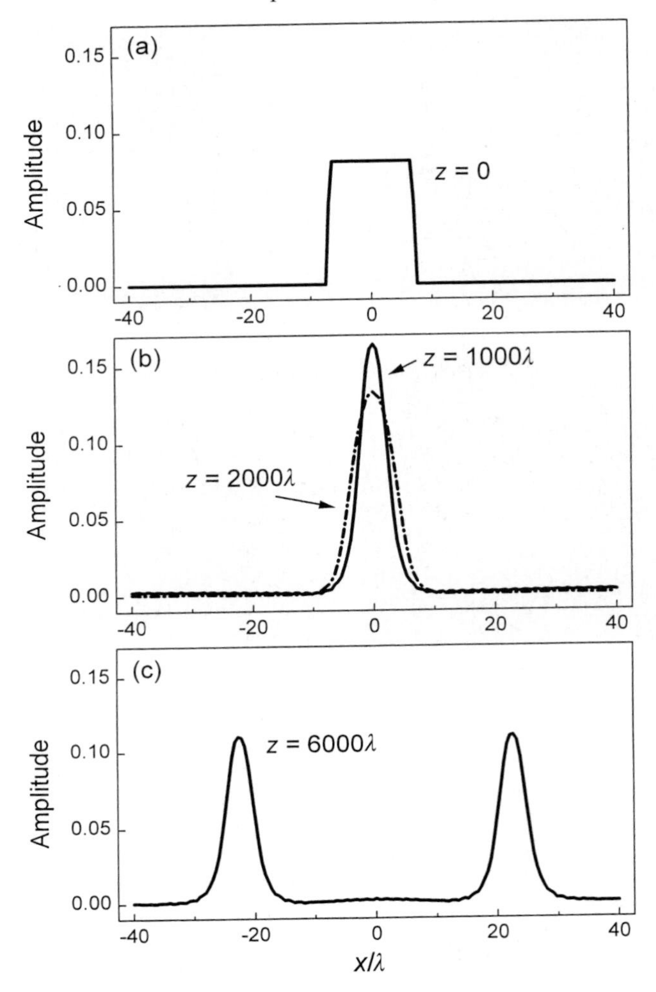

Figure 23.10 Plots of amplitude distribution at various cross-sections of the Y-splitter depicted in Figures 23.8 and 23.9. (a) The initial distribution at $z = 0$ is uniform, having a circular cross-section of diameter 14λ. (b) At $z = 1000\lambda$ (solid line), the end of the single-mode input channel, the beam is confined to the core region. At $z = 2000\lambda$ (broken line), just before branching, the broadened beam is seen to fit into the wider channel. (c) Emerging from the guide at $z = 6000\lambda$ are two identical beams.

tion of the guide is long enough, the back and forth coupling between the two channels may be repeated many times. In this region of strong coupling, the lowest-order modes of the guide are the even and odd modes depicted in Figure 23.12. Because these modes travel at different speeds, their relative phase $\phi_1 - \phi_2$ varies with distance along the guide. The beam resides entirely in channel 1 or channel 2 when $\phi_1 - \phi_2$ is 0° or 180°. When $\phi_1 - \phi_2 = \pm 90°$

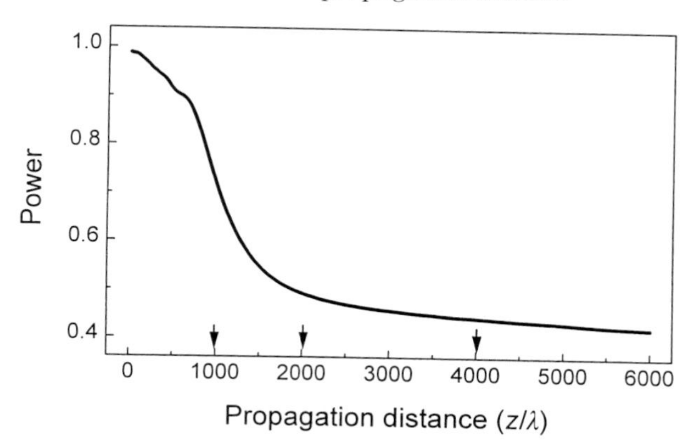

Figure 23.11 Power content of the beam versus z/λ in the BPM simulation of the Y-splitter depicted in Figures 23.9 and 23.10. The arrows indicate the beginning of the split section, the end of the split section, and the location where the split channels stop receding from each other and become parallel.

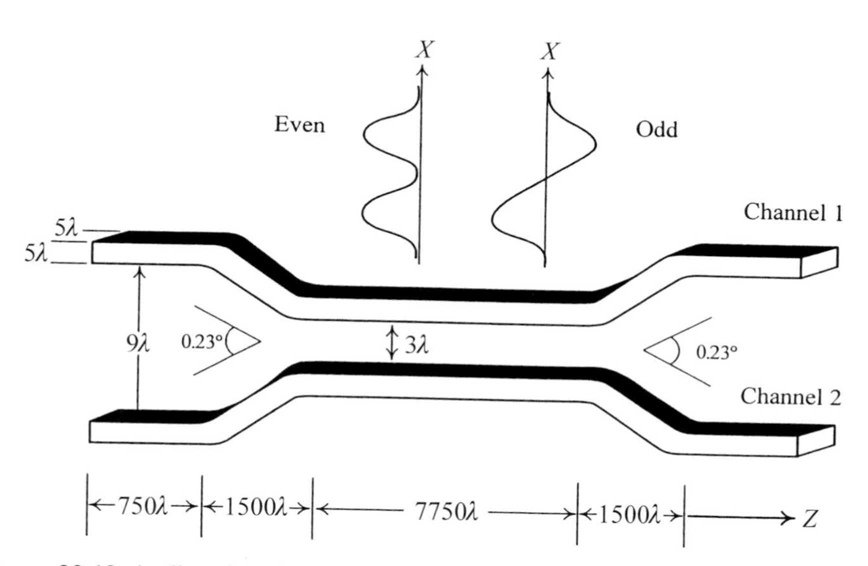

Figure 23.12 A directional coupler allows the coupling of light between adjacent waveguides. Each guide's cross-section in this example is a $5\lambda \times 5\lambda$ square, and both channels are embedded in a cladding of lower refractive index. Between $z = 0$ and $z = 750\lambda$ the separation between the guides is fixed at 9λ; it then decreases continuously to 3λ by $z = 2250\lambda$, and remains fixed at 3λ until $z = 10^4\lambda$. In the coupling region, where the guides are close together, the lowest-order modes of the two channels (taken together as a single waveguide) are the displayed even and odd modes.

the two channels contain equal amounts of light, albeit with a 90° relative phase. Eventually the two channels recede from each other, and the beam stays in the guide in which it was residing just before the separation.

Figure 23.13(a) displays the phase masks used in our BPM simulation of the directional coupler shown in Figure 23.12. The initial intensity distribution produced by a $0.2NA$ lens at the entrance to channel 1 is shown in Figure 23.13(b). Figure 23.14 shows several intensity plots at various cross-sections of the guide, demonstrating the transfer of light between the two channels. Figure 23.15 is a plot of the phase distribution at a location along the guide where the two channels carry equal amounts of optical power; the plot indicates the existence of a 90° phase difference between the two channels. Figure

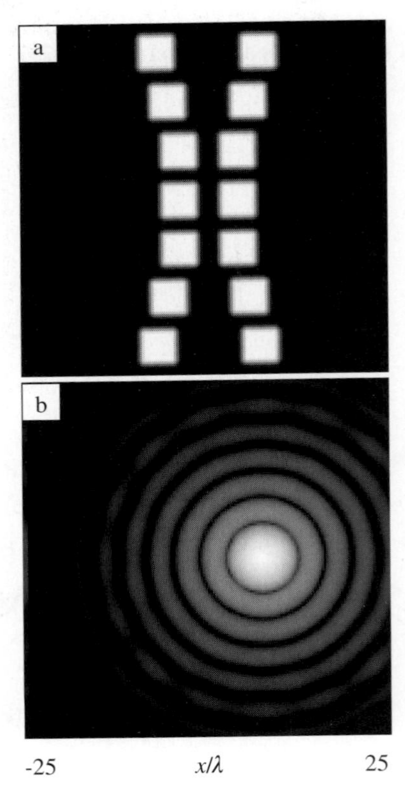

Figure 23.13 (a) Phase masks used in the BPM simulation of the directional coupler of Figure 23.12. Each mask, which consists of a pair of $5\lambda \times 5\lambda$ square apertures, imparts a phase shift of 7.5° to the beam at the end of each propagation step of $\Delta z = 2.5\lambda$. (b) Logarithmic plot of the intensity distribution created by a $0.2NA$ lens at the entrance to channel 1. The wavelength λ is that inside the cladding material, and the effect of losses incurred upon reflection from the front facet of the waveguide is included in this picture.

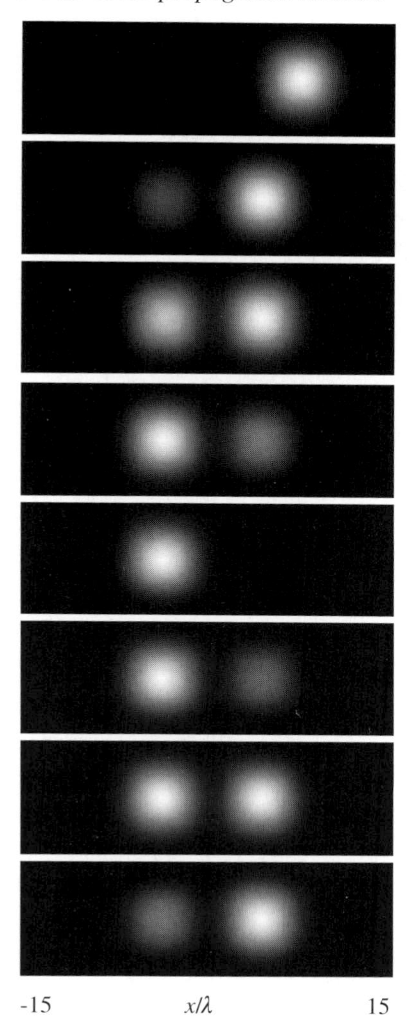

Figure 23.14 Computed plots of intensity distribution at various cross-sections along the directional coupler depicted in Figures 23.12 and 23.13. From top to bottom: $z/\lambda = 750$, 2250, 2500, 2750, 3250, 3750, 4000, 4250.

23.16 shows the amplitude distributions at several cross-sections of the guide. Figure 23.17 shows the power content of each channel versus z, indicating several oscillations of the power between the two channels.

Multimode interference device

Figure 23.18 is a diagram of a multimode interference (MMI) device used as a three-way power splitter. This device consists of an input channel, a wide

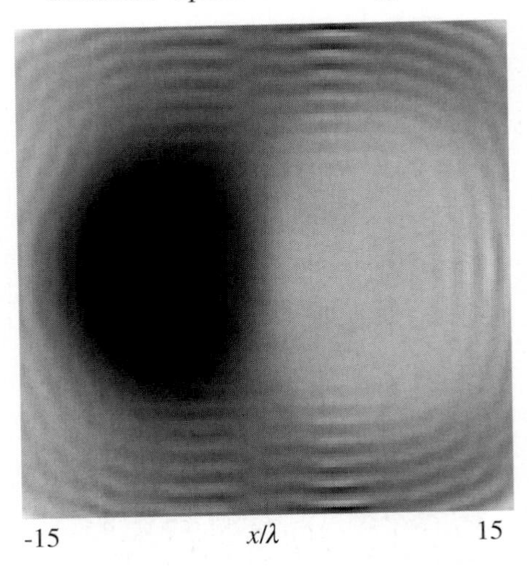

Figure 23.15 The computed phase distribution at $z = 4000\lambda$ in the BPM simulation of the directional coupler depicted in Figure 23.14, showing a 90° phase difference between the two channels at this location.

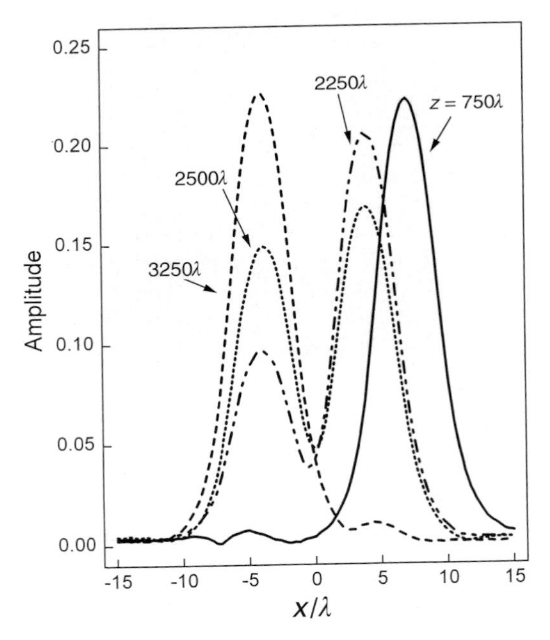

Figure 23.16 Plots of the light amplitude distribution at various cross-sections of the directional coupler depicted in Figures 23.12–23.15. At $z = 750\lambda$ (solid line) a single guided mode is established in channel 1. At $z = 2250\lambda$ (broken and double dotted line) the two channels have come close to each other, and some of the light has already leaked into channel 2. At $z = 2500\lambda$ (dotted line) the power contents of the two channels are nearly equal. At $z = 3250\lambda$ (broken line) the beam has all but moved into channel 2.

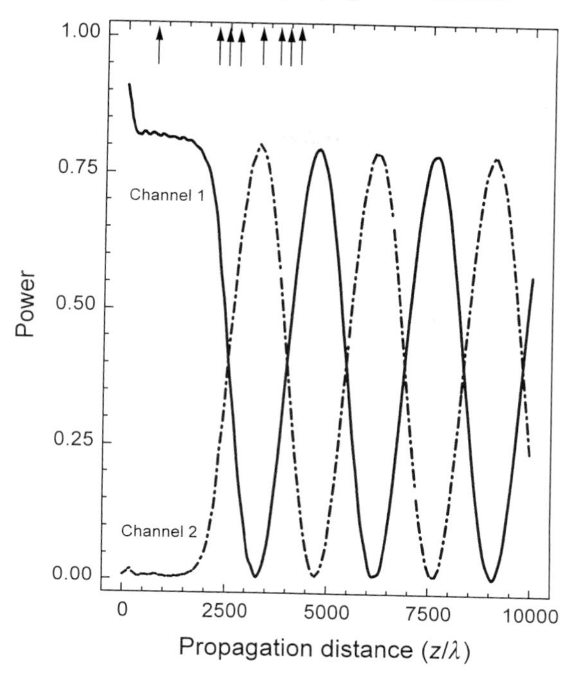

Figure 23.17 Power content versus z for each channel in the directional coupler of Figure 23.12. The incident focused laser beam loses about 4% of its power upon entering the front facet of the guide, and another 14% while establishing itself in channel 1. When the two channels slowly approach each other, the total power in the guide does not change appreciably, but it begins to couple out of channel 1 and into channel 2. By the time the separation of the channels has reached 3λ, a fraction of the beam already resides in channel 2. The oscillation of optical power between the two channels continues as long as they remain close to each other. The arrows at the top of the figure mark the locations of the intensity plots of Figure 23.14.

(multimode) section, and three output channels. The single-mode input guide carries the incident beam to the multimode region, where the beam suddenly expands, exciting the various modes of the broad waveguide. The ensuing interference among these modes creates periodic patterns of intensity and phase at specific locations along the Z-axis. This behavior is reminiscent of the Talbot effect, and in fact its explanation rests on the same principles (see chapter 17, "The Talbot effect"). At a particular distance L from the port of entry, the beam breaks up into several bright spots of equal intensity. If access channels are placed at this location they carry away the resulting isolated beams.[5,6]

Figure 23.19 shows the phase masks used in the BPM simulation of the MMI device depicted in Figure 23.18. At the top of the figure is the cross-section of the $5\lambda \times 5\lambda$ input channel (length 750λ), in the middle is the

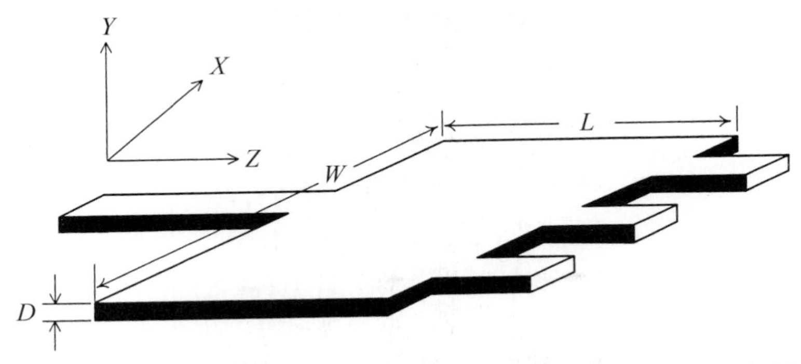

Figure 23.18 In a multimode interference (MMI) device the beam carried by a single-mode channel suddenly expands into a broad, multimode section of length L, width W, and thickness D. The many modes of the broad waveguide thus excited propagate at different speeds along the Z-axis, their interference giving rise to complex patterns of intensity distribution confined within the guide's cross-section in the XY-plane. Access guides placed at the end of the multimode section carry away the concentrated optical energy localized in isolated bright spots at $z = L$.

$45\lambda \times 5\lambda$ multimode section of the guide (length 3000λ), and at the bottom are the cross-sections of the three $5\lambda \times 5\lambda$ output channels (length 1250λ). Computed intensity profiles at several cross-sections of this device are shown in Figure 23.20. Depending on the distance from the port of entry, the guide's width W, and the wavelength λ, interference among the excited modes can give rise to a number of different intensity patterns. In the present example, the chosen parameters of the multimode section ($L = 3000\lambda$, $W = 45\lambda$) result in a three-way splitting of the input optical power. The computed intensity pattern at the end of the output channels appears in the bottom frame of Figure 23.20.

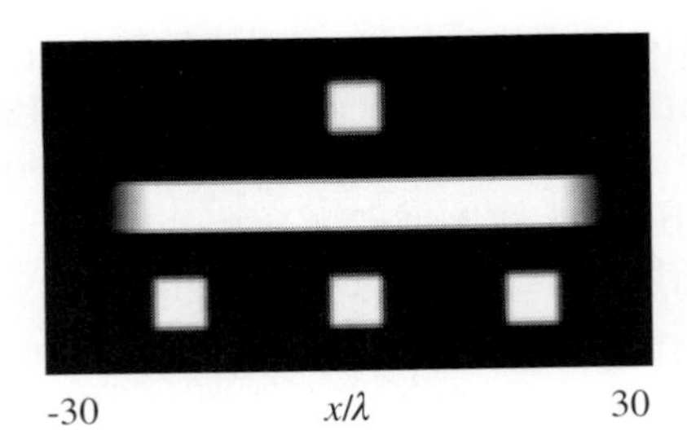

Figure 23.19 Phase masks used in the BPM simulation of the 1×3 splitter depicted in Figure 23.18. Each mask imparts a phase shift of $7.5°$ to the beam at the end of each propagation step of $\Delta z = 2.5\lambda$.

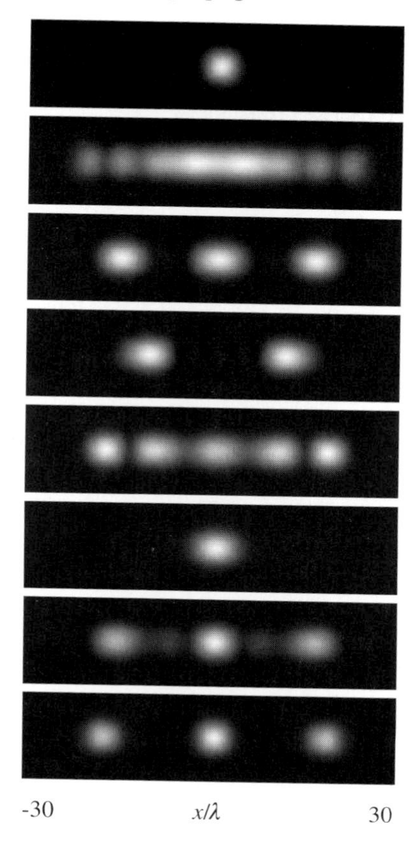

Figure 23.20 Computed plots of intensity distribution in the MMI device of Figures 23.18 and 23.19, showing (from top to bottom) the single-mode beam in the input channel just before entering the multimode section at $z = 0$, the distribution of light in the multimode region at $z/\lambda = 250$, 750, 1125, 1775, 2250, and at the end of the multimode section at $z/\lambda = 3000$. The bottom frame shows the intensity distribution emerging from the three output channels. (The initial distribution entering the input channel at $z = -750\lambda$ was uniform and had a circular cross-section of diameter 10λ.)

References for chapter 23

1 M. D. Feit and J. A. Fleck, Computation of mode properties in optical fiber waveguides by the propagating beam method, *Applied Optics* **19**, 1154 (1980); Analysis of rib waveguides and couplers by the propagating beam method, *J. Opt. Soc. Am. A* **7**, 73–79 (1990).

2 T. Tamir, ed., *Guided-wave Optoelectronics*, 2nd edition, Springer-Verlag, Berlin, 1990.

3 D. Marcuse, *Theory of Dielectric Optical Waveguides*, 2nd edition, Academic Press, New York, 1991.

4 C. R. Pollock, *Fundamentals of Optoelectronics*, R. D. Irwin, Chicago, 1995.

5 O. Bryngdahl, Image formation using self-imaging techniques, *J. Opt. Soc. Am.* **63**, 416–419 (1973).
6 R. Ulrich, Image formation by phase coincidences in optical waveguides, *Optics Communication* **13**, 259–264 (1975).

24

Michelson's stellar interferometer

The essential idea behind the stellar interferometer is that of a double-slit interferometer, such as that shown in Figure 24.1. This type of instrument dates back to 1868 when Fizeau[1] proposed using it to measure the diameters of the fixed stars. Some modern textbooks[2] describe the stellar interferometer in the language of coherence theory, which tends to obscure its fundamental simplicity. This chapter attempts to present the original concept in its simplest form while providing a historical perspective.

The double-slit interferometer

With reference to Figure 24.1, let us assume that a quasi-monochromatic point source of wavelength λ is placed at the origin of the XY-plane, which is the focal point of the collimator lens. The beam emerges from the lens collimated along the optical axis, effectively placing the point source at infinity. A double-slit mask blocks most of the light, allowing only the rays within two narrow slits to pass through to the focusing lens. The slits have a separation d along the X-axis, their widths being inconsequential as long as a sufficient amount of light gets through and a reasonable number of fringes appear at the observation plane. The focusing lens of focal length f brings together the rays that emerge from the two slits, causing them to interfere and produce a fringe pattern. The simple geometrical construction in Figure 24.2 shows that at the observation plane the fringe period p may be written as[3,4]

$$p \approx f\lambda/d. \tag{24.1}$$

Figure 24.3 shows computed plots of intensity distribution in the system of Figure 24.1 for a point source centered on the optical axis; Figure 24.3(a) is the distribution immediately after the slits, while Figures 24.3(b) and 24.3(c)

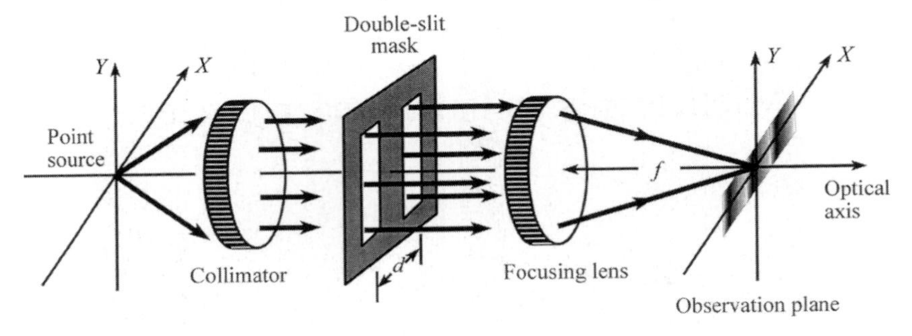

Figure 24.1 Schematic diagram of a double-slit interferometer. The collimator lens has $NA = 0.8, f = 6000\lambda$. The mask has two slits each of width 500λ, the centres of which are separated by $d = 6500\lambda$. The focusing lens has $NA = 0.008$ and $f = 6 \times 10^5\lambda$.

present the intensity pattern within the observation plane. For the assumed system parameters ($f = 6 \times 10^5\lambda$, $d = 6500\lambda$) the fringe period found from Eq. (24.1) is $p \approx 92.3\lambda$, in agreement with the simulated results.

Next, assume that a second point source is placed in the focal plane of the collimator lens, slightly displaced from the first one located at the origin. Assume further that the two sources, although both quasi-monochromatic with wavelength λ, are completely uncorrelated and independent, so that their radiation may be considered to be spatially incoherent. The collimated beam arriving at the plane of the slits from this second source will make a small angle ψ with the optical axis. As shown in Figure 24.4, this angle causes the phase of the light arriving at the two slits to differ by $\Delta\phi$ where[3,4]

$$\Delta\phi \approx 2\pi\psi d/\lambda. \tag{24.2}$$

A 2π phase difference at the slits corresponds to a fringe translation by one period at the observation plane. Thus the two sets of fringes arising from the

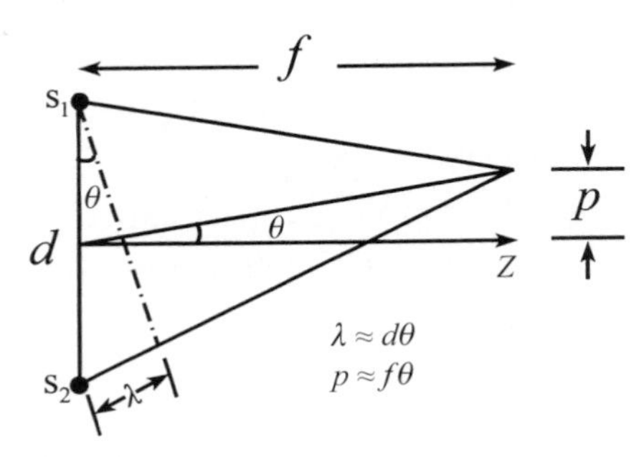

Figure 24.2 Geometrical construction showing the relation between the fringe period p, the distance d between the slits, the focal length f of the focusing lens, and the wavelength λ of the light.

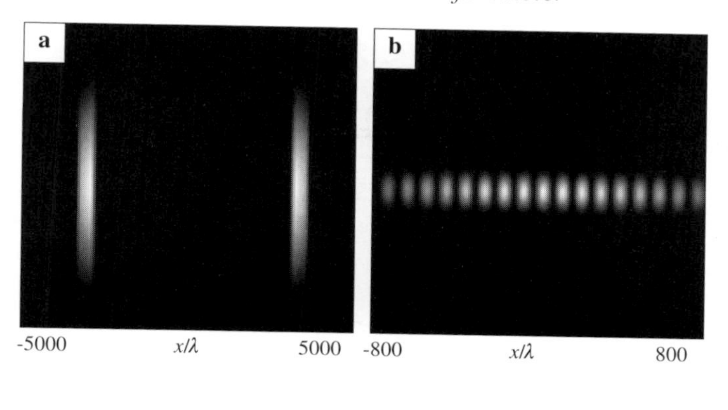

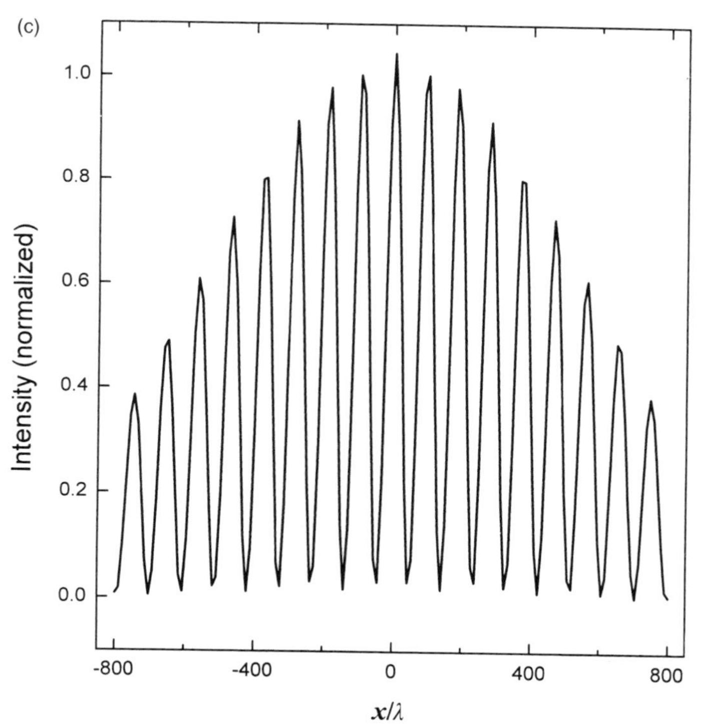

Figure 24.3 Results of a computer simulation involving a single point source, centered on the optical axis of the system of Figure 24.1. (a) Intensity distribution immediately after the mask. (b) The fringe pattern at the focal plane of the focusing lens. (c) Cross-section of the fringe pattern along the X-axis.

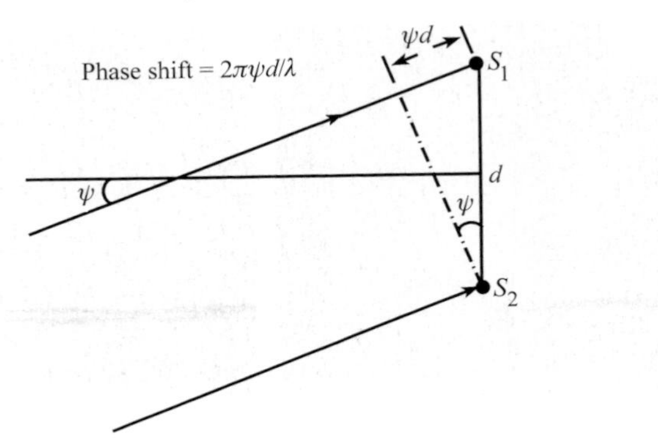

Figure 24.4 A collimated beam arriving at the double-slit mask at an angle ψ relative to the optical axis will exhibit a phase difference of $\Delta\phi$ between the two slits.

two point sources will be shifted relative to each other by an amount that will depend on ψ. Since the point sources are uncorrelated, it is their corresponding *intensity* distributions at the observation plane that will be added together. This results in a ψ-dependence of the fringe visibility V, a quantity defined by Michelson as

$$V = (I_{\max} - I_{\min})/(I_{\max} + I_{\min}), \tag{24.3}$$

where $I_{\min}$ and $I_{\max}$ are the minimum and maximum values of intensity within one fringe period.

As an example, let us assume that two point sources of equal strength, separated by 0.5λ along the X-axis, are placed in the focal plane of the collimator lens of Figure 24.1. The angular separation between the two sources as viewed from the plane of the slits, therefore, is $\psi = 17.2$ seconds of arc.[5] In the absence of the double-slit mask, the images of the two point sources will be unresolvable at the observation plane; see Figure 24.5(a). With the mask in place, however, the plots of intensity distribution in Figures 24.5(b), (c) indicate that the fringe visibility will be substantially altered from that for a single point source; the latter can be deduced from Figure 24.3(c) to be 100%. Close inspection of the fringes, therefore, enables one to infer the presence of a second source of light in the system. As a practical matter, one may adjust the distance d between the two slits until the phase shift $\Delta\phi$ of Eq. (24.2) becomes equal to π, at which point the two fringe systems will be shifted by half a period. Under these conditions the maxima of one set of fringes will overlap the minima of the other, resulting in a complete "washing out" of the interference pattern. Equation (24.2) can then be used to determine the angular separation ψ between the two point sources from the knowledge of the slit separation d that resulted in minimum

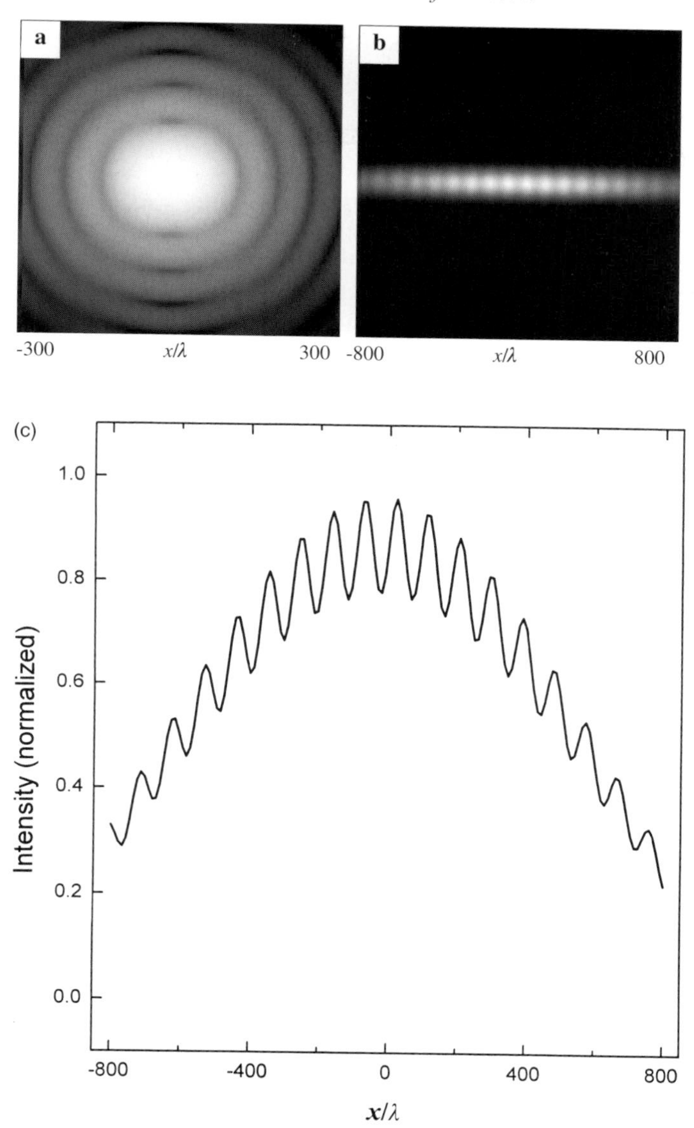

Figure 24.5 Results of a simulation of the system of Figure 24.1 involving two independent point sources, one centered on the optical axis (i.e., at the origin), the other shifted by 0.5λ along the X-axis. (a) Logarithmic plot of the intensity distribution at the observation plane in the absence of the double-slit mask. (b) Fringe pattern at the observation plane with the double-slit mask present. (c) Cross-section of the fringe pattern along the X-axis.

fringe visibility. (Note that changing d will have the undesirable effect of changing the fringe period p according to Eq. (24.1), but, as long as the fringes remain visible to the observer, this change should be inconsequential.)

Dependence of fringe visibility on d

Of course one may not know a priori whether the source is an extended object (such as a large star) or consists of a number of distinct point sources (e.g., a double star) and, in the latter case, whether the point sources are of equal intensity. It turns out that a measurement of fringe visibility V as a function of the separation d between the slits can provide ample information about the intensity distribution of the source. A pair of equal-intensity stars, for instance, will make the visibility versus d a periodic function, whereas the more-or-less uniform disk of a giant star will give rise to an oscillating $V(d)$ whose magnitude declines with increasing d. Calculations show that in the former case the first zero of $V(d)$ appears at $d = 0.5\lambda/\psi$, whereas in the latter the first zero occurs at $d = 1.22\lambda/\psi$; here ψ is the angle subtended by the diameter of the giant star's disk.[2–4]

In any event, it is clear that a measurement of fringe visibility for several different separations of the slits will provide much information about the distribution of intensity at the source. As another example, we show in Figure 24.6 the case of three point sources of equal intensity placed at $x = -0.25\lambda$, 0, and $+0.25\lambda$ in the system of Figure 24.1. Again the image obtained at the observation plane without the double-slit mask does not resolve the sources of light, but the fringe visibility obtained as a function of d carries enough information to allow one to make a fairly accurate statement about the distribution of intensity at the source.

A historical perspective

R. W. Wood[6] sums up the origins of the interferometer: "This method was proposed by Fizeau[1] in 1868 for measuring the diameters of the fixed stars. In 1874 Stefan made an attempt to carry out Fizeau's plan, placing two slits in front of the objective of the Marseilles telescope, the largest available at the time. The fringes remained visible even when the slits were separated by the full diameter of the objective. In 1890 Michelson measured the diameters of the four moons of Jupiter, using the 36 inch telescope of the Lick observatory.[7] The method can also be used for determining the distance between the components of a double star.

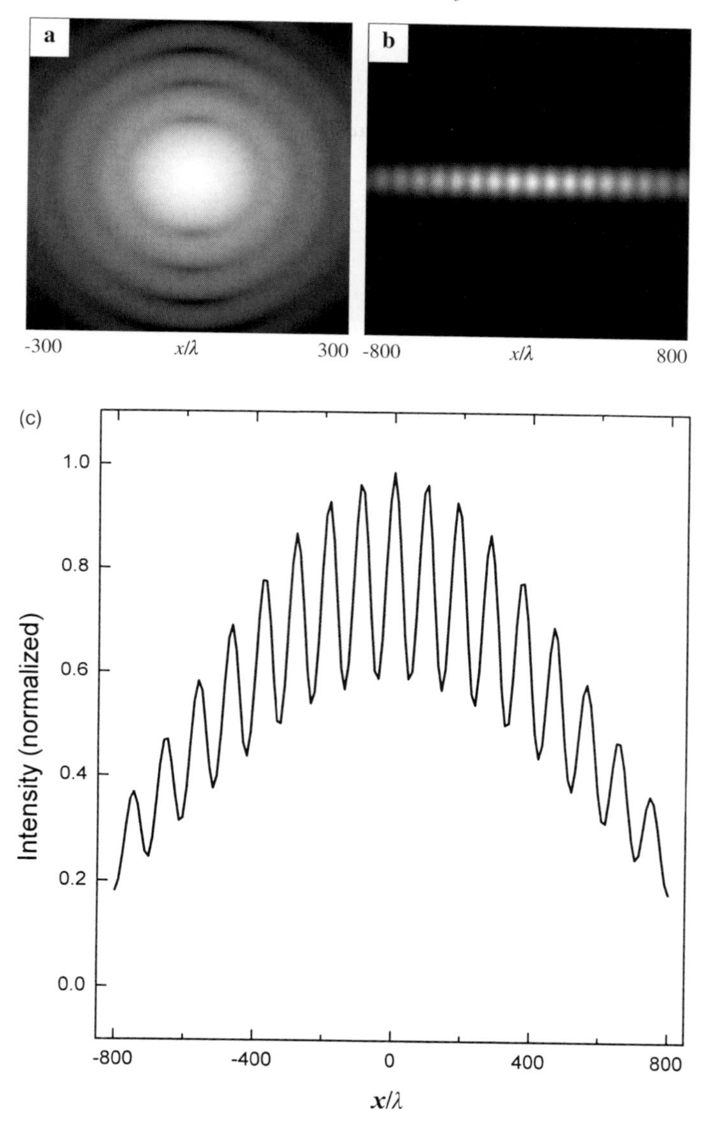

Figure 24.6 Results of a simulation of the system of Figure 24.1 involving three independent point sources, one centered on the optical axis, the others shifted by $\pm 0.25\lambda$ along the X-axis. (a) Logarithmic plot of the intensity distribution at the observation plane in the absence of the double-slit mask. (b) Fringe pattern at the observation plane with the double-slit mask present. (c) Cross-section of the fringe pattern along the X-axis.

Albert Abraham Michelson (1852–1931) was born in what was then Germany (now Poland) and emigrated with his family to the United States in 1855. He became professor of physics at the Case School of Applied Science (Cleveland, Ohio), then at Clark University (Worcester, Massachusetts), and then at the University of Chicago. In 1907 he became the first American to receive a Nobel prize; the prize citation reads: "For his optical precision instruments and the spectroscopic and meteorological investigations carried out with their aid." (Photo: courtesy of AIP Emilio Segré Visual Archives.)

"In 1920 Michelson took up the problem of the determination of stellar diameters.[8] Even the great 100 inch telescope of the Mount Wilson Observatory is not large enough to allow of a sufficient separation of the slits; consequently Michelson designed a 'periscopic' arrangement of four mirrors, the two outer ones, twenty feet apart, reflecting the light to two inner ones which in turn reflected the beams down upon the mirror of the 100 inch telescope. The mirrors were mounted on a metal beam attached to the top of the telescope tube. The instrument was constructed in collaboration with F. G. Pease of the Mount Wilson Observatory."

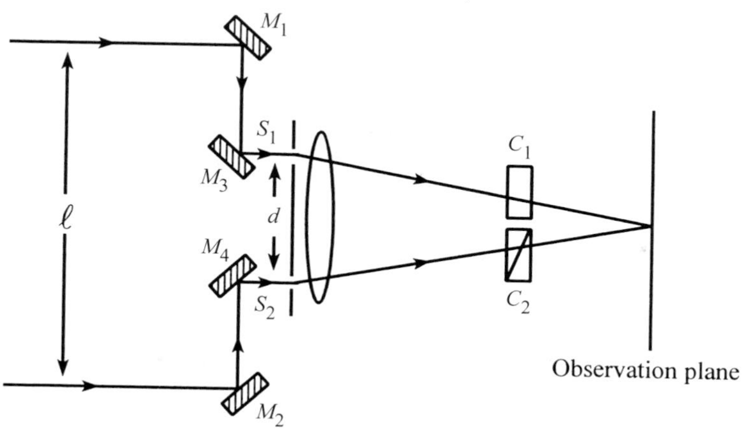

Figure 24.7 Michelson's stellar interferometer. The apertures S_1 and S_2 are fixed, and the light reaches them after reflection at mirrors M_1, M_2, M_3 and M_4. The inner mirrors M_3 and M_4 are fixed, while the outer mirrors M_1 and M_2 can be moved symmetrically in the direction joining S_1 and S_2. If the optical paths $M_1M_3S_1$ and $M_2M_4S_2$ are maintained equal, the optical path difference for light from a distant point source is the same at S_1 and S_2 as at M_1 and M_2, so that the outer mirrors play the part of the movable apertures in the Fizeau method. A plane-parallel glass plate C_1, which can be inclined in any direction, is used to maintain the geometrical pencils from S_1 and S_2 in coincidence in the focal plane. A second plane-parallel glass plate C_2, of variable thickness, is used to compensate inequalities of the optical paths $M_1M_3S_1$ and $M_2M_4S_2$. (Adapted from Born and Wolf.[3])

A schematic diagram of the stellar interferometer constructed by Michelson (and mentioned by R. W. Wood in the preceding paragraph) is shown in Figure 24.7. In this instrument the distance ℓ between mirrors M_1 and M_2 was varied to effect a change of fringe visibility; one must therefore substitute ℓ for d in Eq. (24.2) in order to make it applicable to the new instrument. The fringe period p, however, is still determined by the distance d between the slits, and Eq. (24.1) applies to Michelson's interferometer without any modifications. Thus there is the further advantage that the fringe spacing remains constant as the separation of the movable mirrors is varied. The interferometer was mounted on the 100 inch reflecting telescope of the Mount Wilson Observatory in California, which was used because of its mechanical strength. The apertures S_1 and S_2 were 114 cm apart, giving a fringe spacing of about 20 μm in the focal plane. The maximum separation of the outer mirrors was 6.1 m, so that the smallest measurable angular diameter (with $\lambda = 550$ nm) was about 0.02 seconds of arc.[3]

Again quoting R. W. Wood:[6] "The bright star Betelgeuse was the first investigated. This star shows evidence of its diameter with the 100 inch telescope if a canvas cover is placed over the instrument, provided with two holes

7 inches in diameter and 94 inches apart, the diffraction disk of the star being crossed with *faint* interference bands. If either hole is covered the bands disappear. If the telescope is pointed at Rigel, however, the bands are clear and strong, showing that its angular diameter is smaller than that of Betelgeuse. With the twenty-foot interferometer the bands disappeared entirely in the case of Betelgeuse when the mirrors were separated by a distance of 120 inches, while Rigel showed very distinct bands. The angular diameter of Betelgeuse was computed as 0.047 seconds of arc. From the known distance of the star [determined by triangulation], its actual diameter was calculated as 250 million miles [i.e., 300 times the diameter of the sun] or greater than the earth's orbit about the sun [180 million miles across]. Its diameter has been found to vary, however, for at times the mirrors must be separated by a distance of 14 feet before the fringes disappear. Antares was found to be still larger, having a diameter of 400 million miles. The minimum angular diameter measurable with the 20 foot instrument is 0.024 seconds of arc."

The majority of stars are either too distant or too small for the Michelson interferometer to measure their diameter. For example, at the distance of the nearest star (Alpha Centauri) the sun's disk would subtend an angle of only 0.007 seconds of arc, and to observe the first disappearance of the fringes a mirror separation of 20 m would be necessary. The construction of such a large interferometer would be a difficult undertaking because of the requirement of rigid mechanical connection between the collecting mirrors and the eyepiece.[3] In recent years, the method of Hanbury Brown and Twiss as well as extensions of Michelson's method to radio astronomy have been used for measurements of some of the smaller astronomical objects.[2–4]

References for chapter 24

1 H. Fizeau, *C. R. Acad. Sci.* Paris **66**, 934 (1868).
2 For example, L. Mandel and E. Wolf, *Optical Coherence and Quantum Optics*, Cambridge University Press, London, 1995.
3 M. Born and E. Wolf, *Principles of Optics*, 6th edition, Pergamon Press, Oxford, 1980.
4 M. V. Klein, *Optics*, Wiley, New York, 1970.
5 One degree is 60 minutes, and one minute is 60 seconds of arc. One second of arc is the angle subtended by a small coin at a distance of about 3.5 km.
6 R. W. Wood, *Physical Optics*, third edition, reprinted by the Optical Society of America, 1988.
7 A. A. Michelson, *Phil. Mag.* **30**, 1 (1890); A. A. Michelson, *Nature* (London), **45**, 160 (1891).
8 A. A. Michelson, *Astrophys. J.* **51**, 257 (1920); A. A. Michelson and F. G. Pease, *Astrophys. J.* **53**, 249 (1921).

25

Bracewell's interferometric telescope

There are countless suns and countless earths all rotating around their suns in exactly the same way as the seven planets of our system. We see only the suns because they are the largest bodies and are luminous, but their planets remain invisible to us because they are smaller and non-luminous. The countless worlds in the universe are no worse and no less inhabited than our Earth.

Giordano Bruno (1584) in De L'Infinito Universo E Mondi

In 1978 Ronald Bracewell of Stanford University proposed the use of a nulling interferometer to cancel the image of a bright star in order to observe the relatively faint planets which might be in orbit around the star.[1] This idea, which has been expounded and further extended by others,[2–4] is presently the most promising method of detecting terrestrial planets (i.e., small, rocky planets similar to Venus, Earth, and Mars) orbiting in habitable zones around our neighboring stars. Because atmospheric turbulence distorts the stellar wavefronts and limits the resolution of ground-based observations, an interferometric telescope capable of detecting planets in other solar systems must, of necessity, be stationed in space. The National Aeronautics and Space Administration (NASA) is currently working on a program called the Terrestrial Planet Finder (TPF), and has tentatively scheduled the launch of a nulling interferometer into orbit in about the year 2010.[5]

Nulling interferometer

Figure 25.1 is a diagram of a basic Bracewell telescope intended for operation in the infrared range of wavelengths $\lambda \sim 7 - 20\,\mu\text{m}$. The reason for working in the infrared is that the expected brightness of the star in this region is only $\sim 10^6$ times that of the planet, which is much better than the $\sim 10^9$ brightness ratio in the visible. Moreover, several signature absorption lines corresponding to ozone, water vapor, methane, and carbon dioxide reside in this band, which can be exploited in the spectroscopic analysis of these planets to determine whether they harbor life as we know it.[5]

In the following discussion we confine our attention to a single infrared wavelength of $\lambda = 10\,\mu\text{m}$, even though the interferometric telescope can operate over a fairly broad range of wavelengths. We assume the primaries each have an aperture diameter $D_p = 1$ m and focal length $f_p = 2.5$ m. (The angu-

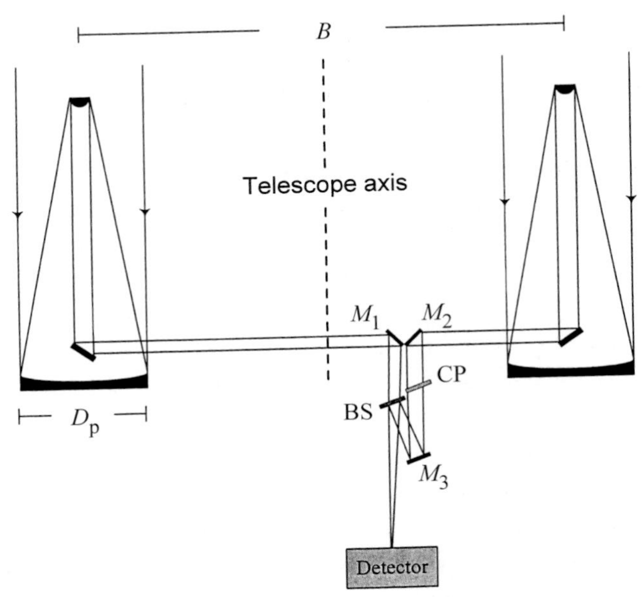

Figure 25.1 (adapted from reference 4). Diagram of a nulling interferometric telescope. The primary mirrors have diameter D_p and baseline B. Unmatched reflections are made at nearly normal incidence to minimize polarization differences. The folded beams are combined at the beam-splitter (BS), which is designed for equal transmission and reflection in the desired range of wavelengths. The beam from the left-hand side, after crossing the axis of the telescope, is folded down at M_1 and transmitted by BS before coming to a focus. The beam from the right-hand side is folded downward at M_2 before reaching the axis of the telescope, and then passes through a compensator plate CP and is reflected back up at M_3 to equalize the path lengths before being reflected from the underside of BS. An achromatic 180° phase difference is realized by balancing a slight difference in the air path with the path difference between BS and CP, fine-tuned by a slight rotation of CP.

lar resolution of the individual mirrors is thus $\sim \lambda/D_p = 10^{-5}$ radians.) The assumed baseline (i.e., center-to-center separation of primary mirrors) is $B = 5\,\text{m}$.† With the telescope pointing at a star, an angular separation of $\phi = 10^{-6}$ radians between the star and its planet results in a relative phase of $2\pi B\phi/\lambda = 180°$ between the light rays arriving at the two mirrors from the planet. (1 µrad ≈ 0.2 arcsec is the separation between the Sun and the Earth

†These parameters, chosen for the sole purpose of demonstrating the basic concepts, are not representative of the planned systems. A typical design under consideration by the TPF program, for example, has four primary mirrors, each ~ 2.5 m in diameter and separated by ~ 100 m baselines. It is envisioned that these free-flying mirrors would collect and forward the beam of light to a local combiner and controller unit (also free-flying). The planned system will be capable of executing nulling interferometry over the broad band of $\lambda = 7$–$20\,\mu\text{m}$. The adjustment of mirror positions and their distances from each other as well as from the combiner would allow the configuration to be optimized on the spot in accordance with the characteristics of the particular solar system under consideration.[5]

observed from $\sim$ 16 light years away.) The separation of the planet from its parent star in this case is an order of magnitude below the resolution of the individual mirrors, yet the assumed nulling interferometer is capable of detecting the planet in the vicinity of the star.

The secondary mirrors in the system of Figure 25.1, placed at $z = 25$ cm before the primary focus, bring the reflected beam to a final focus at $z' = 5$ m in front of the secondary. These negative mirrors, designed for a 20 : 1 conjugate ratio, have aperture diameter $D_s = 10$ cm, focal length $f_s = -26.32$ cm, and magnification $M_s = 20$. The focused cone of light emerging from each secondary is an $f/50$ beam (i.e., numerical aperture $NA' = 0.01$), giving rise to an Airy disk diameter of $1.22\lambda/NA' = 1.22$ mm at the image plane.

The light from the planet, entering the primary at the oblique angle of $\phi = 10^{-6}$ radians, emerges from the secondary at $\phi' = 10\phi$. (The secondary is ten times closer than the primary to the virtual image of the sky at the primary's focal plane.) The final image of the planet, therefore, is shifted by $\Delta r = z'\phi' = 50\,\mu$m from the image of the star at the center of the image plane. This separation, being more than an order of magnitude below the Airy disk diameter of $1220\,\mu$m, is clearly insufficient to resolve the planet's image from the parent star's, confirming once again the inadequacy of the individual mirrors for the task.

The case against a conventional telescope

Even a conventional (filled-aperture) space telescope 25 m in diameter will fail to detect the planet in the preceding example. The problem in this case is not resolution but photon noise. The image of the star, being about $\sim 10^6$ times brighter than that of the planet, floods the detectors and obscures the planet's signal. The nulling interferometer, however, yields an acceptable signal-to-noise ratio at the detector output by canceling the light of the star arriving from the two mirrors while, at the same time, enhancing the image of the planet by constructive interference. Not only does the nulling interferometer eliminate the complete Airy pattern of the star (i.e., the central disk as well as the rings), it does so without requiring any significant displacement of the planet's Airy pattern. What is important for the nulling interferometer is not how much the two Airy disks in the image plane are separated from each other but how much the wavefront arriving from the planet at one mirror is delayed relative to the time of arrival of the same wavefront at the other. This delay or phase shift, being a function of the baseline B, is independent of the mirror diameter D_p.

Destructive and constructive interference

As a specific example consider the system of Figure 25.1 with the aforementioned parameters. The beam-splitter (BS) is an important component of this system; to simulate its behavior we used a six-layer stack on a 1mm substrate, as shown in Figure 25.2. The alternate layers are high- and low-index dielectrics, their thicknesses chosen to yield a 50/50 beam-splitter at the operating wavelength of $\lambda = 10\,\mu m$. The top of the substrate is anti-reflection coated with a low-index layer to minimize undesirable reflections. This stack design, although adequate for demonstration purposes, is not suitable for broadband applications requiring cancellation of the star light with high accuracy. Such applications require alternative designs or more complex multilayer stacks.

Figure 25.3 shows computed images of a single star obtained with the above telescope. The Airy pattern in Figure 25.3(a) is obtained when the light from one arm of the telescope is blocked. When both channels are open and properly balanced, destructive interference at the beam-splitter yields the null image in Figure 25.3(b); here the peak intensity is only $\sim 1.4 \times 10^{-4}$ that in Figure 25.3(a). The weak residual image of the star in Figure 25.3(b) is due to a slight imbalance of the two channels brought about by the beam-splitter's minute departure from the ideal 50/50 ratio. Although this four-orders-of-magnitude reduction of intensity in the null image is sufficient for the present discussion, it is totally unacceptable for the observation of actual terrestrial planets. Because the radiation levels of these planets are expected to be at least a million times weaker than their parent star's, it is

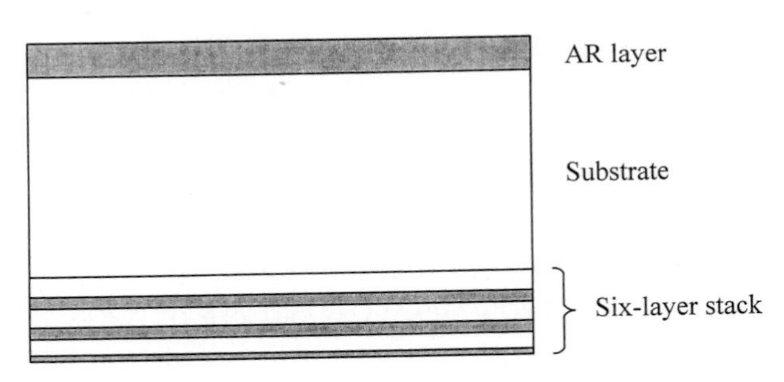

Figure 25.2 A simple design for the beam-splitter (BS) in the system of Figure 25.1, having a 50/50 reflection to transmission ratio at $\lambda = 10\,\mu m$. The 1 mm-thick substrate has refractive index $n = 2$, and is antireflection coated (AR) on the top surface with a $t = 1.785\,\mu m$ layer having $n = 1.4$. Deposited on the substrate bottom is a six-layer stack. Numbered in increasing order starting at the substrate interface, these layers have the following parameters: layers 1, 3, 5, $t = 1.7\,\mu m$, $n = 1.5$; layers 2, 4, $t = 1.25\,\mu m$, $n = 2.0$, layer 6, $t = 0.475\,\mu m$, $n = 2.0$.

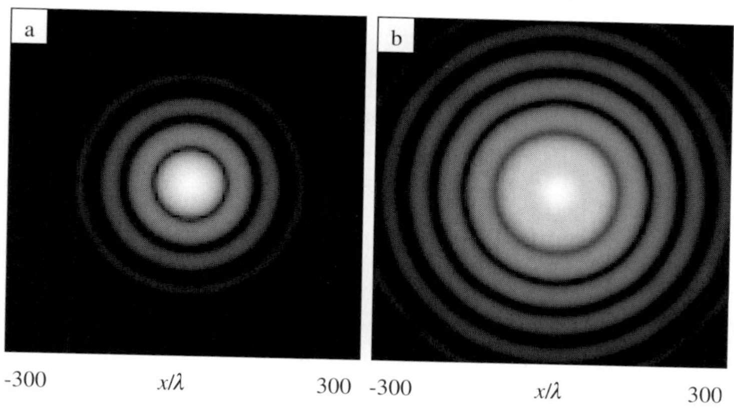

Figure 25.3 Logarithmic intensity distributions in the image plane corresponding to a single star with no planets: (a) when the light from either arm of the interferometer is blocked; (b) with both channels open and the path lengths properly balanced to allow interferometric cancellation of the star's image. The ratio of the peak intensity in (b) to that in (a) is 1.4×10^{-4}. The non-zero values of the residual intensity in (b) are due to imperfect balance between the two channels.

imperative to design the telescope components, with much higher accuracy, for a maximum rejection of the star light.

Consider next a planet only 100 times weaker than its star, with an angular separation of 25 μrad. With one channel of the telescope blocked the image in Figure 25.4(a) is obtained, whereas in a balanced interferometer one obtains the image in Figure 25.4(b). It is obvious in this example that the single-channel output, corresponding to a conventional telescope's image, shows a faintly visible planet next to a bright star, whereas in the interferometric image the light from the star is all but eliminated. Note that for clarity of presentation we have chosen a relatively bright planet with a large separation from its parent star (25 μrad ~ 5 arcsec). Both these assumptions are much too optimistic and, in practice, one must substantially improve the sensitivity of the assumed telescope in order to detect terrestrial planets in our neighboring solar systems.[5]

The fringe pattern and the spinning telescope

Figure 25.5 shows a Bracewell telescope oriented with its baseline in the XY-plane at an angle θ from X, pointing at a star along the Z-axis. Each point in the star's neighborhood may be identified either by its angular coordinates

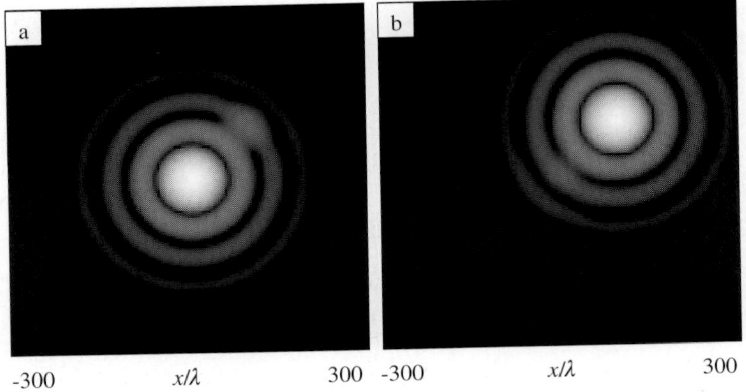

Figure 25.4 Images of star and planet, when the assumed brightness of the planet is 1% of the star's and their angular separation is 5.16 arcsec. (a) When either beam is blocked the intensity distribution is essentially that of the star; the planet is barely visible. (b) With both channels open and the phase difference between them adjusted to 180°, the bright star is canceled out and the planet becomes visible. The center-to-center spacing between the images of the star and the planet is 1.25 mm, and the ratio of the peak intensity in (b) to that in (a) is 0.038.

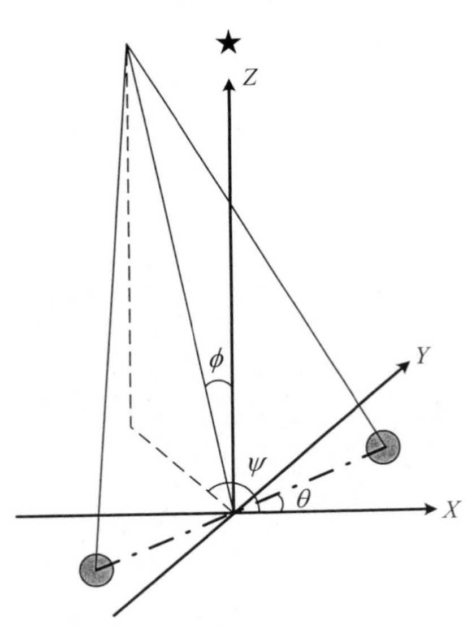

Figure 25.5 The Bracewell telescope, with its baseline in the XY-plane and oriented at an angle θ from X, targeting a star along the Z-axis. The points in the vicinity of the star are identified by their polar and azimuthal coordinates ϕ, ψ. The wavefront, arriving from an oblique direction, reaches one mirror later than the other, producing a path-length difference of $B \sin\phi \cos(\psi - \theta)$.

ϕ, ψ or by the Cartesian coordinates x, y of its image in the telescope. The image location is related to the polar coordinates through the equation

$$(x, y) = M_s f_P \tan\phi(\cos\psi, \sin\psi). \tag{25.1}$$

Here f_P is the focal length of the primary mirror and M_s is the magnification of the secondary. For the light arriving at the two primaries from the direction (ϕ, ψ) in the sky the relative phase is

$$\Delta\Phi = 2\pi(B/\lambda)\sin\phi\cos(\psi - \theta). \tag{25.2}$$

The two arms of the telescope are adjusted in such a way that when $\Delta\Phi = 0$ the two beams interfere destructively whereas a 180° phase shift results in constructive interference. The corresponding light amplitude at the image plane is thus given by

$$A = \tfrac{1}{2}A_0[1 - \exp(\mathrm{i}\Delta\Phi)], \tag{25.3}$$

and the resulting intensity may be written

$$|A|^2 = |A_0|^2 \sin^2(\tfrac{1}{2}\Delta\Phi) = |A_0|^2 \sin^2[\pi(B/\lambda)\sin\phi\cos(\psi - \theta)]. \tag{25.4}$$

Figure 25.6 is a gray-scale plot of $|A/A_0|^2$ in the image plane of the telescope (black and white represent 0 and 1, respectively). Each point (x, y) in this plane corresponds to a point (ϕ, ψ) in the sky in accordance with Eq. (25.1); it is also assumed that the primary mirrors are separated by $B = 5\,\mathrm{m}$ along the $\theta = 45°$ line. The field of view is centered at $(x, y) = (\phi, \psi) = (0, 0)$, which is the target star's location. Only a circle of radius 100λ within the field of view – corresponding to $0 \le \phi \le 20\,\mu\mathrm{rad}$ – is shown in Figure 25.6, but the

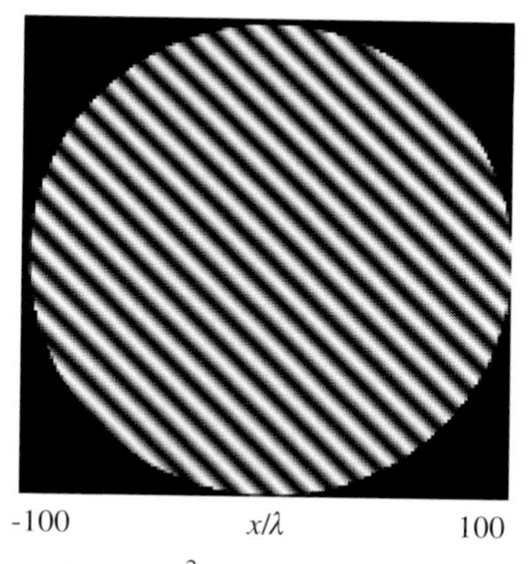

Figure 25.6 Gray-scale plot of $|A/A_0|^2$ (Eq. 25.4) in the image plane of the telescope (black and white are represented by 0 and 1, respectively). The primary mirrors are separated by $B = 5\,\mathrm{m}$ along the $\theta = 45°$ line; $\lambda = 10\,\mu\mathrm{m}$.

same pattern could extend over a much larger patch of sky around the targeted star.

Any planet (or other source of radiation) located in the bright fringes of Figure 25.6 will produce a bright Airy pattern in the image plane at that location. However, planets located in the dark fringes disappear from the image because destructive interference cancels them out. If the telescope is rotated around the Z-axis while maintaining a tight fix on the target star, θ will change continuously and the pattern of Figure 25.6 will rotate around its center. The image of a planet within the field of view, however, will remain fixed while the fringes rotate. The planet's image thus waxes and wanes as the bright and dark fringes cross it one after the other. The number of times that the planet's image appears and disappears in a single revolution of the telescope depends on the polar coordinate ϕ of the planet; specifically, the frequency of the planet's signal at the detector output increases in proportion to its separation ϕ from its parent star. In this way it is possible to modulate the signal of a given planet and, by integration over time, to reduce noise components residing outside the specific frequency of the planet's signal.[1]

Interplanetary dust and zodiacal light

In addition to the Sun and the nine planets and their moons, our solar system is home to countless rocks, pebbles, and dust particles floating in interplanetary space. The light of the Sun scattered from these dust particles (the so-called zodiacal light) will enter a space-based telescope and create a background noise. A similar diffuse radiation from the targeted solar system (exo-zodiacal light) will also be imaged as a fairly uniform distribution across the telescope's field of view.[4,5]

It is true, of course, that the ideal image of a broad, uniform source of light in the Bracewell telescope should resemble the striped pattern of Figure 25.6. However, the Airy disk produced by the finite aperture of each mirror is typically many times larger than the fringe spacing in Figure 25.6, and, therefore, the image of the zodiacal light will be the convolution between the Airy pattern of Figure 25.3(a) and the stripes of Figure 25.6. The zodiacal emissions, therefore, appear as a fairly uniform distribution in the image plane of the telescope. The shot noise from this captured background radiation is mainly responsible for the unavoidable noise in the photodetector output, its elimination requiring the spinning of the telescope followed by integration of the signal over time, as discussed in the preceding section.

Effect of star's finite diameter

In the presence of pointing errors or when the star has a finite angular diameter, the star light leaks out of the null and swamps the planet's image. Figure 25.7(a) shows the computed image of a star having an angular diameter of 0.05 arcsec, obtained in the nulling interferometer of the previous examples. To simulate this finite-size star we assumed 25 equally bright point sources spread over the surface of the star and superimposed the intensities of their Airy patterns at the image plane. The peak intensity of this image is about 230 times stronger than that in Figure 25.3(b), which was obtained under identical conditions except for neglect of the star's diameter. It is clear that the interferometer's null must be made broader if such effects of the finite diameter (as well as any pointing errors) are to be avoided. A proposed solution to this problem involves the use of several telescopes instead of just two, as in the original Bracewell concept. With the beams from four or more telescopes combined in a nulling interferometer, it is possible to broaden the central null of the fringe pattern.[2,3]

Achromatic path-length equalization

The compensator plate CP in the system of Figure 25.1 is used to balance the path lengths of the two interferometer arms over a range of wavelengths. In

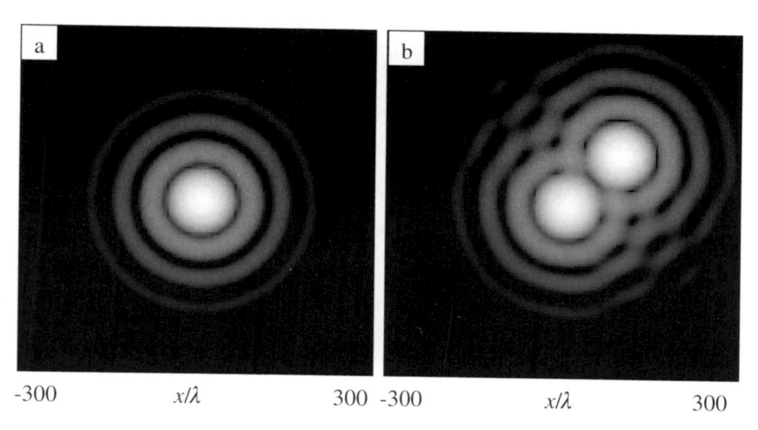

Figure 25.7 (a) Null image of a star of finite diameter (0.05 arcsec). The peak intensity is about 230 times greater than that in Figure 25.3(b). (b) Image of the finite-diameter star and its planet. This image should be compared with Figure 25.4(b), which was obtained under identical conditions except for neglect of the angular diameter of the star. The peak intensities of the planet and star in the present image are nearly the same.

the particular case studied in reference 4, CP was 42 μm thicker than BS, and achromaticity was achieved for $\lambda = 10-14\,\mu$m.

Let d_1 and d_2 be the optical path lengths of the two channels in air, and denote by t_1 and t_2 the thicknesses of CP and BS, respectively. These plates are made of the same material, whose refractive index within the wavelength range of interest may be approximated by $n(\lambda) \approx \alpha + \beta\lambda$ (α and β are material constants). Also, one must take into account the 90° phase shift introduced by the (symmetric) beam-splitter between the reflected and transmitted beams. The overall optical phase difference between the two channels is thus given by

$$\begin{aligned}\Delta\Phi &= \tfrac{1}{2}\pi + 2\pi[d_1 + t_1 n(\lambda) - d_2 - t_2 n(\lambda)]/\lambda \\ &\approx \tfrac{1}{2}\pi + 2\pi\{(t_1 - t_2)\beta + [d_1 - d_2 + (t_1 - t_2)\alpha]/\lambda\}. \qquad (25.5)\end{aligned}$$

In this equation the first bracketed term can be chosen to yield a 90° phase shift by selecting the plate thicknesses such that $(t_1 - t_2)\beta = \frac{1}{4}$. The second bracketed term is dependent on λ and must therefore be set to zero. Since $t_1 - t_2$ is already fixed, elimination of the second term requires an adjustment of $d_1 - d_2$, the path-length difference in air. In practice these adjustments are made iteratively by changing $d_1 - d_2$ while rotating CP by small amounts until the desired null is achieved.

References for chapter 25

1 R. N. Bracewell, Detecting nonsolar planets by spinning infrared interferometer, *Nature* **274**, 780–781 (1978).
2 J. R. P. Angel and N. J. Woolf, searching for life on other planets, *Scientific American* **274**, 60–66 (April 1996).
3 N. Woolf and J. R. Angel, Astronomical searches for Earth-like planets and signs of life, *Ann. Rev. Astron. Astrophys.* **36**, 507–537 (1998).
4 P. M. Hinz *et al.*, Imaging circumstellar environments with a nulling interferometer, *Nature* **395**, 251–253 (1998).
5 J. R. Angel *et al.*, TPF: Terrestrial Planet Finder, JPL publication 99-3, May 1999. For more information visit the worldwide web at http://tpf.jpl.nasa.gov.

26

Scanning optical microscopy[†]

The diffraction-limited focusing of a laser beam to either explore or modify a surface is the basis of several important technologies. Examples include scanning optical microscopy, optical disk data storage, and laser printing. The size of the focused spot and the corresponding depth of focus are important factors in determining the performance characteristics of these systems. In this chapter we examine methods of forming the focused spot, and clarify the relation between spot size and depth of focus.

Principle of operation

The essential features of a scanning optical microscope are shown in Figure 26.1. A laser beam is sent through an objective lens to form a focused spot on the sample. Ideally, the objective is corrected for all aberrations, yielding a diffraction-limited focused spot. The light reflected from the sample returns through the objective and is redirected by the beam-splitter to a detection module. The detection module may be designed to monitor the power, the phase, or the polarization state of the returning beam. The electrical signal $S(x, y)$ produced by the detector is thus representative of the small area of the sample illuminated by the focused spot at and around the point (x, y). The sample is moved to different locations by the XY stage on which it is mounted; the signal $S(x, y)$, plotted against the sample's position, yields an image of the sample's surface over the desired area.

[†] The coauthors of this chapter are Lifeng Li and Wei-Hung Yeh.

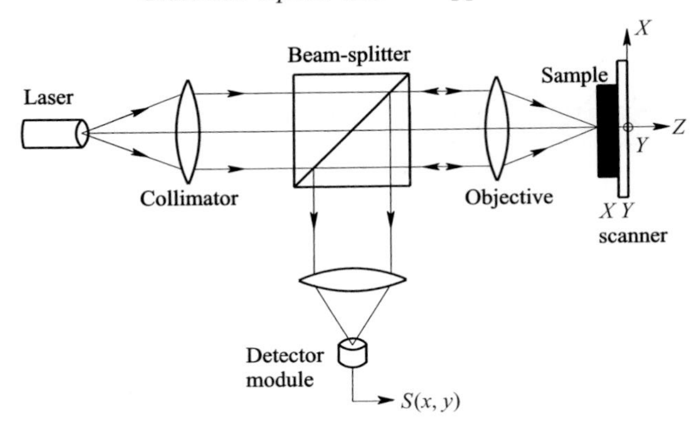

Figure 26.1 Schematic diagram of a scanning optical microscope. The objective lens focuses the laser beam at the point (x, y) on the sample. The XY scanner on which the sample is mounted moves the sample in small steps along both X- and Y-directions, covering the area of interest. At each point the reflected light is picked up by the detector module and converted to a signal $S(x, y)$. A plot of $S(x, y)$ constitutes the image of the scanned area.

Spot size at best focus

The most important component of any optical microscope is its objective lens. The quality of the focused spot produced by the objective determines the resolution of the images obtained, so it is important to have a very small, aberration-free spot at the focal plane of the objective. Figure 26.2, a schematic drawing of an objective lens, defines some of its important characteristics. The converging cone of light has a half-angle θ. The numerical aperture NA of the lens is defined in terms of this half-angle and the refractive index n of the medium in which the sample is immersed:

$$NA = n \sin \theta. \tag{26.1}$$

When the sample is in air ($n = 1$) the numerical aperture is less than unity. However, if the sample is embedded in a liquid or solid of refractive index $n > 1$, the numerical aperture can be as large as n.

The diameter D of an aberration-free focused spot is given by diffraction theory as[1]

$$D \approx \lambda_0 / NA \tag{26.2}$$

where λ_0 is the vacuum wavelength of the laser beam. The above equation gives only a rough estimate of the spot diameter, the exact value depending on how the diameter is defined [e.g., the diameter of the first dark ring of the Airy disk, the full width at half maximum (FWHM) of the intensity distribu-

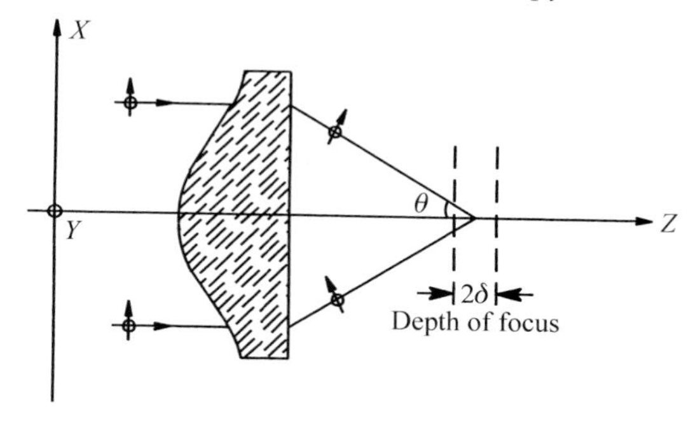

Figure 26.2 A polarized beam of light is brought to diffraction-limited focus by a microscope objective lens. Since scanning microscopy is typically done with a monochromatic laser beam, chromatic aberrations of the lens are of no concern. Bending of the polarization vector, however, is significant and must be taken into consideration. The half-angle θ of the focused cone is used to define the NA-value of the lens. The depth of focus is within $\pm\delta$ of the focal plane. For a high-NA singlet, such as the plano-convex lens shown here, diffraction-limited performance over a flat field can be achieved only with an aspheric surface. This particular lens, designed for operation at $\lambda_0 = 633$ nm, has the following set of parameters: $n = 1.806092$, $R_c = 0.9846$ mm, $K = -1.00938$, $A_4 = 6.16672 \times 10^{-2}$, $A_6 = 1.42948 \times 10^{-2}$, $A_8 = -2.14376 \times 10^{-2}$, $A_{10} = 8.12147 \times 10^{-3}$, aperture radius = 1 mm, thickness = 1.142 mm. The lens NA-value is 0.615 and its focal length is 1.2315 mm.

tion, etc.], on the distribution of light at the entrance pupil of the lens (e.g., uniform, truncated Gaussian, etc.), and on the state of polarization of the laser beam. The proportionality constant between D and λ_0/NA is typically between 0.5 and 1.5, depending on the circumstances.

Figure 26.3 shows plots of intensity distribution at the focal plane of the 0.615NA objective shown in Figure 26.2. The incident beam is assumed to be

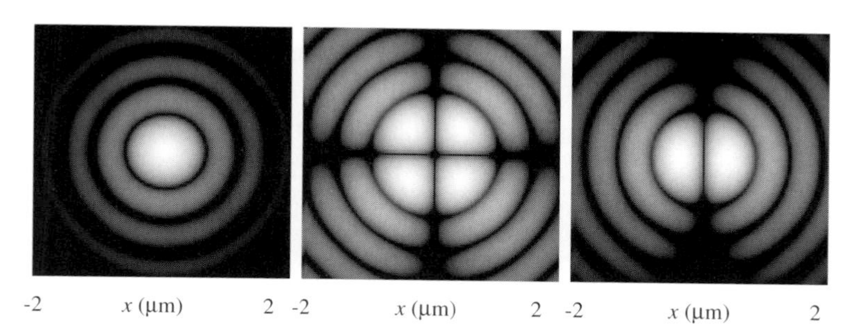

Figure 26.3 Logarithmic plots of intensity distribution at the focal plane of the 0.615NA objective shown in Figure 26.2. The incident beam is uniform and has linear polarization along the X-axis. From left to right: X-, Y-, and Z-components of polarization at best focus. The integrated intensities of the three components are in the ratios 1 : 0.002 : 0.113.

uniform and linearly polarized along the X-axis. The bending of the rays by the lens produces E-field components along the Y- and Z-axes as well; the distributions of these components, which carry only a small fraction of the total optical energy, are shown in Figure 26.3(b), (c). The logarithmic scale of these plots enhances the rings of light around the central bright spot; in fact these rings are typically weak and do not contribute much to the scanning signal. The central bright spot in Figure 26.3(a) is the most important contributor to the signal, but for accurate measurements the effects of the entire focused spot should be taken into consideration.

Depth of focus

Another important characteristic of the focused spot is its depth of focus. Typically, for high-NA objectives the range over which the spot size can be considered to be small is quite limited. As shown in Figure 26.2, if the sample moves by $\pm\delta$ along the Z-axis, deviations from perfect focus may be tolerable; for larger movements, the quality of the scanning signal suffers. The order of magnitude of the depth of focus is given by the theory of diffraction as $\delta/\lambda \approx \pm(D/\lambda)^2$, which is an expression for the Rayleigh range[2] of the beam in a medium in which the wavelength is λ. This expression may be written as

$$\delta \approx \pm D^2/\lambda. \tag{26.3}$$

The proportionality constant between δ and D^2/λ depends on the performance criteria of the system and may be anywhere in the range 0.1 to 1. For the $0.615NA$ lens of Figure 26.2, plots of total intensity distribution (i.e., the X-, Y-, and Z- components of polarization combined) at several distances from focus are shown in Figure 26.4. At best focus a small elongation of the spot along the X-axis may be observed. This is characteristic of the focused spots obtained with linearly polarized light at high NA: the spot is always elongated along the direction of incident polarization. The FWHM of the spot at this point is 0.57 μm along X and 0.51 μm along Y. Equations (26.2) and (26.3) predict $\delta \approx \pm\lambda_0/NA^2 = \pm 1.67$ μm, in agreement with the distributions of Figure 26.4. The spot diameter is substantially enlarged if the depth of focus is exceeded.

Oil immersion objective

To obtain improved resolution one may use an oil-immersion objective. As shown in Figure 26.5, the front element of this type of lens is in contact with a fluid having a specific refractive index n. (The front element is typically an

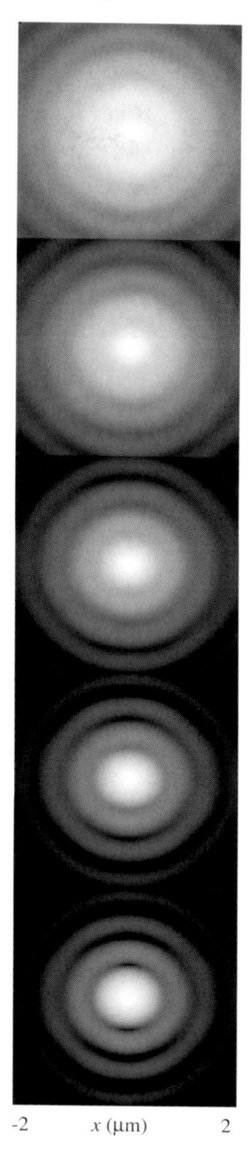

Figure 26.4 Logarithmic plots of total intensity distribution at and near the focus of the 0.615NA objective shown in Figure 26.2. From top to bottom $\Delta z = 2$ μm, 1.5 μm, 1 μm, 0.5 μm, and 0. Because of the symmetry between the two sides of focus, the distributions for $\pm\Delta z$ are the same. At best focus the spot's FWHM is 0.57 μm along X and 0.51 μm along Y.

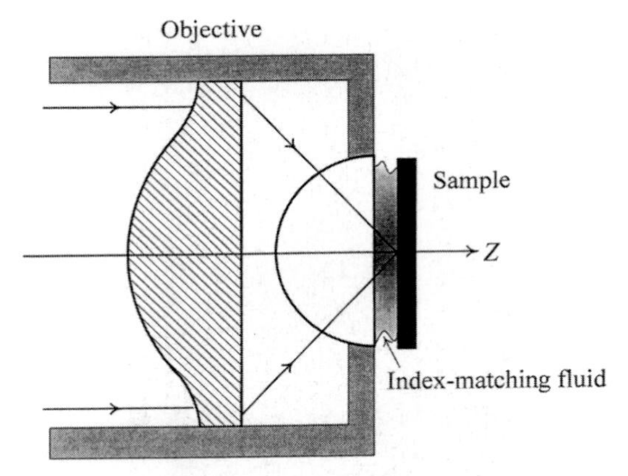

Figure 26.5 An oil-immersion objective focuses the beam onto the sample through an index-matched fluid of refractive index n. The fluid is in contact with both the sample and the front element of the lens. The rays that emerge from the objective do not bend on their way to the sample, thus forming a high-NA cone of light. For a given half-angle θ of the cone, the NA of an oil immersion objective is superior to that of an air-incidence objective by a factor n.

aplanatic sphere; for a discussion of aplanatism see chapter 1, "Abbe's sine condition".) The front element of the lens, the fluid, and the cover plate protecting the sample (if any) should all have the same or nearly the same refractive index. Thus, upon emerging from the objective the rays go directly to the sample's surface without further bending. Under such conditions the wavelength of the light within the immersion oil is reduced by a factor n, in consequence of which the effective NA of the lens increases by the same factor. Equations (26.1)–(26.3) apply to this case as well, showing that for a given cone angle θ both the spot size D and the depth of focus δ shrink by a factor n, compared with an objective designed for operation in air.

For an oil-immersion lens having $\sin\theta = 0.615$ and $n = 2$, Figure 26.6 shows plots of the total intensity distribution at 1 μm defocus (top) and at best focus (bottom). Compared to Figure 26.4, which corresponds to the same value of θ in air, it is apparent that both the spot diameter and the depth of focus have decreased by a factor $n = 2$.

Line scans across a grating

Figure 26.7 shows the cross-section of a diffraction grating. The grating is coated with a thick layer of gold, n and k are 0.14 and 3.37, respectively; the grating has a groove depth 170 nm and a period 1.5 μm, of which 0.5 μm is the groove width, 0.66 μm is the land width, and the remaining 0.34 μm is

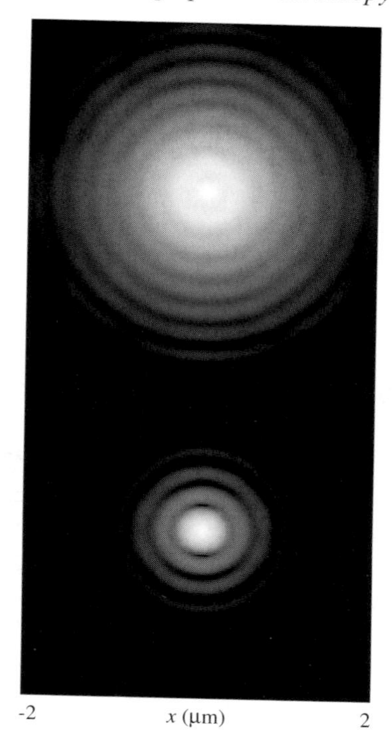

Figure 26.6 Logarithmic plots of intensity distribution at and near the focus of an oil-immersion objective. The objective consists of the $0.615NA$ lens of Figure 26.2 in conjunction with a hemispherical glass cap. Both the cap and the immersion oil have index $n = 2$, resulting in an overall NA of 1.23. Top: Δz is 1 μm away from the focal plane. Bottom: the position of best focus; FWHM = 0.28 μm along X, 0.25 μm along Y.

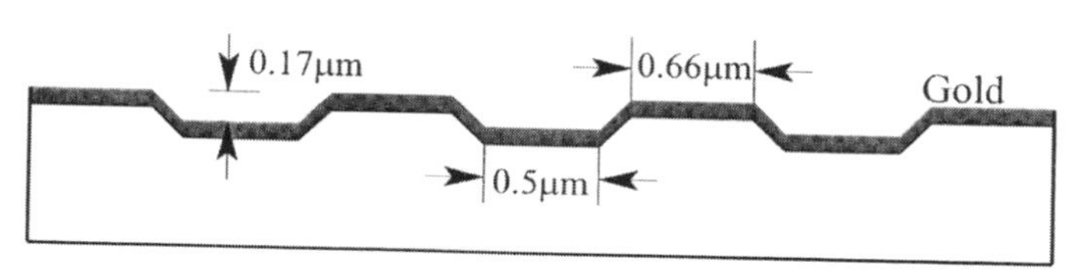

Figure 26.7 Cross section of a diffraction grating used in computer simulations (period 1.5 μm). The gold coating is thick enough to prevent the light from penetrating through to the other side.

taken up by the two side walls, pitched at 45°. For the purpose of imaging this grating, the assumed detector module in the system of Figure 26.1 is a split detector, oriented with its splitting line parallel to the grooves. As will be described below, the outputs S_1, S_2 of the split detector may be combined in different ways to yield the scanning signal.

Figure 26.8 shows plots of a single line-scan of the grating in the direction perpendicular to the grooves; the scalar theory of diffraction has been used to compute these plots. The dashed curves correspond to a 0.6*NA* air-incidence objective, while the solid curves represent a 1.2*NA* oil-immersion objective. The scans in Figure 26.8(a), obtained by adding S_1 and S_2, represent the total optical power returning from the sample. The monitored signal in Figure 26.8(b) is the so-called push–pull signal, $(S_1 - S_2)/(S_1 + S_2)$, which is sensitive to the position of the groove edges. Clearly, the oil-immersion objective with its superior *NA*-value provides a better resolution in both cases.

The origin of the push–pull signal used for sensing the groove edges may be understood by considering Figure 26.9, which shows the intensity distribution in the exit pupil of the objective lens for three cases: from top to bottom, the spot is focused on the land center, on the groove edge, and on the groove center. The symmetry of this so-called "baseball pattern" is such that, with the beam focused on the land or on the groove, the split detector receives equal amounts of light on both its halves. However, on the groove edge the diffraction orders appearing on one side of the baseball pattern have a different phase from those appearing on the opposite side and, therefore, the asymmetry between the two halves of the baseball pattern yields a fairly large differential signal. Since these calculations are based on the scalar theory of diffraction, anomalous effects due to surface plasmon excitation and dependence on the beam's polarization state are not observed. Such effects will show up later in our full vector diffraction calculations.

Focusing through a cover plate

At times it is necessary to observe a sample through a transparent cover plate. Biological samples, for instance, are usually prepared between a pair of thin glass plates, and the storage layer of compact disks is protected from dust and fingerprints by a plastic substrate 1.2 mm thick. In either case the objective lens must be corrected for the specific thickness and refractive index of the cover plate.

As shown in Figure 26.10, a cone of light focused through a parallel plate becomes compressed toward the optical axis, its value of $\sin\theta$ shrinking by the refractive index n of the plate. At the same time, the wavelength of the

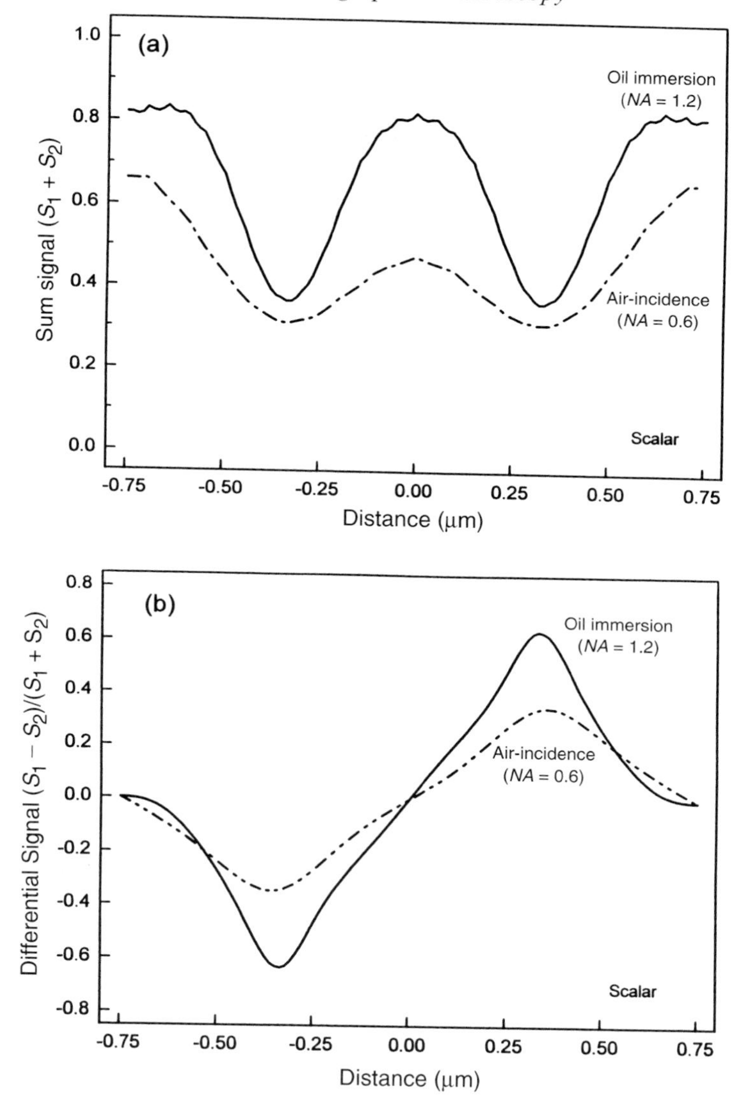

Figure 26.8 Scalar diffraction theory applied to the grating of Figure 26.7 yields single-line scans in the direction perpendicular to the grooves. The scanned period extends from the center of the land at −0.75 μm to the center of the adjacent land, at +0.75 μm, with the groove center at 0. The broken line corresponds to a 0.6*NA* air-incidence objective, while the solid line represents a 1.2*NA* oil-immersion objective. The detector module consists of a split detector aligned with the grooves, yielding signals S_1 and S_2. (a) Sum signal scans corresponding to the total reflected power. (b) Differential signal scans, corresponding to the "push–pull" method.

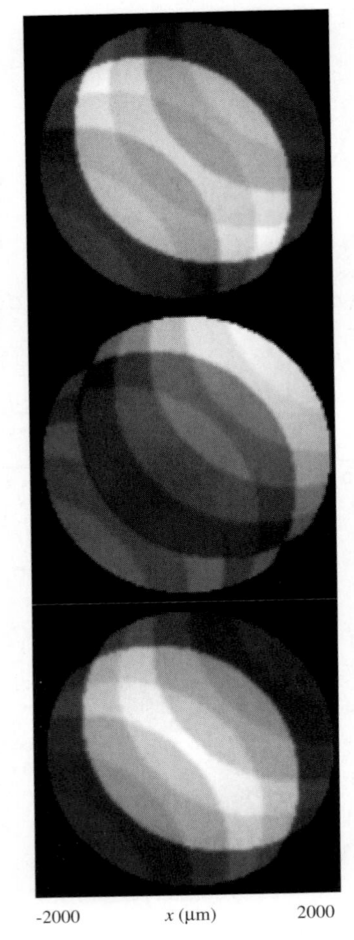

Figure 26.9 Computed baseball patterns at the exit pupil of the 1.2NA oil-immersion objective during the scans depicted in Figure 26.8. From top to bottom, the focused spot is on the land center, on the groove edge, and on the groove center.

light inside the plate also shrinks by the same factor, to give $\lambda = \lambda_0/n$. The net effect is that the spot diameter D does not change as a result of focusing through the cover plate. However, Eq. (26.3) implies that the depth of focus will improve. This would be true, of course, if one interpreted the depth of focus as the depth of the sample interrogated by the focused beam while the sample remained at rest. But what happens if one moves the sample in the $\pm Z$-direction and determines the distance Δz over which the image of the sample remains sharp? One finds in the latter case that focusing through the cover plate does not improve the depth of focus at all. In other words, the

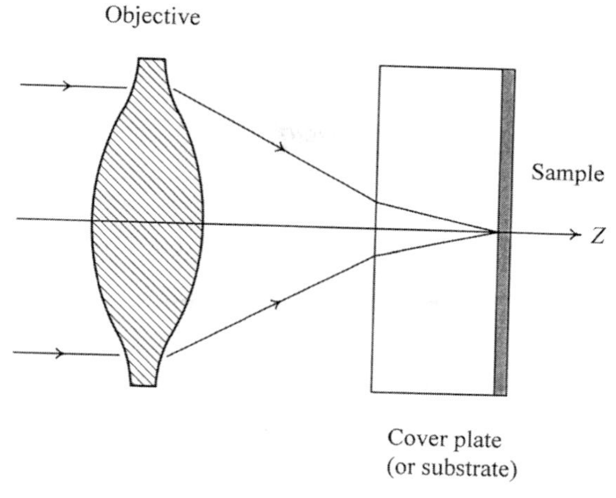

Figure 26.10 Focusing through a transparent cover plate of refractive index n. The cone angle shrinks by a factor of n, but the spot size and the depth of focus are not affected.

depths of focus with and without the cover plate are exactly the same. (Keep in mind that the objective lens is corrected for each case separately.)

The reason for the above apparent discrepancy is as follows. If one moves the sample and the cover plate together by Δz along the positive Z-axis, the top of the cover plate also moves away from the lens by the same distance. Consequently the focused spot recedes from the sample's surface by $n\Delta z$, which is greater than the actual travel of the sample. (This analysis, which ignores residual spherical aberrations, is quite straightforward and requires only the use of Snell's law and simple geometry. It also applies to the case where the sample and the cover plate move along the negative Z-axis.) Thus, as long as the lens remains stationary while the sample and the cover plate travel together along Z, the cover plate does not increase, nor does it decrease, the depth of focus. Aside from protecting the sample, focusing through the cover plate has no obvious advantages.

The solid immersion lens

A transparent hemisphere of refractive index n may be placed over the sample in such a way as to bring the cone of light to focus at the center of the hemisphere, as shown in Figure 26.11. The use of this type of hemisphere, often referred to as a solid immersion lens (SIL),[3] improves the resolution of the system by a factor of n. To establish smooth and seamless contact without the use of an index-matching fluid, the bottom of the hemisphere and the top

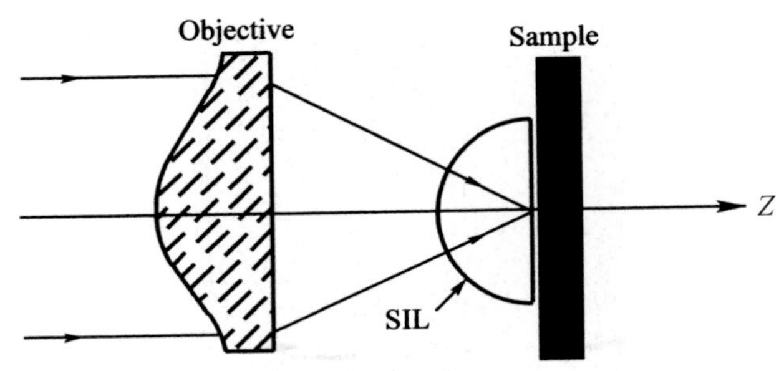

Figure 26.11 Focusing through a solid immersion lens (SIL) of refractive index n. The spot size shrinks by a factor of n, but, assuming the SIL and the sample move together along Z, the depth of focus remains the same.

of the sample must both be flat and free from dirt, dust and scratches. The resolution gain thus achieved is a consequence of the fact that in going through the hemisphere the cone angle θ remains the same while the wavelength of the light shrinks by a factor of n.

What is remarkable about the SIL is that, unlike in oil-immersion microscopy, the depth of focus does not suffer as a result of the improved resolution. As long as the SIL and the sample move together along Z, whether towards or away from the objective, the bending of the light rays at the spherical surface of the SIL (governed by Snell's law) makes the focused spot move in the direction of the sample, thereby helping to increase the depth of focus. The net effect is that the depth of focus of the system remains the same whether or not the SIL is placed on the sample. Figure 26.12 shows computed plots of intensity distribution at the sample's surface when the assembly of the sample and the SIL travels by distances $\Delta z = 2$ μm, 1 μm, and 0 away from the position of best focus. Comparing Figure 26.12 with Figure 26.4, one concludes that the use of the SIL has reduced the spot diameter by a factor of $n = 2$ but has not changed the system's depth of focus.

Effect of the air gap

In applications of SIL to microscopy, where the sample is stationary, the SIL and the sample remain in contact, keeping the width of the air gap at $W_g = 0$. However, in optical disk systems, where the disk spins under the SIL at a rapid rate, a small air gap develops between the bottom of the SIL and the top of the disk surface.[4] Under such circumstances the light must jump through the gap in order to interact with the storage layer of the disk. This

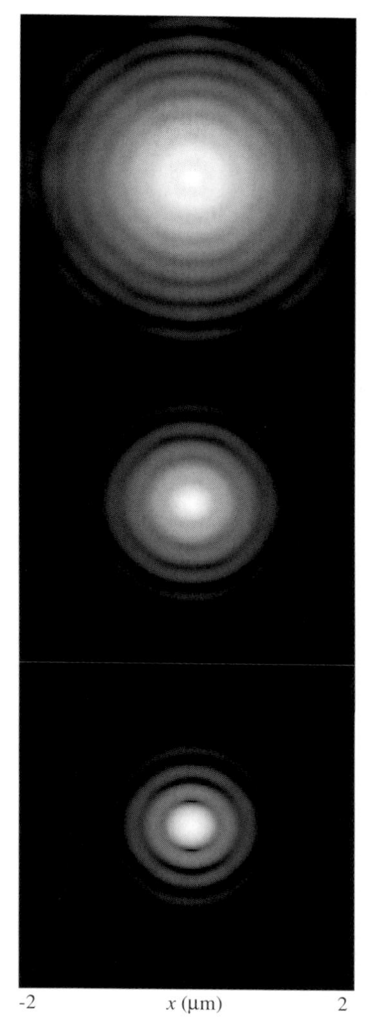

Figure 26.12 Logarithmic plots of intensity distribution when a SIL (radius = 0.5 mm, $n = 2$) is placed in front of a 0.615NA objective lens. From top to bottom, the SIL and the sample, moving together along the Z-axis, deviate from the position of best focus by 2 μm, 1 μm, and 0. At best focus the spot's FWHM is 0.28 μm along X and 0.25 μm along Y. The effective NA is 1.23, but the depth of focus is the same as it was prior to inserting the SIL.

is not a serious problem for those rays that propagate along the Z-axis, or at a small inclination with respect to it, since they are readily transmitted through the bottom of the SIL. However, for those rays that make a large angle with the Z-axis, the Fresnel transmission coefficients become small; in particular, when the incidence angle exceeds the critical angle of total internal reflection the transmissivity drops to zero.[1] Fortunately, the phenomenon of frustrated total internal reflection allows photons to tunnel through the gap and reach the storage layer of the disk. For this to happen efficiently, the gap width W_g must be a small fraction of λ_0. (See chapter 19, "Some quirks of total internal reflection".)

The effects of the air gap on the signal level can be seen in Figure 26.13, which shows results of computer simulations based on the full vector theory of diffraction.[5,6] The grating used in these simulations is that of Figure 26.7, the assumed refractive index of the SIL is $n = 2$, and the incident beam is assumed to be linearly polarized. The direction of incident polarization is parallel to the grooves in Figures 26.13(a), (b), and perpendicular to the

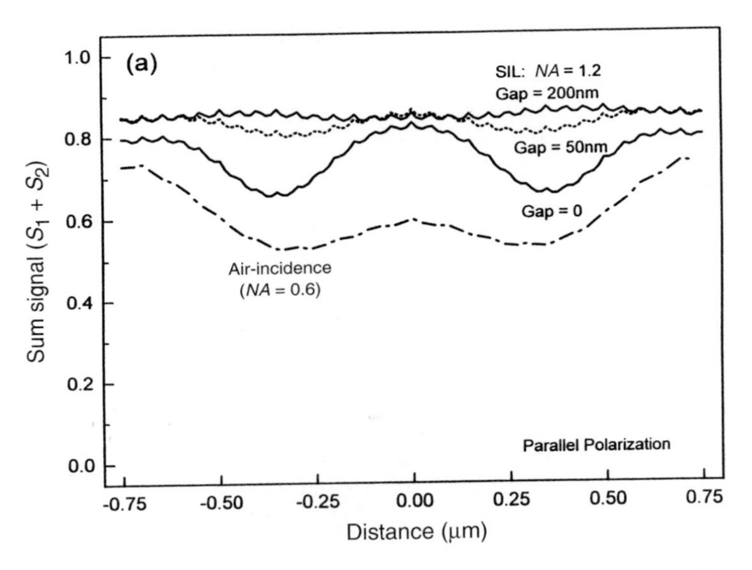

Figure 26.13 Computed line-scans in the direction perpendicular to the grooves of the grating of Figure 26.7, based on vector diffraction theory. The scanned range extends from the center of a land at −0.75 μm to the center of an adjacent land at +0.75 μm. The detector module consists of a split detector aligned with the grooves, yielding the signals S_1, S_2. (a), (c) Sum signal scans corresponding to the total reflected power collected by the detector: upper solid line, gap-width $W_g = 200$ nm; dotted line, $W_g = 50$ nm, lower solid line, $W_g = 0$. The small oscillations riding over the signals are caused by numerical errors. (b), (d) Differential signal scans corresponding to the push–pull method of detection: solid line of smaller amplitude, $W_g = 200$ nm; dotted line, $W_g = 50$ nm, solid line of greater amplitude, $W_g = 0$. In each figure the broken-and-dotted curve is obtained in the absence of the SIL with a 0.6NA objective lens, while the other three curves correspond to a SIL with $n = 2$ and an overall objective NA of 1.2. The linear incident polarization in (a) and (b) is parallel to the grooves, while in (c) and (d) it is perpendicular to the grooves.

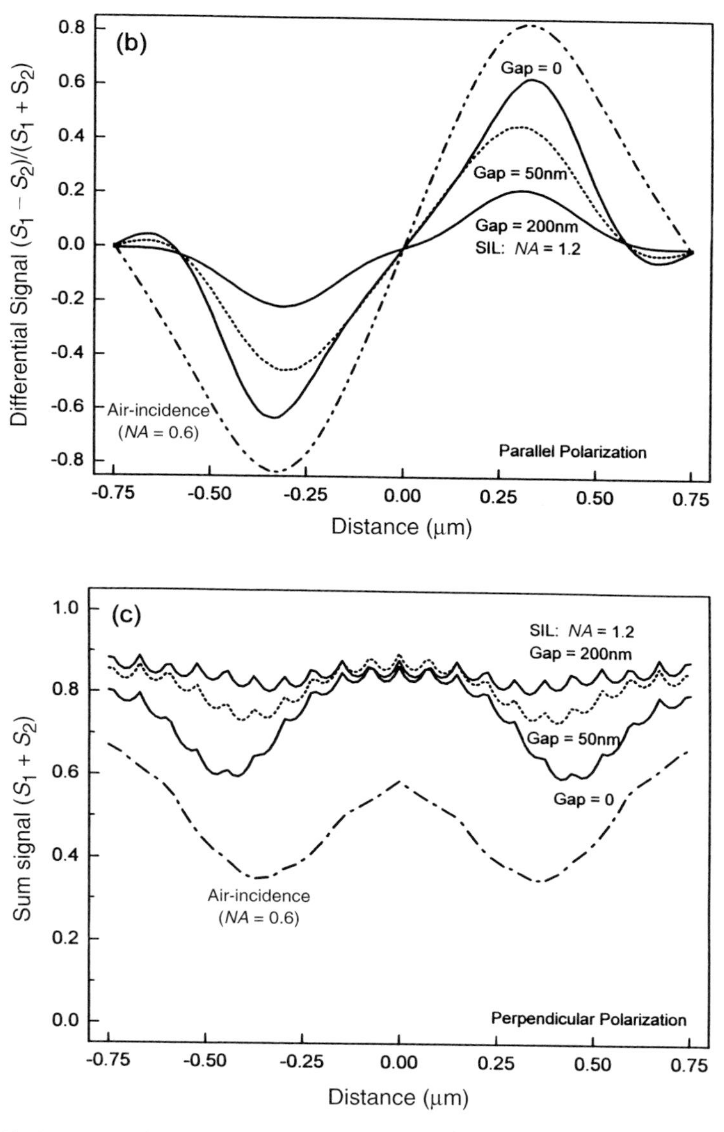

Figure 26.13 (*continued*)

grooves in Figures 26.13(c), (d). Several line-scans across a single period of the grating are shown; the plots in Figures 26.13(a), (c) correspond to the total returned optical power while those in Figures 26.13(b), (d) represent the differential (or push–pull) signal.[5] The signal amplitude is highest when $W_g = 0$, but it has dropped considerably at $W_g = 50$ nm and even further

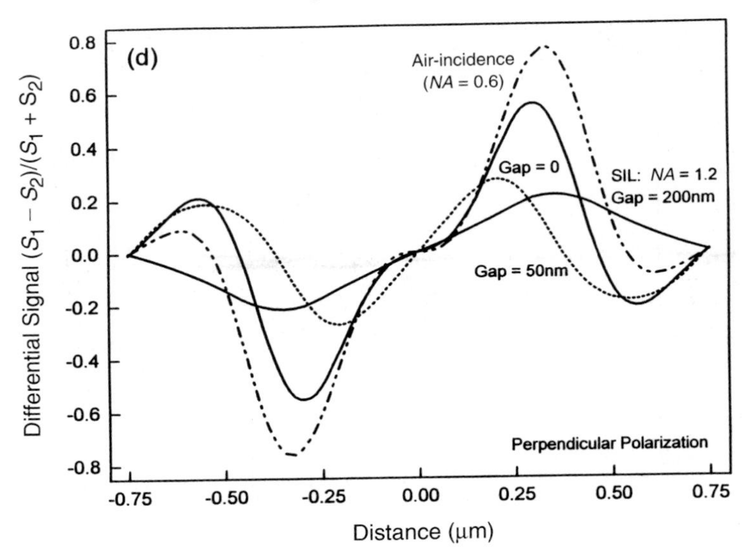

Figure 26.13 (*continued*)

at $W_g = 200$ nm. In practice a gap-width below about $\lambda_0/10$ is usually acceptable; going beyond this value causes a sharp reduction in the signal level.

The scanning signals are sensitive to the direction of incident polarization. In general the two polarization directions parallel and perpendicular to the grooves are not equivalent and yield different results, as may be readily observed in Figure 26.13. To emphasize further the significance of polarization, Figure 26.14 shows the light intensity pattern at the exit pupil of the objective lens for polarization directions parallel and perpendicular to the grooves, all other things being kept equal. The two baseball patterns show clear differences.

The super SIL

It is well known that a converging cone of light aimed at a point a distance nR below the center of a glass sphere of radius R and refractive index n comes to diffraction-limited focus within the sphere at a distance of R/n below the center.[1] This fact has been exploited in the design of the super SIL shown in Figure 26.15. The effective NA of the objective thus increases by a factor n^2, not only because the wavelength within the super SIL is shortened by a factor n but also because the super SIL increases the sine of the cone angle θ by a factor n. The super SIL is aplanatic, and placing it in front of any aplanatic objective renders the combination aplanatic as well.

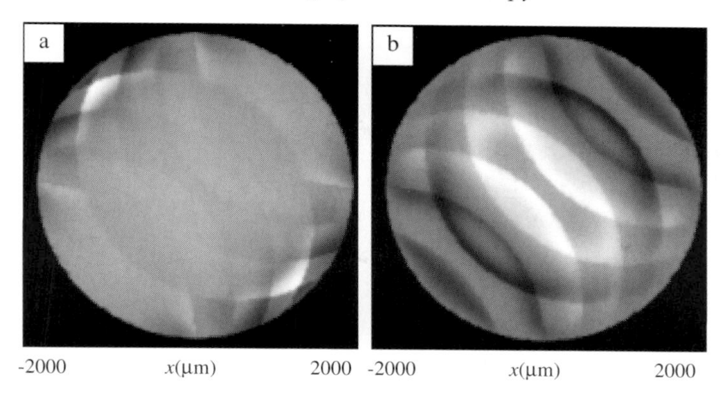

Figure 26.14 Computed baseball patterns at the exit pupil of a $0.6NA$ objective lens when the grating of Figure 26.7 is placed beneath a SIL of refractive index $n = 2$ (total $NA = 1.2$). The focused spot is at the center of the land and the assumed gap width is 100 nm. In (a) the polarization vector is parallel to the grooves, while in (b) it is perpendicular to the grooves.

The full factor n^2 mentioned above may not be realized in practice, because the bending of the rays within the super SIL works only up to a point, stopping when the marginal rays become orthogonal to the Z-axis. If the objective happens to have a large NA to begin with, the super SIL can only increase the value of its $\sin\theta$ up to 1, at which point the remaining rays will miss the super SIL. The other improvement by a factor of n, however, is always realized in practice because the wavelength always shrinks by this factor.

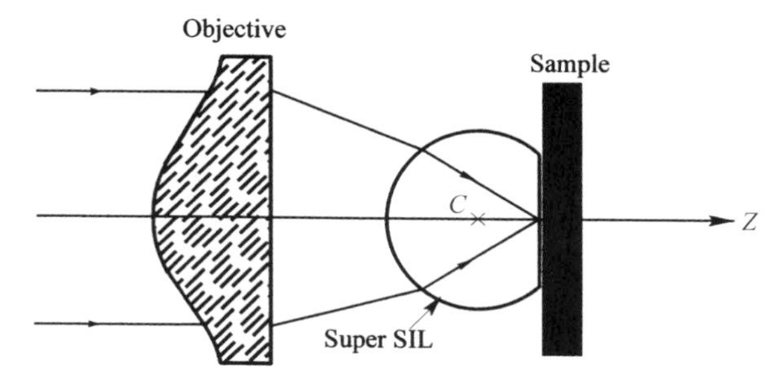

Figure 26.15 Focusing through a "super SIL" of refractive index n and radius R. In the absence of this SIL the cone of light from the objective comes to focus at a distance of nR from the center C of the sphere. The bending of the rays at the sphere's surface shifts the focal point to a distance of R/n from C, without introducing any aberrations.

To see the effect of the super SIL on the focused spot, computed light intensity distributions at and near the focus of a $0.4NA$ objective are shown in Figure 26.16. The FWHM spot diameter at best focus is 0.84 μm along X and 0.8 μm along Y, while the depth of focus according to Eqs. (26.2) and (26.3) is around ±4 μm. When a super SIL of index $n = 2$ is placed in front of this objective the plots of Figure 26.17 are obtained. Clearly the spot size has shrunk by n^2, but the depth of focus is nearly the same as it was before the super SIL was introduced. Once again, it is observed that focusing through the super SIL does not reduce the depth of focus, as long as the sample and the super SIL move together along the Z-axis. (This statement ignores the effects of a small amount of spherical aberration introduced by the departure of the super SIL from its ideal location.)

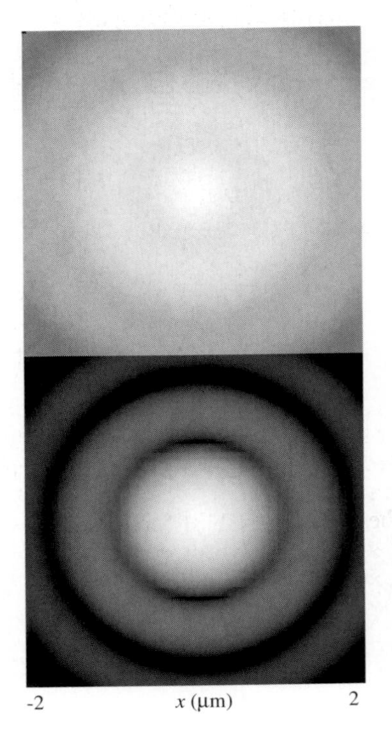

Figure 26.16 Logarithmic plots of intensity distribution at and near the focus of a $0.4NA$ objective lens operating at $\lambda_0 = 633$ nm. Top: $\Delta z = 5$ μm defocus. Bottom: best focus ($\Delta z = 0$). The spot's FWHM at best focus is 0.84 μm along X and 0.80 μm along Y.

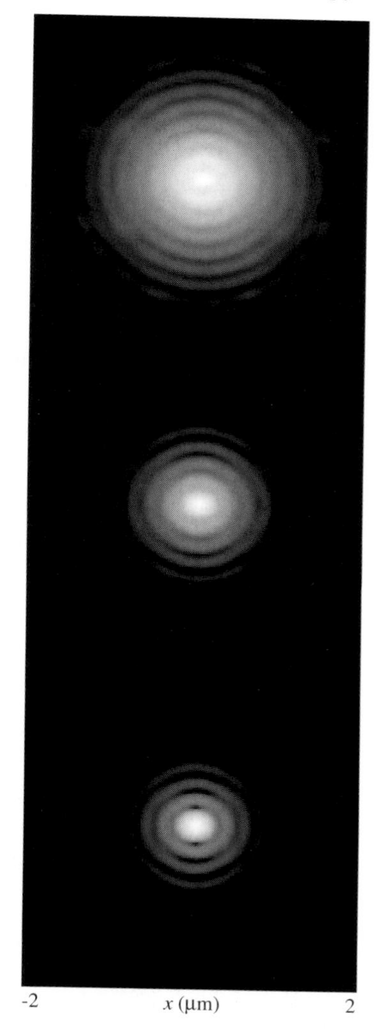

Figure 26.17 Logarithmic plots of intensity distribution obtained when a super SIL ($R = 0.5$ mm, $n = 2$) is placed in front of the $0.4NA$ objective depicted in Figure 26.16. From top to bottom, the super SIL and the sample, moving together along the Z-axis, deviate from the position of best focus by $\Delta z = 5$ μm, 2.5 μm, and 0. At best focus the FWHM of the spot is 0.24 μm along X and 0.20 μm along Y. The effective NA is 1.6, and the spot size has shrunk accordingly, but the depth of focus is essentially the same as it was without the super SIL.

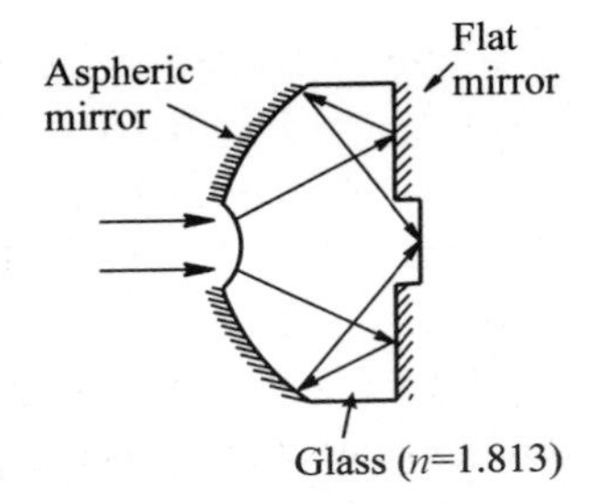

Figure 26.18 A catadioptric optical element molded from glass of refractive index $n = 1.813$. The spherical entrance facet has a radius of curvature $R_c = 0.7$ mm and an aperture radius of 0.41 mm. Once inside the glass, the light rays are reflected first from the flat aluminized surface and then from the aspheric surface (also aluminized), before arriving at the flat exit surface of the plateau. The aspheric parameters are as follows: $R_c = 2.5308$ mm, $K = -1.7076$, $A_1 = 0.01233$, $A_2 = 0.209 \times 10^{-3}$, $A_3 = 0.4476 \times 10^{-4}$, $A_4 = 0.8797 \times 10^{-5}$, aperture radius = 1.8 mm. The vertex of the aspheric surface is at $z = 0$, that of the spherical surface is at $z = 0.3$ mm, the flat mirror is at $z = 1.5$ mm, and the exit facet is at $z = 1.8$ mm.

A catadioptric SIL

A design that combines the objective and the SIL into one catadioptric element is shown in Figure 26.18.[7] (A catadioptric element is one that involves both the reflection and the refraction of light.) A collimated beam enters the concave facet of the lens, is reflected first at a flat internal mirror, then at an aspheric internal mirror, and is finally brought to focus at the bottom of a plateau that is in contact (or near contact) with the surface of the sample

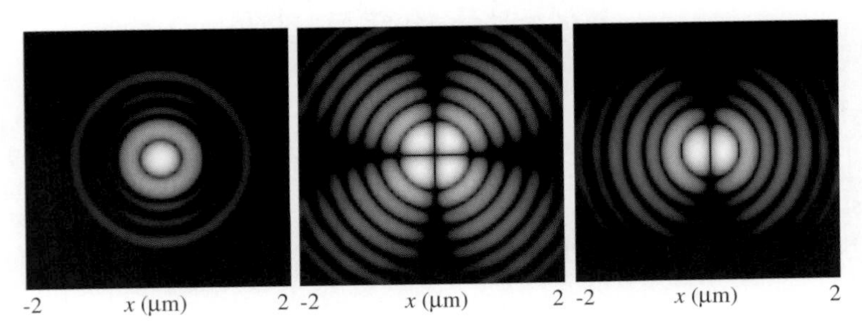

Figure 26.19 Logarithmic plots of the intensity distribution at the focal plane of the catadioptric lens of Figure 26.18. The incident beam is collimated and linearly polarized along the X-axis. From left to right are shown the X-, Y-, and Z- components of polarization. The integrated intensities of the three components are in the ratios 1 : 0.003 : 0.128. The effective NA-value of the lens is 1.1, but its annular shape of aperture gives rise to a spot size slightly less than that of the Airy disk. The enhanced rings are also caused by the annular shape of the aperture.

under investigation. This particular lens, which is also aplanatic, has a reasonably large field of view, with $NA = 1.1$. Because the central portion of the incident beam is not used, the lens effectively has an annular aperture, which makes the central spot even smaller than the Airy disk but at the expense of increasing the brightness of the rings. Figure 26.19 shows plots of intensity distribution at the focus of the lens for the three components of polarization. The FWHM of the total intensity distribution is 0.3 μm along X and 0.26 μm along Y. Depth of focus is not a very useful concept for this particular element because the incident beam is collimated.

References for chapter 26

1 M. Born and E. Wolf, *Principles of Optics*, 6th edition, Pergamon Press, Oxford, 1980.

2 A. E. Siegman, *An Introduction to Lasers and Masers*, McGraw-Hill, New York, 1971.

3 S. M. Mansfield, W. R. Studenmund, G. S. Kino, and K. Osato, *Opt. Lett.* **18**, 305–307 (1993).

4 B. D. Terris, H. J. Mamin, and D. Rugar, *Appl. Phys. Lett.* **65**, 388–390 (1994).

5 For the computations that led to Figures 26.13 and 26.14, the reflection coefficients of the grating were first calculated using DELTA, a vector diffraction code developed by Lifeng Li. These coefficients were subsequently imported to DIFFRACT™ where they were combined to represent the effects of a focused beam.

6 Lifeng Li, Multilayer-coated diffraction gratings: differential method of Chandezon *et al.* revisited, *J. Opt. Soc. Am.* A**11**, 2816–2828 (1994).

7 C. W. Lee *et al.*, Feasibility study on near field optical memory using a catadioptric optical system, *Optical Data Storage Conference*, Aspen, Colorado, May 1998.

27

Zernike's method of phase contrast

Frederik (Fritz) Zernike (1888–1966)

Zernike invented the phase-contrast microscope in 1935, and was awarded the 1953 Nobel prize in physics for this achievement.[1] In an ordinary optical microscope, an object that imparts a phase modulation to the incident light will produce only a faint image. This faint image may be attributed to the diffraction of a small amount of the light out of the entrance pupil of the objective lens. To improve this image, Zernike in effect extracted a reference beam from the light collected by the objective lens and produced an interferogram of the object at the image plane of the microscope, thus converting phase information into amplitude (or intensity) modulation.

The principles of operation of the phase-contrast microscope are by now fully understood.[1–6] Both spatially coherent and spatially incoherent light may be used in this type of microscopy. For best results, a quasi-monochromatic light source with a reasonable coherence time must be employed. Our goal in the present chapter is to give a simple explanation of the main ideas behind the method and to provide a pictorial survey of this important branch of modern optical microscopy.

The phase-contrast microscope

The diagram in Figure 27.1 shows the main elements of a phase-contrast microscope. The light source may be a coherent source (e.g., a laser) or an incoherent one (e.g., a tungsten lamp or an arc lamp); monochromaticity may be achieved by means of a colored glass filter. The condenser lens projects the source onto the object, whose image is formed by the objective lens. Although the system depicted in Figure 27.1 appears as transmissive, it could just as well represent the unfolded view of a reflective system. In the latter case, the condenser and the objective lens are physically the same element, and a means of separating the incident path from the reflected path (such as a beam-splitter) must be provided.

The main difference between an ordinary microscope and a phase contrast microscope is the presence, in the latter, of a spatial filter (or mask) within the rear focal plane of the objective lens (see Figure 27.1). To appreciate the action of this filter, we note that the light emerging from the object over its XY-plane has a complex-amplitude distribution that may be assumed to be proportional to $\exp[i\phi(x, y)]$. Here $\phi(x, y)$ is the phase distribution imparted to a uniform incident beam upon transmission through (or reflection from) a sample that may have surface-relief structure and non-uniform thickness, or perhaps even an inhomogeneous refractive index profile.

Assuming that $\phi(x, y)$ is sufficiently small, we may use a Taylor series expansion to arrive at the following approximation:

$$\exp[i\phi(x, y)] \approx 1 + i\phi(x, y). \tag{27.1}$$

In the Fourier domain, the first term in the above expansion, being the constant or d.c. term, appears at the center of the plane of spatial frequencies.

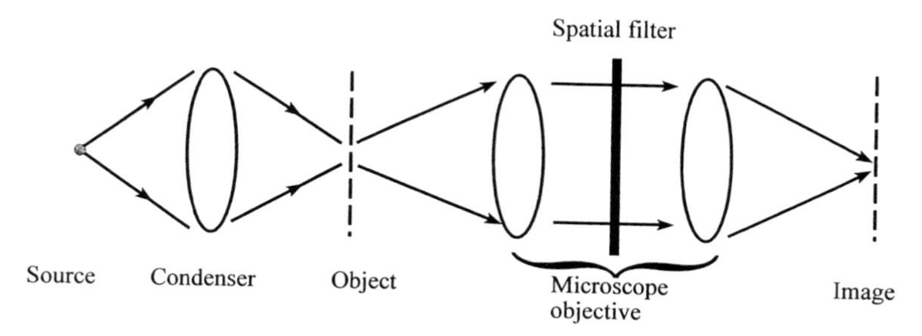

Figure 27.1 Schematic diagram of a simple phase-contrast imaging system. The light source is projected by the condenser lens onto a phase object, allowing the objective lens to form an image of this object at the image plane. The main component is the mask in the Fourier plane, which imparts a uniform phase shift (and possibly some amplitude attenuation) to the undiffracted component of the beam.

In the rear focal plane of the objective lens, therefore, the d.c. term appears as a bright spot centered at and around the optical axis. Zernike realized that by placing a 90° phase shift on this d.c. term (i.e., multiplying it by i), he could bring it in phase with the second term in Eq. (27.1). In this way he enabled beams corresponding to the two terms in the above expansion to interfere with each other when they overlapped within the image plane of the system. The primary function of the spatial filter, therefore, is to delay, by one quarter of a wavelength, the central region of the beam within the rear focal plane of the objective lens.

The source and the illumination optics

Two types of illumination will be considered. To provide collimated coherent illumination we assume that a monochromatic laser beam is brought to focus on the object by a condenser lens of a very small numerical aperture (NA). Figure 27.2(a) is the logarithmic intensity distribution at the object plane produced by a $0.03NA$ condenser. This distribution has the shape of an Airy pattern, with a central lobe diameter of $1.22\lambda/NA \approx 41\lambda$, where λ is the wavelength of the light source. Since the objects of interest will be small compared to the Airy disk diameter, and since they will be placed near the center of the Airy disk, this illumination qualifies as coherent, fairly uniform, and nearly collimated.

The second type of illumination to be considered is incoherent. Our concern, of course, is solely with spatial incoherence. However, to ensure that the phase is meaningfully defined throughout the system and that the coherence time is long enough for interference to occur, we assume a quasi-monochromatic source with a sufficiently narrow bandwidth. With this type of illumination, the source can be modeled as a collection of independent point sources extending over the luminous area of the lamp. We compute the image obtained with each such point source independently, and add up the *intensities* of the resulting images to obtain the final image.

The imaging optics

The objective lens used in the simulations described below is free from aberrations and, therefore, its performance is diffraction-limited. The objective is a finite-conjugate lens with a numerical aperture of 0.25 (on the side of the object), a focal length of 5000λ, and a magnification of 10.

The object used throughout this chapter is a transparent piece of flat glass or plastic, embossed with seven marks of various sizes and shapes, as shown

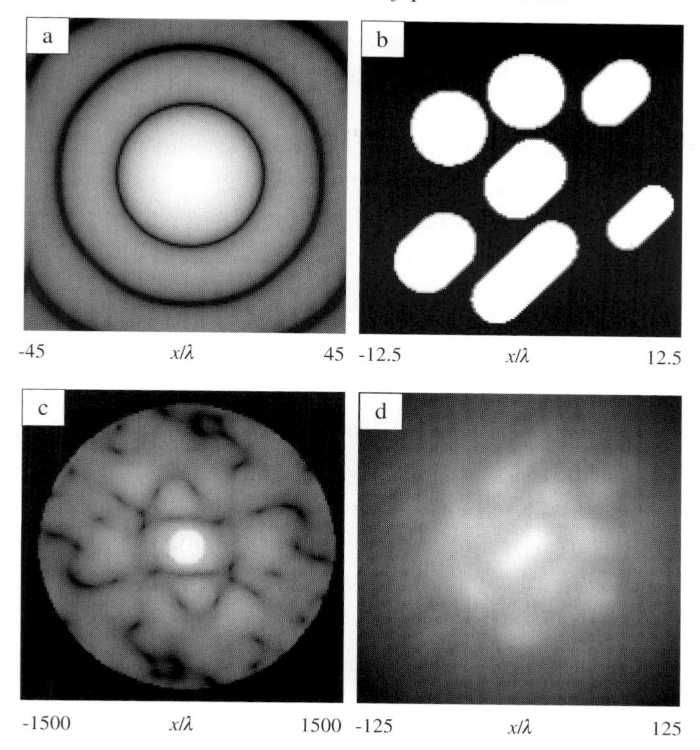

Figure 27.2 Computed distributions at various cross-sections of the system of Figure 27.1 (without the phase-contrast mask) for the case of coherent illumination. (a) Logarithmic plot ($\alpha = 4$) of the intensity distribution at the object plane, obtained when a collimated, coherent source is brought to focus by a $0.03NA$ condenser lens. (b) Pattern of phase objects (marks) with different sizes and separations, on a uniform background. The transmissivity is 100% over the entire area of this object, but the marks impart to the incident beam a 36° phase shift (i.e., one-tenth of a wavelength) relative to the background. (c) Logarithmic plot ($\alpha = 6$) of the intensity distribution at the exit pupil of the objective lens. (d) Distribution of intensity in the image plane of the system in the absence of a phase-contrast filter; the outlines of the marks are barely visible in this image.

in Figure 27.2(b). The largest mark is 10λ long and the smallest mark is 3λ wide. These marks are large enough to yield a reasonably clear image with both coherent and incoherent illumination, in conjunction with an appropriate phase-contrast filter. All the marks impart to the incident beam a phase shift of 36° relative to the background (corresponding to an optical path-length difference of $\lambda/10$).

For coherent illumination of the object by the beam depicted in Figure 27.2(a), the logarithmic plot of intensity distribution at the Fourier plane is shown in Figure 27.2(c). The bright central spot in this figure is the d.c. term mentioned earlier. Note that the cutoff point of this logarithmic plot is at $\alpha = 6$ and, therefore, the light diffracted by the object and spread throughout the aperture of the objective lens is quite weak. In the absence of any phase-contrast mechanism the computed image of the object is as shown in Figure 27.2(d). This obviously is a very poor image, one in which the boundaries of the marks are barely perceptible. We will see below how the action of the phase-contrast filter dramatically improves the quality of this image.

Contrast enhancement with coherent illumination

With a disk-shaped spatial filter (diameter = 550λ) placed in the Fourier transform plane of the object, the image shown in Figure 27.3 will be obtained. The filter in this case is a simple 90° phase-shifter, affecting the bright, central region of the beam shown in Figure 27.2(c). The images of the marks are now clearly visible, but the contrast is not remarkable.

To study the effect of amplitude filtering on image quality, we replace the phase-shifting filter with one that simply blocks the central region of the beam within the Fourier plane. The resulting image is shown in Figure 27.4. Note that, by eliminating the d.c. component of the phase modulation function $\phi(x, y)$, use of this filter has emphasized the boundaries of the marks.

The best choice for the phase-contrast filter is generally a phase/amplitude mask that shifts the d.c. component of the beam by 90°, while attenuating its amplitude to bring it in line with the magnitude of $\phi(x, y)$. Figure 27.5 shows

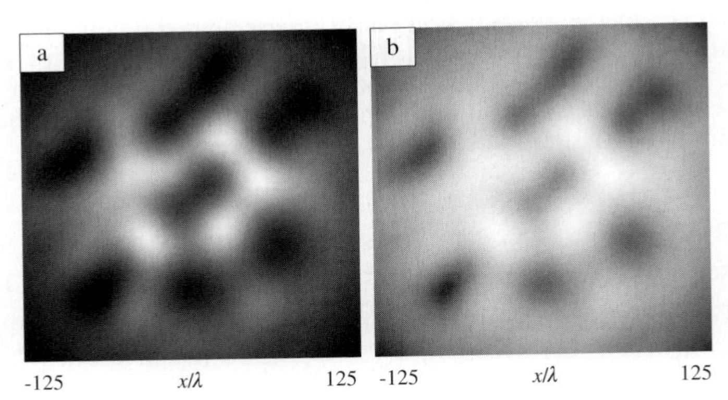

Figure 27.3 Image of the phase object of Figure 27.2(b), obtained with the coherent illumination of Figure 27.2(a) when a phase-contrast mask is placed in the Fourier plane. The mask is a small disk of radius 275λ, imparting a +90° phase shift to the central region of the beam. (a) Intensity distribution in the image plane. (b) Same as (a) but on a logarithmic scale ($\alpha = 1.65$).

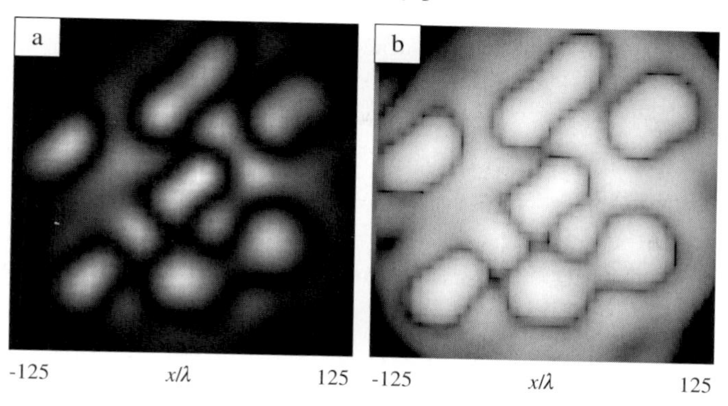

Figure 27.4 Image of the phase object of Figure 27.2(b), obtained with the coherent illumination of Figure 27.2(a), when an amplitude mask is placed in the Fourier plane. The mask, a small disk of radius 275λ, blocks the central region of the beam. (a) Intensity distribution in the image plane. (b) Same as (a) but on a logarithmic scale ($\alpha = 3$).

the image obtained with a filter that cuts the amplitude in half while shifting the phase by 90°. The resulting contrast enhancement is quite impressive.

Finally we consider the effect of changing the phase shift from +90° to −90°. This is shown in Figure 27.6, where the images of the marks are now brighter than their background. A similar situation will arise, of course, if instead of reversing the sign of the phase at the filter we reverse the phase of the marks at the object. In practice, most phase objects contain a number of

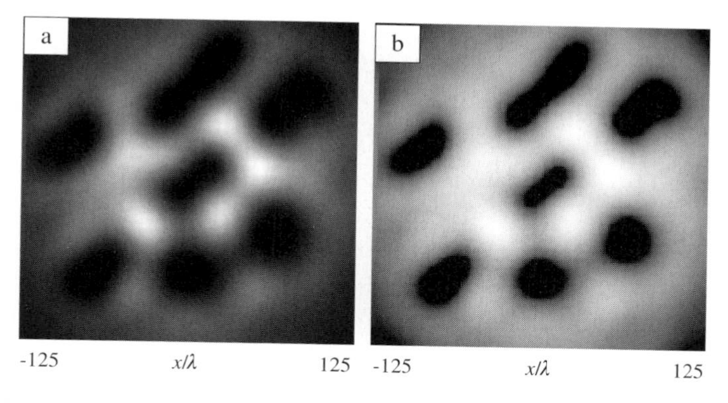

Figure 27.5 Image of the phase object of Figure 27.2(b) obtained with the coherent illumination of Figure 27.2(a) when a phase/amplitude mask is placed in the Fourier plane. The mask, a small disk of radius 275λ, imparts a +90° phase shift to the central region of the beam while attenuating its amplitude by 50%. (a) Intensity distribution in the image plane. (b) Same as (a) but on a logarithmic scale ($\alpha = 1.65$).

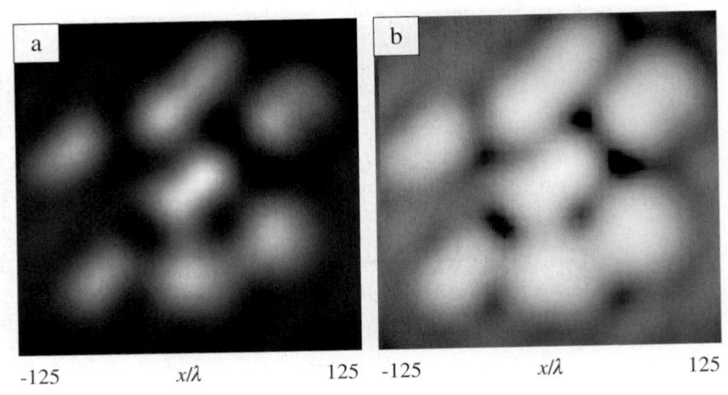

Figure 27.6 Same as Figure 27.5 except for the phase shift of the mask, which is $-90°$ in the present case. (a) Intensity distribution in the image plane. (b) Same as (a) but on a logarithmic scale ($\alpha = 2$).

positive as well as negative features, and their images will appear to be darker than the background in some regions and brighter in other regions.

Contrast enhancement with incoherent illumination

To obtain better resolution in optical microscopy one must illuminate the object with a cone of light (as opposed to a cylindrical collimated beam). This point was discussed in chapter 5, "Coherent and incoherent imaging". The best results are typically achieved when the numerical apertures of the illumination cone and the objective lens are identical because, under such circumstances, twice as many spatial frequencies of the object are captured by the objective lens. It turns out that the cone of light does not have to be solid in order to achieve high resolution; the same benefits also derive from a hollow cone of light.

In Figure 27.7(a) we show the annular source of incoherent light that is used in our calculations to illuminate the condenser lens. In this annulus there are 36 independent "point sources", which provide a good approximation to a homogeneous ring of incoherent light. A $0.25NA$ condenser lens projects the annulus to a bright spot at its focal plane, as shown in Figure 27.7(b). This spot is large enough to cover the phase object depicted in Figure 27.2(b).

The logarithmic plot of intensity distribution at the Fourier plane, Figure 27.7(c), shows a bright annulus as well as a fairly uniform disk of diffracted light within the exit pupil of the objective lens. Evidently, the phase-contrast

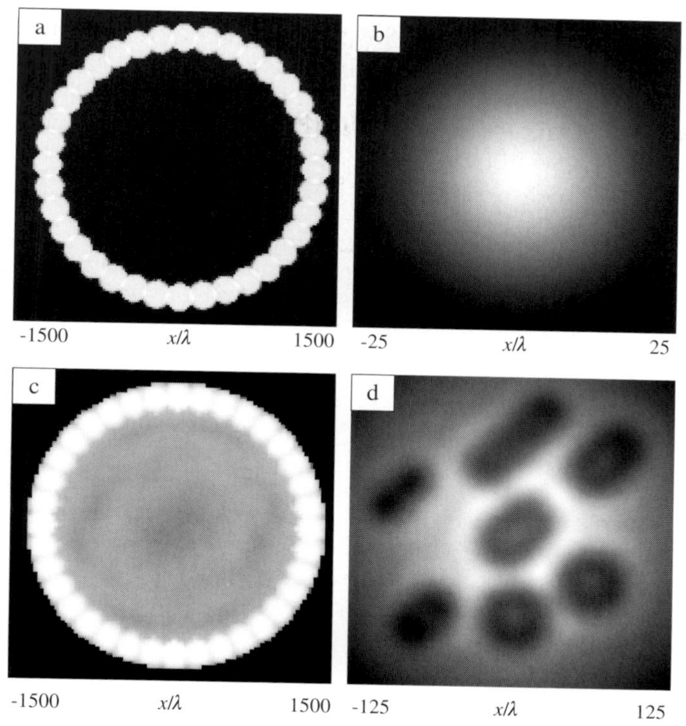

Figure 27.7 Imaging of the phase object of Figure 27.2(b), obtained with an incoherent, annular illuminator. (a) The simulated homogeneous, annular light source consists of 36 independent, quasi-monochromatic point sources. These point sources are arranged uniformly around the circumference of the entrance pupil of the $0.25NA$ condenser lens. (b) Computed intensity distribution at the focal plane of the condenser, which is also the location of the object. (c) Distribution of the logarithm of intensity ($\alpha = 6$) at the exit pupil of the $0.25NA$ objective lens. The annular phase mask placed at this pupil has a width of 300λ, it imparts a $+90^\circ$ phase shift and a 50% (amplitude) attenuation to the beam at the outer periphery of the exit pupil. (d) Computed intensity distribution at the image plane of the system.

filter must also be in the form of an annular ring, covering the circumference of the objective's exit pupil and capable of delivering a 90° phase shift as well as a reasonable attenuation factor to the incident beam.

The resulting image shown in Figure 27.7(d) is obviously of high quality, both in terms of resolution and contrast.

References for chapter 27

1 F. Zernike, *Z. Tech. Phys.* **16**, 454 (1935); *Phys. Z.* **36**, 848 (1935); *Physica* **9**, 686, 974 (1942).
2 M. Françon, Le contraste de phase en optique et en microscopie, *Revue d'Optique*, Paris (1950).
3 A. H. Bennett, H. Jupnik, H. Osterberg, and O. W. Richards, *Phase Microscopy*, Wiley, New York, 1952.
4 F. D. Kahn, *Proc. Phys. Soc. B* **68**, 1073 (1955).
5 M. Born and E. Wolf, *Principles of Optics*, 6th edition, Pergamon Press, Oxford, 1980.
6 M. V. Klein, *Optics*, Wiley, New York, 1970.

28

Polarization microscopy

The state of polarization of a given beam of light is modified upon reflection from (or transmission through) an object. The resulting change in polarization state conveys information about the structure and certain physical properties of the illuminated region. Polarization microscopy is a variant of conventional optical microscopy that enables one to monitor these changes over a small area of a specimen. Such observations then allow the user to identify and analyze the specimen's structural and other physical features.[1,2]

Traditionally, observations with a polarization microscope have been categorized "orthoscopic" or "conoscopic." Orthoscopic observations involve direct imaging of the sample itself, thus allowing one to view the indentations, striations, variations of optical activity and birefringence, etc., over the sample's surface. Conoscopic observations, however, involve illuminating a crystalline surface with a cone of light and then imaging the exit pupil of the objective lens. This mode of observation is used in characterizing the crystal's ellipsoid of birefringence and identifying its optical axes.

The polarization microscope

Figure 28.1 is a simplified diagram of a polarization microscope. The light source is typically an extended white light source, such as a halogen lamp or an arc lamp. The collected and collimated beam from the source is linearly polarized as a result of passage through a polarizer. In metallurgical microscopes, such as the one shown here, the objective lens is used both for illuminating the sample and for collecting the reflected light. Typically the source is imaged onto the entrance pupil of the objective lens, which provides for maximum light-collection efficiency while producing a highly defocused image of the source at the sample.[1,3] Any non-uniformities of the source

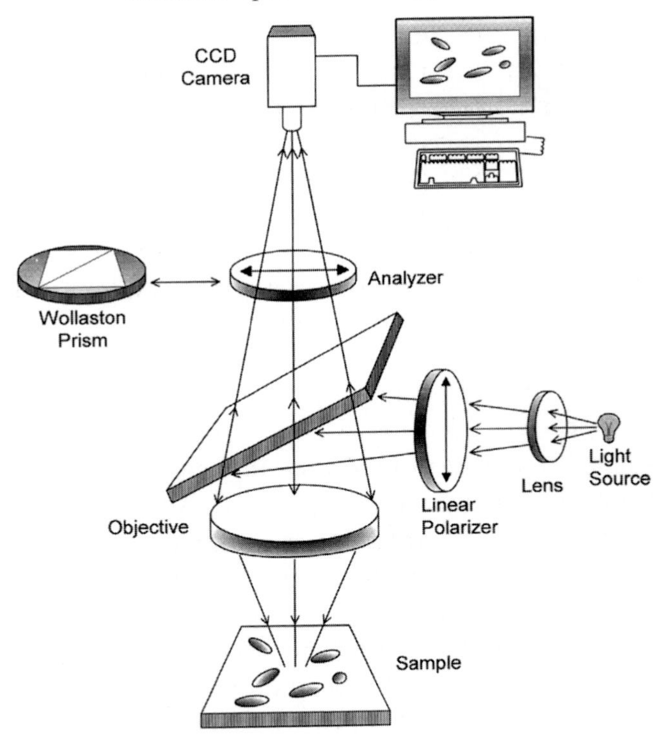

Figure 28.1 Diagram of a conventional polarization microscope. The spatially incoherent light source is linearly polarized and imaged onto the entrance pupil of the objective lens. The reflected light returns through the objective and, after passage through the analyzer, arrives at the image plane. The analyzer is in a rotatable mount, and its transmission axis is adjusted to yield maximum image contrast. If the analyzer is replaced with a Wollaston prism, two images will appear, side by side, on the camera's CCD plate. The computer downloads both images simultaneously and subtracts one from the other in order to produce a differential image.

are thereby averaged, to yield a more uniform light intensity distribution at the sample's surface.

Although the source is spatially incoherent, the projected beam at the sample's surface is, in general, partially coherent. As for the degree of temporal coherence of the light source, it does not play a role in polarization microscopy and is, therefore, ignored throughout this chapter. All one needs to assume is that the light source is quasi-monochromatic, with a bandwidth that is sufficiently narrow to allow one to restrict attention to a single wavelength. The bandwidth must be wide enough, however, to render the source spatially incoherent. (An extended but purely monochromatic source is, of necessity, spatially coherent because the radiated fields from any two locations on the source maintain their relative phase at all times.)

Throughout this chapter we assume a quasi-monochromatic source of wavelength λ_0, consisting of a fixed number of independent and mutually incoherent point sources arranged on a tightly packed square lattice. The contribution of each such point source to the final image is computed independently of those of all the other point sources. The sum of the *intensity* distributions thus produced at the image plane by the individual point sources constitutes the image of the object. This method of computing the image takes full account of the partial spatial coherence of the illuminating beam without ever having to introduce the corresponding correlation functions explicitly.

The light reflected from the sample is collected by the microscope's objective lens, then passed through another linear polarizer (usually referred to as the analyzer), and finally brought to focus at the image plane. This image plane coincides with the front focal plane of the eyepiece (not shown) or the plane of the detectors within a TV camera. Modern optical microscopes are usually equipped with a charge-coupled device (CCD) camera, which picks up the image and displays it on a computer monitor. The possibility of digital image processing afforded by this electronic acquisition allows new methods of microscopy, such as the differential method to be described shortly.

The analyzer is rotated about the optical axis until its transmission axis is crossed (or nearly crossed) with that of the polarizer. The image contrast is primarily determined by the action of the object on the state of polarization of the incident beam. In regions where the sample does not affect the polarization, the reflected light is blocked by the analyzer, making the corresponding regions of the image dark. However, in those regions that rotate the polarization vector, a fraction of the light goes through the analyzer, the transmitted optical power being proportional to the degree of rotation of the polarization as well as to the actual reflectivity of the sample at the given spot. The resulting image thus provides a map and a measure of the ability of the sample to rotate the direction of incident polarization at its various locations. This has been the basis of orthoscopic polarization microscopy for many years. The conoscopic approach, which involves the imaging of the exit pupil of the objective lens, will be discussed towards the end of this chapter.

The four-corners problem

A limitation of polarization microscopy is rooted in the fact that the beam's state of polarization is affected by ordinary reflections and refractions at the various surfaces throughout the optical path.[1,4,5] This usually results in polarization rotation and/or ellipticity in the four corner areas of the objective's exit pupil, as shown in Figure 28.2. The four-corners problem allows transmission of spurious light through the analyzer, thereby reducing the contrast of the image. When the problem is caused by reflections and refractions at the various surfaces of the objective (or condenser) lens, a viable solution is to use a specialty objective that incorporates a half-wave plate in the midst of its optical train.[1,6] The half-wave plate rotates the polarization direction by 90°, allowing the four-corner rotations before and after the plate to cancel each other out. This solution was offered by objective-lens manufacturers in the early days, before the advent of powerful antireflection coatings. Nowadays the various surfaces of the objective and the condenser are antireflection coated, and the four-corners problem caused by these surfaces is negligible.

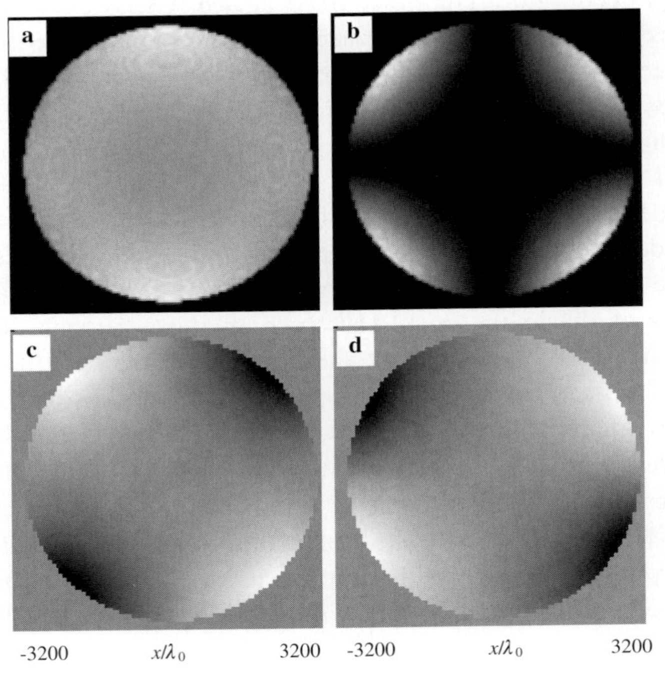

Figure 28.2 Various distributions of the reflected light at the exit pupil of the objective when a single monochromatic point source is used to illuminate the sample. The intensity plots in (a) and (b) correspond respectively to the components of polarization parallel and perpendicular to the polarizer's transmission axis. The polarization rotation angle ρ is depicted in (c) and the polarization ellipticity η is shown in (d). The gray-scale of the latter plots depicts positive values of ρ and η as bright and negative values as dark.

The problem still remains, however, that Fresnel's reflection coefficients at the sample's surface differ for p- and s-polarized rays, causing a polarization rotation problem that is aggravated with increasing angle of incidence. Moreover, if the sample is observed through a birefringent substrate, the resulting polarization variations over the beam's cross-section give rise to spurious light transmission through the analyzer, which, once again, reduces the image contrast.[5] These problems can no longer be solved by the incorporation of a half-wave plate within the objective lens, because they are sample dependent. The differential method of microscopy described below solves the four-corners problem by splitting the spurious light between two images of the sample and then eliminating it by subtracting one image from the other.

Differential method†

A simple modification of the conventional microscope of Figure 28.1 involves replacing the analyzer with a Wollaston prism. The Wollaston splits the image of the sample into two and transmits both images, side by side, to the camera. With the transmission axes of the Wollaston fixed at 45° relative to the polarizer's axis, the unrotated light is split equally between the two images. When there is polarization rotation, however, one image receives more light than the other, the sense of rotation of the polarization determining which image gets the larger share. The two images are then subtracted from each other (within the computer) to produce a single differential image of the sample. The differential image is superior in many respects to the conventional image, as will be seen in the examples that follow. The main advantage of differential polarization microscopy is that it does not suffer from the four-corners problem. Another advantage is that a map of reflectivity variations across the sample can be readily constructed by adding the two images together; normalizing the differential image by the sum image then provides a pure map of polarization rotation at the sample.

The sample

In general, the polarization image of a sample is mixed with its other images, say, those produced by reflectivity variations or optical phase variations across the sample. To avoid such complications, we consider a smooth sam-

† To the author's best knowledge the concept of differential polarization microscopy has not been described previously in the technical and patent literature and may therefore be novel.

ple having uniform amplitude and phase reflectivity everywhere, but one that rotates the polarization of the incident beam as a result of optical activity. A perpendicularly magnetized thin-film sample provides a good example in this case. By changing the direction of magnetization (from up to down) in different locations, one can create a pattern of magnetic domains such as that shown in Figure 28.3. Here the smallest domain (shown at the center) is one wavelength in diameter. The black and white regions are magnetized in opposite directions and rotate the incident (linear) polarization by $+0.5^\circ$ and -0.5°, respectively.

The material of the sample used in the following examples is assumed to have complex index of refraction $(n, k) = (3.35, 4.03)$ which gives it a reflectivity of 62% at normal incidence. At oblique incidence the Fresnel reflection coefficients for p- and s-polarized light differ from each other, thus inducing some rotation and ellipticity into the reflected polarization state. For instance, at a 53° angle of incidence, the linear polarization of a ray originally directed at 45° with respect to the p-direction rotates by 7.4° and acquires 8.7° of ellipticity. This change of the polarization state upon reflection is caused solely by the Fresnel coefficients of the sample, independently of its optical activity.

Low-resolution imaging

Figure 28.4 shows computed images, both conventional and differential, of the magnetic marks of Figure 28.3 obtained with a $50\times$, $0.4NA$ objective. In

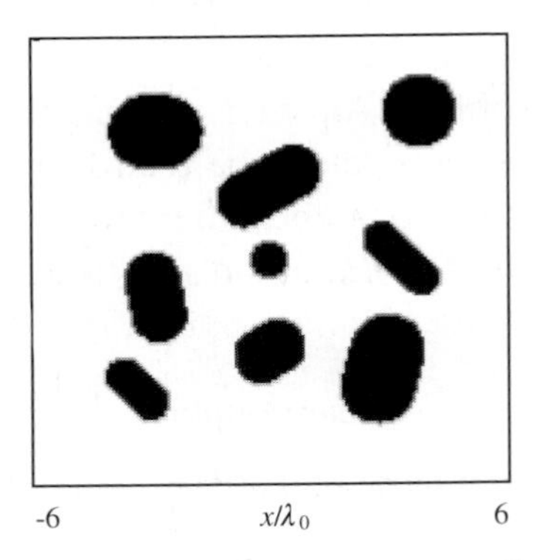

Figure 28.3 Pattern of magnetic domains on a perpendicularly magnetized sample. The magnetic material rotates the polarization of a linearly polarized beam at normal incidence by ± 0.5. The domains are chosen to represent a wide range of sizes and shapes; the smallest domain appearing in the center is one wavelength (λ_0) in diameter.

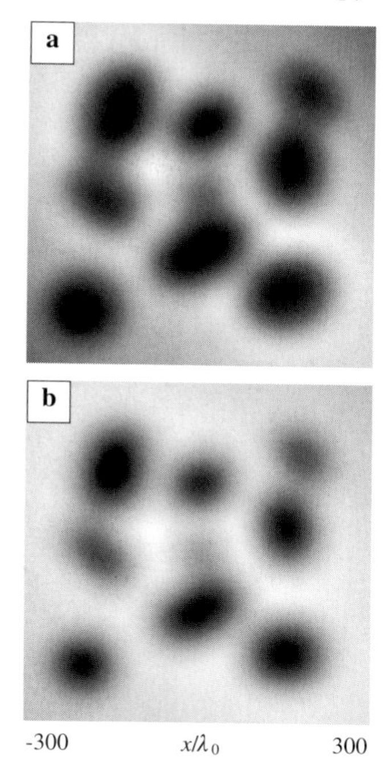

Figure 28.4 Images of the sample of Figure 28.3 in a polarization microscope having a 50×, 0.4*NA* objective lens. (a) Conventional image obtained with the analyzer set 0.5° away from extinction. (b) Differential image obtained with the Wollaston prism.

these calculations the source was defocused by a distance of $35\lambda_0$ below the object plane, and the images from a total of 361 point sources were superimposed to simulate the (spatially incoherent) light source. For the conventional image shown in Figure 28.4(a) the analyzer axis was set 0.5° away from the cross position, nearly the optimum setting for achieving maximum contrast in this case. (The contrast may be reversed by rotating the analyzer to the opposite side of the cross position.) The resolution of these images is not great, as evidenced by the near-disappearance of the small mark in the center. The contrast, however, is quite good, and there is little difference between the conventional and differential methods of imaging. The reason is that at 0.4*NA* the half-angle of the focused cone of light is only 23.6°, which is not large enough to cause a significant four-corners problem.

High-resolution imaging

Obtaining images with high resolution requires a high-*NA* objective lens. Figure 28.5 shows both conventional (a), (b) and differential (c), (d) images

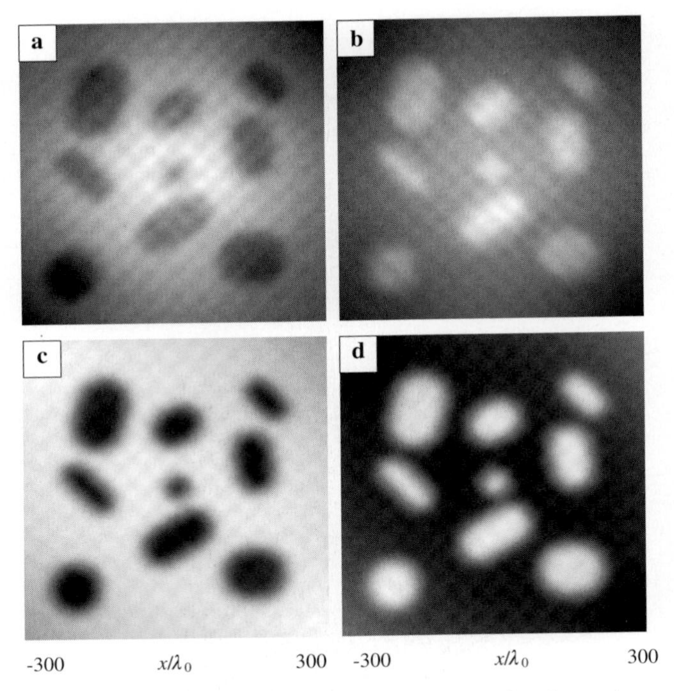

Figure 28.5 Images of the sample of Figure 28.3 in a polarization microscope having a 50×, 0.8NA objective lens. (a) Conventional image obtained with the analyzer set +1.5° away from extinction. (b) Same as (a) but now the analyzer is set −1.5° from extinction to reverse the contrast. (c) Differential image. (d) Same as (c) but with the order of subtraction reversed.

of the sample of Figure 28.3 obtained with a 50×, 0.8NA objective. The images on the left show dark domains on a bright background, while the reverse-contrast counterpart of each image is shown to its right. In these calculations the source was defocused by a distance of $10\lambda_0$ below the object plane, and the images from a total of 361 point sources were superimposed to simulate the (spatially incoherent) light source. Inspection of Figure 28.5 reveals that the resolution has improved over that of Figure 28.4. The contrast, however, is quite poor for the conventional images in Figures 28.5(a), (b), even though the analyzer has been set optimally at 1.5° from the crossed position. This poor contrast is a manifestation of the four-corners problem. In comparison, the differential images of Figures 28.5(c), (d) show excellent contrast, which is not surprising considering that the four-corners contributions to individual images (before subtraction) are identical and can therefore be removed by subtraction.

To gain a better appreciation of the four-corners problem, consider the intensity distribution at the plane of the sample, Figure 28.6, corresponding to a single point source defocused by $10\lambda_0$. Although the incident beam entering the objective lens is linearly polarized along the X-axis, the defocused spot, in consequence of the bending of the rays by the lens, contains all three components of polarization, along the X-, Y-, and Z- axes; these are shown respectively from top to bottom in Figure 28.6. The peak intensities of

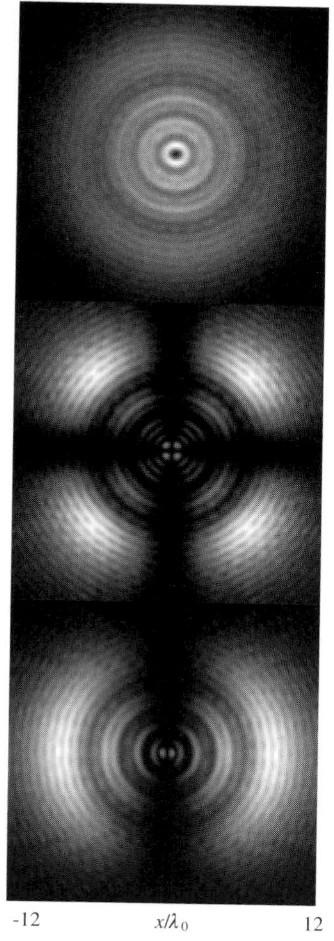

Figure 28.6 Distribution of incident intensity at the plane of the sample corresponding to a single point source defocused by $10\lambda_0$ through a $0.8NA$ objective. The incident beam entering the lens is linearly polarized along the X-axis. Top to bottom: intensity distributions corresponding to polarization components along the X-, Y-, and Z- axes.

the three components in Figure 28.6 are in the ratios $I_x : I_y : I_z =$ $1 : 0.007 : 0.185$. Upon reflection from the sample the distributions remain qualitatively the same, but the peak-intensity ratios change to $1 : 0.017 : 0.142$. Thus the relative content of the Y-component increases upon reflection while that of the Z-component decreases. When this distribution returns to the objective lens, it gives rise to patterns of intensity and polarization similar to those shown in Figure 28.2. At the exit pupil the values of the polarization rotation angle ρ range from -7.0° to $+8.1^\circ$, while the polarization ellipticity η ranges from -8.8° to $+8.6^\circ$. The slight asymmetry between positive and negative values is caused by the presence of magnetization in the sample. In the absence of magneto-optical activity, ρ and η vary between $\pm 7.4^\circ$ and $\pm 8.7^\circ$, respectively.

Substrate birefringence

Sometimes it is necessary to observe a sample through an intervening medium, such as a coating layer or a substrate. If this medium happens to be birefringent, it creates a four-corners problem of its own.[5] As a typical example, assume that the sample of Figure 28.3 is coated with a birefringent layer 500 nm thick whose principal refractive indices along the coordinate axes are $(n_x, n_y, n_z) = (1.5, 1.6, 1.7)$. For this sample, conventional microscopy yields the image shown in Figure 28.7(a), while differential microscopy produces the normal and reverse-contrast images of Figures 28.7(b), (c). Clearly, in the presence of birefringence differential polarization microscopy is far superior to the conventional method. For this sample, the reflected polarization pattern at the exit pupil for a single illuminating point source (see Figure 28.2) exhibits ρ-values ranging from -20.4° to $+22.0^\circ$, and η-values ranging from -23.3° to $+23.0^\circ$. In the absence of magnetic activity ρ and η would vary between $\pm 21.3^\circ$ and $\pm 23.2^\circ$, respectively.

Conoscopic observations

The system depicted in Figure 28.8 captures the essence of conoscopic polarization microscopy. Here a coherent, monochromatic beam of light is linearly polarized and sent through an objective lens to be focused on a birefringent crystal. The reflected light is re-collimated by the objective and observed after going through a crossed analyzer. For the specific example described below, the objective's NA-value is 0.375 and its focal length f is $20\,000\lambda_0$. The sample is in the XY-plane, the Z-axis being perpendicular to its surface. The crystal slab's thickness is $430\lambda_0$, its principal refractive indices are

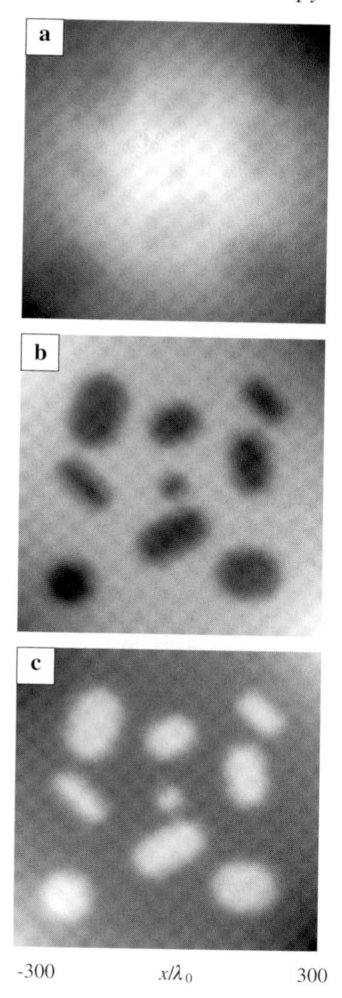

Figure 28.7 Images of the sample of Figure 28.3, coated with a birefringent layer and placed in a microscope having a $50\times$, $0.8NA$ objective. (a) Conventional image, obtained with the analyzer set optimally at 5° away from extinction. (b) Differential image. (c) Same as (b) but with the order of subtraction reversed.

$(n_x, n_y, n_z) = (1.686, 1.682, 1.531)$, and its ellipsoid of birefringence is rotated around the Z-axis by 13°.

The computed intensity distribution at the observation plane of Figure 28.8 is shown in Figure 28.9(a), and the corresponding logarithmic plot appears in Figure 28.9(b). Within the focused cone there are two rays that propagate along the two optical axes of the crystal; these rays return without any change in their state of polarization and are therefore blocked by the analyzer. There are also groups of rays whose polarization vectors undergo

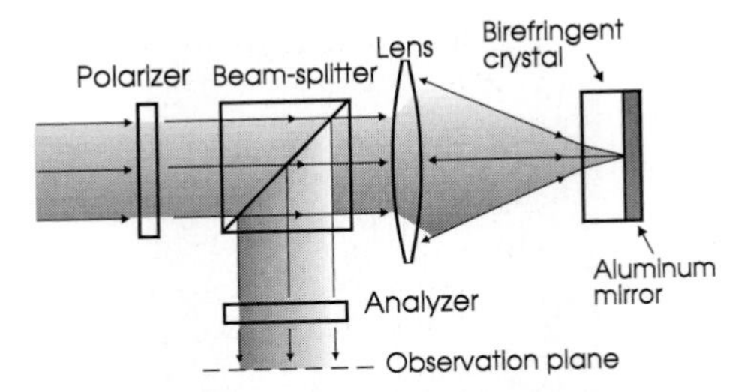

Figure 28.8 Schematic diagram of a simplified conoscopic microscope. The double passage of the focused beam through the birefringent crystal causes varying degrees of polarization rotation over the beam's cross-section. The crossed analyzer converts these rotations into an intensity pattern.

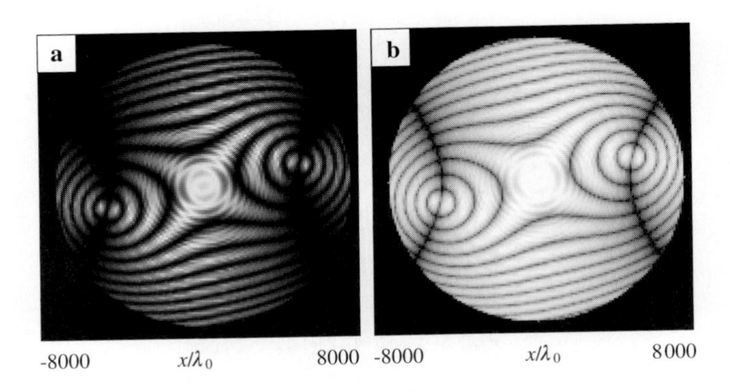

Figure 28.9 (a) Intensity and (b) logarithmic intensity distributions at the observation plane in the system of Figure 28.8 with a biaxially birefringent crystal.

rotation by integer multiples of 180° in double passage through the slab. These rays are also blocked by the analyzer, giving rise to the various dark regions in the intensity patterns of Figure 28.9. A systematic analysis of the exit-pupil distribution can, therefore, provide detailed information about the sample's ellipsoid of birefringence.

References for chapter 28

1 S. Inoué and R. Oldenbourg, Microscopes, in *Handbook of Optics*, Vol. II, second edition, McGraw-Hill, New York, 1995.

2 J. R. Benford and H. E. Rosenberger, Microscopes, in *Applied Optics and Optical Engineering*, Vol. IV, ed. R. Kingslake, Academic Press, New York, 1967.

3 M. Born and E. Wolf, *Principles of Optics*, 6th edition, Pergamon Press, Oxford, 1980.

4 H. Kubota and S. Inoué, Diffraction images in the polarizing microscope, *J. Opt. Soc. Am.* **49**, 191–198 (1959).
5 Y. C. Hsieh and M. Mansuripur, Image contrast in polarization microscopy of magneto-optical disk data-storage media through birefringent plastic substrates, *Applied Optics* **36**, 4839–4852 (1997).
6 J. R. Benford, Microscope objectives, in *Applied Optics and Optical Engineering*, Vol. III, ed. R. Kingslake, Academic Press, New York, 1965.

29

Nomarski's differential interference contrast microscope

George Nomarski invented the method of differential interference contrast for the microscopic observation of phase objects in 1953.[1–3] The features on a phase object typically modulate the phase of an incident beam without significantly affecting the beam's amplitude. Examples include unstained biological samples having differing refractive indices from their surroundings, and reflective (as well as transmissive) surfaces containing digs, scratches, bumps, pits, or other surface-relief features that are smooth enough to reflect specularly the incident rays of light. A conventional microscope image of a phase object is usually faint, showing at best the effects of diffraction near the corners and sharp edges but revealing little information about the detailed structure of the sample.[4]

Nomarski's method creates two slightly shifted, overlapping images of the same surface. The two images, being temporally coherent with respect to one another, optically interfere, producing contrast variations that contain useful information about the phase gradients across the sample's surface. In particular, a feature that has a slope in the direction of the imposed shear appears with a specific level of brightness that is distinct from other, differently sloping regions of the same sample.[4–6]

The Nomarski microscope uses a Wollaston prism in the illumination path to produce two orthogonally polarized, slightly shifted bright spots at the sample's surface. Upon reflection from (or transmission through) the sample, the two beams are collected by the objective lens, then sent through the same (or, in the case of a transmission microscope, a similar) Wollaston prism, which recombines the two beams by sliding them back over each other. The two beams subsequently arrive coincidentally in the image plane of the microscope, but the two images of the sample which they carry will be relatively displaced. A linear analyzer, placed after the Wollaston prism in the reflected (transmitted) path, brings the polarization vectors of the two images into alignment, enabling the two to interfere with each other. A sheared interferogram of the sample's surface is thus formed at the image plane of the microscope.

Wollaston prism

Because Nomarski's method of microscopy is fundamentally dependent on the action of the Wollaston prism, a brief description of this polarizing beam-splitter is in order. The Wollaston prism, depicted in Figure 29.1, consists of two cemented wedges from the same uniaxial birefringent crystal (e.g., quartz or calcite). The individual wedges are precisely cut and polished, then aligned with their optic axes orthogonal to each other.[4] In Figure 29.1 the optic axis of the upper wedge is horizontal within the plane of the page, while that of the lower wedge is perpendicular to the plane. The crystal's ordinary and extraordinary refractive indices, n_o and n_e, interact with the E-field components perpendicular and parallel to the optic axis, respectively.

The incident beam, in general, has both s- and p-components of polarization. In going through the upper half of the Wollaston, the p-component interacts with n_e and the s-component with n_o, but the propagation direction remains the same for both the p- and s- beams. In the lower half the roles of n_o and n_e are exchanged, with the result that the p-component is deflected to

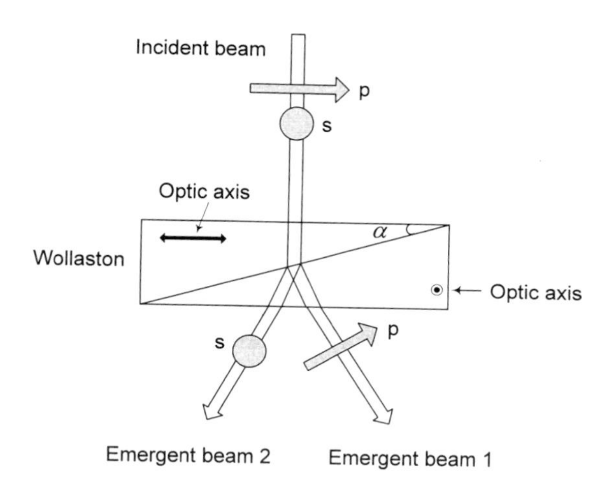

Figure 29.1 The Wollaston prism consists of two cemented wedges of the same uniaxial birefringent crystal, aligned with their optic axes in different directions. The incident beam, with its p- and s- components of polarization, is split at the interface between the wedges. Emerging from the Wollaston are two orthogonally polarized beams that propagate in different directions.

one side and the s-component to the other (one beam enters a denser, the other a rarer medium). The angular separation of the beams is further enhanced by Snell's law when they exit the prism. Emerging from the Wollaston, therefore, are two beams, propagating in different directions and having mutually orthogonal directions of polarization.

Figure 29.2 shows a thin bundle of rays arriving at a Wollaston prism and splitting into two orthogonally polarized beams. The p- and s- beams go through a microscope objective and illuminate the sample in two small, slightly displaced patches that cover the objective's field of view. Upon reflection from the sample the beams return through the objective and come together again as they emerge from the Wollaston. Note that, in a round trip through this system, the optical path lengths of the p- and s- beams will be the same only if the Wollaston is centered on the Z-axis. In particular, if the Wollaston is translated along the X-axis then, during a round trip, one beam sees a longer optical path than the other. The relative phase of the p- and s- beams, referred to as the bias phase ϕ_B, can therefore be adjusted by sliding the Wollaston along the X-axis. Note that, for a given lateral position

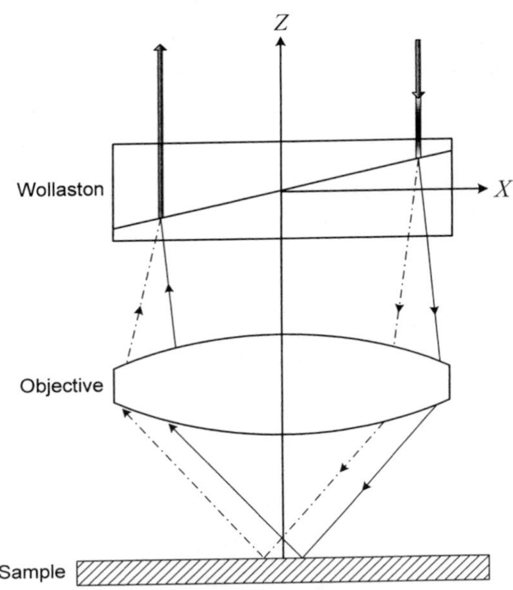

Figure 29.2 A bundle of rays entering a Wollaston prism is split into p- and s-polarized beams. The beams go through a microscope objective and illuminate the sample in two small, slightly displaced patches that cover the objective's field of view. Upon reflection from the sample, the beams return through the objective and come together as they exit the Wollaston prism. The bias phase ϕ_B between the two beams may be adjusted by sliding the Wollaston in the horizontal direction.

of the Wollaston, the bias phase ϕ_B is constant for all the ray bundles that go through the system: it is independent of their initial distance from the Z-axis.

Assuming $\alpha = 0.84^\circ$ for the wedge angles and $n_o = 1.54467$, $n_e = 1.55379$ for the ordinary and extraordinary refractive indices of the crystal (quartz), the angular separation of the two beams emerging from the Wollaston (in the forward path) will be 0.0153°. For an objective lens having $f = 3750\lambda$, where λ is the wavelength of the quasi-monochromatic light source, this angular separation results in one λ of displacement between the two spots that illuminate the sample. Moreover, for every lateral shift by 100λ of the Wollaston, there occurs a bias phase $\phi_B = 19.26^\circ$ between the p- and s-beams in a double pass through the system. So, for example, if the lateral shift is 1870λ then one beam will be retarded by a full 2π relative to the other.

Differential interference contrast microscope

Figure 29.3 is a diagram of an epi-illumination Nomarski differential interference contrast microscope. For the computer simulations reported in this chapter the spatially incoherent light source is assumed to be quasi-monochromatic (wavelength λ), consisting of 529 point sources arranged in a square array. These point sources are projected onto the mid-plane of the Wollaston prism, which sits at the entrance pupil of the objective lens. The entrance pupil being at the back focal plane of the objective, uniform illumination at the sample's surface is achieved (Köhler illumination). The illumination is called "critical" if the source is imaged directly onto the sample. In practice Köhler illumination is preferred over critical illumination because of its superior uniformity, but coherence-related properties of the system (such as resolution) are not affected by this choice of illumination. In this chapter, for reasons having to do with nuances of the computer simulation, we have chosen to illuminate the sample with a somewhat defocused image of the source.

The polarizer renders the illuminating beam linearly polarized, and the Wollaston prism, whose axes are at 45° relative to the transmission axis of the polarizer, creates two orthogonally polarized, slightly displaced patches of light at the sample. The light reflected from the sample returns through the objective and the Wollaston but, as it arrives at the crossed analyzer, its two components of polarization are no longer in phase. The phase difference between the p- and s- beams at this point is $\phi_B + \Delta\phi$, where ϕ_B is the constant bias phase produced by the Wollaston's displacement from the center and $\Delta\phi$ is the imparted phase retardation at the sample's surface. The amount of light that gets through the analyzer depends on the above phase shift, with more

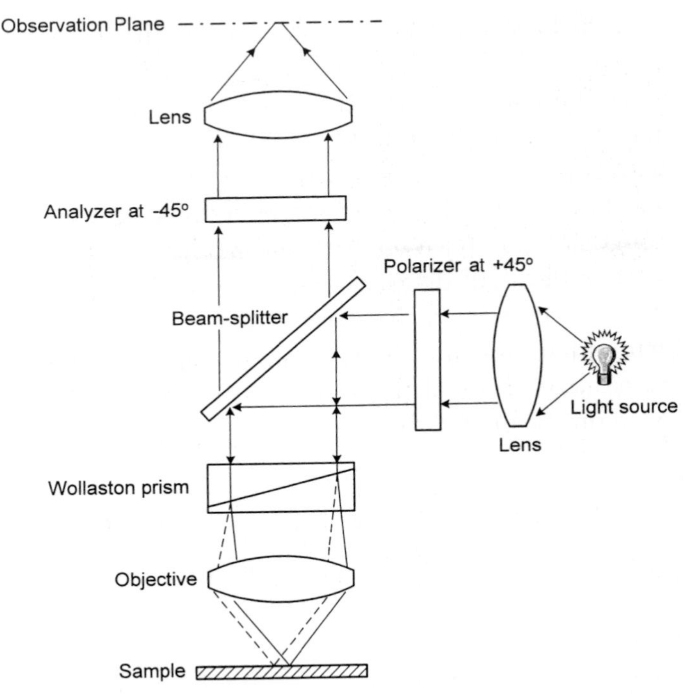

Figure 29.3 Schematic diagram of an epi-illumination Nomarski microscope. The spatially incoherent light source is quasi-monochromatic (wavelength λ), the polarizer renders the illuminating beam linearly polarized, and the Wollaston prism, with axes at 45° to the direction of incident polarization, creates two slightly displaced, orthogonally polarized patches of light at the sample. The light reflected from the sample returns through the objective and the Wollaston, arriving at the crossed analyzer with its two components of polarization relatively phase-shifted. The light that gets through the analyzer forms an image of the sample at the observation plane.

light going through as the phase shift increases from 0° to 180°. Each bright point within the light source illuminates the entire field of view of the objective and creates an image at the observation plane. The various point sources thus create overlapping images, which add up in intensity by virtue of the (spatial) incoherence of the light source.

Examples

Figure 29.4(a) shows the distribution of phase on a uniformly reflecting surface having several sphero-cylindrical pits with varying depths. The nose feature has a depth of 0.5λ, and the mouth, eyes, and eyebrows are respectively 0.25λ, 0.375λ, and 0.75λ deep. The computed image of this phase

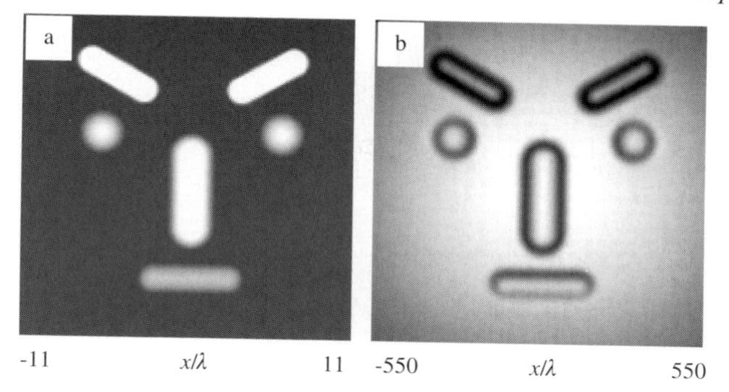

Figure 29.4 (a) The distribution of phase at an object's surface and (b) the distribution of intensity in the image of the same object, as observed in a conventional optical microscope. In (a) the various features of the "face" have the same reflectance but different depth, resulting in phase modulation of the incident light. The nose, mouth, eyes, and the eyebrows are respectively 0.5λ, 0.25λ, 0.375λ, and 0.75λ deep. The image in (b) is formed by a $0.8NA$, $50\times$ objective. The simulated light source consisted of 529 spatially incoherent point sources, each defocused by 10λ above the sample's surface. The observed contrast is purely due to diffraction effects, as the phase object does not give rise to any contrast in geometric-optical terms.

object in a conventional optical microscope (i.e., like that in Figure 29.3 but without the polarizer, analyzer, and Wollaston) is shown in Figure 29.4(b). Note that diffraction of light from the edges of the various features of the face creates dark borders in the corresponding image regions, but this conventional image lacks information about the slope and depth distribution within those features.

The computed Nomarski image of the phase object of Figure 29.4(a), obtained with one λ of sheer along the X-axis, is shown in Figure 29.5. The intensity distribution in the image plane is shown in Figure 29.5(a), while a logarithmic plot of intensity (resembling an over-exposed photographic plate) is shown in Figure 29.5(b). In these calculations the assumed bias phase $\phi_B = 0$; this results in identical image brightness for regions with equal but opposite slopes, and also yields a completely dark image background. Since the assumed shear in Figure 29.5 is along the X-direction, vertical features (such as the nose) are clearly visible in the Nomarski image, while horizontal features (such as the mouth) are hidden. The reverse is true when the shear is along the Y-axis, as in Figure 29.6, where horizontal features become visible while vertical features disappear.

Figure 29.7 shows the Nomarski image of the object in Figure 29.4(a), but with a bias phase $\phi_B = 90^\circ$. The background of the image is now bright, because the analyzer no longer blocks the light reflected from flat regions of the sample. Moreover there is an asymmetry between regions with positive

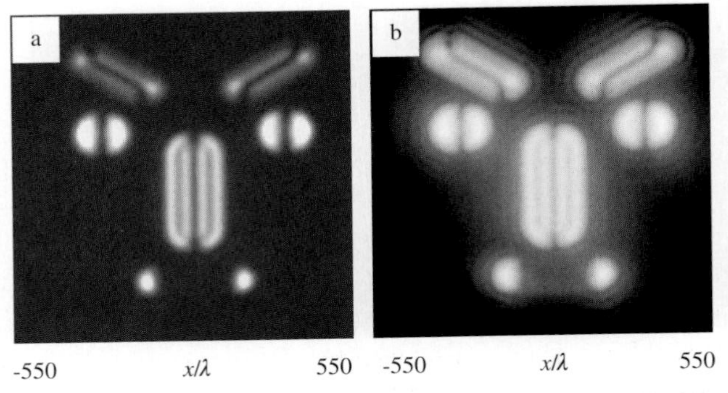

Figure 29.5 Nomarski images of the phase object in Figure 29.4(a), when the Wollaston produces one λ of shear along the X-axis. The microscope is that shown in Figure 29.3, having a 50×, 0.8NA objective, and the Wollaston's horizontal position is adjusted for $\phi_B = 0°$. (a) Intensity distribution in the image plane; (b) logarithmic plot of the intensity distribution.

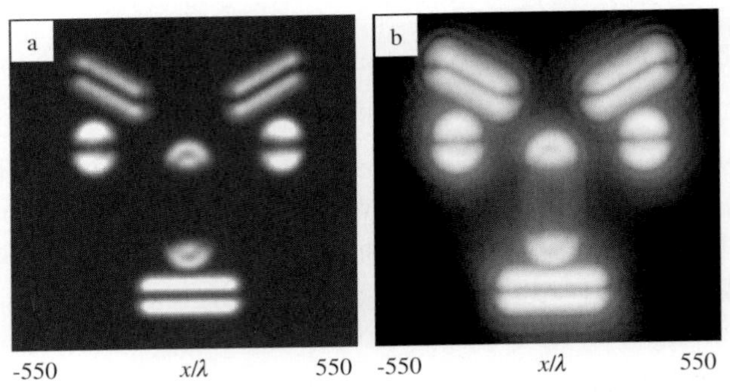

Figure 29.6 Same as Figure 29.5, except for the direction of shear, which is along the Y-axis.

Figure 29.7 Nomarski image of the phase object in Figure 29.4(a), when the Wollaston produces one λ of shear along the X-axis. The microscope is that shown in Figure 29.3, having a 50×, 0.8NA objective, and the Wollaston's horizontal position is adjusted for $\phi_B = 90°$.

and negative slope, as can be seen by comparing the right and left sides of the nose feature.

Another example of a phase object is shown in Figure 29.8(a). Here a ridge having height λ runs along the 45° direction in the XY-plane. The two edges of the ridge have differing slopes, the lower edge being 4λ wide while the upper edge is 2λ wide. In the middle of the ridge there is a pit of depth λ in the shape of a football stadium. The conventional image of this sample is shown in Figure 29.8(b). Again diffraction from the various edges renders certain features visible in the image, but specific information about the slopes is lacking. In contrast, two Nomarski images of the same object obtained with one λ of horizontal shear are shown in Figure 29.9. The bias phase $\phi_B = 0°$ in Figure 29.9(a), whereas $\phi_B = 90°$ in Figure 29.9(b). Different slopes produce different intensity levels in these images. Also note that the symmetry present in Figure 29.9(a) between equal but opposite slopes is broken in Figure 29.9(b), where $\phi_B \neq 0°$.

Practical considerations

The back focal plane of high-NA objectives is usually inaccessible from outside the lens, so the Wollaston prism cannot be directly inserted at the entrance pupil. By choosing a somewhat different orientation for the optic

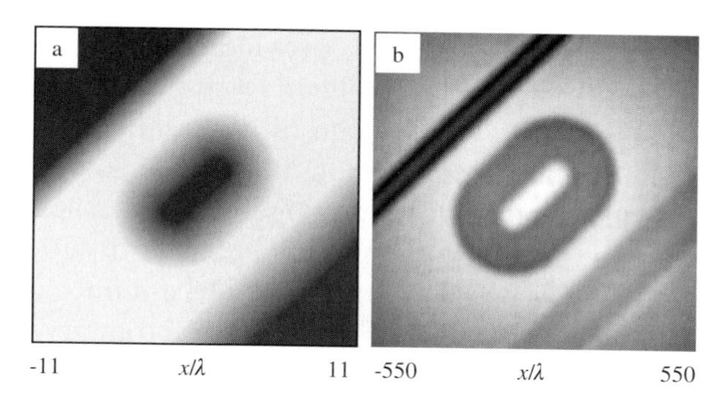

Figure 29.8 (a) Phase object and (b) its conventional microscope image. The object consists of a ridge with a height of λ, running at 45° to the X- and Y-axes, and a pit in the middle of the ridge whose depth is also λ. The ridge's side-walls have different slopes: the lower wall is 4λ wide, while the upper wall is 2λ wide. The flat-bottomed pit has the shape of a football stadium. The image in (b) is formed through a 50×, 0.8NA microscope objective. The simulated light source consisted of 529 spatially incoherent point sources, each defocused by 10λ above the sample's surface. The observed image contrast is purely due to diffraction effects, as the phase object does not give rise to any contrast in geometric-optical terms.

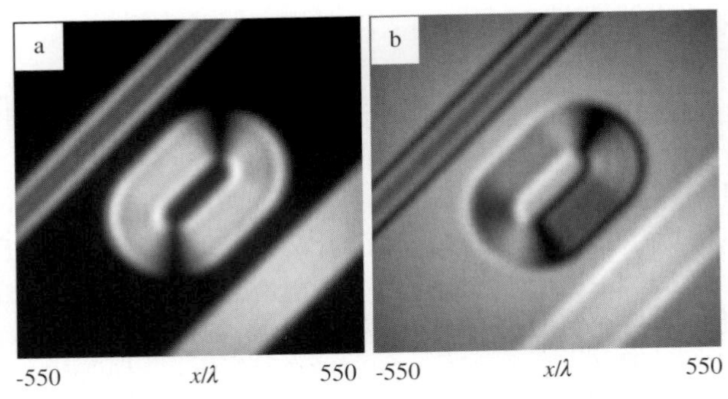

Figure 29.9. Nomarski images of the phase object of Figure 29.8(a), when the Wollaston produces one λ of shear along the X-axis. The microscope is that shown in Figure 29.3, having a 50×, 0.8NA objective. The Wollaston's horizontal position is adjusted to yield a bias phase ϕ_B between the p- and s-polarized beams. (a) $\phi_B = 0°$, (b) $\phi_B = 90°$.

axes of the crystal wedges, Nomarski modified the Wollaston prism in such a way that the p- and s- beams appeared to be separating from each other in a plane external to the prism.[3] In this way the light source could be imaged onto the entrance pupil of the objective through the Nomarski-modified Wollaston prism, allowing both Köhler illumination and the separation and recombination of the p- and s- beams at the entrance pupil.

Another practical consideration involves the use of broadband light sources. The sources used in practice are not always monochromatic and, in fact, may have a fairly broad spectrum. The analysis offered in this chapter applies to multi-color sources as well, provided that the individual wavelengths are treated independently and their corresponding images are eventually superimposed. In any given region of the sample, interference causes certain colors to fade while strengthening others. The color or hue observed through a broadband Nomarski microscope at a given location is thus a qualitative measure of the slope of the sample at that location. For quantitative measurements, however, it is best to use quasi-monochromatic light in conjunction with some form of phase-shifting interferometry.[7–9] This may be achieved, for instance, by sliding the Wollaston prism along the shear direction while monitoring (with a CCD camera) the variations in intensity at specific locations of the image.

References for chapter 29

1 G. Nomarski, Diapositif interferentiel à polarisation pour l'étude des objects transparents ou opaques appartenant à la classe des objects de phase, French patent No. 1059 124, 1953.

2 G. Nomarski, Microinterféromètre différential à ondes polarisées, *J. Phys. Radium* **16**, 9S–11S (1955).
3 R. D. Allen, G. B. David, and G. Nomarski, The Zeiss–Nomarski differential interference equipment for transmitted light microscopy, *Z. Wiss. Mikroskopie* **69** (4), 193–221 (1969).
4 M. V. Klein, *Optics*, Wiley, New York, 1970.
5 S. Inoué and R. Oldenbourg, *Microscopes*, chapter 17 in *Handbook of Optics*, Vol. II, McGraw-Hill, New York, 1995.
6 M. Pluta, *Advanced Light Microscopy*, Vol. 2: *Specialized Methods*, Elsevier, Amsterdam; Polish Scientific Publishers, Warszawa, 1989.
7 D. L. Lessor, J. S. Hartman, and R. L. Gordon, Quantitative surface topography determination by Nomarski reflection microscopy. I. Theory, *J. Opt. Soc. Am.* **69**, 357–366 (1979).
8 J. S. Hartman, R. L. Gordon, and D. L. Lessor, Quantitative surface topography determination by Nomarski reflection microscopy. II. Microscope modification, calibration, and planar sample experiments, *Applied Optics* **19**, 2998–3009 (1980).
9 W. Shimada, T. Sato, and T. Yatagai, Optical Surface Microtopography using phase-shifting Nomarski microscope, *SPIE* **1332**, *Optical Testing and Metrology*, 525–529 (1990).

30

The van Leeuwenhoek microscope

Antoni van Leeuwenhoek (1632–1723), a fabric merchant from Delft, the Netherlands, used tiny glass spheres to study various microscopic objects at high magnification with surprisingly good resolution. A contemporary of Sir Isaac Newton, Christiaan Huygens, and Robert Hooke, he is said to have made over 400 microscopes and bequeathed 26 of them to the Royal Society of London. (A handful of these microscopes are extant in various European museums.) Using his single-lens microscope, van Leeuwenhoek observed what he called animalcules – or micro-organisms, to use the modern terminology – and made the first drawing of a bacterium in 1683. He kept detailed records of what he saw and wrote about his findings to the Royal Society of London and the Paris Academy of Science. His contributions have made him the father of scientific microscopy.[1–3]

van Leeuwenhoek was an amateur in science and lacked formal training. He seems to have been inspired to take up microscopy by Robert Hooke's illustrated book, *Micrographia*, which depicted Hooke's own observations with the microscope. In basic design, van Leeuwenhoek's instruments were simply powerful magnifying glasses, not compound microscopes of the type used today. An entire instrument was only 3–4 inches (8–10 cm) long, and had to be held up close to the eye; its use required good lighting and great patience.[4] van Leeuwenhoek devised tiny, double-convex lenses to be mounted between brass plates. Through them, he was able to peer at objects mounted on pinheads, magnifying them up to 300 times, a power that far exceeded that of early compound microscopes.

Compound microscopes had been invented around 1595. Several of van Leeuwenhoek's contemporaries, notably Robert Hooke in England and Jan Swammerdam in the Netherlands, had built compound microscopes and were making important discoveries with them. However, because of various technical difficulties, early compound microscopes were not practical for magnifications beyond 20× or 30×. van Leeuwenhoek's skill at grinding lenses, together with his naturally acute eyesight and great care in adjusting the lighting, enabled him to build microscopes with clearer and brighter images than any of his contemporaries could achieve.

van Leeuwenhoek used his invention to confirm the discovery of capillary systems, to describe the life cycle of ants, and to observe plant and muscle tissue, protozoa and bacteria, and the spermatozoa of insects and humans. In 1673, van Leeuwenhoek began writing letters to the newly formed Royal Society of London, describing his findings – his first letter contained some observations on the stings of bees. For the next 50 years he corresponded with the Royal Society; his letters, written in Dutch, were translated into English or Latin and printed in the *Philosophical Transactions of the Royal Society*, and often reprinted separately. His experiments with microscope design and function made him an international authority on microscopy, and in 1680 he was made a Fellow of the Royal Society.

It is suspected that van Leeuwenhoek produced his lenses by chipping away the excess glass from the thickened droplet that forms on the bottom of a blown-glass bulb. These lenses probably had a thickness of ~ 1 mm and a radius of curvature of ~ 0.75 mm. They had superior magnification and resolution when compared to other microscopes of the time. The Utrecht museum has one of van Leeuwenhoek's microscopes in its collection. This amazing instrument has a magnification of about 275× with a resolution approaching one micron (in spite of a scratch on the lens).[5]

Towards the end of his life van Leeuwenhoek wrote: ". . . my work, which I've done for a long time, was not pursued in order to gain the praise I now enjoy, but chiefly from a craving after knowledge, which I notice resides in me more than in most other men. And therewithal, whenever I found out anything remarkable, I have thought it my duty to put down my discovery on paper, so that all ingenious people might be informed thereof."

Elementary optics of glass spheres

Figure 30.1 shows a ray of light parallel to the optic axis at height h, going through a glass sphere of radius R and refractive index n. The angle of incidence on the sphere is denoted by θ, and the refracted ray inside the glass makes an angle θ' with the surface normal. According to Snell's law, $\sin\theta = n\sin\theta'$, and from simple geometry

$$CA = R\sin\theta/\sin(2\theta - 2\theta'). \qquad (30.1)$$

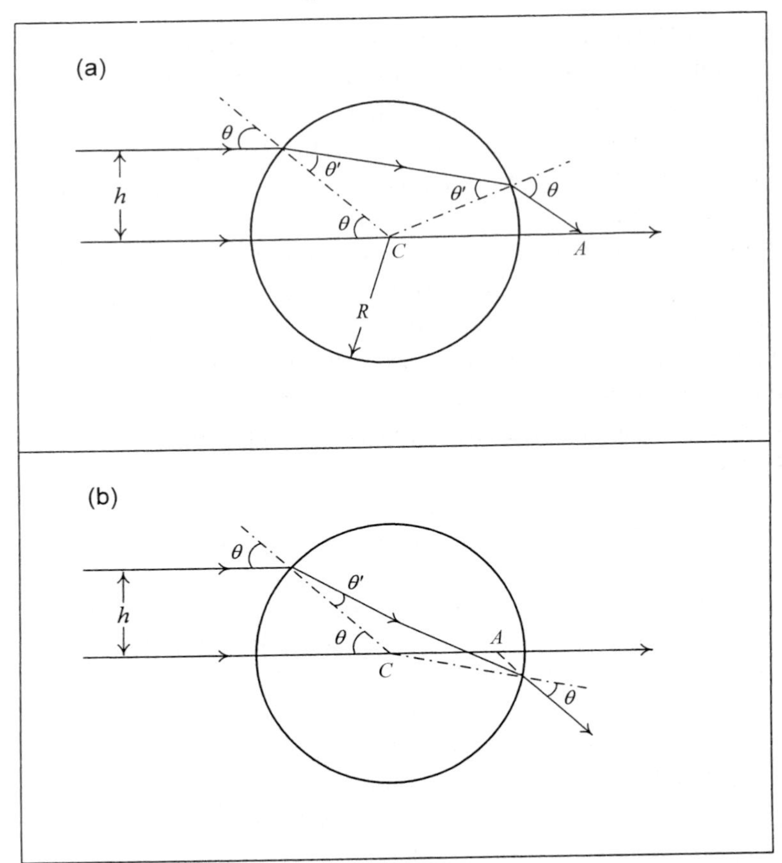

Figure 30.1 A ray of height h traveling parallel to the optic axis is refracted by a glass sphere of radius R and refractive index n. Upon emerging from the sphere, the ray crosses the optic axis at point A. When h becomes very small the point A approaches the paraxial rear focus F' of the lens. In (a) $n < 2.0$ and the emergent ray crosses the axis outside the sphere, whereas in (b), where $n > 2.0$, only the backward extension of the ray crosses the axis. (When $n = 2.0$ the paraxial rays come to focus on the rear facet of the sphere.)

When the ray height h is much smaller than the radius R of the sphere, the angles θ and θ' will be small, in which case the small-angle approximation yields

$$CA \approx nR/[2(n-1)]. \tag{30.2}$$

Thus, for example, if $n = 1.5$ then the paraxial focus of the lens is at a distance $CA = 1.5R$ from the lens center, or if $n = 2$ then the paraxial focus coincides with the rear vertex of the sphere, that is, $CA = R$.

Depending on the values of n and h, the proper path of the ray may be that shown in Figure 30.1(a) or (b), but equations (30.1) and (30.2) apply to both cases. The paraxial focus, of course, is relevant only for rays with a small height h; when h increases beyond the paraxial regime, the point A moves closer to the center C, giving rise (for a beam of wide cross-section) to spherical aberrations.

Confining our attention to a glass sphere having $R = 1$ mm and $n = 1.5$ – typical of what Van Leeuwenhoek used for his microscopes – we suppose that a point source of light is placed at the front (paraxial) focus F of the lens, as in Figure 30.2. A ray that leaves the source at an angle ϕ relative to the optic axis will emerge parallel to the axis only in the paraxial regime, i.e., when ϕ is small. For larger values of ϕ the emergent ray crosses the optic axis at the point A, where

$$CA = R\sin\theta/\sin(2\theta - 2\theta' - \phi). \tag{30.3}$$

Here ϕ and θ are related through $\sin\phi = R\sin\theta/FC$. Thus a point source located at the front focus F and radiating into a reasonably large cone will produce a real image on the opposite side at some finite distance from C. To be sure, this image has a certain amount of spherical aberration and, to obtain a good image, one must limit the angular range of the cone of light accepted by the lens. This may be achieved by closing down the aperture stop, which may be located either on the object side or the image side of the

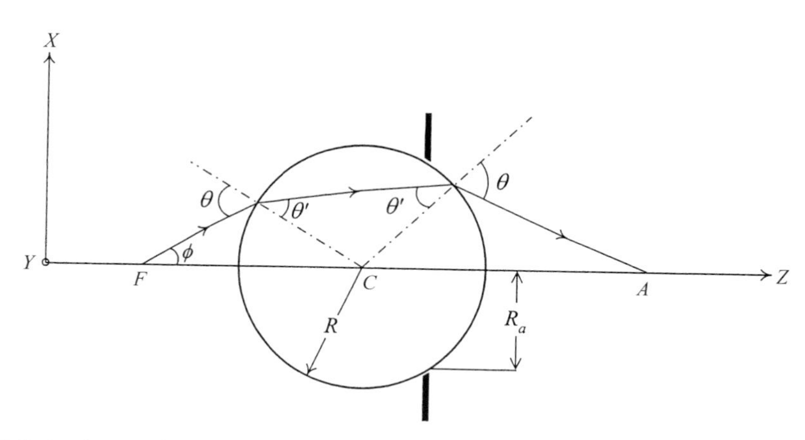

Figure 30.2 A glass sphere of radius $R = 1$ mm and refractive index $n = 1.5$. The aperture stop, of radius R_a, is also the exit pupil of the lens in this case. A monochromatic point source ($\lambda = 0.5$ μm) placed at the paraxial front focus F is approximately imaged to the point A, at a finite distance from the lens center.

lens. In Figure 30.2 the stop is in the image space and may thus be referred to as the exit pupil of the lens.

Figure 30.3 shows computed distributions pertaining to the system of Figure 30.2. The point source is located at the paraxial focus of the lens ($R = 1\,\text{mm}, n = 1.5, CF = 1.5\,\text{mm}$), and the assumed radius of aperture $R_\text{a} = 0.55$ mm. Figure 30.3(a) shows that the emergent intensity at the exit pupil is somewhat brighter near the rim compared with that at the center of the aperture. Figure 30.3(b), a plot of phase distribution at the exit pupil (minus the curvature), shows a significant amount of spherical aberration. (The curvature of the emergent beam has been removed from the phase plot; only the residual aberrations are shown.) The emergent beam comes to best focus at a distance $CA = 27.36$ mm behind the lens. Figure 30.3(c), a logarithmic plot of intensity distribution in the plane of best focus, also shows the substantial rings of light caused by spherical aberration. These clearly indicate that the image quality of a wide-aperture system would be poor.[4]

When the aperture is further closed down to $R_\text{a} = 0.4$ mm the distributions of Figure 30.4 are obtained. The intensity distribution at the exit pupil is now fairly uniform, and the phase plot shows convergent behavior towards the point of best focus at $CA = 59.3$ mm behind the lens. (Notice that in Figure 30.4(b), unlike Figure 30.3(b), the curvature has not been subtracted from the phase plot.) The best-focused spot is shown in Figure 30.4(c). In addition to a relatively small spherical aberration, this system also has a fairly large field of view, as may be inferred from the plots of Figure 30.5. Here a number of identical point sources are placed in the front focal plane of the lens, and

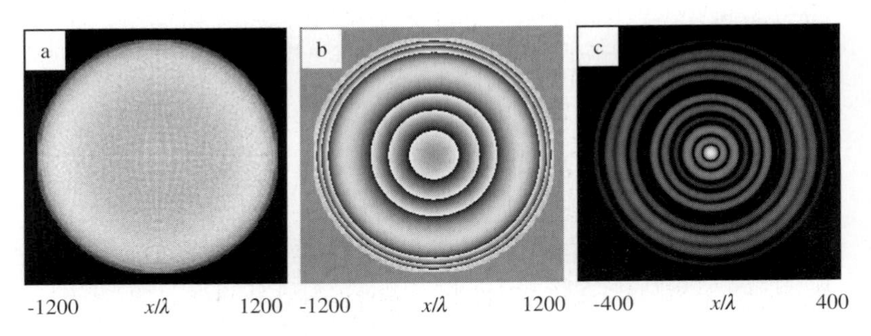

Figure 30.3 Various distributions in the system of Figure 30.2 when $R_\text{a} = 0.55$ mm. (a) Emerging intensity distribution at the exit pupil. (b) Distribution of residual phase at the exit pupil when the curvature of the emergent beam is taken out (r.m.s. aberrations $= 0.96\lambda$). The gray-scale encodes values of phase from -180° (black) to $+180^\circ$ (white). (c) Logarithmic plot of intensity in the plane of best focus, located a distance of 27.36 mm from the lens center. The logarithmic scale emphasizes the weak rings.

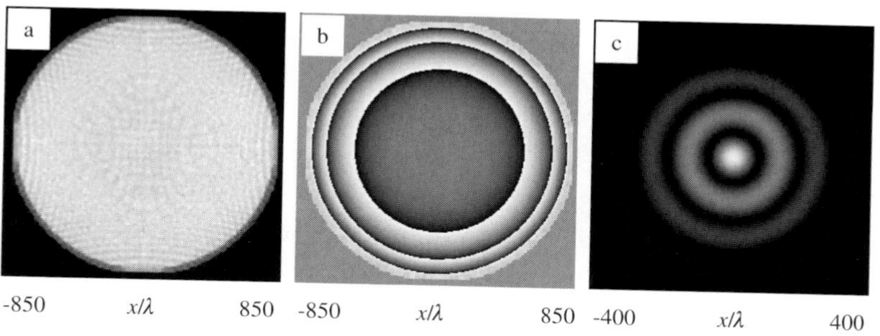

Figure 30.4 Various distributions in the system of Figure 30.2 when $R_a = 0.4$ mm. (a) Intensity distribution at the exit pupil. (b) Total phase distribution at the exit pupil; the r.m.s. value of residual aberrations (with the curvature taken out) is 0.22λ. The gray-scale encodes values of phase from -180° (black) to $+180^\circ$ (white). (c) Intensity distribution in the plane of best focus, located a distance of 59.3 mm from the lens center.

their corresponding images are computed in the plane of best focus, at $CA = 59.3$ mm. All imaged points show spherical aberration similar to that of the central spot, but there is very little coma and astigmatism, owing to the fact that the system is essentially monocentric.

Glass sphere as a magnifier

Up to this point we have studied the properties of real images formed by point sources placed in the (paraxial) focal plane of a spherical lens. Now we

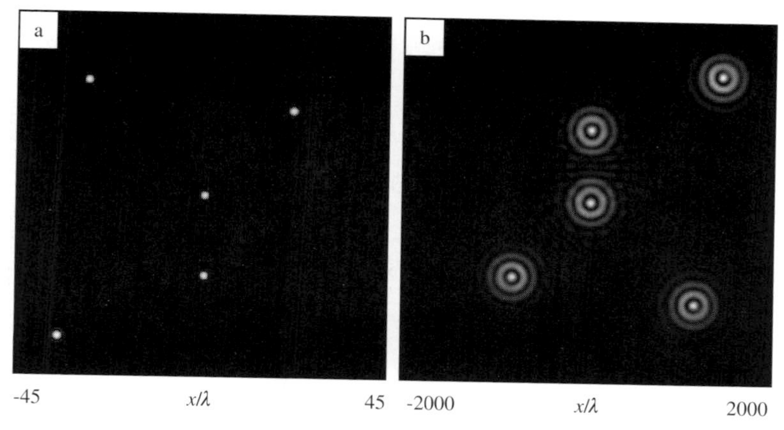

Figure 30.5 Five point sources placed in the front focal plane of the spherical lens shown in Figure 30.2. The exit-pupil radius $R_a = 0.4$ mm, and the best image (with $\sim 45\times$ magnification) appears in a plane 59.3 mm away from the lens center. (a) Intensity distribution in the object plane. (b) Intensity distribution in the image plane. All imaged points show spherical aberration, but there is very little coma or astigmatism.

will consider the spherical lens as a magnifying glass, placing the object somewhat closer to the lens than its front focus and examining the properties of the virtual image thus formed.

The diagram of Figure 30.6 is a representation of a Van Leeuwenhoek microscope with a spherical glass lens having $R = 1$ mm, $n = 1.5$. To achieve high-resolution imaging with this system the aperture is closed down to $R_a = 0.25$ mm, and the object is displaced from the paraxial focus F by 20 μm towards the lens. The observer's eye is placed very close to the lens, so that the pupil of the eye essentially coincides with the exit pupil of the lens.

The object used in the following calculations is shown in Figure 30.7. This is a transmissive object with several micron-sized features that impart phase and amplitude modulation to the incident beam. With this object we demonstrate both coherent and incoherent imaging through the system of Figure 30.6. The illumination in both cases is monochromatic at a wavelength $\lambda = 0.5$ μm, although white light or other broadband sources can also be used to illuminate the object. The simplicity of this single-lens microscope keeps chromatic aberrations to a minimum.[1]

In the case of coherent imaging, the incident beam is collimated, uniform, and propagates along the Z-axis. The computed distributions of intensity and phase at the exit pupil of the lens are shown in Figure 30.8. The intensity plot in Figure 30.8(a) is drawn on a logarithmic scale to emphasize the spatial-frequency content of the image-carrying beam. It is found numerically that the best focus of this system is at a distance $CA = -316$ mm from the lens

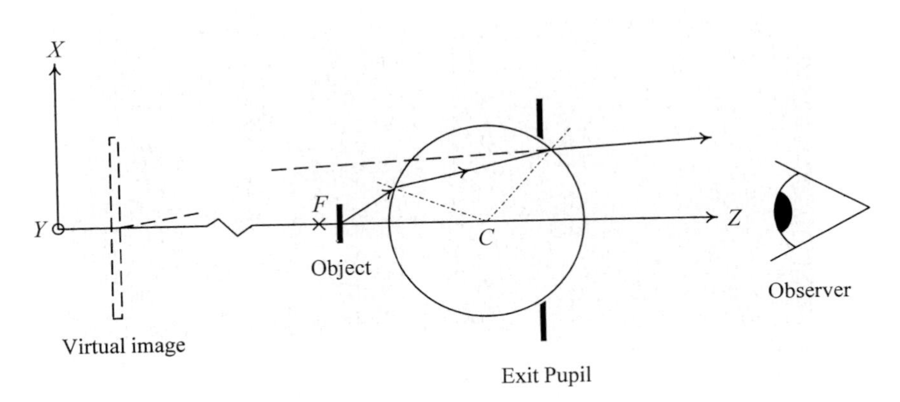

Figure 30.6 The simulated Van Leeuwenhoek microscope. The lens radius $R = 1$ mm, its refractive index $n = 1.5$, the object is 20 μm to the right of the paraxial focus F (i.e., 0.48 mm away from the lens), and the exit-pupil radius $R_a = 0.25$ mm. The virtual image, formed 316 mm to the left of the lens center, can be comfortably viewed when the eye is placed at or near the exit pupil.

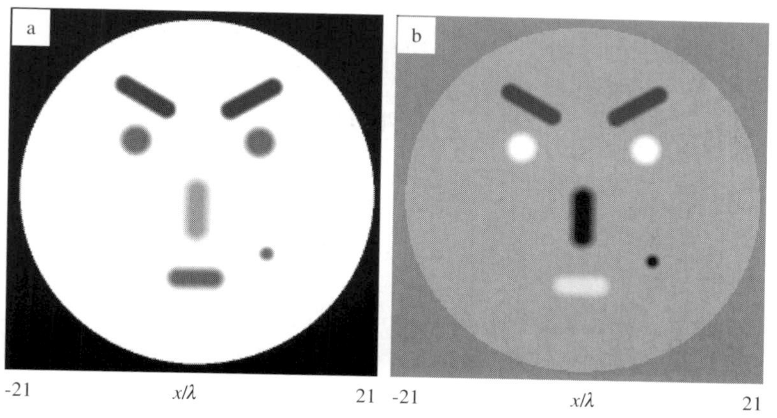

Figure 30.7 Distributions of (a) intensity and (b) phase immediately in front of the object. The object is trans-illuminated with a uniform, coherent, and monochromatic plane wave $\lambda = 0.5$ μm. The smallest feature in the lower right-hand side is 1 μm in diameter. The phase values in (b) range from $-144°$ (black) to $+108°$ (white).

center; the computed coherent image at this distance, having a magnification close to 200, is shown in Figure 30.9(a).

To compute the incoherent image, we illuminate the object with 225 monochromatic point sources ($\lambda = 0.5$ μm, $NA = 0.15$), and superimpose the resulting intensity distributions in the image plane obtained for individual point sources. Figure 30.9(b) is the computed incoherent image of the object shown in Figure 30.7 through the system of Figure 30.6. The magnification is

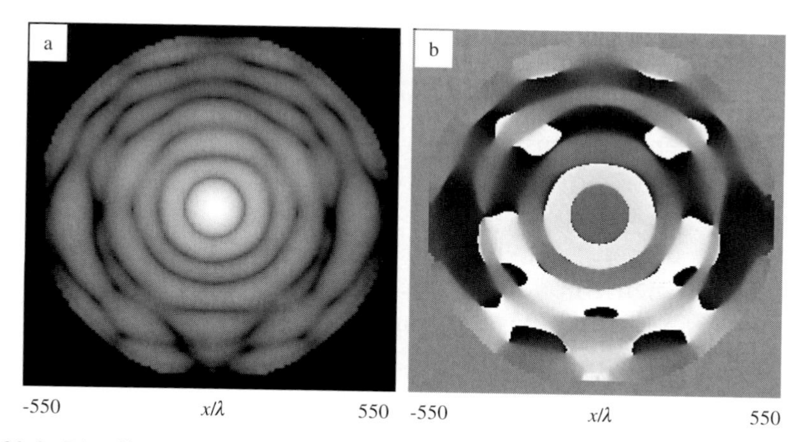

Figure 30.8 Distributions of (a) intensity and (b) phase at the exit pupil of the microscope of Figure 30.6 with the coherently illuminated object of Figure 30.7. The intensity is shown on a logarithmic scale to emphasize its weak regions. The phase ranges from $-180°$ (black) to $+180°$ (white).

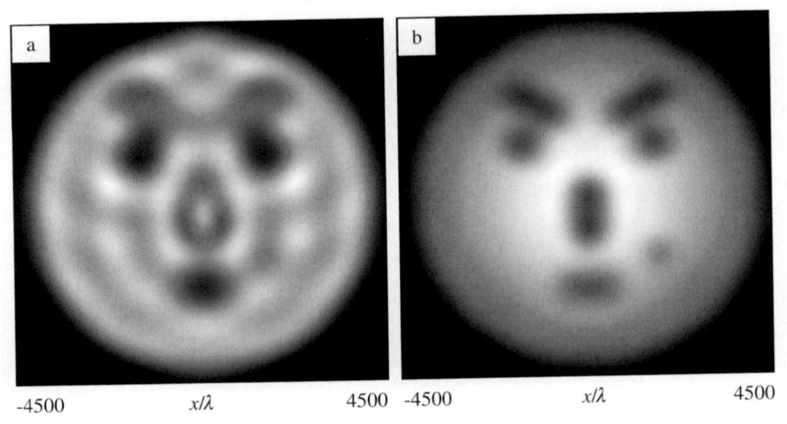

Figure 30.9 Distributions of intensity in the virtual image seen through the microscope of Figure 30.6 with the object of Figure 30.7. The image in (a) is computed for a coherent, monochromatic beam of light normally incident on the object. The incoherent image in (b) is obtained by illuminating the object with 225 point sources through a $0.15NA$ condenser lens. These virtual images have a magnification of $\sim 200\times$ and appear at a distance of 316 mm behind the lens center

about 200, and the image exhibits a fairly accurate reproduction of the various features present in the object, except perhaps the spot 1 μm in diameter on the lower right-hand side. Thus the microscope depicted in Figure 30.6, having numerical aperture $NA \approx 0.16$, is nearly diffraction-limited, over at least a 20 μm field of view, with a resolution of ~ 2 μm at $\lambda = 0.5$ μm. (In reality, the field of view of the microscope is several times greater than that demonstrated in this particular example.)

Method of computation

The results presented in this chapter were obtained by a combination of ray-tracing and diffraction calculations. The light emanating from the object was propagated to the vicinity of the lens using far-field (Fraunhofer) diffraction formulas. The complex-amplitude distribution at this point was converted into a set of geometric-optical rays, using the local Poynting vector to represent the ray. The rays were traced from the entrance pupil to the exit pupil of the lens using standard methods of ray-tracing. At the exit pupil the ray magnitude and phase information was converted into a complex wavefront, and the wavefront was propagated to the image plane using near-field (Fresnel) diffraction formulas.

Other applications of glass spheres

Glass balls have found application in other areas as well. A simple method of coupling the light from a diode laser (or a light-emitting diode) into an optical fiber uses a spherical glass ball between the source and the fiber's entrance facet. This may not be the most efficient coupling mechanism, but it is simple, inexpensive, and easy to implement in conjunction with multimode fibers. Tiny glass beads are often mixed with ordinary paint for use on the streets, on automobile license plates, etc., to enhance retro-reflectivity. My colleague Stephen Jacobs of the University of Arizona has made a fused silica ball six inches in diameter, through which one can look toward the sun and observe beautiful optical phenomena.[6] Looking through this glass sphere, one cannot help but remember that Nature has employed spherical droplets of water to create the magnificent rainbow.[7,8]

References for chapter 30

1 B. J. Ford, The earliest views, Scientific American, 50–53, April 1998.
2 B. J. Ford, *Leeuwenhoek Legacy*, Bristol, Biopress; London, Farrand Press; 1991.
3 L. Yount, *Antoni van Leeuwenhoek: First to See Microscopic Life*, Enslow Publishers, 1996.
4 J. A. Mahaffey, Making Leeuwenhoek proud: building simple microscopes, *Opt. & Phot. News* **10**, 62–63, March 1999.
5 These historical anecdotes have been compiled from information available on the worldwide web. See, for example, encarta.msn.com, www.hcs.ohio-state.edu, www.letsfindout.com, www.feic.com, www.ucmp.berkeley.edu, www.utmem.edu.
6 S. F. Jacobs and S. C. Johnston, Unusual optical effects of a solid glass sphere, *Opt. & Phot. News* **8**, 44–45, October 1997.
7 H. M. Nussenzveig, The theory of the rainbow, Scientific American, 116–127, April 1977.
8 C. B. Boyer, *The Rainbow, From Myth to Mathematics*, Sagamore Press, Thomas Yoseloff, New York, 1959.

31

Projection photolithography[†]

Photolithography is the technology of reproducing patterns using light. Developed originally for reproducing engravings and photographs and later used to make printing plates, photolithography was found ideal in the 1960s for mass-producing integrated circuits.[1] Projection exposure tools, which are now used routinely in the semiconductor industry, have continually improved over the past several decades in order to satisfy the insatiable demand for reduced feature size, increased chip size, improved reliability and production yield, and lower overall cost. High-numerical-aperture lenses, short-wavelength light sources, and complex photoresist chemistry have been developed to achieve fabrication of fine patterns over fairly large areas. Research and development efforts in recent years have been directed at improving the resolution and depth of focus of the photolithographic process by using phase-shifting masks (PSMs) in place of the conventional binary intensity masks (BIMs). In this chapter we describe briefly the principles of projection photolithography and explore the range of possibilities opened up by the introduction of PSMs.

Basic principles

Figure 31.1 is a diagram of a typical projection system used in optical lithography. A quasi-monochromatic, spatially incoherent light source (wavelength λ) is used to illuminate the mask. Steps are usually taken to homogenize the source, thus ensuring a highly uniform intensity distribution at the plane of the mask. The condenser stop may be controlled to adjust the degree of coherence of the illuminating beam; this control of partial coher-

[†]The coauthor of this chapter is Rongguang Liang.

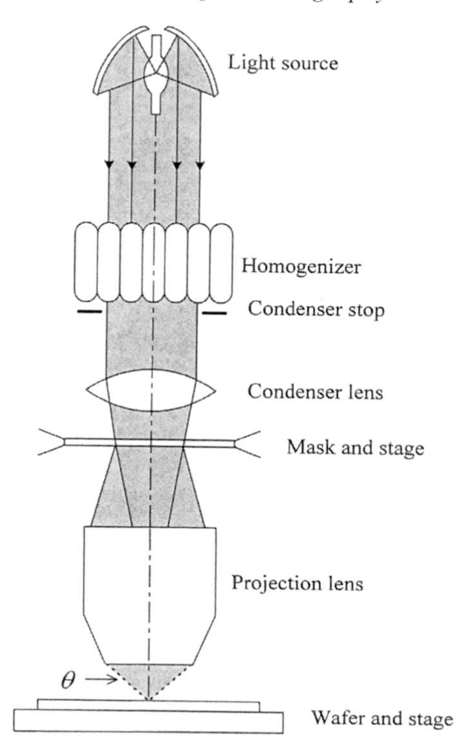

Figure 31.1 Essential elements of a photolithographic "stepper" used for exposing semiconductor wafers. The condenser stop controls the degree of coherence of the illumination. The numerical aperture NA_0 of the projection lens is defined as $\sin\theta$, where θ is the half-angle of the cone subtended by the clear aperture of the projection lens at the wafer. The uniformly illuminated mask is imaged onto the wafer with a magnification M that is typically around 1/5.

ence is especially important when PSMs are used to improve the performance of optical lithography beyond what is achievable with the traditional BIMs.

The light transmitted through the mask is collected by the projection lens, which images the mask onto the wafer, typically with a magnification $M = 1/5$. Thus, if the numerical aperture of the projection lens is defined as $NA_0 = \sin\theta$, its angular aperture on the mask side will be $\sin\theta' = MNA_0$. If the condenser's numerical aperture NA_c happens to be much less than $\sin\theta'$ then the illumination is coherent, while if $NA_c \geq \sin\theta'$ then the illumination is essentially incoherent. In practice the ratio $\sigma = NA_c/(MNA_0)$ is used as a measure of the incoherence of illumination. For example, if $M = 1/5$ and $NA_0 = 0.6$, then $NA_c = 0.084$ yields $\sigma = 0.7$, while $NA_c = 0.06$ yields $\sigma = 0.5$. For a given projection lens, therefore, the incoherence of illumination is proportional to the condenser's stop diameter.[1–3]

Over the past decade, photolithographic systems have evolved through several generations. The wavelength of the light source has steadily decreased from 365 nm (i-line of mercury) to 257 nm (high-pressure mercury arc lamp) to 248 nm (KrF laser), and is presently at 193 nm (ArF excimer laser). The numerical aperture NA_0 of the projection lens, having increased from its value of ~ 0.16 in the early days to ~ 0.6 in present-day systems, is likely to increase still further. The illumination systems have also improved, taking advantage of off-axis illumination and related configurations.[1,2] Other improvements have occurred in the area of photoresists and the control of their exposure and development processes and also in the control of the flatness of the wafer, which reduces the need for a large depth of focus, etc.

These topics are beyond the scope of the present chapter, and we refer the interested reader to the published literature for further information.[1–6] In the remainder of this chapter we present computed images of various masks obtained in a typical projection system ($NA_0 = 0.6$, $M = 1/5$) and compare the resulting image contrasts and resolutions.

PSM versus BIM

Traditional "binary intensity" masks (BIMs) consist of opaque chromium lines on transparent glass substrates; these masks modulate the intensity of the incident light without affecting its phase. Modern masks have begun to take advantage of optical phase by changing the thickness of the transparent regions of the mask, either by depositing additional transparent material where needed or by removing a thin layer from the substrate at specific locations, thereby selectively adjusting the transmitted optical phase.[1,2]

The basic idea of an optical phase-shifting mask for lithography originated in the early 1980s with M. D. Levenson[6] in the US and, independently and almost simultaneously, with M. Shibuya[7] in Japan. Figure 31.2 shows several different mask designs that exploit optical phase to improve the resolution of the photolithographic process. In addition to improved resolution, these PSMs also increase the effective depth of focus and provide a wider process window (i.e., a wider range of acceptable focuses and exposures).[1]

Alternating-aperture phase-shifting mask

Consider the simple mask consisting of three bright lines on a dark background shown in Figure 31.3. Each bright line is 3λ wide, and the separation

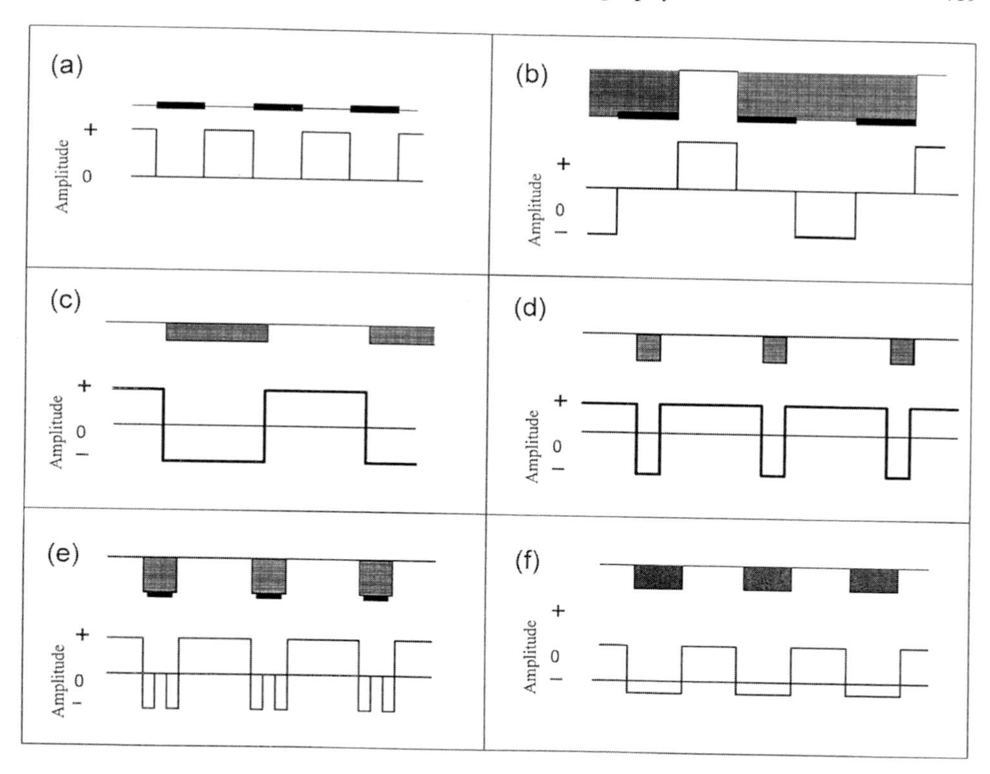

Figure 31.2 Several mask structures and, below each structure, the corresponding E-field patterns immediately after transmission through the mask. (a) Conventional transmission mask. (b) Alternating-aperture phase mask with etched substrate. (c) A chromeless phase-edge mask produces dark lines in the image solely through destructive interference at the phase transitions. (d) A shifter–shutter mask is similar to (c) except that each dark line is produced by a pair of adjacent phase-edges. (e) A rim-shifter mask contains chrome lines bracketed by 180° phase-edges. (f) An attenuated phase-shift mask; here the shaded regions represent partially transmissive material with a 180° phase shift. (Adapted from reference 1.)

between adjacent lines is also 3λ. (Note that these are the mask dimensions; at the wafer the features are demagnified by a factor $1/M = 5$.) We assume two different designs for the mask. In the first, the mask is a conventional BIM, the same phase being imparted to the light transmitted through each aperture. In the second, the mask is a PSM in which the upper and lower bright lines are phase-shifted by 180° relative to the central bright line. In Figures 31.4(a), (b) we compare the intensity patterns of the images obtained

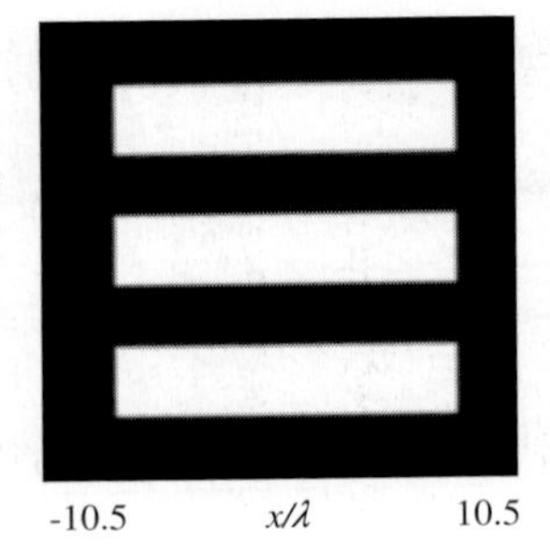

Figure 31.3 A simple mask containing three transparent apertures on an opaque background. The apertures as well as the spaces between apertures are 3λ wide. When the apertures impart a uniform phase to the transmitted beam, the mask is a BIM. When the upper and lower apertures impose on the transmitted beam a 180° phase shift relative to the middle aperture, the mask is an alternating-aperture PSM.

at the wafer for these two types of mask. The assumed projection system is that of Figure 31.1, with $NA_0 = 0.6$, $M = 1/5$, and $\sigma = 0.7$. Clearly the PSM is better at resolving the dark spaces between adjacent bright lines. For direct comparison, a cross-section through these two intensity distributions is shown in Figure 31.4(c). Increasing the coherence of the illumination by closing down the aperture of the condenser to $\sigma = 0.5$ improves the image contrast of the PSM but degrades that of the BIM image, as can be readily observed in Figures 31.4(d)–(f).

Isolated bright line

As our second example we consider the case of an isolated bright line. Figures 31.5(a), (b) show respectively a BIM and a PSM for a line of width 4λ. (Again, this is the dimension at the mask; the projected line at the wafer is only 0.8λ wide.) The PSM of Figure 31.5(b) contains two 0.8λ-wide side-riggers, each imparting a 180° phase shift to the incident beam relative to the central bright line.[4]

Figures 31.6(a), (b) show the computed intensity patterns at the wafer for the two masks, and Figure 31.6(c) shows cross-sections of both patterns (the assumed coherence factor σ is 0.7). The side-riggers produce small bumps in the intensity pattern of the PSM, but these are usually below the resist threshold and are not printed. In Figure 31.6 it can be seen that the computed image of the bright line using the PSM is about 10% narrower than that obtained with the BIM. This modest reduction in the printed line-width can be slightly improved upon if the side-riggers' location and width are properly optimized and also if the condenser stop is further closed down to increase

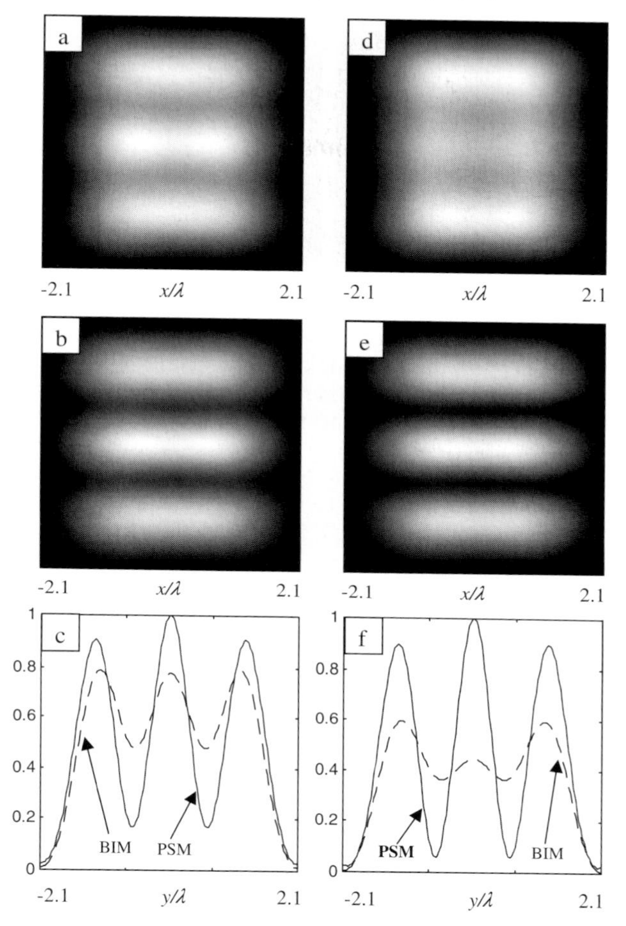

Figure 31.4 Computed plots of intensity distribution at the wafer for the mask of Figure 31.3 placed in the system of Figure 31.1 ($NA_0 = 0.6$, $M = 1/5$). (a) Image of the BIM obtained with $\sigma = 0.7$. (b) Image of the PSM obtained with $\sigma = 0.7$. (c) Cross-sections of the intensity patterns for the BIM (broken line) and the PSM (solid line). (d)–(f) Same as the patterns in the left-hand column, but for $\sigma = 0.5$.

the coherence of illumination (σ-values as low as 0.3 have been suggested in the literature[4,5]).

Contact hole

Figure 31.7(a) shows a simple $4\lambda \times 4\lambda$ square aperture on a dark background. This feature has uniform phase across the aperture and, therefore, represents the BIM for a contact hole. A corresponding PSM for the same hole is shown in Figure 31.7(b). Here four side-rigger lines of width 0.5λ and

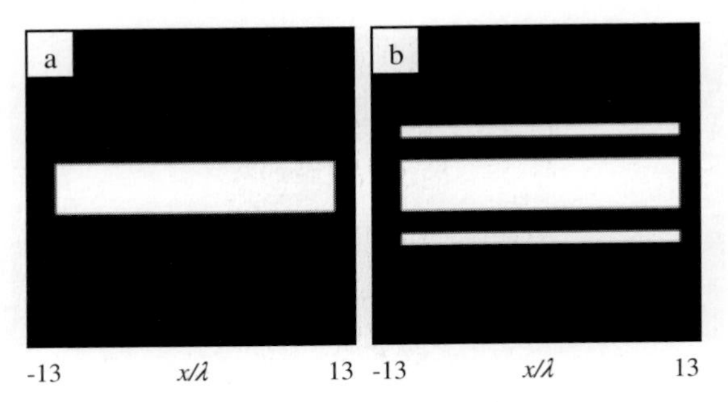

Figure 31.5 Masks designed for creating an isolated bright line at the wafer. (a) BIM containing a 4λ-wide line on an opaque background. (b) PSM featuring the same 4λ-wide line flanked by a pair of 0.8λ-wide side-riggers. Each side-rigger imparts to the incident beam a 180° phase shift relative to the central line. The separation between the central line and each side-rigger is 2λ.

180° phase shift (relative to the central aperture) are placed around the hole.[4] The computed intensity patterns of the images of these masks at the wafer appear in Figures 31.8(a), (b), respectively. The side-rigger features are too small to be printed, but their destructive interference with the central aperture results in a smaller projected hole, as revealed in the cross-sectional intensity profiles at the wafer shown in Figure 31.8(c). As before, the printed feature size can be further optimized by adjusting the dimensions of the side-riggers as well as by closing the condenser stop to reduce the value of σ.

More complicated patterns

Figure 31.9(a) shows a mask with five transparent apertures. The widths of line (bright) and space (dark) on this mask are both equal to 4.8λ. If the mask is used without any phase shifts, the intensity pattern of Figure 31.9(b) will be obtained at the wafer. Placing 180° phase-shifters on alternate bright apertures results in the image intensity distribution shown in Figure 31.9(c). Two different cross-sections of these patterns are also given in Figures 31.9(d), (e). In this case of relatively large features, there are apparently no significant differences between a BIM and a PSM.

With shrinking feature size, however, the advantages of the PSM become apparent. Figure 31.10 is the counterpart of Figure 31.9 for the case where the line- and space-widths (at the mask) are both reduced to 3λ. The BIM is now seen to yield a fairly low-contrast image at the wafer, while the PSM provides better resolution and sharper contrast. Reducing the feature size still further to 2.4λ (at the mask) results in the patterns of Figure 31.11. Here the

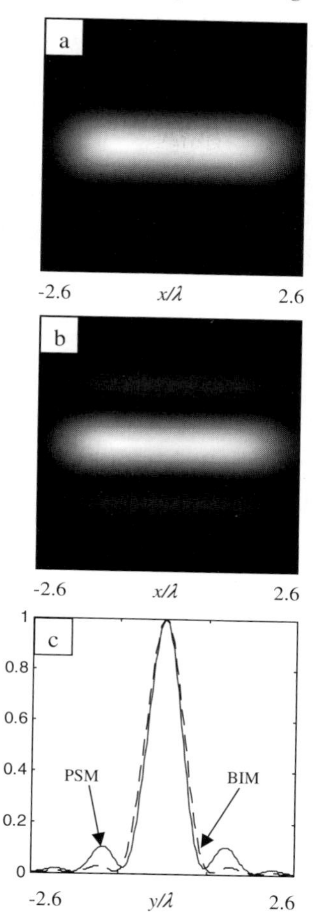

Figure 31.6 Computed intensity patterns at the wafer for the masks of Figure 31.5 in the system of Figure 31.1 ($NA_0 = 0.6, M = 1/5, \sigma = 0.7$). (a) Using the BIM; (b) using the PSM; (c) the cross-sections of the intensity patterns in the images of the BIM (broken line) and the PSM (solid line).

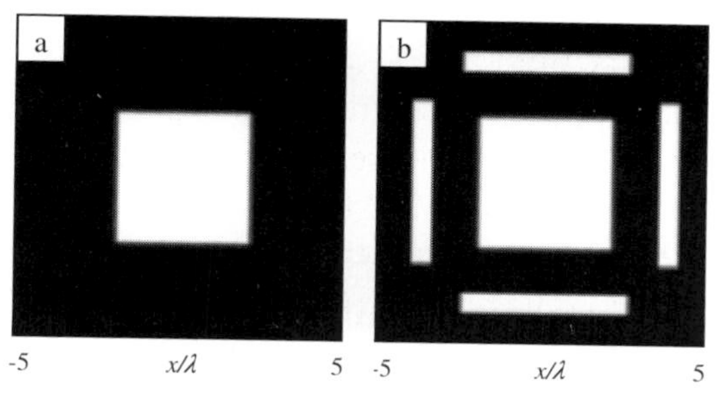

Figure 31.7 Mask patterns for creating a contact hole. (a) BIM containing a $4\lambda \times 4\lambda$ square aperture on an opaque background. (b) PSM featuring the same $4\lambda \times 4\lambda$ aperture surrounded by 0.5λ-wide side-riggers. Each side-rigger imparts to the incident beam a 180° phase shift relative to the central aperture.

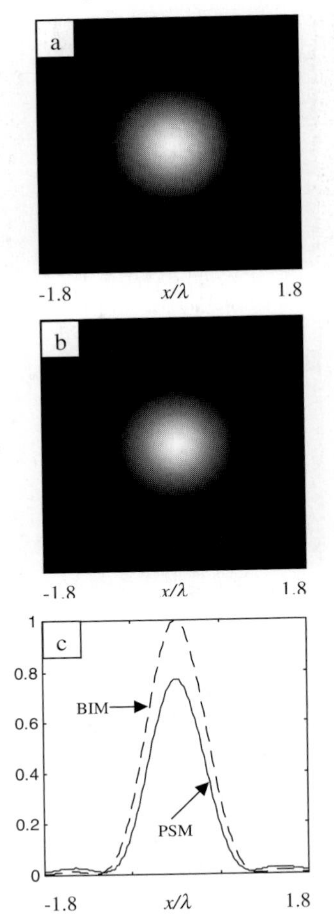

Figure 31.8 Computed intensity patterns at the wafer for the masks of Figure 31.7 in the system of Figure 31.1 ($NA_0 = 0.6, M = 1/5, \sigma = 0.7$). (a) Using the BIM; (b) using the PSM; (c) the cross-sections of the intensity patterns in the images of the BIM (broken line) and the PSM (solid line).

PSM still performs reasonably well, while the image quality of the BIM has been substantially degraded.

Phase-shifters on a transparent background

As a final example, consider the fully transparent (i.e., chromeless) PSM shown in Figure 31.12(a). Each of the three rectangular features on this mask is 4λ wide and is phase-shifted by 180° relative to the background. Also, the spaces separating adjacent rectangular features are each 4λ wide. The computed intensity distribution at the wafer in a system having $NA_0 =$

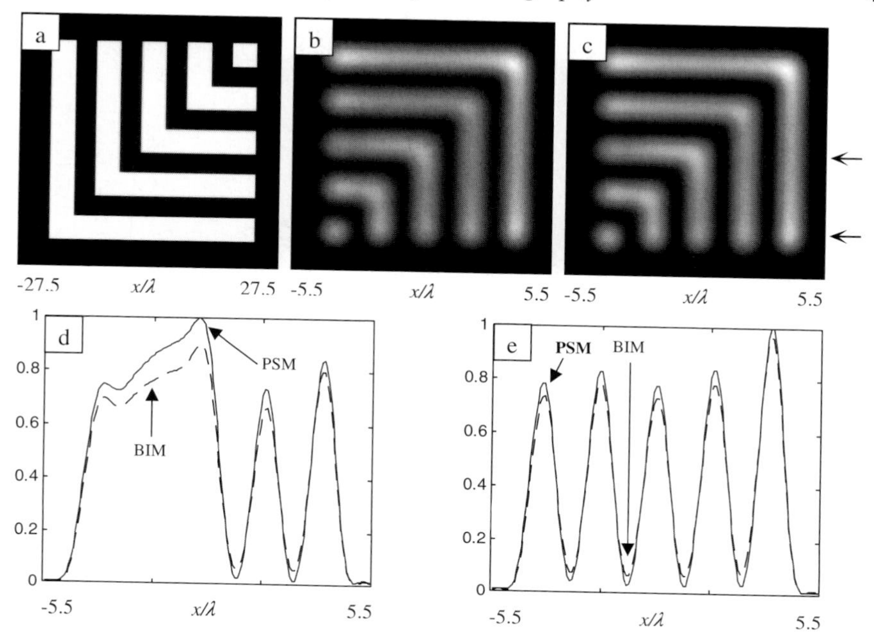

Figure 31.9 (a) Mask pattern containing five transparent apertures on an opaque background. The lines and spaces are all 4.8λ wide. When used as a BIM, all apertures impart the same uniform phase to the incident beam. When used as a PSM, the apertures are alternately phase-shifted by 0° and 180°. The assumed projection-system parameters are $NA_0 = 0.6$, $M = 1/5$, $\sigma = 0.7$. (b) Computed intensity pattern in the image of the BIM. (c) Computed intensity pattern in the image of the PSM; the arrows mark the cross-sections displayed in (d) and (e). (d) Cross-sectional plots of intensity distributions in the images of the BIM (broken line) and the PSM (solid line). (e) A different cross-section of the two images.

0.6, $M = 1/5$, $\sigma = 0.7$ is shown in Figure 31.12(b), and a cross-sectional view is provided in Figure 31.12(c). Depending on the intended application, this image may or may not be acceptable. For instance, suppose the long edges of the rectangular features of the mask are meant to produce dark lines at the wafer. This they do quite well, as is evident from the presence of four horizontal dark lines in Figure 31.12(b). However, if the ends of these dark lines are required to be disconnected from each other, then the PSM has failed in providing the necessary isolation. The problem is rooted in the sharp 0°–180° phase-edge occurring at the short end of each rectangular feature. This problem can be remedied in principle by softening the phase transition at these short ends by providing a gradual transition from 180° to 120° to 60° and eventually to 0°. Such phase stair-steps, however, are usually impractical because they are costly and, moreover, they produce masks that are difficult

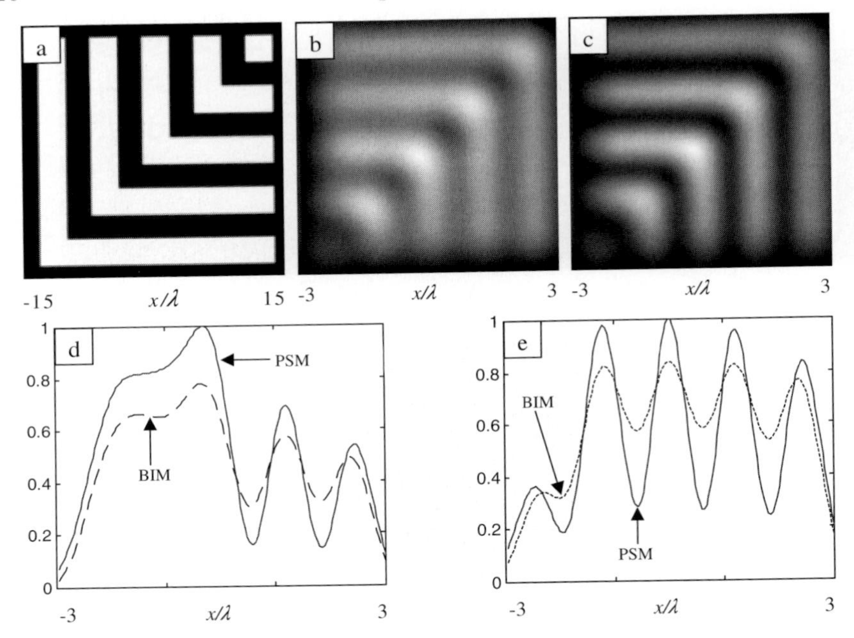

Figure 31.10 Same as Figure 31.9 but for smaller mask features. The lines and spaces on the mask are now 3λ wide.

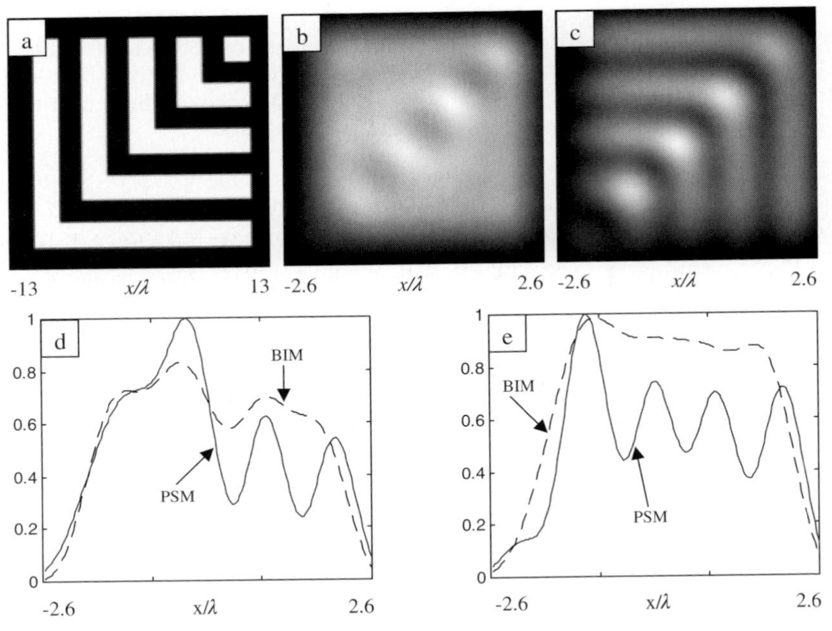

Figure 31.11 Same as Figure 31.9 but for very small mask features. The lines and spaces on the mask are now 2.4λ wide.

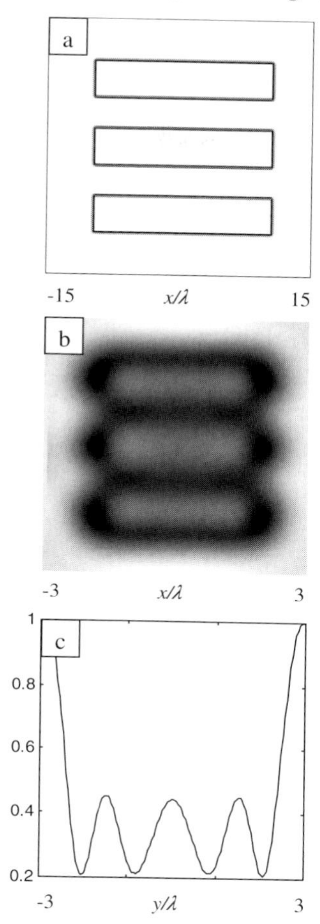

Figure 31.12 (a) Transparent PSM containing three rectangular regions of width 4λ, each imparting a 180° phase shift to the incident beam. Like the background, the spaces between adjacent apertures (also 4λ wide) are fully transparent and impart a 0° phase to the beam. (b) Computed intensity distribution in the image plane of the system of Figure 31.1 having $NA_0 = 0.6$, $M = 1/5$, $\sigma = 0.7$. (c) Central cross-section of the intensity pattern of the image seen in (b).

to inspect and to repair. In today's practice, such unwanted dark lines are erased by a second exposure through a different mask.

Concluding remarks

Incorporating the advantages of optical phase in the design, manufacture, and testing of photomasks is still very much a research topic; many potential benefits of the PSM await to be realized. The type of PSM in common use today is the attenuated PSM depicted in Figure 31.2(f), where the traditional opaque chrome is replaced by a material that transmits ~8% with a 180° phase shift. This is useful for printing bright spaces and contact holes, and has essentially replaced the shifter–shutter type of mask (see Figure 31.2(d)). Also, the more recent high-transmission tri-tone PSM, where the phase-shifted material transmits ~18% and there is a separately patterned opaque layer, has superseded the rim-shifters (see Figure 31.2(e)).[8]

References for chapter 31

1 M. D. Levenson, Wavefront engineering for photolithography, *Physics Today*, 28–36, July 1993.
2 M. D. Levenson, Extending the lifetime of optical lithography technologies with wavefront engineering, *Jpn. J. Appl. Phys.* **33**, 6765–6773 (1994).
3 M. D. Levenson, Wavefront engineering from 500 nm to 100 nm CD, in *Emerging Lithographic Technologies, SPIE* **3048**, 2–13 (1997).
4 T. Terasawa, N. Hasegawa, T. Kurosaki, and T. Tanaka, 0.3-micron optical lithography using a phase-shifting mask, *SPIE* **1088**, 25–33 (1989).
5 N. Hasegawa, T. Terasawa, T. Tanaka, and T. Kurosaki, Submicron optical lithography using phase-shifting mask, *Electro-chem. Ind. Phys. Chem.* **58**, 330–335 (1990).
6 M. D. Levenson, N. S. Viswanathan and R. A. Simpson, Improving resolution in photolithography with a phase-shifting mask, *IEEE Trans. Electron Devices* **ED-29**, 1828–1836 (1982).
7 M. Shibuya, Projection master for transmitted illumination, Japanese Patent Gazette # Showa 62-50811, application dated 9/30/80, issued 10/27/87.
8 M. D. Levenson, private communication.

32
The Ronchi test

In the 1920s Vasco Ronchi developed the well-known method of testing optical systems now named after him.[1,2] The essential features of the Ronchi test may be described by reference to Figure 32.1. A lens (or more generally, an optical system consisting of a number of lenses and mirrors) is placed in the position of the "object under test". The lens is then illuminated with a beam of light, which, for the purposes of the present chapter, will be assumed to be coherent and quasi-monochromatic. These restrictions on the beam may be substantially relaxed in practice.[3]

The lens brings the incident beam to a focus in the vicinity of a diffraction grating, which is placed perpendicular to the optical axis, i.e., the Z-axis. The grating, also referred to as a Ronchi ruling, may be as simple as a low-frequency wire grid or as sophisticated as a modern short-pitched, phase/amplitude grating. The position of the grating should be adjustable in the vicinity of focus, so that it may be shifted back and forth along the optical axis. The grating breaks up the incident beam into multiple diffracted orders, which will subsequently propagate along Z and reach the lens labeled "pupil relay" in Figure 32.1.

The pupil relay may simply be the lens of the eye, which projects the exit pupil of the object under test onto the retina of the observer. Alternatively, it may be a conventional lens that creates a real image of the exit pupil on a screen or on a CCD camera.

The diffracted orders from the grating will be collected by the relay lens and, within their overlap areas, will create interference fringes characteristic of the aberrations of the optical system under consideration. By analyzing these fringes, one can determine the type and, with some effort, the magnitude of the aberrations present at the exit pupil of the system.

The above description of the Ronchi test relies on its modern interpretation; this is based on our current understanding of physical optics and the

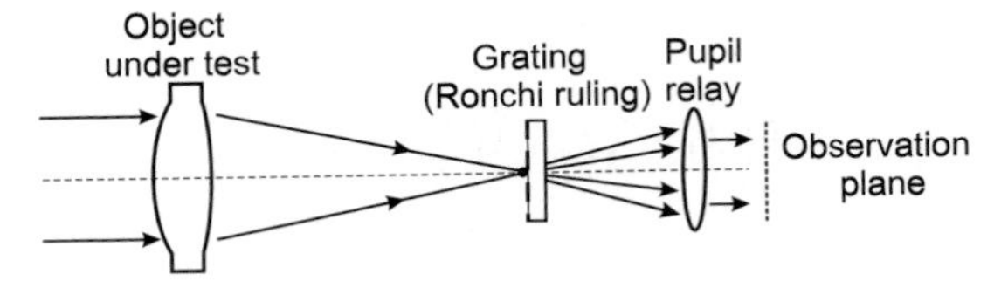

Figure 32.1 A beam of coherent, quasi-monochromatic light is brought to focus by an optical system that is undergoing tests to determine its aberrations. A diffraction grating, placed perpendicular to the optical axis in the vicinity of focus, breaks up the incident beam into several diffraction orders. The diffracted orders propagate, independently of each other, and are collected by a pupil relay lens, which forms an image of the exit pupil of the object under test at the observation plane.

theory of diffraction gratings. Historically, however, the gratings used in the early days were quite coarse, and the results obtained with them required no more than a simple geometric-optical theory for their interpretation. Typically, one would place the eye at the focus of the lens and hold a grating (e.g., a wire grid) in front of the eye, moving the grating in and out until a clear pattern became visible. At this point the beam would be illuminating several of the wires simultaneously. By looking through the grating and observing the shadows that the wires cast on the exit pupil, one could determine the type of aberration present in the system. The coarseness of the grating, of course, caused several of the diffracted orders (as we understand them today) to overlap each other, thus resulting in reduced contrast and smearing of the patterns near the boundaries. These problems were eventually overcome when finer gratings became available and the diffraction theory of the Ronchi test was better understood.

Choosing an appropriate grating

For best results the pitch of the grating should be chosen such that, as shown in Figure 32.2, no more than two diffraction orders will overlap at any given point. To determine the appropriate grating period P, one needs to know the wavelength λ_0 of the beam used for testing, and the numerical aperture NA of the focused cone of light. (By definition, $NA = \sin\theta$, where θ is the half-angle subtended by the exit pupil of the lens at its focal point. If the lens under test is being used at full aperture, NA will also be equal to 0.5 divided by the lens's f-number.) To avoid multiple overlaps among diffracted orders, the angle between adjacent orders must exceed the focused cone's half-angle. Now, it is well known in the theory of diffraction gratings that, at normal incidence, $\sin\theta_n = n\lambda_0/P$ where n, an integer, is the order of diffraction and θ_n is the corresponding deviation angle from the surface normal. Therefore, we arrive

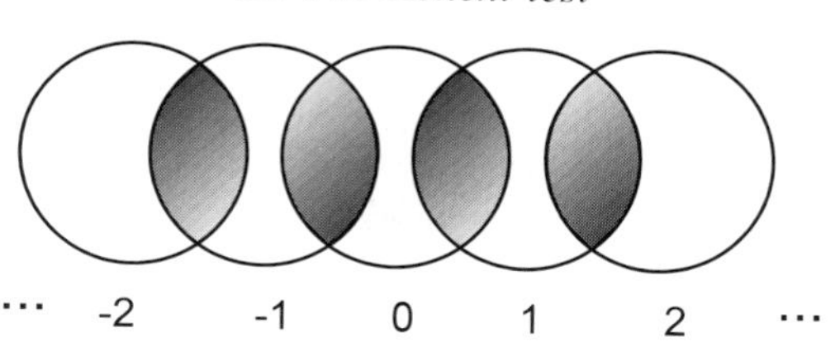

Figure 32.2 Several diffracted orders in the far field of the grating of Figure 32.1. When the grating's period is chosen properly, each diffracted order (i.e., emergent cone of light) will overlap only with its nearest neighbors. Except for a lateral shift in position, the various orders are identical, carrying the amplitude and phase distribution of the beam as it appears at the exit pupil of the object under test.

at the conclusion that P should be less than or equal to λ_0/NA. For example, assume that the lens under test has a numerical aperture $NA = 0.5$. Then, if the grating period is chosen to be $2\lambda_0$, each diffracted order will deviate from the zero-order beam by 30°, making the +first-order beam just touch the −first-order beam in the far field.

Figure 32.3 shows the computed intensity distribution at the observation plane of an aberration-free system in which the relay lens has the same numerical aperture as the lens under test ($NA = 0.5$). This equality of the numerical apertures means that only the zeroth-order diffracted beam will be fully transmitted to the observation plane. Of the ±first-order beams, only those portions that overlap the zero order will reach the observation plane. The period of the grating in this example has been a little less than λ_0/NA, leaving a small gap between +first order and −first order. The absence of aberrations means that the phase distribution over the cross-sections of the

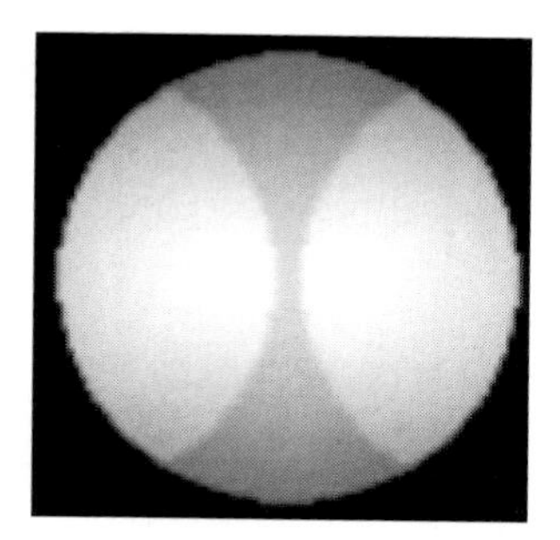

Figure 32.3 Distribution of intensity at the observation plane of Figure 32.1 in the absence of aberrations. The pupil relay lens is chosen to have the same numerical aperture as the object under test, thereby limiting the collected light to the zeroth-order beam and to those portions of the ±first-order beams that overlap the zeroth-order beam.

various diffracted orders is uniform and, therefore, no interference fringes are to be expected.

Ronchigrams for primary or Seidel aberrations

Figure 32.4 shows the computed patterns of intensity distribution at the observation plane of Figure 32.1 corresponding to different types of primary (Seidel) aberrations of the lens. For these calculations we fixed the distance between the lens under test and the relay lens and then placed the grating at the paraxial focus of the converging wavefront. The pattern in Figure 32.4(a) was obtained when we assumed the presence of three waves of curvature (or defocus) at the exit pupil of the lens. Different amounts of defocus would create essentially the same pattern, albeit with a different number of fringes. In Figure 32.4(b) we observe the fringes arising from the presence of three waves of third-order spherical aberration in the test system. The shapes of these fringes depend not only on the magnitude of the aberration but also on the position of the grating relative to the focal plane. (We will have more to say about this point later.) Figure 32.4(c) shows the fringes that would arise when three waves of primary astigmatism are present. When the orientation

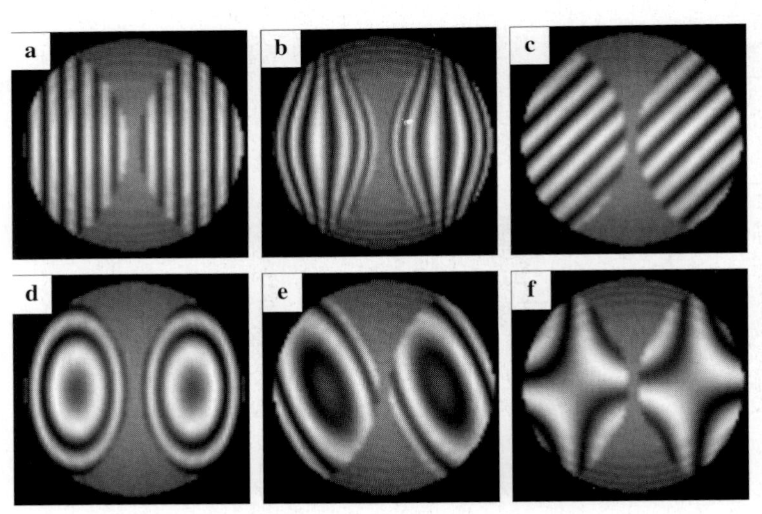

Figure 32.4 Computed plots of intensity distribution at the observation plane of Figure 32.1. The lens under test is assumed to have three waves of primary (Seidel) aberrations; the grating is at the nominal focal plane of the lens. (a) Defocus, (b) spherical aberration, (c) astigmatism oriented at 45°, (d) coma at 0°, (e) coma at 45°, (f) coma at 90°.

of the astigmatism changes, the fringes will remain straight lines but their orientation within the observation plane will change accordingly.

The last three frames in Figure 32.4 represent the effects of third-order coma. A change in orientation of this aberration causes the interference pattern to change drastically. Figures 32.4(d)–(f) correspond to three waves of coma oriented at 0°, 45°, and 90°, respectively.

Sliding the grating along the optical axis

A change in the position of the grating relative to the focal plane influences the observed fringe pattern. We limit our discussion to the case of spherical aberration, although similar analyses could be performed for other aberrations. Assuming three waves of spherical aberration as before, we obtain the patterns displayed in Figure 32.5 as we slide the grating along the optical axis in the system of Figure 32.1. Once again, we have taken the lens under test to have $NA = 0.5$ and $f = 6000\lambda_0$. The paraxial focus of the lens under test coincides with the front focal point of the relay lens, and the grating is shifted by different amounts Δz relative to this common focus. Frames (a)–(f) in

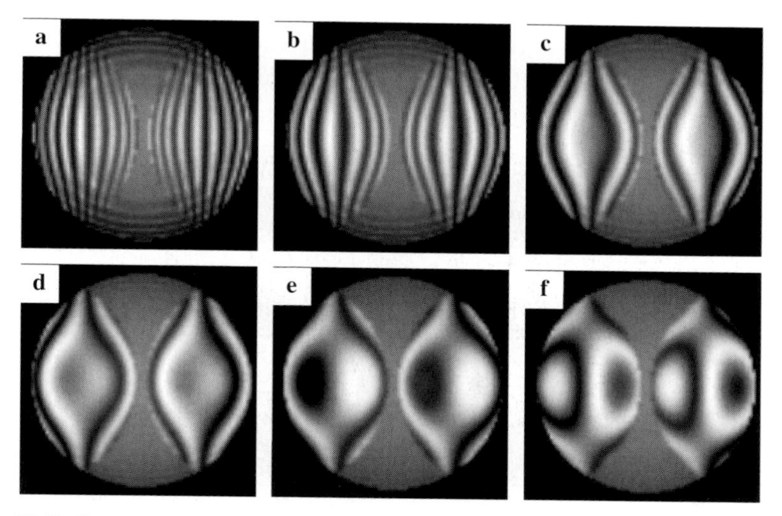

Figure 32.5 Computed plots of intensity distribution at the observation plane of Figure 32.1, showing the patterns obtained by sliding the grating along the optical axis. The lens under test ($NA = 0.5, f = 6000\lambda_0$) is assumed to have three waves of primary spherical aberration, and its paraxial focus is coincident with the focal point of the relay lens. The grating is moved along the optical axis by an amount Δz relative to the (common) focal plane; positive distances are towards the marginal focus. (a) $\Delta z = -10\lambda_0$, (b) $\Delta z = 0$, (c) $\Delta z = 10\lambda_0$, (d) $\Delta z = 15\lambda_0$, (e) $\Delta z = 20\lambda_0$, (f) $\Delta z = 25\lambda_0$.

Figure 32.5 correspond to different values of Δz, starting at $\Delta z = -10\lambda_0$ in (a) and moving forward to $\Delta z = +25\lambda_0$ in (f). In the process, as the grating moves through paraxial focus and towards marginal focus, we observe a rich variety of patterns that aid us in determining the nature and the magnitude of the aberration.

To be sure, the Ronchi test is not the only scheme used during the fabrication and evaluation of optical systems; several other tests exist and their relative merits have been expounded in the literature.[3] It is useful here to examine some of these alternative methods and to compare the resulting patterns (interferograms or otherwise) with those obtained with the Ronchi test.

Testing by interfering with a reference plane wave

Figure 32.6 shows the schematic diagram of a Mach–Zehnder interferometer, which is one among many that can be used to evaluate the aberrated wavefronts directly. In this system a coherent monochromatic beam of light is sent through the lens under test, is collected and recollimated by a well-corrected lens, and is made to interfere with a reference beam that has been split off the incident wavefront. The flat mirror shown in the lower left side of the interferometer is mounted on a tip–tilt stage that allows the introduction of a small amount of tilt in the reference beam. Figure 32.7 shows the computed patterns of intensity distribution at the observation plane of the Mach–

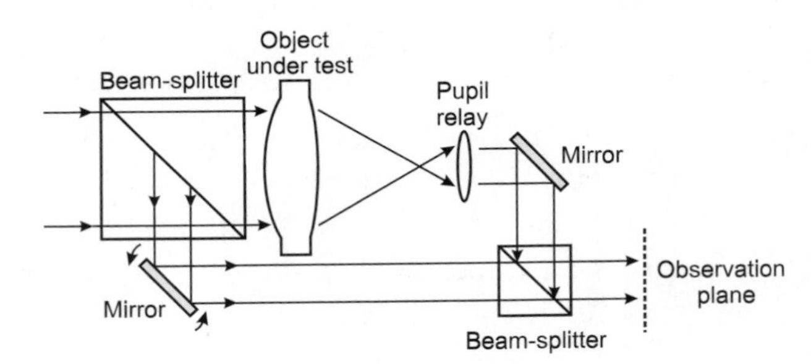

Figure 32.6 Schematic diagram of a Mach–Zehnder interferometer that might be set up for a direct measurement of wavefront aberrations. The pupil relay lens (itself free from aberrations) forms at the observation plane an image of the exit pupil of the lens under test. A fraction of the incident beam is diverted from its original path and sent to the observation plane by means of the various mirrors and beam-splitters. The observed fringes are characteristic of the aberrations present at the exit pupil of the lens under test. A small tilt of the mirror shown at the lower left side of the figure would introduce a linear phase shift on the reference beam. This tilt is generally useful in producing signature fringe patterns at the observation plane.

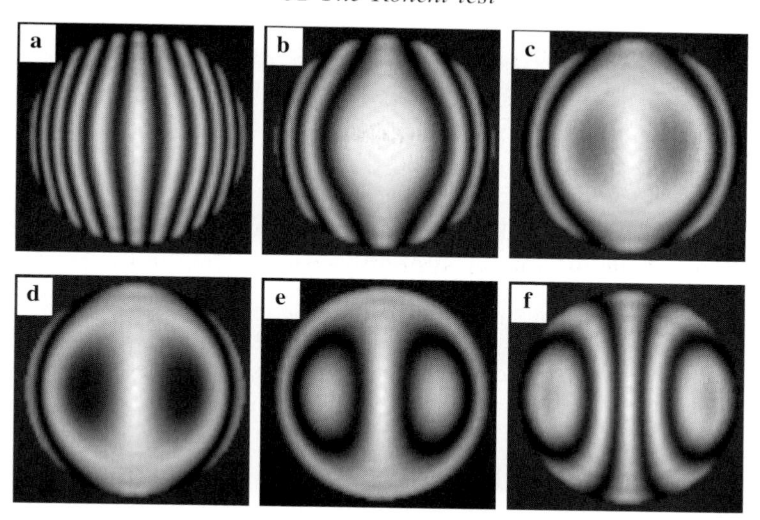

Figure 32.7 Computed plots of intensity distribution at the observation plane of Figure 32.6. The lens under test ($NA = 0.5, f = 6000\lambda_0$) is assumed to have three waves of primary coma, and its nominal focus is coincident with the focal point of the relay lens. The tilt angle ψ of the reference beam increases progressively from (a) to (f): (a) $\psi = -0.1°$, (b) $\psi = 0°$, (c) $\psi = 0.05°$, (d) $\psi = 0.07°$, (e) $\psi = 0.1°$, (f) $\psi = 0.18°$.

Zehnder interferometer corresponding to three waves of primary coma. In obtaining the various frames of Figure 32.7 we have fixed all the system parameters and only varied the tilt of the reference beam. Note that the characteristic fringes of coma in Figure 32.7 are quite different from those of coma in the Ronchi test, shown in Figures 32.4(d)–(f). Incidentally, the patterns of Figure 32.7 show similarities with the Ronchigrams of spherical aberration displayed in Figure 32.5. This is not a coincidence; it is rooted in the algebraic forms of the aberration function for third-order coma ($\rho^3 \cos\phi$) and spherical aberration (ρ^4) and also in the fact that a Ronchigram, being a kind of shearing interferogram (albeit with a large shear), is related to the derivative of the wavefront aberration function.

Knife-edge and wire tests

A schematic diagram of the knife-edge method of testing optical systems is shown in Figure 32.8. A geometric-optical interpretation of this test suffices for most practical purposes: the knife-edge blocks different groups of rays in its various positions along the optical axis, allowing the remaining rays to reach the observation plane.[3] Another method of testing, known as the wire

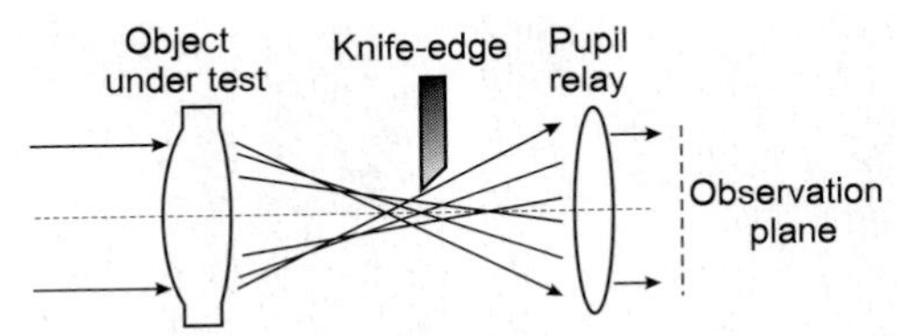

Figure 32.8 In the knife-edge test a certain region in the vicinity of focus is blocked by a knife-edge; the nature and the magnitude of the aberrations are then inferred from the resulting patterns of intensity distribution at the observation plane. (The knife-edge may be moved both along and perpendicular to the optical axis.) The wire test is similar to the knife-edge test except that a fine wire is used instead, to block certain groups of rays.

test, is quite similar to the knife-edge method, being obtained from it by substituting for the knife-edge a length of fine wire.[3]

Since the grating in the Ronchi test may be thought of as a series of parallel knife-edges or, more aptly, a series of parallel wires, it should not come as a surprise that similarities exist between Ronchigrams and the patterns observed in these other tests. In fact, early attempts at explaining the results of Ronchi's method were based on geometrical optics, and considered the grating as a set of parallel wires whose shadows produced the observed patterns.[4] We will not delve into these matters, but simply draw the reader's attention to Figures 32.9 and 32.10, where we show several computed patterns of intensity distribution for the knife-edge and wire tests, respectively.

The results of the simulated knife-edge test depicted in Figure 32.9 assume a laser as the light source. Consequently, frames (a) and (b) of Figure 32.9 exhibit several dark lines which, with a less coherent light source, would have been absent. The results of the simulated wire test shown in Figure 32.10 assume an extended light source, since the small amount of spherical aberration present in the system under consideration would render the test useless with a wire, which fine as it may be, will still be wider than the focused spot produced by a laser beam. Note the similarities between the patterns of Figures 32.9 and 32.10 on the one hand, and those of Figures 32.5(d)–(f) on the other.

Extensions of the Ronchi test

Several modifications and extensions of the Ronchi test have appeared over the years, and have helped to solve specific problems in testing of optical systems.[3] As an example we mention the double-frequency grating lateral-shear interferometer invented by James Wyant in the early 1970s.[5] The grat-

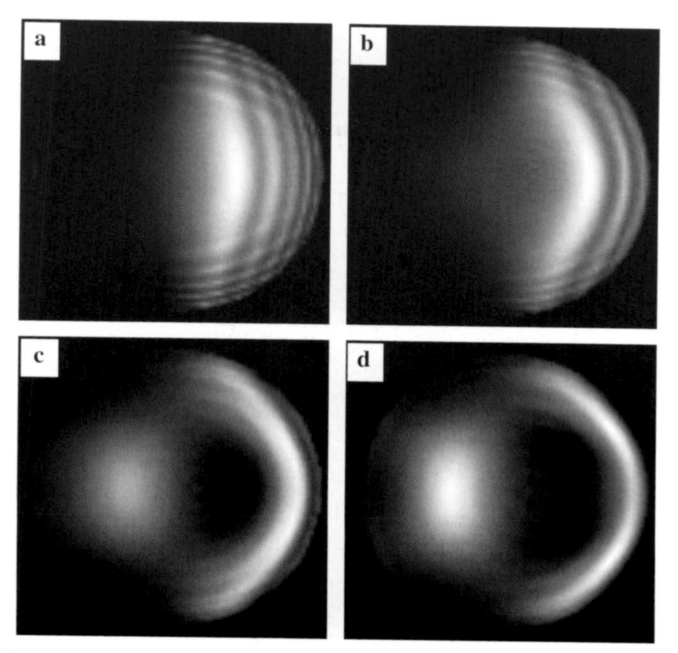

Figure 32.9 Computed plots of intensity distribution at the observation plane of Figure 32.8 corresponding to the knife-edge test carried out with a laser beam. The lens under test ($NA = 0.5, f = 6000\lambda_0$) and the pupil relay lens ($NA = 0.5$) are assumed to be fixed in their respective positions, while the knife-edge moves along the optical axis. (The tip of the knife remains on the axis at all times.) The lens under test is assumed to have three waves of primary spherical aberration. In frames (a) to (d) the distance of the knife-edge from paraxial focus $\Delta z = -15\lambda_0$, 0, $+15\lambda_0$, and $+20\lambda_0$, respectively. (Positive distances are in the direction of the marginal focus.)

ing in this device has two slightly different frequencies, which give rise to two +first-order beams as well as two −first-order beams; the beams in each pair are slightly shifted relative to each other. Moreover, the (average) pitch of the grating is such that there is no overlap between the zeroth, +first and −first orders. Consequently, interference occurs between the two +first-order beams (and, likewise, between the two −first-order beams). One can thus obtain an arbitrarily small lateral shear of the wavefront under test and use the results to achieve accurate quantitative measurements.

A two-dimensional version of the double-frequency grating has also been employed to generate lateral wavefront shear simultaneously along the X- and Y- axes. (Remember that beam propagation is along Z and, therefore, X and Y are orthogonal axes in the plane of the grating.) In the absence of a two-dimensional grating, one must rotate a one-dimensional grating by 90° to obtain wavefront shear first along the X- and then along the Y-axis.

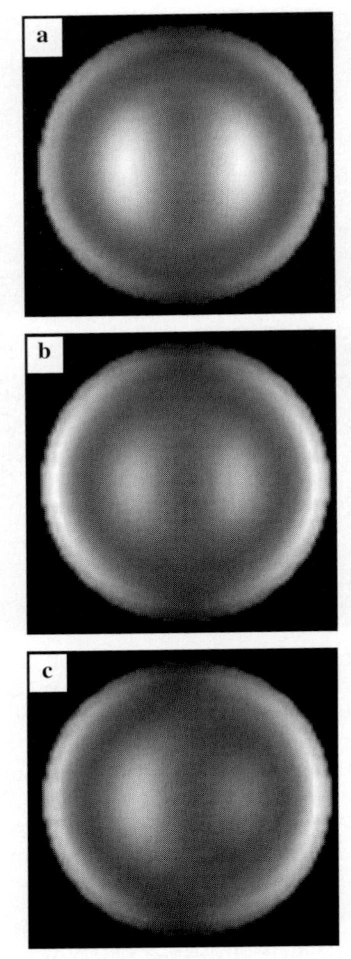

Figure 32.10 Computed plots of intensity distribution at the observation plane of Figure 32.8 corresponding to the wire test with an extended, quasi-monochromatic light source. The lens under test ($NA = 0.5, f = 6000\lambda_0$) has three waves of primary spherical aberration. The assumed wire diameter is $15\lambda_0$, which is comparable to the size of the image of the extended light source, as measured in the vicinity of focus. In (a) the wire is centered on axis and is $25\lambda_0$ away from paraxial focus (in the direction of the marginal focus). In (b) the wire is again centered on the axis, but is $20\lambda_0$ away from paraxial focus. In (c) the wire has been shifted $0.5\lambda_0$ off-axis while its distance from paraxial focus remains at $20\lambda_0$.

References for chapter 32

1 V. Ronchi, Le Frange di Combinazioni Nello Studio delle Superficie e dei Sistemi Ottici, *Riv. Ottica Mecc. Precis.* **2**, 9 (1923).
2 V. Ronchi, Due Nuovi Metodi per lo Studio delle Superficie e dei Sistemi Ottici, *Ann. Sc. Norm. Super. Pisa* **15** (1923).
3 D. Malacara, ed., *Optical Shop Testing*, second edition, Wiley, New York, 1992.
4 G. Toraldo di Francia, Geometrical and interferential aspects of the Ronchi test, in *Optical Image Evaluation*, National Bureau of Standards Circular 526, issued April 29, 1954.
5 J. C. Wyant, Double frequency grating lateral shear interferometer, *Appl. Opt.* **12**, 2057 (1973).

33

The Shack–Hartmann wavefront sensor

Roland Shack invented the device now known as the Shack–Hartmann wavefront sensor in the early 1970s.[1,2] This sensor, which in recent years has been commercialized, measures the phase distribution over the cross-section of a given beam of light without relying on interference and, therefore, does not require a reference beam.

The standard method of wavefront analysis is interferometry, where one brings together on an observation plane the beam under investigation (hereinafter the test beam) and a reference beam in order to form tell-tale fringes.[3] The trouble with interferometry is that it requires a reference beam, which is not always readily available. Moreover, the coherence length of the light used in these measurements must be long compared with the path-length difference between the reference and test beams. Thus, when the available light source happens to be broad-band, it becomes difficult (though by no means impossible) to produce high-contrast fringes. The Shack–Hartmann instrument solves these problems by eliminating altogether the need for the reference beam.

Wavefront analysis by interferometry

Before embarking on a discussion of the Shack–Hartmann wavefront sensor, it will be instructive to describe the operation of a conventional interferometer. Consider, for instance, the system of Figure 33.1, where a spherical mirror is under investigation. While grinding and polishing the glass blank, the optician frequently performs this type of test to determine departures of the surface from the desired figure. A point source reflected from a 50/50 beam-splitter is used to illuminate the test mirror. Before arriving at the mirror, however, the beam is partially reflected from the spherical surface of a plano-convex lens attached to the front facet of the beam-splitter cube (i.e., the spherical cap). The center of curvature of this spherical cap is at C, which is also the virtual image of the point source in the beam-splitter's half-silvered mirror. The light reflected from the spherical cap (and focused at C) forms the reference beam. (Incidentally, this interferometer was also invented by Roland Shack in the 1970s, and is now known as the Shack cube.[4])

Note in Figure 33.1 that the pinhole is placed directly on the face of the beam-splitter to eliminate possible aberrations of the beam upon entering and exiting the cube. The reflectivity of the spherical cap is about 4%, which is similar to that of the uncoated test mirror. The equal-strength test and reference beams thus produce a high-contrast fringe pattern. Figure 33.2(a) shows a typical phase distribution over the cross-section of a test beam reflected from a mirror having several waves of aberration. The computed interference pattern between this and an equal-strength reference beam is shown in Figure 33.2(b). Needless to say, the fringe contrast is excellent and the observed fringes may be related directly to the wavefront aberrations. In general, the coherence length of the light source must be long enough to

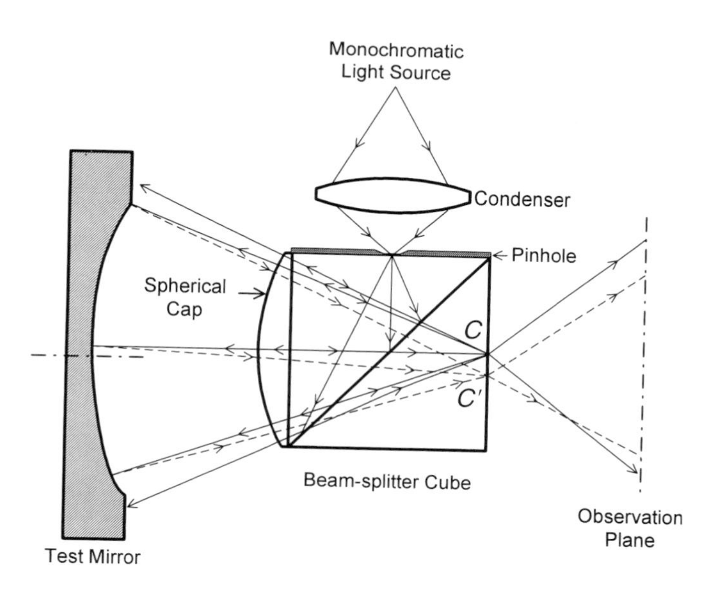

Figure 33.1 The Shack cube is used here to measure the surface quality of a spherical mirror. The cube is a 50/50 beam-splitter capped by an index-matched plano-convex lens. The light from the point source is partially reflected from this spherical cap, producing a reference beam that comes to focus at C. The beam that passes through the cap illuminates the test mirror, then returns and crosses the cube and is focused at C'. The interference pattern between the test and reference beams is viewed at the observation plane. The cube's axis is slightly displaced from the axis of the mirror in order to separate C from C', which is needed for producing straight-line fringes.

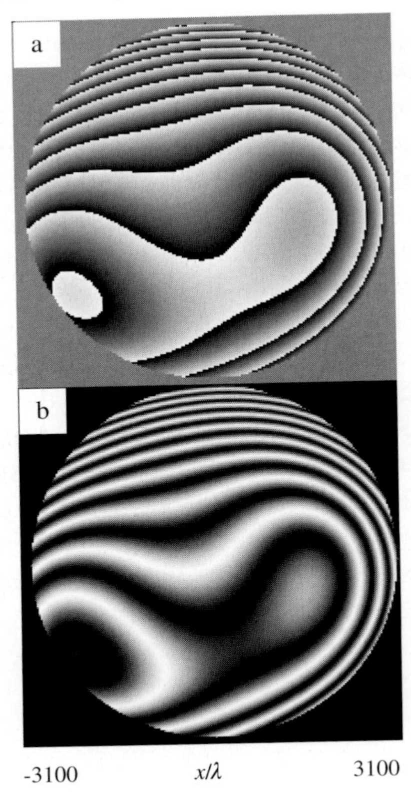

Figure 33.2 (a) A typical phase distribution in the cross-section of a monochromatic test beam (wavelength λ). The gray-scale covers the range from -180° (black) to $+180^\circ$ (white). (b) The interference pattern obtained by adding the test beam in (a) to a reference plane wave of equal magnitude.

ensure that, at the observation plane, the test and reference beams remain mutually coherent. For testing small mirrors having a short focal length (say, less than 10 cm) a single radiation line of an arc lamp may suffice, but for larger mirrors a long-coherence-length laser is usually necessary.

In practice, the center of curvature of the test mirror is slightly displaced from C, as shown in Figure 33.1, so that the rays bouncing off the mirror and arriving at the exit facet of the cube would converge not to C, but to a nearby point C'. This small lateral displacement of the test beam relative to the reference beam produces straight-line fringes in the interferogram at the observation plane. Such fringes are very sensitive to small aberrations of the mirror, and their deviation from linearity can be related easily to minute surface errors. If the errors are large, however, there is no need for straight-line fringes, and the center of the test mirror can coincide with C.

The combination of the cube beam-splitter and the spherical cap may be considered a thick lens. This lens projects a real image of the test mirror in the space behind the cube. The best place to observe the fringes, therefore, is at the location of this image, where the fringes are localized on the mirror, and the observer can readily identify areas that need further grinding and polishing. Another advantage is that scratches and dust particles on the mirror come to focus at its image, thus eliminating spurious fringes of the scattered light that downgrade the quality of the interferograms obtained at other locations in the image space. (The spherical cap, of course, must be kept clean at all times to prevent dust particles that have collected there from producing their own spurious fringes.)

The test mirror depicted in Figure 33.1 does not have to be spherical, but may be a mild paraboloid or hyperboloid whose center of curvature is, as before, placed at or near the point *C*. The departure of the mirror's figure from sphericity imparts a certain amount of spherical aberration to the test beam, which may be calculated in advance. The optician then looks for aberrations above and beyond this expected amount of spherical aberration in order to determine the necessary corrections.

Large telescope mirrors may also be tested with a Shack cube, but they require the use of an additional lens system known as a null-corrector.[3] A telescope's primary mirror is generally a large paraboloid or hyperboloid designed for operation at "infinite conjugate", that is, it brings the collimated beam of a distant star to focus within the mirror's focal plane. Testing such a large mirror with a collimated beam is impractical, however, and its actual departure from sphericity is too severe to be simply subtracted from the interferograms obtained in the system of Figure 33.1. In such situations, a null-corrector is designed to cancel the spherical aberrations of a test beam originating from a point source located at the mirror's center of curvature. When a properly calibrated null-corrector is inserted between the Shack cube and the test mirror, the observed interferogram registers only the departure of the mirror from its desired figure.

The Shack–Hartmann wavefront sensor

Figure 33.3 shows a schematic of the Shack–Hartmann wavefront sensor. This device is in many ways superior to conventional interferometers: since it does not require a reference beam, it is simpler to align and to operate and, because it does not rely on interference, it may be used with a white light source. At the heart of the instrument is a lenslet array and a charge-coupled device (CCD) camera. The lenslets are identical and have a fairly

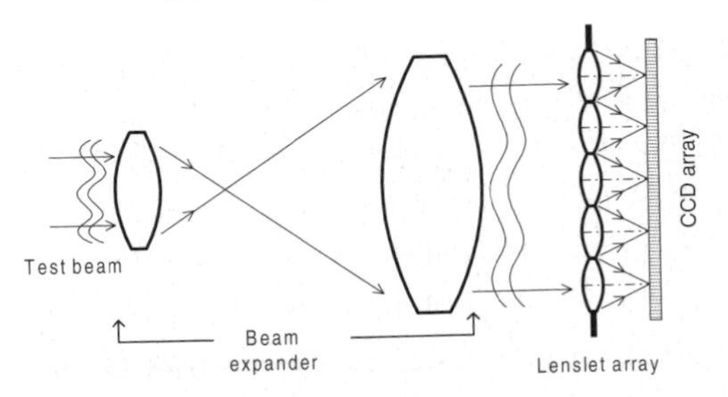

Figure 33.3 The basic elements of a Shack–Hartmann wavefront sensor. Upon entering the system, the (aberrated) test beam is processed to yield a collimated beam with a diameter that matches the dimensions of the lenslet array. Each lenslet captures a fraction of the beam and brings it to focus within its focal plane, where a CCD camera monitors the intensity distribution. The CCD has many more pixels than there are lenslets, so the location of each focused spot is determined simply by identifying the illuminated pixel within the relevant sub-array of pixels. The complete wavefront may be reconstructed from knowledge of the positions of individual focused spots.

large f-number, and the CCD detector, whose number of pixels is much greater than the number of lenslets, is placed at the focal plane of the lenslet array.

Upon entering the system, the (aberrated) test beam is collimated and expanded or reduced, as necessary, to match the dimensions of the lenslet array. The array typically consists of $n \times n$ identical lenslets, where n is around 100. Each lenslet thus acts on a small patch of the wavefront and brings it to focus within its focal plane. Located at this focal plane is the CCD array with $m \times m$ pixels, where m is typically around 1000. Consequently, each focused spot is assigned an exclusive sub-array of the CCD containing $(m/n) \times (m/n)$ pixels. If the patch of the wavefront captured by a given lenslet happens to have uniform phase, it forms a bright spot at the focal point of the lenslet, that is, at the center of the corresponding CCD sub-array. However, if the phase distribution at the lenslet's pupil happens to be non-uniform, the focused spot appears at a different location in the focal plane (but still within the associated CCD sub-array).

Figure 33.4 shows the computed distribution of intensity at the focal plane of a 6×6 lenslet array illuminated by the test beam of Figure 33.2(a). It may be verified that the center of each focused spot is shifted away from its respective frame's center by an amount proportional to the slope of the corresponding segment of the incident wavefront.

In practice, individual lenslets are very small compared to the test beam's diameter. Consequently, the incident phase distribution at the pupil of any

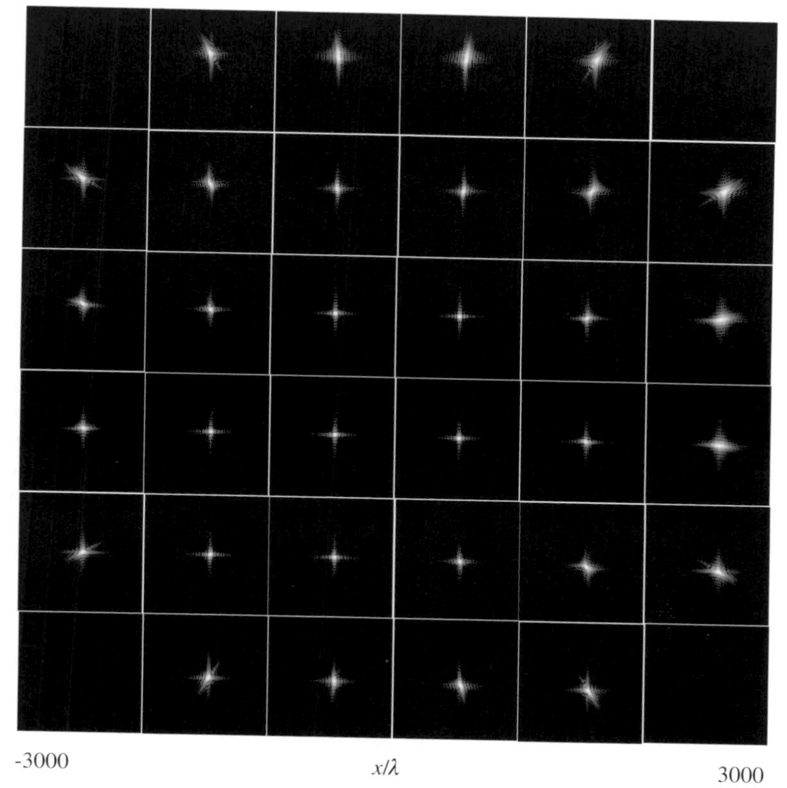

Figure 33.4 Intensity distribution at the focal plane of a 6×6 lenslet array, when the incident beam is assumed to have the phase distribution of Figure 33.2(a). Each square lenslet is $1000\lambda \times 1000\lambda$ in size, and has focal length $f = 25\,000\lambda$. The logarithmic plot of intensity shown here reveals the fine detail of the distribution at the CCD array. In practice, the fine detail is rather faint and only the center of each spot is detected by the CCD.

given lenslet may be approximated by a linear function of the pupil coordinates. It is thus clear that the aberrations of each focused spot are negligible; all one needs to know is the shift of the spot from the focal point of its associated lenslet, which is readily obtained by examining the CCD's output. So long as the focused spots remain within their allotted sub-array of detectors, the system can compute the local slope of the wavefront at the entrance pupil of each and every lenslet. The local slopes are then patched together to reconstruct the complete phase distribution over the cross-section of the test beam.

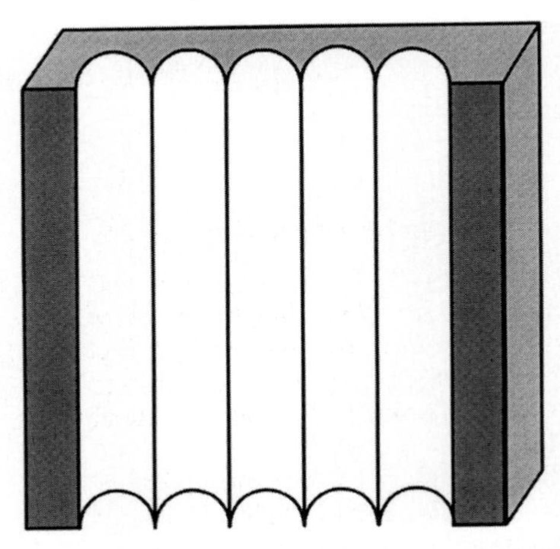

Figure 33.5 A piece of flat glass on which identical grooves were carved served as a mold for the early lenslet arrays. Two such pieces were prepared and placed face-to-face across a plastic sheet at right angles to each other. The assembly was then heated in an oven to transfer the pattern of the mold to the plastic sheet. The 1 mm-wide grooves had a depth of only a few micrometers.

Historical notes

The predecessor to the Shack–Hartmann sensor was Hartmann's screen test, which used an array of holes in place of the lenslets.[3,5–7] Shack realized the advantages of using a lenslet array and set out to fabricate one, since no such array with the characteristics he desired was available at the time. He made a mold by using a cutting tool to carve parallel grooves in a piece of flat glass, as shown in Figure 33.5. Two such pieces of grooved glass, oriented at right angles to each other, were clamped to an acrylic sheet and heated in an oven to mold convex ribs on each side of the acrylic sheet, thus forming an array of crossed cylindrical lenses. The first such array had 50×50 lenslets, each with an area of 1×1 mm^2 and a focal length of 150 mm.

Before the advent of CCD detectors in the 1980s, wavefront analysis was done by examining a photographic plate exposed to the array of focused spots. The plate was also exposed (simultaneously and through the same array of lenslets) to a parallel, aberration-free reference beam. The spots formed by this reference beam marked the center of each frame, thus providing reference points for measuring the displacement of the spots formed by the test beam. The tedious task of exposing and developing the photographic plate, followed by painstakingly measuring the positions of individual spots, was rewarding nonetheless; it allowed astronomers to measure the aberrations of their telescopes in the field using unfiltered star light. Even atmospheric turbulence did not pose a serious problem for this method, since its effects were simply averaged over during the relatively long exposure time of the photographic plate.

References for chapter 33

1 R. V. Shack and B. C. Platt, Production and use of a lenticular Hartmann screen (abstract only), *J. Opt. Soc. Am.* **61**, 656 (1971).
2 R. Riekher, *Fernrohre und ihre Meister*, Verlag Technik, Berlin, 1990.
3 D. Malacara, *Optical Shop Testing*, second edition, Wiley, New York, 1992.
4 R. V. Shack and G. W. Hopkins, The Shack interferometer, *SPIE* **126**, *Clever Optics*, 139–142 (1977).
5 J. Hartmann, Bemerkungen uber den Bau und die Justirung von Spektrographen, *Z. Instrum.* **20**, 47 (1900).
6 J. Hartmann, Objektivuntersuchungen, *Z. Instrum.* **24**, 33 (1904).
7 R. Kingslake, The absolute Hartmann test, *Trans. Opt. Soc.* **29**, 133 (1927–1928).

34
Ellipsometry

The goal of ellipsometry is to determine the optical and structural constants of thin films and flat surfaces from the measurements of the ellipse of polarization in reflected or transmitted light.[1–5] In the absence of birefringence and optical activity a flat surface, a single-layer film, or a thin-film stack may be characterized by the complex reflection coefficients $r_p = |r_p| \exp(i\phi_{rp})$ and $r_s = |r_s| \exp(i\phi_{rs})$ for p- and s-polarized incident beams, as well as by the corresponding transmission coefficients $t_p = |t_p| \exp(i\phi_{tp})$ and $t_s = |t_s| \exp(i\phi_{ts})$.[6]

Strictly speaking, an ellipsometer is a device that measures the complex ratios r_p/r_s and/or t_p/t_s. The amplitude ratios are usually deduced from the angles ψ_r and ψ_t, which are defined by $\tan \psi_r = |r_p|/|r_s|$ and $\tan \psi_t = |t_p|/|t_s|$. In practice, measuring the individual reflectivities $R_p = |r_p|^2$, $R_s = |r_s|^2$ or transmissivities $T_p = |t_p|^2$, $T_s = |t_s|^2$ does not require much additional effort. Measuring the individual phases, of course, is difficult, but the relative phase angles $\phi_{rp} - \phi_{rs}$ and $\phi_{tp} - \phi_{ts}$ can be readily obtained by ellipsometric methods. The values of R_p, R_s, $\phi_{rp} - \phi_{rs}$, ψ_r, T_p, T_s, $\phi_{tp} - \phi_{ts}$ and ψ_t may be measured as functions of the angle of incidence, θ, or as functions of the wavelength of the light, λ, or both.

The results of ellipsometric measurements are fed to a computer program that searches the space of unknown parameters to find agreement between the measured data points and theoretical calculations.[5] The unknown parameters of the sample usually include thickness, refractive index, and absorption coefficient of one or more layers. In general, the larger is the collected data set, the more accurate will be the estimates of the unknown parameters or the greater will be the number of unknowns that can be estimated. The relationship between the measurables and the unknowns is usually nonlinear, and there is no a priori guarantee that the various measurements on a given sample are independent of each other, nor that a given set of measurements is sufficient for determining the unknowns. Powerful numerical algorithms exist that search the space of unknown parameters and find estimates that closely reproduce the measured data.

The nulling ellipsometer

Figure 34.1 is the diagram of a conventional nulling ellipsometer.[4,5] The quasi-monochromatic light of wavelength λ enters a rotatable polarizer, whose transmission axis may be oriented at an arbitrary angle ρ_p relative to the X-axis. The polarizer's output is thus a collimated, linearly polarized beam of light with an adjustable E-field orientation. This beam goes through a quarter-wave plate (QWP) whose fast and slow axes are fixed at $\pm 45°$ to the X-axis. (The QWP imparts a relative 90° phase shift to the E-field components along its axes.) The beam emerging from the QWP has equal ampli-

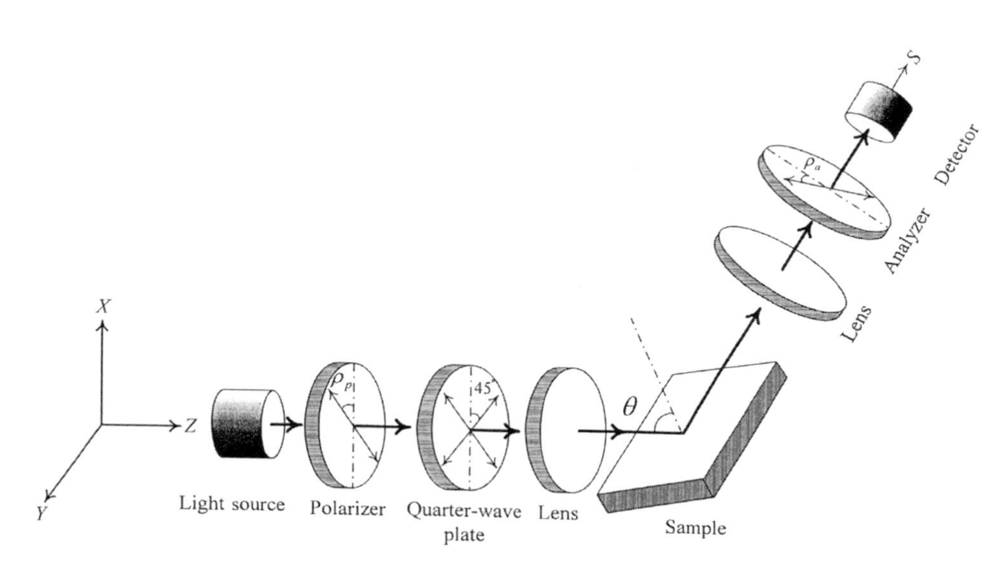

Figure 34.1 In a nulling ellipsometer the collimated beam of light emerging from the source (wavelength λ) is linearly polarized along the direction ρ_p by a rotatable polarizer. The quarter-wave plate's axes are typically at $\pm 45°$ to the XZ-plane of incidence. Thus the beam incident on the sample has equal amounts of p- and s-polarization, the relative phase between these two components depending on ρ_p. Reflection from the sample induces a phase shift $\phi_{rp} - \phi_{rs}$ between the p- and s-components, which may be cancelled out by adjusting the polarizer's orientation. Subsequently, the analyzer in the detection arm is rotated to extinguish the light transmitted to the detector. In the null condition, the value of ρ_p yields the sample's phase shift $\phi_{rp} - \phi_{rs}$ while the analyzer angle ρ_a yields the ellipsometric parameter ψ_r, which is related to the amplitude ratio $|r_p|/|r_s|$ of the reflection coefficients.

tudes along X and Y, that is, $|E_x| = |E_y|$. The phase difference between these E-field components is adjustable in accordance with the following relation: $\phi_x - \phi_y = 2(\rho_p - 45°)$.

Reflection from the sample imparts a phase difference $\phi_{rp} - \phi_{rs}$ to the p- and s- components of the beam, which may be cancelled out by properly selecting the polarizer angle ρ_p. At this point the reflected beam is linearly polarized, its E-field components along X and Y being proportional to $|r_p|$ and $|r_s|$, respectively. In the reflected path the analyzer, whose transmission axis is also adjustable, is rotated through an angle $\rho_a = \tan^{-1}(|r_p|/|r_s|) = \psi_r$ to block the light that would otherwise reach the detector. Thus by measuring the values of ρ_a and ρ_p that null the detector's signal, one obtains the amplitude ratio $|r_p|/|r_s|$ and the relative phase $\phi_{rp} - \phi_{rs}$ of the sample's reflection coefficients.

Measuring the sample reflectivities R_p, R_s using a nulling ellipsometer is straightforward; all one needs to do is monitor the detector signal S at $\rho_a = 0°$ and $90°$. Calibration requires removing the sample and aligning the arms of the ellipsometer with each other (i.e., $\theta = 90°$), in which case the light from the source goes through the entire system and yields a detector signal corresponding to a 100% sample reflectivity. Optical power fluctuations could be countered by splitting off a small fraction of the beam at the source and monitoring its variations with an auxiliary detector. The signal from the auxiliary detector is subsequently used to normalize the reflectivity signals.

Needless to say, the same types of measurement as discussed above, when performed on the transmitted beam, yield the values of T_p, T_s, $\phi_{tp} - \phi_{ts}$ and ψ_t.

Thin film on transparent substrate

Figure 34.2 shows a sample consisting of a thin absorbing layer on a glass substrate. To allow the transmitted beam to exit the substrate without a change in its state of polarization, and also to eliminate spurious reflections, an antireflection-coated hemispherical substrate is assumed. The film, which is 25 nm thick, has a complex index of refraction $n + ik = 4.5 + 1.75i$, and the substrate's refractive index is $n_0 = 1.5$. Computed values of the sample's reflection and transmission characteristics at $\lambda = 633$ nm and $\theta = 60°$ are: $R_p = 29.63\%$, $R_s = 74.83\%$, $\phi_{rp} - \phi_{rs} = 3.95°$, $\psi_r = 32.18°$, $T_p = 24.13\%$, $T_s = 6.96\%$, $\phi_{tp} - \phi_{ts} = 1.50°$, $\psi_t = 61.76°$.

We now examine the sensitivity of ellipsometric measurements to variations in the sample parameters. For example, if the refractive index n of the film is varied in the range 4.0 to 5.0, the various characteristics of the sample

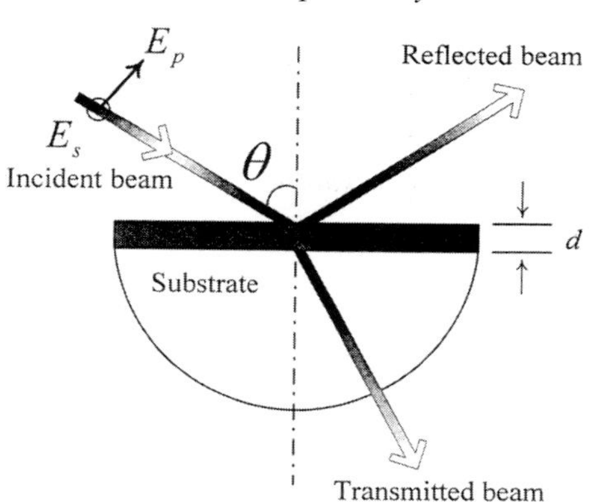

Figure 34.2 A 25 nm-thick film of complex refractive index $n + ik = 4.5 + 1.75i$ is deposited on a hemispherical glass substrate ($n_0 = 1.5$). The probe beam has $\lambda = 633$ nm and is incident at $\theta = 60°$. To avoid complications arising from reflections or losses at the substrate bottom, the hemispherical surface is antireflection coated.

vary as in Figure 34.3. (The variations shown here are relative to the nominal sample characteristics evaluated at $n = 4.5$.) It is seen that R_p, R_s, ψ_r, ψ_t are more sensitive to changes of n than T_p, T_s, $\phi_{rp} - \phi_{rs}$ and $\phi_{tp} - \phi_{ts}$. Similarly, Figure 34.4 shows variations of the sample characteristics with changes in k. Here T_p, $\phi_{rp} - \phi_{rs}$ and $\phi_{tp} - \phi_{ts}$ are seen to be more sensitive to k than R_p, R_s, T_s, ψ_r, ψ_t. Figure 34.5 shows variations of the sample characteristics with a

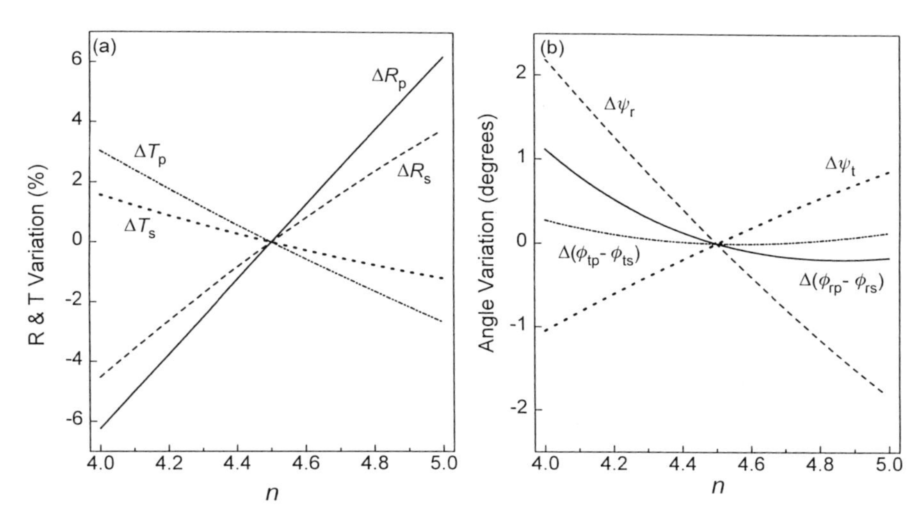

Figure 34.3 Variations of the reflection and transmission characteristics of the sample of Figure 34.2 at $\lambda = 633$ nm, $\theta = 60°$, when the film's refractive index n is varied from 4.0 to 5.0. The changes are relative to the nominal values obtained with $n = 4.5$.

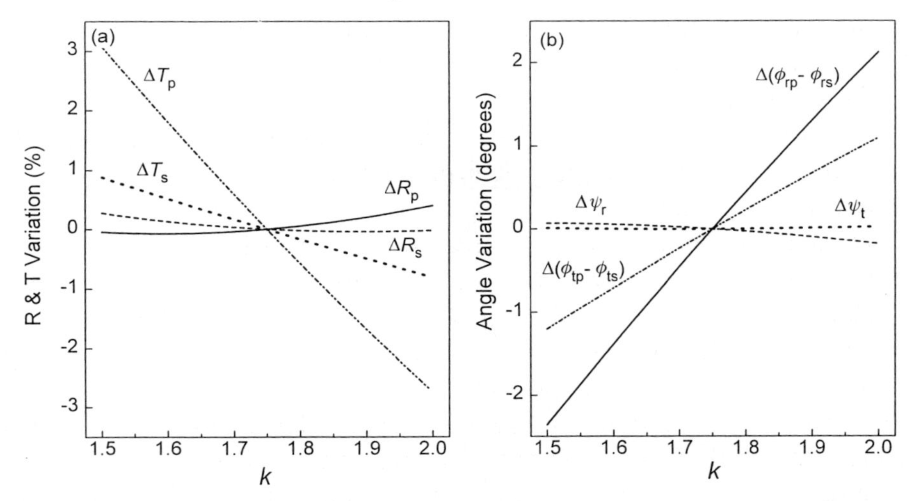

Figure 34.4 Variations of the reflection and transmission characteristics of the sample of Figure 34.2 at $\lambda = 633$ nm, $\theta = 60°$, when the film's absorption coefficient k is varied from 1.5 to 2. The changes are relative to the nominal values obtained with $k = 1.75$.

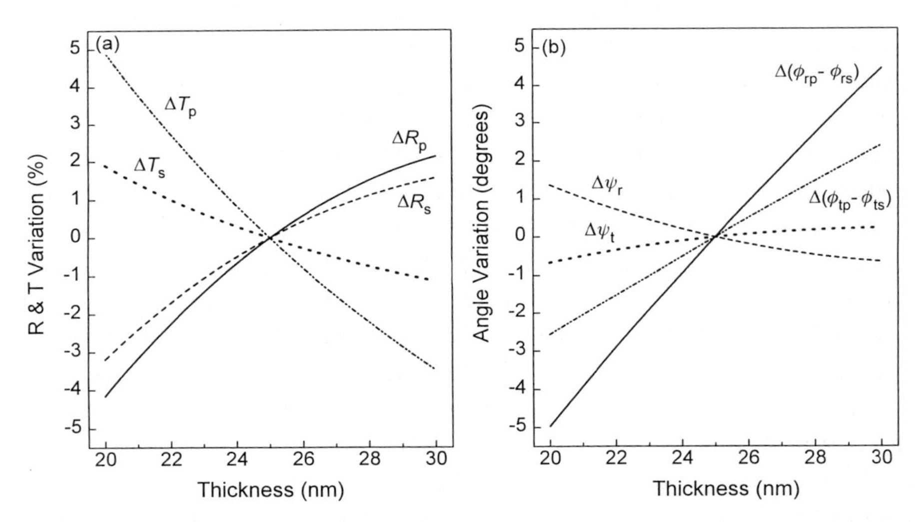

Figure 34.5 Variations of the reflection and transmission characteristics of the sample of Figure 34.2 at $\lambda = 633$ nm, $\theta = 60°$, when the film thickness d is varied from 20 nm to 30 nm. The changes are relative to the nominal values obtained with $d = 25$ m.

changing film thickness d in the range from 20 nm to 30 nm. In this case ψ_t and, to some extent, ψ_r are insensitive to d, but the remaining characteristics are quite sensitive.

When all the components of the system are assumed to be perfect, the ellipsometer is sensitive enough to determine accurately the unknown sample parameters. In practice, however, no measurement system is perfect: the polarizer and the analyzer have a finite extinction ratio, allowing a small fraction of the undesirable E-field component to pass through; the quarter-wave plate's retardation deviates from 90°, and the beam that illuminates the sample is not an ideal plane wave but has a finite diameter. Moreover, when the beam is focused on the sample to provide a reasonable spatial resolution, the focused cone of light contains a range of incidence angles, resulting in measured values that are averages over these angles. One consequence of such system imperfections is that, in the "null condition," a minimum amount of light would still reach the detector. Another consequence is the limited accuracy with which the various reflection and transmission characteristics of the sample are measured.

Performance of the nulling ellipsometer

For the system depicted in Figure 34.1 we show in Figure 34.6 computed plots of the detector signal S versus the angle ρ_a of the analyzer for several values of the polarizer angle ρ_p. The assumed focusing and collimating lenses are identical, having $NA = 0.025$, which corresponds to a 3° focused cone at the sample. In Figure 34.6(a) the assumed system is perfect, while in Figure 34.6(b) errors are incorporated into the various components, namely, the assumed polarizer and analyzer have a 1:100 extinction ratio, the angle of incidence on the sample is $\theta = 61°$, and the QWP's retardation is 87° while its axes are 1° away from the ideal 45° orientation.

The null in Figure 34.6(a) is achieved with $\rho_p = 47°$ and $\rho_a = 32.2°$, yielding $\phi_{rp} - \phi_{rs} = 4°$ and $\psi_r = 32.2°$, as expected. Also the detector signals at $\rho_a = 0°$ and 90° are 0.296 and 0.748, which correspond to the correct values of R_p and R_s. In practice, even in this ideal case with perfect components the exact location of the null may not be easy to determine. This produces a certain degree of inaccuracy, depending on the available signal-to-noise ratio at the detector. In the case of Figure 34.6(b), where the assumed components have substantial errors, the minimum signal occurs at $\rho_p = 54°$ and $\rho_a = 30°$, yielding $\phi_{rp} - \phi_{rs} = 18°$ and $\psi_r = 30°$. The reflectivities in this case (obtained at $\rho_p = 9°$, and $\rho_a = 0°$ and 90°) are $R_p = 0.308$, $R_s = 0.727$. If we consider

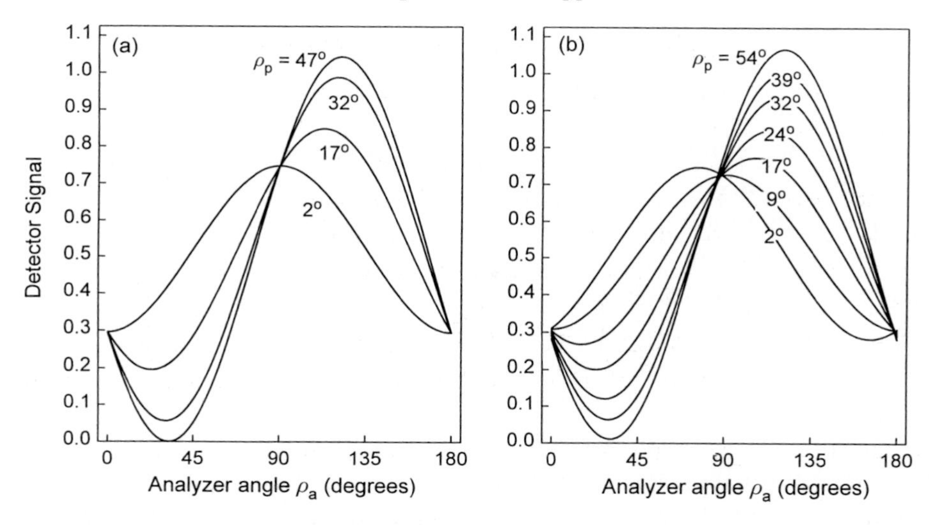

Figure 34.6 The detector signal S versus the orientation angle ρ_a of the analyzer in the nulling ellipsometer of Figure 34.1 with the sample of Figure 34.2. Different curves correspond to different values of the polarizer angle ρ_p. The total optical power of the unpolarized (or circularly polarized) beam emerging from the source is unity, the detector's conversion factor is 4, the incidence angle is $\theta = 60°$, and the focusing and collimating lenses have $NA = 0.025$. In (a) the assumed system is perfect. In (b) there are departures from ideal behavior, namely, the polarizer and analyzer have a 1:100 extinction ratio, the angle of incidence deviates by 1°, and the quarter-wave plate's retardation is 87° while its axes are 1° away from the ideal 45° orientation.

the sensitivity curves in Figures 34.3–34.5, such huge errors are clearly unacceptable.

A more realistic situation might correspond to small system errors; suppose, for instance, that the polarizer and the analyzer have extinction ratios of 1:1000, the angle of incidence on the sample has a 0.25° error ($\theta = 60.25°$), and the QWP's retardation is 90.5° while its axes are misaligned by only 0.25°. In this case the minimum signal occurs at $\rho_p = 49°$ and $\rho_a = 31.8°$, yielding $\phi_{rp} - \phi_{rs} = 8°$ and $\psi_r = 31.8°$. The reflectivities (obtained at $\rho_p = 4°$, and $\rho_a = 0°$ and 90°) are $R_p = 0.291$ and $R_s = 0.757$. It is thus clear that the nulling ellipsometer requires a high degree of accuracy in its components in order to achieve a reasonable level of confidence in its estimates of sample parameters.

Ellipsometry with a variable retarder

Figure 34.7 shows a different kind of ellipsometer, consisting of a fixed polarizer, a variable retarder (e.g., a liquid crystal cell or a photoelastic modulator), and a fixed differential detection module. None of these components needs to be rotated or otherwise adjusted during measurements. The variable retarder provides a range of polarization states at the sample. For instance, the incident beam is p-polarized when the retardation $\Delta\phi$ is $0°$, circularly polarized when $\Delta\phi = \pm 90°$, and s-polarized when $\Delta\phi = 180°$. The detection module consists of a Wollaston prism with transmission axes fixed at $\pm 45°$ to the plane of incidence, followed by a pair of identical photodetectors.

When the relative phase $\Delta\phi$ imparted by the retarder to the incident beam is continuously varied from $0°$ to $360°$, the sum signal $S_1 + S_2$ oscillates between a maximum and a minimum value; these correspond to R_p and

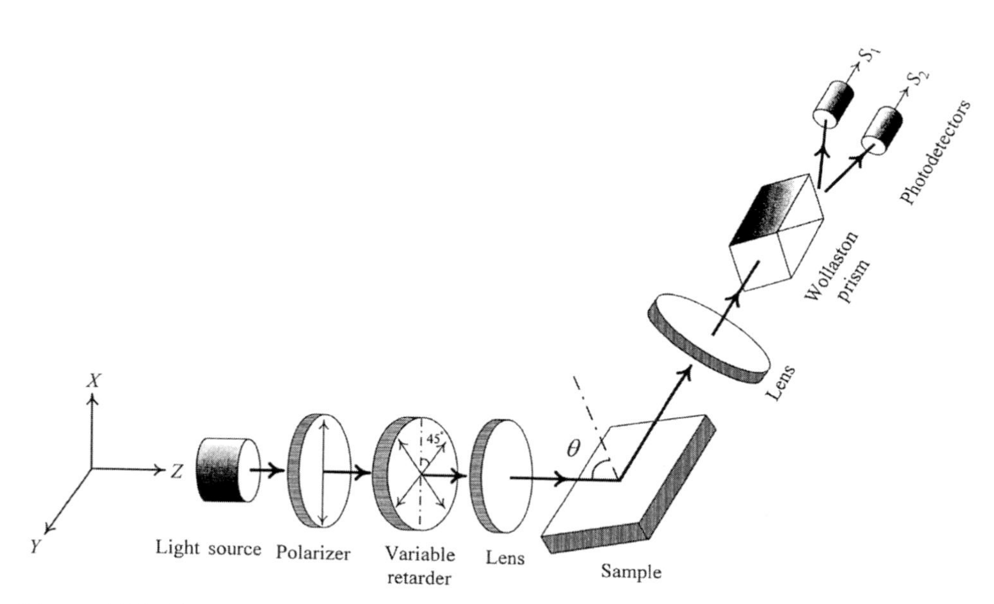

Figure 34.7 Diagram of an ellipsometer based on a variable retarder and a differential detection module. The beam emerging from the polarizer is collimated and linearly polarized along the X-axis. The variable retarder's axes are fixed at $\pm 45°$ to the XZ-plane of incidence, while its phase is varied continuously from $0°$ to $360°$. The light beam is focused on the sample through a low-NA lens, and the reflected beam is recollimated by an identical lens in the reflection path. The reflected beam is monitored by a differential detector consisting of a Wollaston prism (oriented at $45°$ to the plane of incidence) and two identical photodetectors. The sum of the detector signals $S_1 + S_2$ contains information about the sample reflectivities R_p and R_s, while their normalized difference $(S_1 - S_2)/(S_1 + S_2)$ yields the relative phase $\phi_{rp} - \phi_{rs}$.

R_s, although not necessarily in that order. At the same time, the normalized difference signal $(S_1 - S_2)/(S_1 + S_2)$ exhibits a peak-to-valley variation equal to $2\sin(\phi_{rp} - \phi_{rs})$. The system of Figure 34.7 does not provide an independent measure of the other ellipsometric parameter, ψ_r. However, since R_p and R_s are directly measurable, ψ_r is redundant.

In operating the system of Figure 34.7 it is not necessary to know the time-dependence of the retardation $\Delta\phi$, nor in fact does one need to know the specific value of $\Delta\phi$ at any point during the measurement. The maximum and minimum values of the sum signal and of the normalized difference signal contain all the necessary information. Unlike the nulling ellipsometer, this system does not require any adjustment of angles around a broad minimum; therefore, there is much less uncertainty about the measured data points.

For the ideal system depicted in Figure 34.7, Figure 34.8(a) shows computed plots of the sum signal and the normalized difference signal versus the retardation $\Delta\phi$. The maximum and minimum values of the sum signal are 0.748 and 0.296, corresponding to R_s and R_p. The normalized difference signal has a peak-to-valley variation of 0.1375, yielding $\phi_{rp} - \phi_{rs} = 3.94°$.

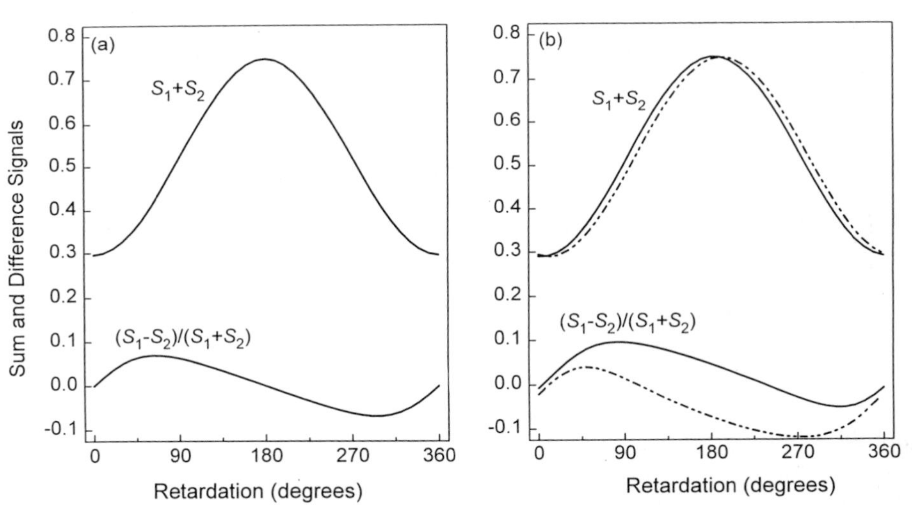

Figure 34.8 Computed plots of the sum and (normalized) difference signals in the system of Figure 34.7 for the sample shown in Figure 34.2. The horizontal axis depicts the relative phase imparted to the beam by the variable retarder. The beam emerging from the polarizer has unit optical power, the detectors' conversion factor is unity, the incidence angle is $\theta = 60°$, and the focusing and collimating lenses have $NA = 0.025$. (a) The assumed system is perfect. (b) Two instances (solid lines, broken lines) of imperfect system behavior.

In Figure 34.8(b) we have assumed some imperfection in the system components. Two cases are examined, one leading to the solid curves and the other to the broken curves. In the former case the polarizer's extinction ratio is 1:1000, the retarder axes are misaligned by 1°, the Wollaston prism has a 1:100 leak ratio between its two channels, and the angle of incidence θ is in error by 0.5°. From the computed sum and difference signals $R_p = 0.290$, $R_s = 0.750$, and $\phi_{rp} - \phi_{rs} = 4.23°$. In the case of the broken curves in Figure 34.8(b) the assumed imperfections are large. Here the polarizer's extinction ratio is 1:100, the retarder's orientation angle is 43°, the angle of incidence is $\theta = 60.5°$, and the Wollaston prism leaks 2% of the wrong polarization into each channel. From the computed sum and difference signals the values of $R_p = 0.290$, $R_s = 0.749$, and $\phi_{rp} - \phi_{rs} = 4.6°$ are obtained. Obviously, the system of Figure 34.7 is quite tolerant of imperfections and misalignments; therefore, it is suitable for accurate determination of the sample parameters.

References for chapter 34

1 A. Rothen, The ellipsometer, an apparatus to measure thicknesses of thin surface films, *Review of Scientific Instruments* **16**, 26–30 (1945).
2 A. B. Winterbottom, Optical methods of studying films on reflecting bases depending on polarization and interference phenomena, *Trans. Faraday Society* **42**, 487–495 (1946).
3 R. H. Muller, Definitions and conventions in ellipsometry, *Surface Science* **16**, 14–33 (1969).
4 R. M. A. Azzam and N. M. Bashara, *Ellipsometry and Polarized Light*, North-Holland, Amsterdam, 1977.
5 R. M. A. Azzam, Ellipsometry, chapter 27 in *Handbook of Optics*, Vol. 2, McGraw-Hill, New York, 1995.
6 O. S. Heavens, *Optical Properties of Thin Solid Films,* Butterworths, London, 1955.

35

Holography and holographic interferometry

Dennis Gabor (1900–1979). His life-long love of physics started at the age of 15. Fascinated by Abbe's theory of the microscope and by Lippmann's method of color photography, he and his brother built up a home laboratory and began experimenting with X-rays and radioactivity. Gabor entered the Technische Hochschule Berlin and acquired a diploma in 1924 and an electrical engineering doctorate in 1927. His thesis work involved the development of high-speed cathode ray oscillographs, in the course of which he built the first iron-shrouded magnetic electron lens. In 1927 he joined Siemens & Halske AG, where he invented a high-pressure quartz mercury lamp, since used in millions of street lamps. With the rise of Hitler in 1933, Gabor left for England and obtained employment with the British firm Thomson-Houston. At Thomson–Houston he developed a system of stereoscopic cinematography, and in his last year there carried out basic experiments in holography. In 1949 he joined the Imperial College of Science and Technology (London) and remained there as Professor of Applied Electron Physics until his retirement in 1967. (Photo: courtesy of AIP Emilio Segré Visual Archives, W. F. Meggers Collection.)

Holography dates from 1947, when the Hungarian-born British scientist Dennis Gabor (1900–1979) developed the theory of holography while working to improve electron microscopy.[1,2] Gabor coined the term "hologram" from the Greek words *holos*, meaning whole, and *gramma*, meaning message. The 1971 Nobel prize in physics was awarded to Gabor for his invention of holography.

Further progress in the field was prevented during the following decade because the light sources available at the time were not truly coherent. This barrier was overcome in 1960, with the invention of the laser. In 1962 Emmett Leith and Juris Upatnieks of the University of Michigan recognized, from their work in side-looking radar, that holography could be used as a three-dimensional visual medium. They improved upon Gabor's original idea by using a laser and an off-axis technique.[3] The result was the first laser transmission hologram of three-dimensional objects. The basic off-axis technique of Leith and Upatnieks is still the staple of holographic methodology. These transmission holograms produce images with clarity and realistic depth, but require laser light to view the holographic image.

The Russian physicist Uri Denisyuk combined holography with Lippmann's method of color photography. In 1962 Denisyuk's approach produced a white-light reflection hologram, which could be viewed in the light from an ordinary light bulb. In 1968 Stephen Benton, then at Polaroid corporation, invented white-light transmission holography.[4–6] This type of hologram can be viewed in ordinary white light and is commonly known as the rainbow hologram. These holograms, which are "printed" by direct stamping of the interference pattern onto plastic, can be mass produced rather inexpensively.[7]

Basic principles

A setup for recording a simple transmission hologram is shown in Figure 35.1. The coherent beam of the laser, after being expanded to cover the area of interest, is split into an object beam and a reference beam. The object beam passes through (or reflects from) the object before arriving at the photographic plate; the reference beam is directed toward the photographic plate at an oblique angle θ. At the XY-plane of the plate the complex-amplitude distribution of the object beam is $A_O(x, y)$. The reference beam's amplitude, $A_R(x, y)$, is proportional to $\exp[i(2\pi/\lambda)(xS_x + yS_y)]$, where S_x and S_y are the direction cosines of the beam. The two beams interfere at the plate, upon which their interference fringes are recorded. When the plate is properly

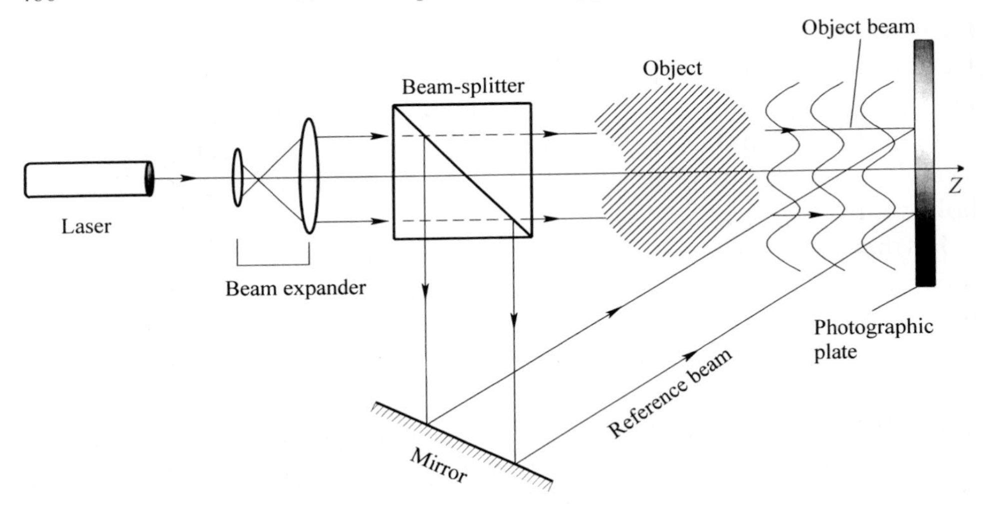

Figure 35.1 The basic optical system used for recording a simple hologram. The laser beam is expanded to accommodate the size of the object. The beam-splitter separates a fraction of the light to be used as a reference beam and sends it along a path that reaches the photographic plate at an oblique angle. The rest of the beam continues along the Z-axis, interacts with the object, and arrives at the photographic plate while carrying the phase/amplitude information about the object. The two beams interfere and the plate records the resulting fringes of the interference pattern. The film is subsequently developed into a positive (or negative) transparency and becomes a permanent record of the object wave.

processed and developed, its amplitude transmissivity $\tau(x, y)$ becomes proportional to the incident intensity pattern, that is,[8–10]

$$\tau(x, y) = I(x, y) = |A_O(x, y) + A_R(x, y)|^2. \tag{35.1}$$

To reconstruct the object wave, the developed plate is returned to its original position and illuminated with the reference beam, as shown in Figure 35.2. The transmitted beam's complex amplitude may thus be written

$$\begin{aligned} A(x, y) &= \tau(x, y) A_R(x, y) \\ &= \{|A_O(x, y)|^2 + |A_R(x, y)|^2\} A_R(x, y) \\ &\quad + |A_R(x, y)|^2 A_O(x, y) + A_R^2(x, y) A_O^*(x, y) \end{aligned} \qquad (35.2).$$

Note in the above equation that $|A_R(x, y)|$ is a constant, independent of x and y, and that $A_R^2(x, y)$ is a plane wave with direction cosines $2S_x$ and $2S_y$. (When θ is small, the propagation direction of this plane wave makes an angle 2θ with the Z-axis.) Thus in addition to the reference beam $A_R(x, y)$ – which is modulated by the squared modulus of the object wave – the

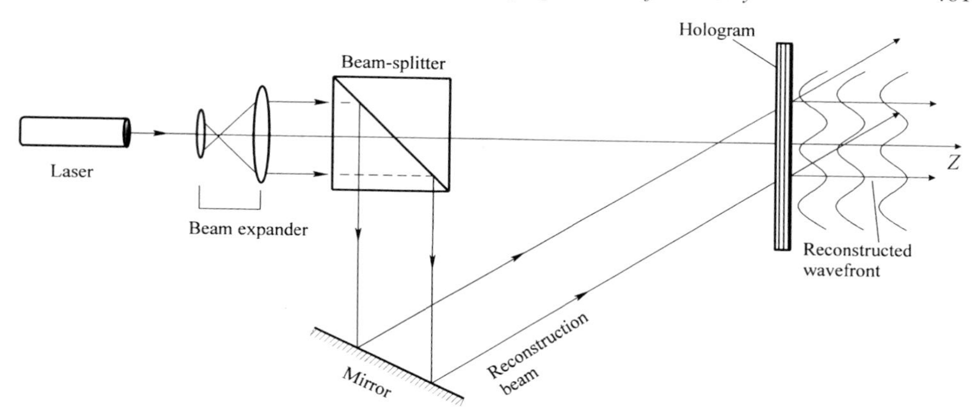

Figure 35.2 To reconstruct the recorded wavefront one places the hologram in front of the same reference beam as used for recording. Upon transmission through the hologram several reconstructed waves emerge. If the hologram is in the same position as it was during recording, the virtual image of the object will be carried by the component of the emergent beam traveling along the Z-axis. However, if the hologram is flipped then a real image of the object emerges along the Z-axis. (The flipping is such that the reconstruction beam becomes the conjugate of the original reference beam with respect to the hologram.)

wavefront emerging from the hologram contains the original object wave $A_O(x, y)$, as well as its complex conjugate $A_O^*(x, y)$. The reconstructed object wave travels in its original direction (i.e., along the Z-axis, in the case of Figure 35.2), but the conjugate wave rides on a plane wave whose deviation angle from the Z-axis is nearly twice that of the original reference beam.

Behind the hologram, the reconstructed object wave yields the virtual image of the recorded object; this image may be viewed through the lens of an eye or photographed through the lens of a camera. The conjugate wave yields a real image of the object, which can be visually inspected or photographed by placing a photographic plate directly in its path. The transmitted portion of the reconstruction beam itself does not carry any useful information and is generally ignored.

Hologram of a simple phase-amplitude object

As an example, consider the phase–amplitude object shown in Figure 35.3. The featureless areas of the face are transparent to the incident light, but the eyes, nose, and mouth alter both the amplitude and the phase of the beam. The eyes are partially transmissive depressions with a 50% transmittance and a maximum phase depth of 5π at the center. The nose and the mouth are also

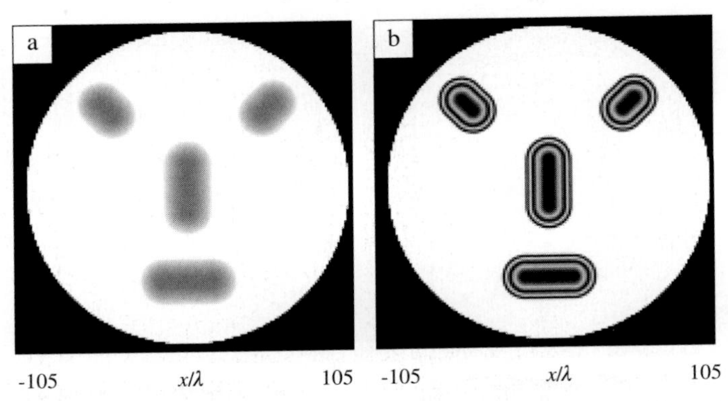

Figure 35.3 This face is a partially transmissive phase/amplitude object. The intensity pattern shown in (a) is obtained when the face is illuminated by a coherent, collimated, and uniform laser beam (i.e., a plane wave). The amplitude transmission coefficient of the facial features (eyes, nose, mouth) is 0.7. The interferogram in (b) is obtained when the transmitted beam is made to interfere with a plane wave. The features of the face modulate the phase of the transmitted beam in a continuous fashion by an amount that rises to 5π at the center of the eyes and falls to -5π at the center of the nose and the mouth.

50% transmissive, but they are raised above the surface of the face and their corresponding phase depth at the center is -5π. Figure 35.3(a) shows the pattern of transmitted intensity for a uniform incident beam. Figure 35.3(b), an interferogram between the beam transmitted through the face and a collinear plane wave, shows the fringes caused by the phase modulation imparted to the beam by the various features of the face.

When a plane wave (wavelength λ) is transmitted through the face at $z = 0$ and propagated to a photographic plate at $z = 3500\lambda$, one obtains the intensity and phase distributions shown in Figures 35.4(a), (b), respectively. Figure 35.4(c) is the interference pattern formed with a reference plane wave traveling at an oblique angle $\theta = 8°$. The photographic plate is exposed to this interference pattern and subsequently developed into a positive transparency, that is, one in which the amplitude transmissivity is proportional to the incident intensity distribution during exposure. This transparency is a coherent-light hologram of the face.

Note in Figure 35.4(c) that the chosen diameter of the reference beam is not large enough to cover the regions of the object wave far away from the Z-axis. This is simply due to the limited computer memory available for these calculations and is not a limitation in holography. Whereas in practice the reference beam is usually large enough to record all significant spatial frequencies of the object onto the hologram, in the present calculations the

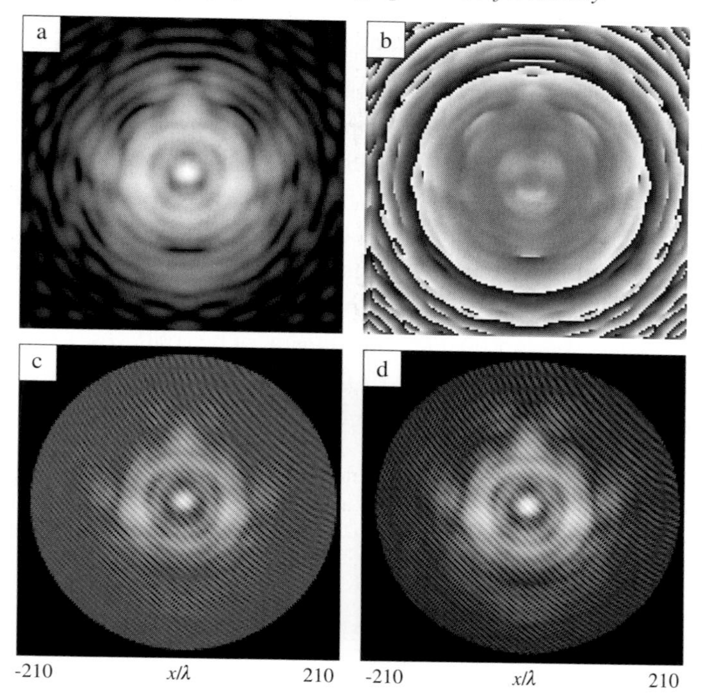

Figure 35.4 A plane wave traveling along the Z-axis and transmitted through the face at $z = 0$ arrives at the photographic plate at $z = 3500\lambda$. (a) Distribution of the logarithm of intensity of the object wave at the plate. (b) Object wave's phase distribution at the plate. (c) Interference pattern (logarithm of intensity) between the object wave and a reference plane wave traveling at $\theta = 8°$ relative to the Z-axis. (d) Distribution of the logarithm of intensity immediately after the hologram, when the exposed plate is developed into a positive transparency and placed in front of the reconstruction beam.

small diameter of the reference beam limits the range of admissible spatial frequencies, resulting in the loss of fine detail in the reconstructed images of the original object.

When the developed hologram is placed in the system of Figure 35.2 and illuminated with the reconstruction beam, the original object wave and its complex conjugate appear among the transmitted waves, in accordance with Eq. (35.2). Figure 35.4(d) shows the transmitted intensity pattern immediately behind the hologram. At this point the overlapping components of the emergent beam are all mixed together and, therefore, difficult to identify separately. Since these components are traveling in different directions, propagation over a short distance is all that is required to disentangle them from each other.

Holographic images of the recorded object

When the above hologram is placed in the same position as during recording and illuminated with the same reference beam (now called the reconstruction beam) one obtains, at $z = 3500\lambda$ behind the hologram, the reconstructed intensity and phase patterns of Figures 35.5(a), (b). The central region of this figure contains the reconstructed object wave, $A_O(x, y)$, carrying the virtual image of the face. The transmitted fraction of the reconstruction beam – modulated by the squared modulus of the object wave – appears to the right and above the central region. The real image of the face – produced by the conjugate wave $A_O^*(x, y)$ – is shifted further off-axis, and appears in the upper right corner of Figure 35.5(a).

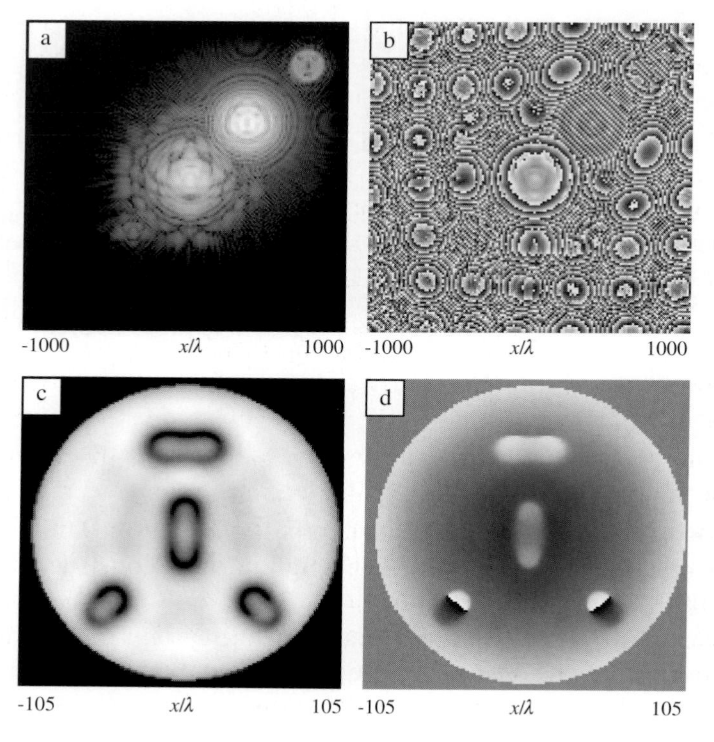

Figure 35.5 The reconstructed wavefront at $z = 3500\lambda$ behind the hologram. The incident beam is the same as the reference beam used in creating the hologram. (a), (b) Distributions of the logarithm of intensity and the phase over the entire reconstructed field. The central region of this field carries a virtual image of the face. (c), (d) Distributions of the logarithm of intensity and the phase in the image plane of a unit-magnification lens that captures the central portion of the field and creates a real image of the face from the reconstructed object wave.

Holographic reconstruction produces not only the amplitude of the original object but also its phase pattern, as is evident from Figure 35.5(b). Unlike regular photography, which maintains a record of the intensity profile but loses all trace of phase, the holographic process preserves both the amplitude and the phase information, and faithfully reproduces the entire object wave upon reconstruction. A comparison of the central regions of Figures 35.5(a), (b) with the original object wave of Figures 35.4(a), (b) might be worthwhile here, although one should note that the reconstructed wave in Figures 35.5(a), (b) is captured at an effective distance of 7000λ from the original object, whereas the patterns of Figures 35.4(a), (b) correspond to a propagation distance of only 3500λ.

To observe the virtual image, one should place an imaging lens in the central region of the field and produce a real image from the reconstructed object wave. (Alternatively, one could propagate the reconstructed object wave backwards in space by 7000λ to reproduce the object wave at its point of origination.) A one-to-one imaging lens ($NA = 0.04$, $f = 3500\lambda$) placed in the central region of Figures 35.5(a), (b) will create an inverted real image of the face at $z = 7000\lambda$ behind the lens. The resulting intensity and phase patterns are shown in Figures 35.5(c), (d). The loss of resolution due to the small size of the hologram is visible at the edges of the various facial features, from which the high-spatial-frequency content of the original face is obviously missing (compare with Figure 35.3(b)).

If the hologram is flipped during playback, the reconstruction beam, being a plane wave in this example, becomes the conjugate of the original reference beam, namely, $A_R^*(x, y)$. (Alternatively, the reference beam may be conjugated and brought in from the opposite side of the hologram.) Under such circumstances the transmitted wave along the original direction of the object wave (i.e., the Z-axis in the present example) becomes the conjugated object wave, $A_O^*(x, y)$, and the reconstructed object wave moves off-axis. This situation is depicted in Figure 35.6, where, after propagating 3500λ beyond the hologram, the various components of the transmitted beam have separated from each other. The intensity distribution in Figure 35.6(a) reveals at the center the real image of the face, slightly to the lower left the directly transmitted reconstruction beam, and close to the lower left corner the beam containing the virtual image. There is also a weaker image of the face on the right-hand side of the real image; this "second harmonic" of the face is created by the nonlinearity of the photographic process. Figures 35.6(c), (d) are close-ups of the intensity and phase patterns in the real image produced by the conjugated object wave.

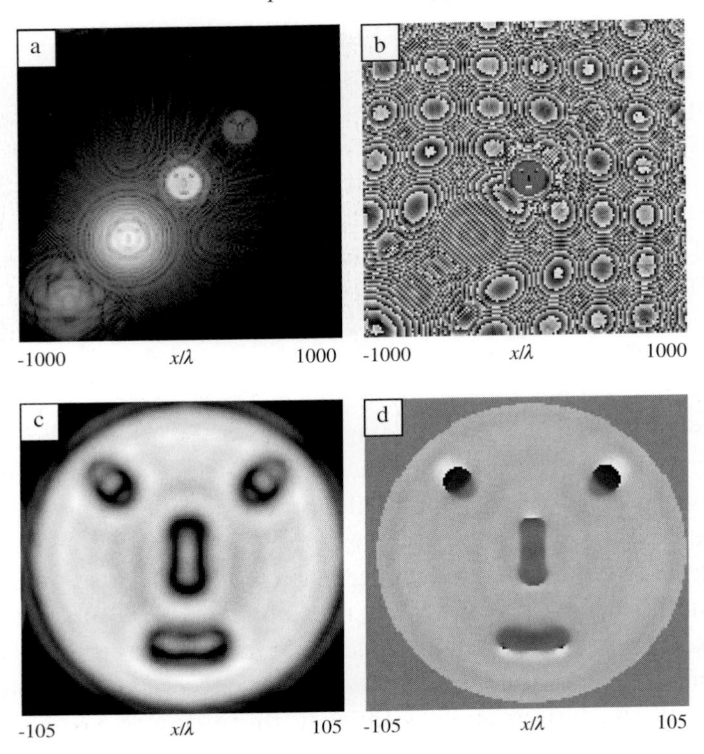

Figure 35.6 The reconstructed wavefront at $z = 3500\lambda$ behind the hologram. The incident beam is the conjugate of the reference beam used in creating the hologram. (a), (b) Distributions of the logarithm of intensity and the phase over the entire reconstructed field. (c), (d) Close-ups of the central region of the reconstructed field, showing the logarithm of intensity and the phase distribution of the real image of the face.

Holographic interferometry

Suppose the face shown in Figure 35.3 is somehow distorted at a later time or has undergone changes in its optical properties such that the beam transmitted through the face has acquired a certain degree of phase modulation. To render this phase modulation visible by converting it to intensity variations, it is necessary to interfere the beam transmitted through the face with a reference beam. If a collinear plane wave is chosen as reference, the resulting interferogram will resemble that in Figure 35.7(a). Here the deformation contours appear as black and white fringes superimposed on the face. One can also see in this figure the fringes caused by the phase structure of the facial features, namely, the eyes, the nose, and the mouth.

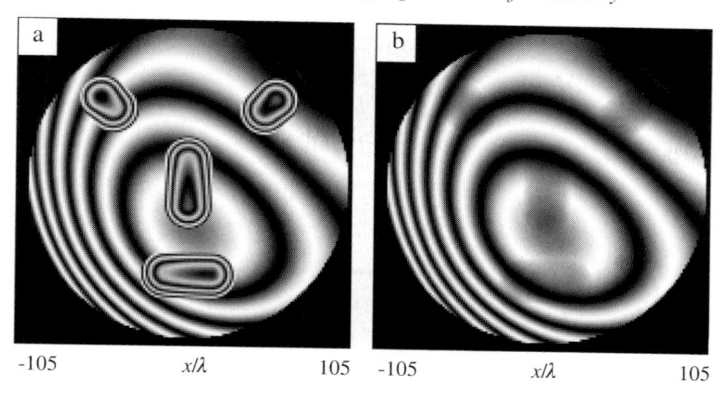

Figure 35.7 Two interferograms of the distorted face. In (a) the reference beam is a plane wave, whereas in (b) the distorted face is made to interfere with its own undistorted version.

An alternative "reference beam" is provided by the original, undistorted wave from the face itself. If the wave transmitted through the distorted face is made to interfere with that from the original face, the resulting fringe pattern will look like that in Figure 35.7(b). Here the features of the face itself do not appear in the interferogram; only the distortion fringes are visible. This is a clear advantage, of course, because one is usually interested in the changes induced in the object, not in the features of the object itself. The problem in most cases, however, is that the distorted and the undistorted objects are not simultaneously available and, therefore, creating an interferogram between the two using traditional methods of interferometry is not a viable option.

Holographic interferometry provides a solution to this problem by allowing the original wavefront, while still available, to be stored on a photographic plate. Later, when the object is distorted, a second recording of its wavefront is made; then the two wavefronts are reconstructed and allowed to interfere with each other. Interestingly enough, these two recordings can be made on the same photographic plate by double exposure. Moreover, the two wavefronts are automatically superimposed during reconstruction.[10] The essential idea behind holographic interferometry may be readily grasped by reference to Eqs. (35.1) and (35.2) above. If the distorted wavefront is denoted by $A'_O(x, y)$, it is clear that, upon reconstructing the double exposure hologram, the emergent object wave will be $A_O(x, y) + A'_O(x, y)$, while the emergent conjugate wave will be $A^*_O(x, y) + A'^*_O(x, y)$. In this way both the virtual image and the real image show fringe patterns corresponding to contours of constant phase shift between the original object and its distorted version.

Figures 35.8(a), (b) show the intensity and phase patterns at the photographic plate corresponding to the distorted face. When this beam is combined with a reference plane wave traveling at 8° to the Z-axis, the fringe pattern of Figure 35.8(c) is obtained. This fringe pattern is recorded on the same film that had previously recorded the hologram of the original face. When the resulting double-exposure hologram is developed into a positive transparency and placed in front of the reconstruction beam, the intensity distribution of Figure 35.8(d) appears immediately behind the hologram.

Assuming that the reconstruction beam is the conjugate of the reference beam used in recording both holograms, the emergent beam along the Z-axis will be the conjugate of the combined object waves, namely, $A_{\mathrm{O}}^{*}(x, y)$ +

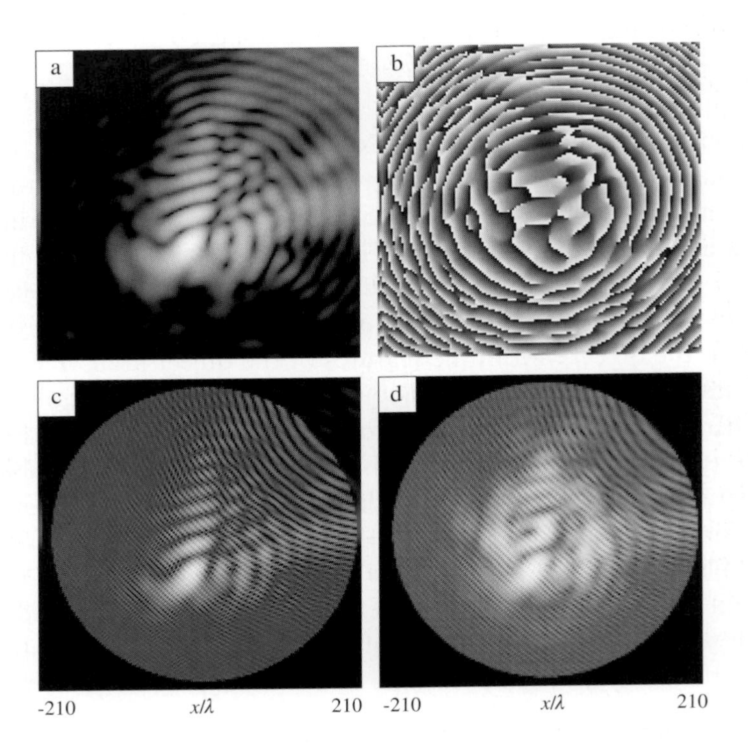

Figure 35.8 A plane wave traveling along the Z-axis and transmitted through the distorted face at $z = 0$ arrives at the photographic plate at $z = 3500\lambda$. In this double-exposure experiment a hologram of the undistorted face has already been recorded on the plate. (a) Logarithmic plot of the object wave's intensity distribution at the plate. (b) The object wave's phase distribution at the plate. (c) Pattern of interference between the object wave and a reference plane wave traveling at $\theta = 8°$ relative to the Z-axis. (d) Distribution of the logarithm of intensity immediately after the hologram, when the twice-exposed film is developed into a positive transparency and placed in front of the reconstruction beam.

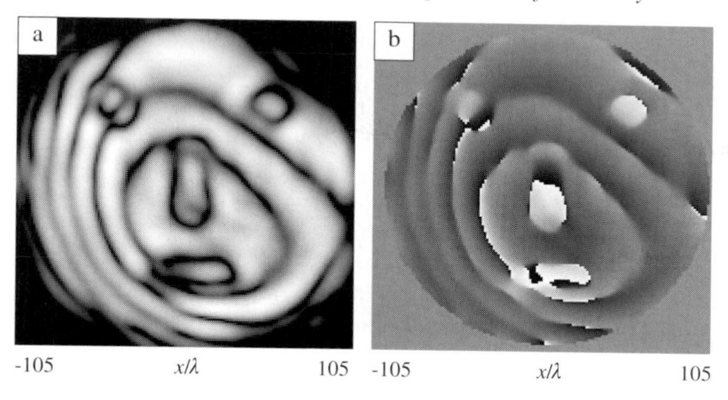

Figure 35.9 The reconstructed wavefront at $z = 3500\lambda$ behind the double-exposure hologram, showing the interference pattern between the real images of the distorted and undistorted face. The incident beam is the conjugate of the original reference beam used in both exposures, and the component of the reconstructed wave traveling along the Z-axis carries the real images. (a) Logarithmic plot of intensity and (b) plot of phase distribution over the area of the real image.

$A'^{*}_{O}(x, y)$. The intensity and phase patterns in Figures 35.9(a), (b) are obtained after propagating the emergent beam a distance of 3500λ beyond the hologram. The fringe pattern caused by the distorted face is clearly visible in this holographic interferogram.

In an ideal situation, where the hologram is large enough to capture all significant spatial frequencies of both object waves, the features of the original object will be invisible in the interferogram. However, in these calculations, the hologram is of necessity small and, therefore, the features are not completely absent from the final image. In any event, if the reference beam is large enough to capture the high-spatial-frequency content of the object waves, the interferogram of Figure 35.9(a) will approach the ideal one shown in Figure 35.7(b).

Real-time interferometry using a holographic image

If a hologram of an object in a given state is made, the reconstructed image can be made to interfere in real time with the "live" images of the same object in different states. Hence deformations that are dynamic in nature can be observed directly. This also provides a natural and very sensitive method of aligning the hologram to the original position after it has been removed for processing.

References for chapter 35

1 D. Gabor, A new microscopic principle, *Nature* **161**, 777–778 (1948).

2 D. Gabor, Microscopy by reconstructed wavefronts, *Proc. Roy. Soc. London A* **197**, 454–487 (1949).

3 E. N. Leith and J. Upatnieks, Reconstructed wavefronts and communication theory, *J. Opt. Soc. Am.* **52**, 1123–1130 (1962).

4 S. A. Benton, Hologram reconstruction with extended incoherent sources, *J. Opt. Soc. Am.* **59**, 1454A (1969).

5 S. A. Benton, The mathematical optics of white light transmission holograms, in *Proceedings of the First International Symposium on Display Holography*, ed. T. H. Jeong, Lake Forest College, July 1982.

6 S. A. Benton, Survey of holographic stereograms, in *Processing and Display of Three-Dimensional Data, SPIE* **367**, 15–19 (1983).

7 The introductory section is adapted from *Holophile, Inc.*'s website at www.holophile. com.

8 J. W. Goodman, *Introduction to Fourier Optics*, McGraw-Hill, New York, 1968.

9 P. Hariharan, *Optical Holography*, Cambridge University Press, UK, 1984.

10 C. M. Vest, *Holographic Interferometry*, Wiley, New York, 1979.

36

Self-focusing in nonlinear optical media†

Self-focusing and self-trapping in nonlinear optical media were discovered soon after the invention of the laser in the early 1960s.[1–5] These phenomena provided an explanation for the appearance of hot spots and associated optical damage in media irradiated by high-power laser pulses. The very high intensities achievable with the laser made it possible to observe these and other nonlinear effects, which depend upon the change in refractive index of the medium in response to the local electric field intensity.

The physics of optical nonlinearity

In a medium exhibiting third-order nonlinearity, the index of refraction n depends on the local E-field intensity $I(x, y, z)$ as follows:[2]

$$n(x, y, z) = n_0 + n_2 I(x, y, z). \tag{36.1}$$

Here n_0 is the medium's background index of refraction (observed at low optical intensities) and n_2 is the nonlinear coefficient of the material. Whereas n_0 is a dimensionless quantity, the nonlinear coefficient n_2 has inverse intensity units, i.e. units of area/power. Several physical mechanisms can cause the refractive index of a given medium to depend on the E-field intensity; notable among them are the anharmonic motion of electrons in crystals, electrostriction, and the molecular orientation known as the Kerr effect.[2] Electrostriction is caused by the volume force of an inhomogeneous electric field within a dielectric medium. The volume force draws the material into the high-field region, increasing its local density and, consequently, its refractive index. Optical glasses such as fused silica exhibit both electronic and electro-

†The coauthor of this chapter is Ewan M. Wright of the Optical Sciences Center, University of Arizona.

strictive nonlinearities, their n_2-values being in the range 5×10^{-16} to 5×10^{-15} cm^2/W. The Kerr effect is observed in materials whose molecules possess anisotropic polarizability and so tend to be aligned by the E-field, thus causing a change in the local refractive index. The liquid carbon disulfide (CS_2), which has a fairly large n_2-value, 2.6×10^{-14} cm^2/W, is a good example of this class of materials.

When n_2 is positive, the index of refraction in regions of high intensity tends to be larger than that in regions where the E-field is weak. Consequently, for an initially collimated and localized beam profile (such as a Gaussian), the wavefront propagating through the medium develops a phase pattern that resembles the curvature of a converging beam. While diffraction effects tend to broaden the cross-section of the beam, wavefront curvature – caused by nonlinearity – attempts to pull the beam towards regions of higher intensity. As long as the nonlinear effect is weak, diffraction predominates; however, as one increases the beam's power a point is reached where the tendency of the beam to become focused balances the effects of diffraction. The beam can then propagate over long distances without any noticeable expansion or contraction. Physically, the field has built an effective waveguide for itself, which enables it to propagate without spreading. This phenomenon, known as self-trapping, occurs at the critical input power $P_{cr} = 0.146\lambda^2/(n_0 n_2)$. Typical values of P_{cr} are 33 kW for CS_2 at $\lambda = 1$ μm, and 0.2–2 MW for common optical glasses in the visible and near-infrared range. Self-trapping is inherently unstable and is readily destroyed by slight perturbations of the wavefront; nonetheless, it is possible to arrange well-controlled experiments to demonstrate the phenomenon.

If the laser power is further increased beyond the threshold of self-trapping, the phenomenon known as self-focusing collapse is observed. In this case, not only does the nonlinear effect counter the natural tendency of the beam to diverge but also it forces the beam to collapse under its own weight and come to a sharp focus (a singularity, in the approximate paraxial theory) within a finite distance.[3] Further increases in laser power break up the beam into multiple filaments, each of which carries enough power to exhibit self-focusing in its own right.

Our goal in this chapter is to demonstrate some interesting examples of self-focusing in nonlinear media, both to elucidate the fundamental physics and to highlight the key effects produced by self-focusing in bulk media.

Gaussian beam profile

Figure 36.1(a) shows the distribution of intensity in the cross-section of a Gaussian beam of wavelength λ and having a 1/e radius 1000λ. The beam is linearly polarized along the X-axis and propagates along the Z-axis. The beam's waist is at $z = 0$, so the phase distribution in this plane is uniform over the beam's cross-section. The full-width at half-maximum (FWHM) intensity of the beam at the waist equals 1177λ, and its peak intensity $I_{\max} = 0.64 I_0$. Here I_0 is an arbitrary scale factor used to normalize all intensity profiles throughout this chapter.

The beam cannot satisfy Maxwell's equations unless it has a component of polarization E_z along the Z-axis; the computed intensity profile $|E_z|^2$ for this Z-component is shown in Figure 36.1(b). For a beam whose cross-section is substantially larger than a wavelength, the power content of E_z is typically much less than that of E_x. For example, in the present case the fraction of the total optical power carried by the Z-component is only 0.25×10^{-7}. We will see below that E_z gains in strength as the beam converges towards focus.

Self-focusing by transmission through a thin slab

Consider the transmission of the Gaussian beam depicted in Figure 36.1 through a thin slab of transparent material. (By thin we mean that the medium thickness is much less than the Rayleigh range of the incident beam, so that diffractive effects in the medium may be neglected.) Let the thickness d and the nonlinear coefficient n_2 of the slab be chosen to yield a phase shift $\Delta\phi = 2\pi n_2 I_{\max} d/\lambda = 10\pi$ at the beam center, where the intensity

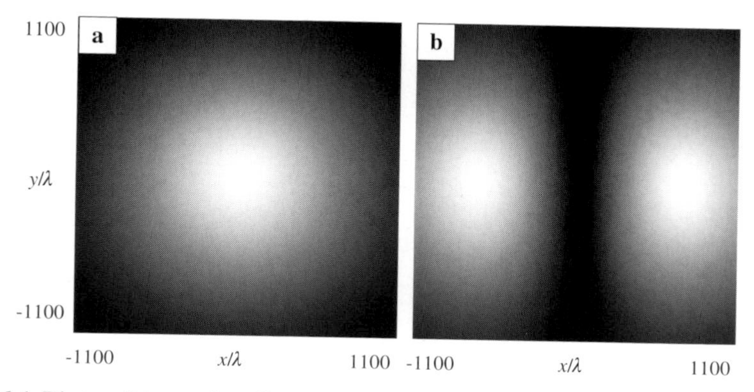

Figure 36.1 Plots of intensity distribution for (a) the X-component and (b) the Z-component of polarization. These plots represent the cross-section of a Gaussian beam having a 1/e radius 1000λ.

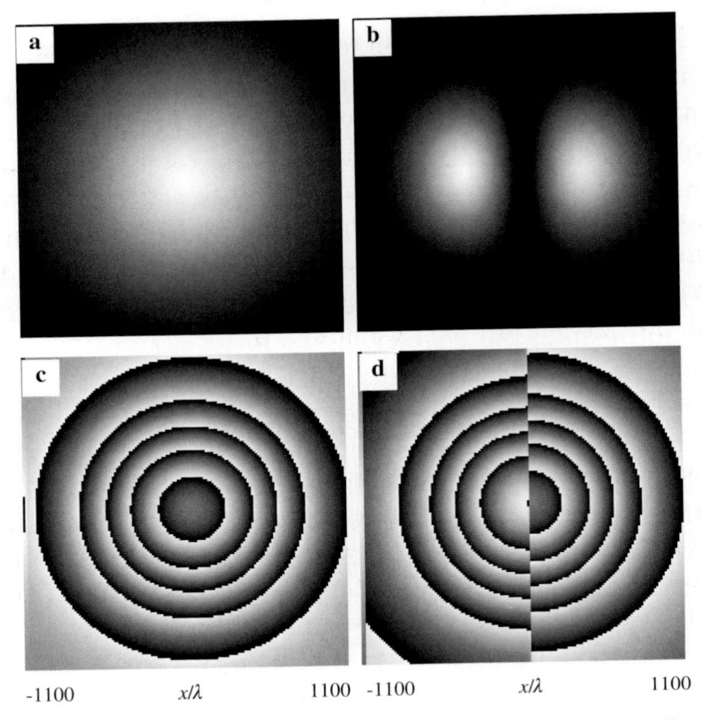

Figure 36.2 The beam of Figure 36.1 goes through a thin slab of a nonlinear material, creating a change in the index of refraction in proportion to its intensity. At the center, where the beam is brightest, the self-induced phase shift is 10π. The intensity of the beam upon emerging from the slab is shown in (a), (b), and its phase distribution in (c), (d). The plots on the left-hand side correspond to the X-component of polarization, while those on the right-hand side represent the Z-component.

is at its peak. Upon transmission through the slab the beam acquires the intensity and phase distributions shown in Figure 36.2. The distribution of $|E_x|^2$ in Figure 36.2(a) is the same as that in Figure 36.1(a), but the intensity profile of E_z in Figure 36.2(b) is somewhat different from that in Figure 36.1(b). The fractional power of the Z-component is now 111×10^{-7}, which, small as it may be, is substantially greater than the corresponding value before entering the nonlinear medium. This behavior may be understood by observing that the emergent beam has acquired a fairly large curvature and, consequently, its polarization vector has bent further toward the Z-axis.

Figure 36.2(c) shows the phase profile of the emergent wavefront for the X-component of polarization. The gray-scale ranges from $-\pi$ (black) to π (white), and the number of rings indicates a total phase shift of 10π from the center to the rim. The phase profile for the Z-component of polarization in Figure 36.2(d) shows, in addition to the curvature, a π phase shift between the right and left halves of the beam. Again this is a simple geometrical consequence of the bending of the rays toward the optical axis.

The above example clearly demonstrates that a nonlinear medium can impart a curvature phase factor to a beam during transmission. When the curvature is negative the beam becomes divergent and expands upon further propagation. Conversely, a positive curvature causes the beam to converge towards a focus. This is the underlying physical mechanism of self-focusing in thick nonlinear media, to which we now turn.

Self-focusing through a thick slab

Let us now consider propagation of the Gaussian beam of Figure 36.1 through a thick slab of a nonlinear material, where the effects of diffraction during propagation within the medium must be retained. For simulation purposes we divide the thick slab into 60 thin slabs (in which we place the nonlinearity), and propagate the beam between pairs of adjacent slabs through a linear medium of refractive index n_0, which fills the gap between the slabs. We choose a separation of 5000λ between adjacent slabs, compute the incident intensity profile at each slab (using Fresnel's diffraction formula), and allow the nonlinear medium of each slab to impart to the beam a phase pattern $\phi(x, y)$ in proportion to the incident intensity distribution $I(x, y)$. The specific phase shift assumed is 5° at the reference intensity of I_0. The above procedure is repeated 60 times for a total propagation distance of $300\,000\lambda$. (This numerical scheme of breaking the propagation into alternate sections of linear propagation followed by a nonlinear phase mask is equivalent to the split-step beam propagation method commonly employed in optics.) For the above choice of parameters the input power is about 20 times greater than the critical power for self-trapping, P_{cr}, and we expect the simulation to display self-focusing collapse.[3]

The results of this simulation appear in Table 36.1 and Figure 36.3. The left-hand column in the figure shows the cross-sectional profile of $|E_x|^2$, while the right-hand column shows the corresponding plots of $|E_z|^2$. From top to bottom, the intensity profiles are obtained after 20, 30, 40, 50, and 60 steps in the simulation. Note that the beam is converging towards a focus and that E_z is becoming stronger as the beam gets smaller. The FWHM of the beam drops from 1177λ in the beginning to 196λ after 60 iterations. The focusing, of course, is not diffraction-limited, because the curvature imparted to the beam by the nonlinear medium does not exactly constitute a spherical wave-

Table 36.1. *Various properties of the beam during propagation through a nonlinear slab. The corresponding intensity profiles are shown in Figure 36.3*

Number of steps	$\frac{I_{\max}}{I_0}$	Fractional power of Z-component	FWHM ($\times\lambda$)
10	0.65	0.29×10^{-7}	1160
20	0.68	0.39×10^{-7}	1103
30	0.76	0.58×10^{-7}	998
40	0.92	0.91×10^{-7}	824
50	1.30	1.52×10^{-7}	545
60	3.10	3.21×10^{-7}	196

front. The departure of the wavefront from perfect sphericity saddles the beam with primary and higher-order spherical aberrations.

It is clear physically that self-focusing collapse cannot proceed indefinitely. Some mechanisms that can arrest the collapse are saturation of the nonlinear refractive-index change, nonlinear absorption arising from multi-photon ionization, and optical breakdown.

Asymmetric intensity profile and self-deflection

Our next example is similar to the previous one, except that now the beam launched into the thick nonlinear medium has an asymmetric profile.[4] The asymmetry is produced by blocking off half the incident Gaussian beam. To maintain the total optical power, we multiply the beam's amplitude by $\sqrt{2}$, thus preserving the integrated intensity over the clear aperture of the beam. As before, the distance between adjacent thin slabs is 5000λ, and the beam is propagated in 60 steps for a total distance of $300\,000\lambda$.

Figure 36.4 shows, from top to bottom, the initial half-Gaussian beam as well as the patterns of intensity distribution within the medium after 20, 40, 50, and 60 propagation steps. Both columns show the profile of $|E_x|^2$, the intensity distribution being on the left-hand side and its logarithm on the right-hand side. (The logarithmic plot enhances weak features of the distribution, just like an over-exposed photograph.)

We note several new features in this example. First, the beam comes to a focus in the narrow dimension before it collapses in the wide dimension. Second, the center of the beam shifts to the right as it propagates. This self-deflection is caused by the prism-like phase factor that the nonlinear medium imparts to the beam.[4] An ideal prism imparts a phase factor that

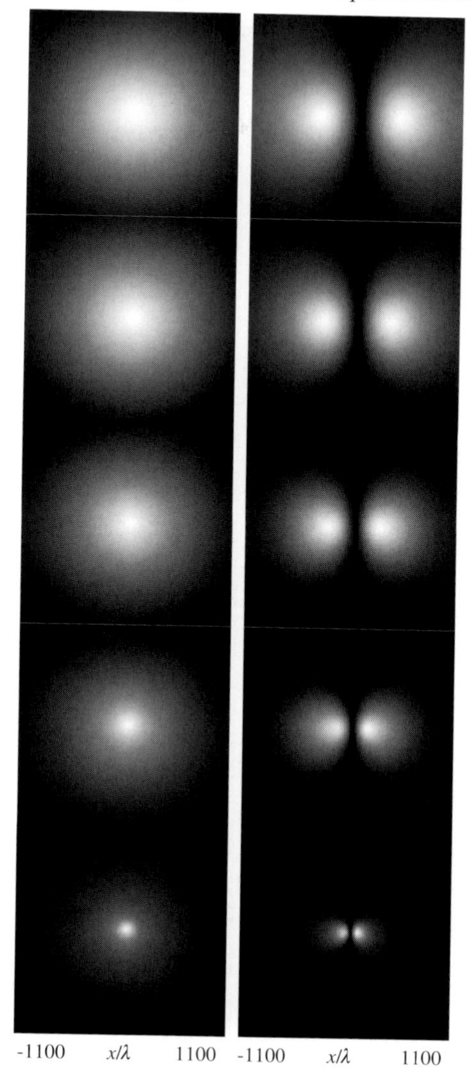

Figure 36.3 Top to bottom: plots of intensity distribution after 20, 30, 40, 50, 60 steps of propagation through a nonlinear medium. The X-component of polarization is on the left, and the Z-component on the right. The incident beam is the Gaussian shown in Figure 36.1.

is linear in the spatial coordinate x, namely, $\exp(i2\pi\sigma x/\lambda)$, deflecting the beam by an angle $\theta = \sin^{-1}\sigma$. One can explain the observed self-deflection in Figure 36.4 by noting the similarity between the ideal phase factor of a prism and the phase factor $\exp[i\phi(x, y)]$ imposed on the half-Gaussian beam by the nonlinear medium.

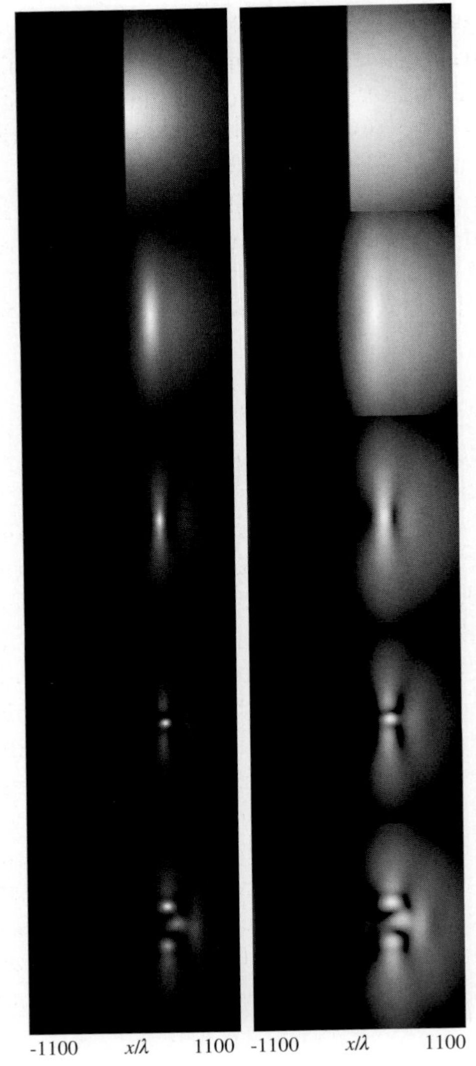

Figure 36.4 Top to bottom: distributions of intensity (left) and logarithm of intensity (right) after 0, 20, 40, 50, 60 propagation steps through a nonlinear medium. The incident beam is the Gaussian of Figure 36.1, with its left half blocked but its intensity doubled to preserve the total power. In units of the reference intensity I_0, the peak intensity $I_{\max}$ starts at 1.28 and increases to 1.51 after 10 steps, 1.95 after 20 steps, 3.24 after 30 steps, 6.67 after 40 steps, and 14.95 after 50 steps, then drops to 7.76 after 60 steps.

Finally, note in Figure 36.4 that the beam breaks up into multiple branches after coming to focus. In practice the intensity at the focal point may be large enough to damage the material. Even if damage does not occur, small material inhomogeneities can cause substantial aberrations, distorting and breaking up the beam in unpredictable ways. The fact that computer simulations also show this type of breakup is due to small numerical errors incurred during computation. Usually these numerical errors are insignificant, but when the intensity begins to build up in the vicinity of a focal point, they cause the breakup of the beam in a random-looking fashion.

Beam filamentation

As mentioned above, if the beam's power is large enough the beam breaks up into many cells, each of which contains several critical powers and comes independently to focus. Our final example concerns a uniform beam of diameter 2000λ with a constant intensity equal to $0.32I_0$ across the aperture. In this simulation we placed 40 thin slabs of a nonlinear material at intervals of $15\,000\lambda$ along the Z-axis. Each slab imparts a phase shift of 15° at the reference intensity of I_0, which is equivalent to an incident optical power of $60P_{cr}$. Shown in Figure 36.5 are the results of simulation after 10, 20, 30, 35, and 40 steps. At first, as a result of diffraction during propagation, the beam breaks up into multiple rings. After 30 iterations the central region of the beam comes to a focus. Afterwards, the central spot goes out of focus, but one of the rings breaks into multiple filaments.[5] Small perturbations are necessary to break up a ring; as mentioned above these are provided by material inhomogeneities in practice, and by small numerical errors inherent to computer simulations in these calculations. The number of filaments depends on the power of the beam as well as on the strength of nonlinearity of the material.

Concluding remarks

Another mechanism that can couple the refractive index to the beam intensity profile is absorption of the light followed by heating and thermal diffusion. Variation of the refractive index in response to thermal expansion (or contraction) of the material is a frequently observed source of nonlinear optical behavior. Thermal effects usually produce negative values of n_2, thus causing defocusing of the beam. Heat diffusion further complicates the relation between $n(x, y, z)$ and $I(x, y, z)$, by removing the local nature of their interdependence. In this chapter we have confined our attention to the simple case

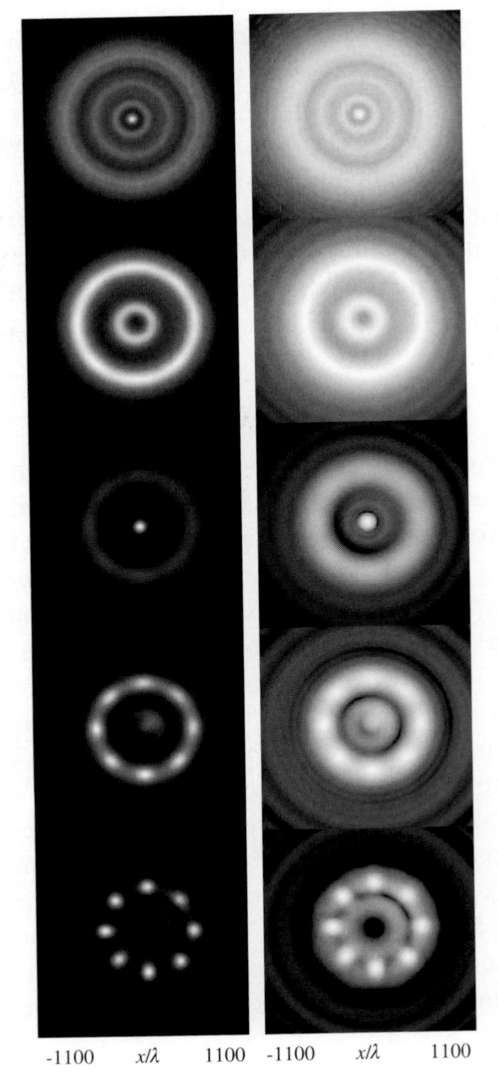

Figure 36.5 Top to bottom: distributions of intensity (left) and logarithm of intensity (right) after 10, 20, 30, 35, 40 propagation steps through a nonlinear medium. The incident beam is uniform, having a circular cross-section of radius 1000λ. In units of the reference intensity I_0, the peak intensity $I_{\max}$ starts at 0.32 and then fluctuates as follows: 1.18 after 10 steps, 0.9 after 20 steps, 7.85 after 30 steps, 3.00 after 35 steps, and 6.87 after 40 steps.

of local nonlinearity with a positive value for n_2 and have shown examples of self-focusing and beam filamentation. Similar studies can be carried out for thermally induced nonlinearities, provided that heat diffusion is taken into consideration properly.

References for chapter 36

1 R. Y. Chiao, E. Garmire, and C. H. Townes, Self-trapping of optical beams, *Phys. Rev. Lett.* **13**, 479–482 (1964).
2 R. W. Boyd, *Nonlinear Optics*, chapter 4, Academic Press, Boston, 1992.
3 P. L. Kelley, Self-focusing of optical beams, *Phys. Rev. Lett.* **15**, 1005–1008 (1965).
4 G. A. Swartzlander and A. E. Kaplan, Self-deflection of laser beams in a thin nonlinear film, *J. Opt. Soc. Am. B* **5**, 765–768 (1988).
5 A. J. Campillo, S. L. Shapiro, and B. R. Suydam, Periodic breakup of optical beams due to self-focusing, *Appl. Phys. Lett.* **23**, 628–630 (1973); also, Relationship of self-focusing to spatial instability modes, *Appl. Phys. Lett.* **24**, 178–180 (1974).

37

Laser heating of multilayer stacks

Laser beams can deliver controlled doses of optical energy to specific locations on an object, thereby creating hot spots that can melt, anneal, ablate, or otherwise modify the local properties of a given substance. Applications include laser cutting, micro-machining, selective annealing, surface texturing, biological tissue treatment, laser surgery, and optical recording. There are also situations, as in the case of laser mirrors, where the temperature rise is an unavoidable consequence of the system's operating conditions. In all the above cases the processes of light absorption and heat diffusion must be fully analyzed in order to optimize the performance of the system and/or to avoid catastrophic failure.

The physics of laser heating involves the absorption of optical energy and its conversion to heat by the sample, followed by diffusion and redistribution of this thermal energy through the volume of the material. When the sample is inhomogeneous (as when it consists of several layers having different optical and thermal properties) the absorption and diffusion processes become quite complex, giving rise to interesting temperature profiles throughout the body of the sample. This chapter describes some of the phenomena that occur in thin-film stacks subjected to localized irradiation. We confine our attention to examples from the field of optical data storage but the selected examples have many features in common with problems in other areas, and it is hoped that the reader will find this analysis useful in understanding a variety of similar situations.

Magneto-optical disk

The cross-section of a quadrilayer magneto-optical (MO) disk, optimized for operation at $\lambda = 400$ nm, is shown in Figure 37.1. (GaN-based semiconductor diode lasers operating at these blue and violet wavelengths are becoming

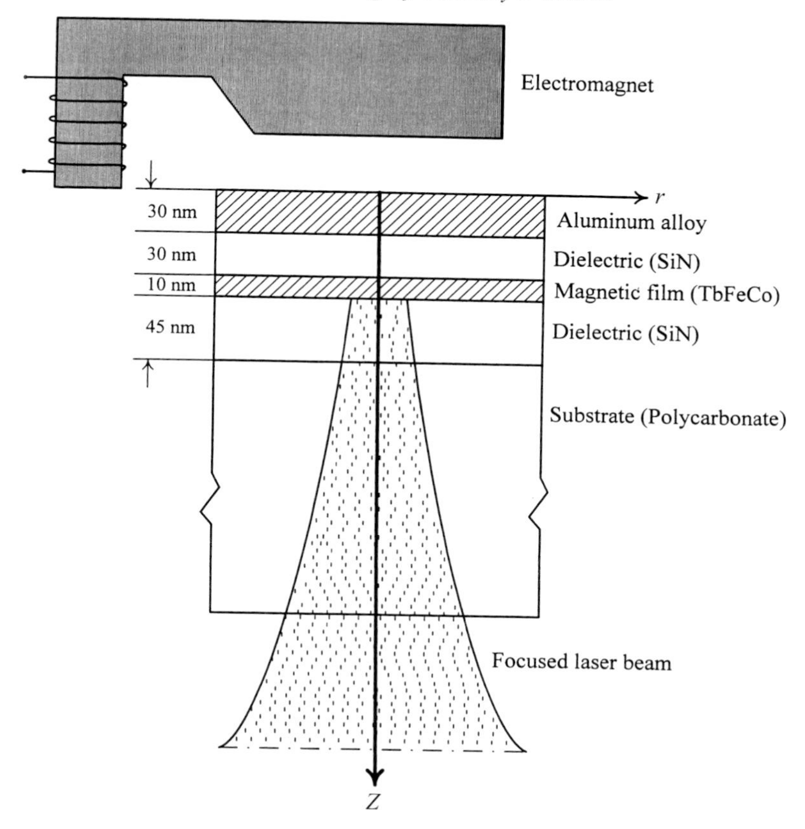

Figure 37.1 Quadrilayer stack of a magneto-optical disk. The electromagnet applies a magnetic field $B_z(t)$ in the Z-direction, which is also the easy axis of magnetization of the magnetic layer. The laser beam, focused on the magnetic film through the substrate, acts as a heat source during recording.

commercially available, and optical disk systems are expected to take advantage of this development by switching to blue or violet lasers within the next two to three years.) The quadrilayer of Figure 37.1 is deposited on a plastic substrate and consists of a thin magnetic film sandwiched between two transparent dielectric layers, capped by a thin layer of an aluminum alloy.[1,2] The optical and thermal constants of the various layers of this stack are listed in Table 37.1.

The focused laser beam arrives at the magnetic layer from the substrate side. This quadrilayer is designed to have a reflectivity of 9%, and has a fairly large polar MO Kerr signal (polarization ellipticity $\eta_K = \pm 1.55^\circ$ and Kerr rotation angle $\theta_K = \pm 0.24^\circ$, where the $\pm$ signs correspond to the up and down directions of magnetization of the storage layer). Aside from contribut-

Table 37.1. *Optical and thermal constants of the various materials used in the calculations*

	Refractive index $n + ik$ ($\lambda = 0.4\,\mu m$)	Dielectric tensor $\varepsilon, \varepsilon'$ ($\lambda = 0.4\,\mu m$)	Specific heat C (J/cm^3 °C)	Thermal conductivity K (J/cm s °C)
Polycarbonate (substrate)	1.6	—	1.4	0.0025
Aluminum alloy	0.50 + 4.85i	—	2.4	0.75
$Tb_{21}Fe_{72}Co_7$ (amorphous ferrimagnet)	2.33 + 3.45i	$\varepsilon = -6.46 + 16.11i$ $\varepsilon' = 0.185 - 0.233i$	2.9	0.10
SiN (dielectric)	2.2	—	2.5	0.030
$Ge_2Sb_2Te_5$ (amorphous)	2.9 + 2.5i	—	1.3	0.002
$Ge_2Sb_2Te_5$ (polycrystal)	2.0 + 3.6i	—	1.3	0.005
ZnS-SiO_2 (dielectric)	2.2	—	2	0.006

ing to the optical properties of the stack, the aluminum layer acts as a heat sink, and the upper dielectric layer is thin enough to provide good thermal coupling between the two metallic layers.[1,2]

Figure 37.2 shows the intensity profile of the focused spot at the storage layer of the disk. The assumed objective lens that brings the laser light to focus in this case is free from all aberrations, is corrected for the thickness of the substrate, and has $NA = 0.8, f = 1.5$ mm. The collimated Gaussian beam entering the lens has 1/e (amplitude) radius $r_0 = 1.2$ mm, which is the same as the radius of the objective's entrance pupil. The distribution of Figure 37.2(a) is displayed on a logarithmic scale to enhance the diffraction rings caused by truncation of the beam at the objective's aperture. The radial profile of the spot, depicted on a linear scale in Figure 37.2(b), reveals that the rings are quite weak, however, and thus incapable of producing much heat at the periphery of the central bright spot.

Figure 37.3 is a plot of the magnitude of the Poynting vector, S, along the Z-axis for a plane wave normally incident on the quadrilayer stack of Figure 37.1 through the substrate.[2] The horizontal axis depicts the distance from the top of the stack. Thus S is seen to be constant in the two dielectric layers ($30 < z < 60$ nm and $70 < z < 115$ nm), indicating no optical absorption in these regions. Most of the absorption takes place in the magnetic film ($60 < z < 70$ nm); a very small fraction of the incident energy goes to the aluminum layer ($0 < z < 30$ nm). The optical energy thus deposited in the magnetic film raises the local temperature immediately, but soon thermal diffusion takes over and carries the heat to other regions of the stack.

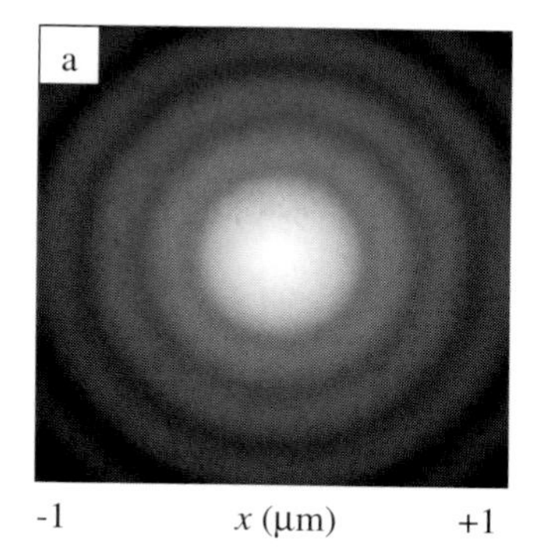

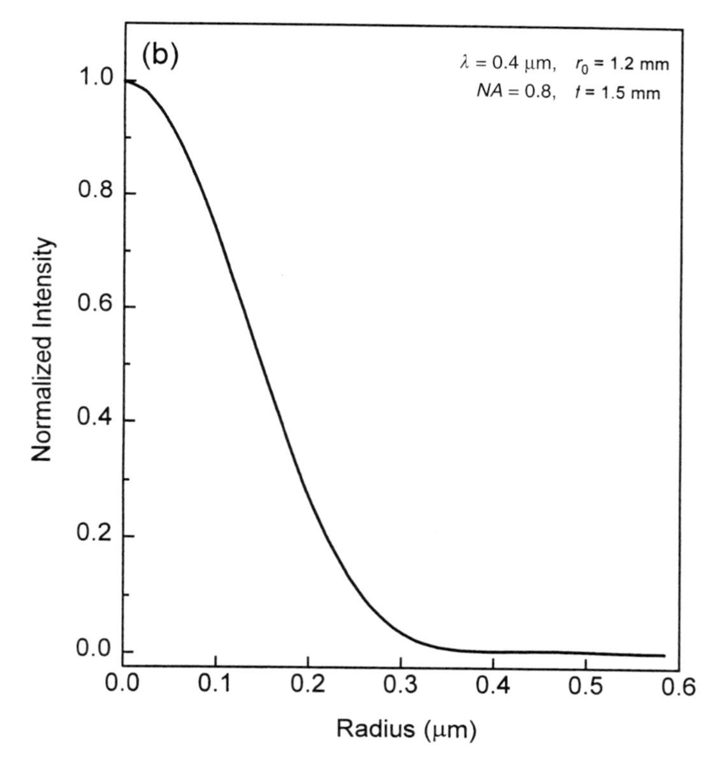

Figure 37.2 Distribution of total E-field intensity, $|E_x|^2 + |E_y|^2 + |E_z|^2$, at the focal plane of a $0.8NA$ objective. The incident Gaussian beam ($\lambda = 0.4\,\mu\text{m}$) is truncated by the lens aperture at its 1/e (amplitude) radius. For simplicity's sake, the beam is assumed to be circularly polarized, so that it would yield a circularly symmetric spot at the focal plane. (a) Logarithmic plot of intensity, showing an Airy disk diameter $\approx 0.6\,\mu\text{m}$ and FWHM $\approx 0.3\,\mu\text{m}$. (b) Radial intensity profile.

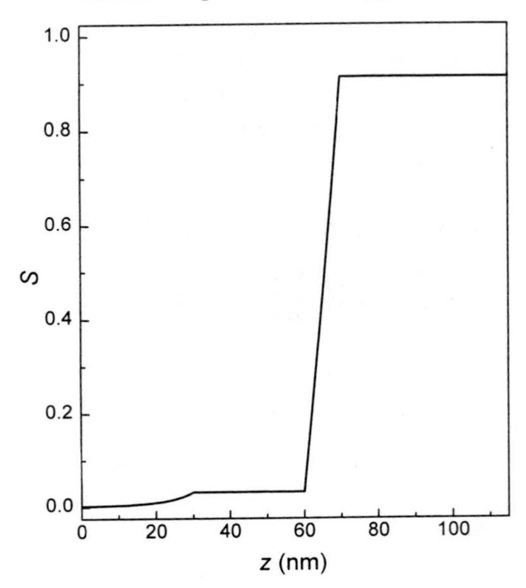

Figure 37.3 The magnitude S of the Poynting vector along the Z-axis, plotted through the thickness of the quadrilayer of Figure 37.1. The incident beam ($\lambda = 0.4\,\mu$m) is assumed to have unit power. Upon entering the stack $S = 0.91$, which indicates that 9% of the incident optical energy is reflected at the substrate interface. Approximately 3% of the energy goes to the aluminum layer and the remaining 88% is absorbed by the magnetic film.

Heat diffusion in the stationary stack

To describe the temperature distribution within the stack, we need to specify the time dependence of the incident laser power, $P(t)$. Figure 37.4 shows three such functions used in the examples throughout this chapter. The first function, $P_1(t)$, is a 1 mW trapezoidal pulse with 55 ns duration and 5 ns rise and fall times. We examine the effect of this pulse on the quadrilayer of Figure 37.1, when the stack is stationary.

Figure 37.5 shows the profiles of temperature versus z at the beam center, $r = 0$; the various curves correspond to different instants of time. Early on, at $t = 10$ ns, the magnetic film is at a relatively high temperature, the aluminum layer has uniform temperature through its thickness, and there is a large thermal gradient between the magnetic film and the aluminum layer. It is through this temperature gradient that heat is transferred from the magnetic layer to the aluminum heat sink.[3] A gradient has been established also in the lower dielectric layer between the magnetic film and the substrate. The temperature in the substrate is seen to decay exponentially with z.

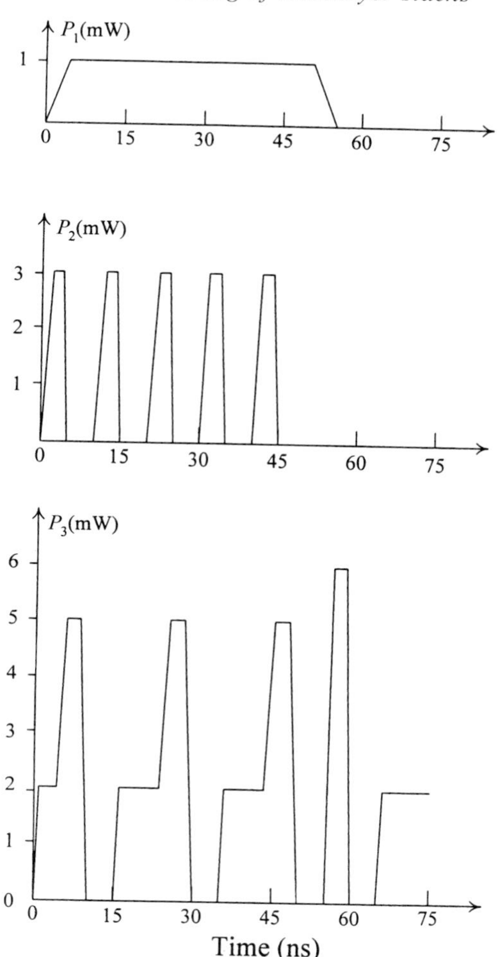

Figure 37.4 The functions representing laser power versus time that are used in the various examples: $P_1(t)$ is a 55 ns trapezoidal pulse with 5 ns rise and fall times; $P_2(t)$ is a sequence of five identical pulses, each with a 5 ns duration, 1 ns rise and fall times, and a center-to-center spacing of 10 ns; $P_3(t)$ is a fairly complex pattern of three-level pulses, used in phase-change recording.

At later times during the heating cycle (i.e., $t = 30$ ns and 50 ns) the patterns are similar to that at $t = 10$ ns, but the temperatures are higher. Once the laser is turned off the temperatures drop abruptly. At $t = 55$ ns, the magnetic film is already cooling down, and the heat is moving to the substrate. The hottest spot at this point is somewhere in the substrate, close to the interface with the lower dielectric layer. At $t = 60$ ns, the cooling has progressed further, and the heat is rapidly spreading through the substrate. By $t = 100$ ns, the temperature everywhere is essentially back to the ambient

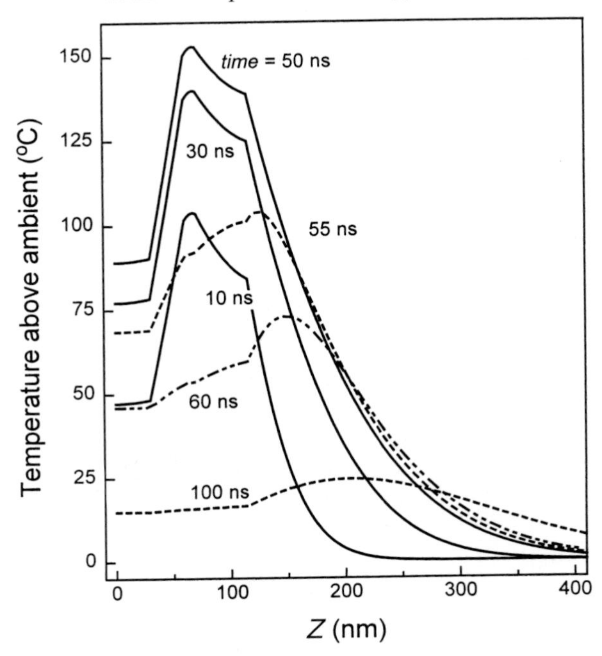

Figure 37.5 Computed temperature profiles along the Z-axis at the beam center, $r=0$, for the stack of Figure 37.1 illuminated by the focused beam of Figure 37.2 and the pulse $P_1(t)$ of Figure 37.4. The profiles at $t=10$, 30, and 50 ns represent the heating-period; cooling-period profiles are shown for $t=55$, 60, and 100 ns. By virtue of its strong absorption of the incident optical energy, the magnetic film is the hottest region during heating. The high thermal conductivity of the aluminum layer gives it a fairly uniform temperature through its thickness. As soon as the laser is turned off, the temperatures drop rapidly, and the peak temperature shifts to the substrate.

temperature. Although only the z-profiles are shown here, it should be remembered that the heat diffuses radially as well; not only does the heat move to the substrate, but also it spreads radially throughout the entire stack.[3]

Next we consider the profiles of temperature versus time in the magnetic layer. Figure 37.6 shows several profiles at different distances r from the beam's center, starting at $r=0$ and increasing in steps of $\Delta r=50$ nm to $r=1$ μm. At $r=0$, the temperature reaches its highest value at the end of the pulse, then decays quickly and, in the span of a few nanoseconds after the laser turn-off, goes down by almost an order of magnitude. At larger radii, the temperature is slow to rise, and it also peaks somewhat after the laser is turned off. The reason for this behavior is that the focused spot at these radii is rather weak, and the heat does not arrive there directly from the laser, but

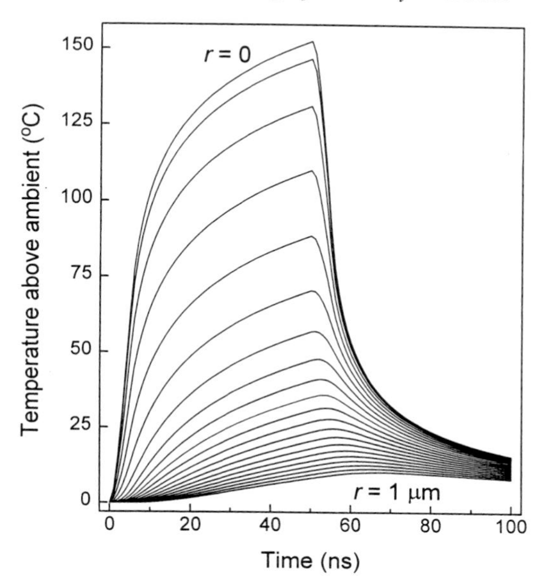

Figure 37.6 Computed profiles of temperature versus time in the magnetic layer under the same conditions as in Figure 37.5. Different curves correspond to different radial distances from the beam center (in steps of $\Delta r = 50$ nm), the largest temperature occurring in the center at $r = 0$ and the lowest temperatures belonging to $r = 1\ \mu$m. As soon as the laser is turned down at $t = 50$ ns temperatures near the beam center drop sharply, but at larger radii, because of radial heat diffusion from the center, T continues to rise for a while after the laser is turned off.

by radial diffusion from the central region, which is under intense illumination.

Recording by magnetic field modulation

The electromagnet (EM) above the quadrilayer stack of Figure 37.1 provides a switched magnetic field $B_z(t)$ between $\pm B_{max}$, thus helping to set the direction of magnetization of the hot spot within the magnetic layer.[4,5] (B_{max} is typically of the order of a hundred oersteds.) To ensure proper alignment of the focused spot with the EM's pole piece, the EM is rigidly attached across the disk to the optical head. (The optical head is the assembly of the laser and other optical, mechanical, and electronic elements that guide the laser beam to the disk and back to the detectors.) The disk moves at a constant velocity V in the space between the EM and the optical head. Typically, it spins at a fixed angular velocity, say, 6000 rpm, which, at a radius of 40 mm on a 3.5 inch diameter disk, corresponds to a linear track velocity $V = 25$ m/s. Since the information track is usually a continuous spiral from the inner to the outer radius of the disk, the combined optical and magnetic head assembly must follow this track by slow, continuous, travel along the disk's radial direction.[1,2]

In a currently popular recording scheme, the laser is pulsed at a fixed rate to produce a sequence of identical hot spots in the magnetic layer.[6] The binary information to be recorded on the disk is fed to the EM, which switches the magnetization of the hot spot between the "up" and "down" stable states. The switching rate must be rapid enough to provide a high data-transfer rate into the recording medium. This requires a compact EM capable of flying very close to the magnetic layer, lest its inductance becomes too large. If the recorded marks are to be 0.25 μm long in the direction of the track, the laser must be pulsed at 10 ns intervals, in which case $B_z(t)$ must switch between $\pm B_{max}$ with rise and fall times of only a few nanoseconds. Such fast magnetic heads are currently at the forefront of conventional magnetic recording technology (i.e., hard-disk drives), but they require further development in order to be suitable for future generations of MO drives.

Consider the quadrilayer disk of Figure 37.1 moving at $V = 25$ m/s under the focused beam of Figure 37.2, modulated with the pulse sequence $P_2(t)$ of Figure 37.4. These five pulses are each 5 ns wide, have 1 ns rise and fall times, and reach a peak power of 3 mW. The assumed ambient temperature is 25 °C. Figure 37.7 shows several isotherms at the critical temperature $T_{crit} = 175$ °C in the magnetic film during the period $0 \leq t \leq 50$ ns. (The maximum temperature of the magnetic film during the same period is $T_{max} = 300$ °C.) To a good approximation, the magnetic dipoles of the storage layer align with the field of the EM in those regions where $T \geq T_{crit}$ but the EM is unable to reorient these dipoles where $T < T_{crit}$.[4–6]

The isotherms in Figure 37.7 are plotted at $\Delta t = 1$ ns intervals whenever $T \geq T_{crit}$ in the MO film. When the temperature is on the rise the isotherms are shown as solid lines, but as broken lines when the temperature is declining. By the end of the fifth pulse the temperature profiles reach the steady state (i.e., there are no significant variations from one set of isotherms to the next). In between the adjacent pulses the temperatures everywhere drop below T_{crit}, so that for about 4 ns before and after each pulse the entire magnetic film is at $T < T_{crit}$. This cooling period is crucial for thermomagnetic recording by magnetic-field modulation, because as long as $B_z(t)$ is saturated at $\pm B_{max}$, the magnetization state of the disk is well defined (either up or down). But when the field is in transition, there is a short time interval during which $B_z(t)$ is weak and, therefore, the magnetization orientation is uncertain. This problem is overcome by keeping the temperature below T_{crit} during the transition period, in which case no changes occur in the disk's

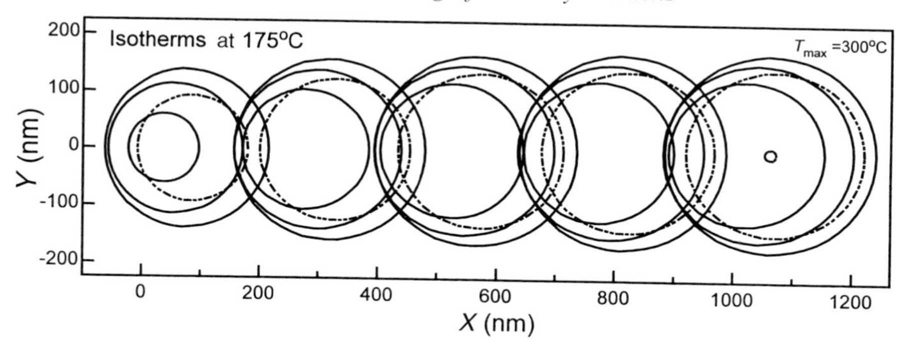

Figure 37.7 Computed isotherms in the magnetic layer of Figure 37.1, when the multilayer is subjected to the focused spot of Figure 37.2 and the pulse sequence $P_2(t)$ of Figure 37.4. The ambient temperature is 25 °C, and the disk moves at $V = 25$ m/s along the X-axis. (In the reference frame of the disk, the focused spot moves from left to right.) The maximum temperature during this period is $T_{\max} = 300$ °C, reached at $t = 44$ ns. All depicted isotherms are at $T = 175$ °C, plotted at $\Delta t = 1$ ns intervals whenever $T \geq 175$ °C. The solid (broken) curves represent the heating (cooling) phase of each pulse. Because of the lateral heat diffusion, each pulse produces slightly larger isotherms than the preceding one, but by the end of the fifth pulse this process reaches a steady state. During the 10 ns period of each pulse the disk moves by $\Delta x = 0.25\,\mu$m, which is the minimum mark-length that can be recorded in this example.

magnetic state and, consequently, the recorded domains acquire sharp boundaries.[6] For these reasons it is imperative to have a quadrilayer design, such as that of Figure 37.1, that cools down below T_{crit} in between adjacent laser pulses.

Phase-change optical recording

The general structure of a phase-change (PC) disk is similar to that of an MO disk. Figure 37.8 shows the cross-section of a quadrilayer PC stack optimized for operation at $\lambda = 400$ nm. The optical and thermal constants of this stack are listed in Table 37.1. The $Ge_2Sb_2Te_5$ material can be switched between amorphous and (poly)crystalline states by the laser beam: melting at $T_{\mathrm{melt}} \approx 625$ °C followed by rapid quenching results in an amorphous mark, whereas annealing for a reasonable length of time above the glass transition temperature ($T_{\mathrm{glass}} \approx 150$ °C) returns the material to the crystalline state.[7–9] The stack shown in Figure 37.8 has reflectivities $R_c = 30\%$ and $R_a = 8\%$ for the crystalline and amorphous phases of the PC film. Note also that the thermal constants for the two phases are somewhat different. In the following analysis we ignore these differences by assuming the PC layer to be crystalline

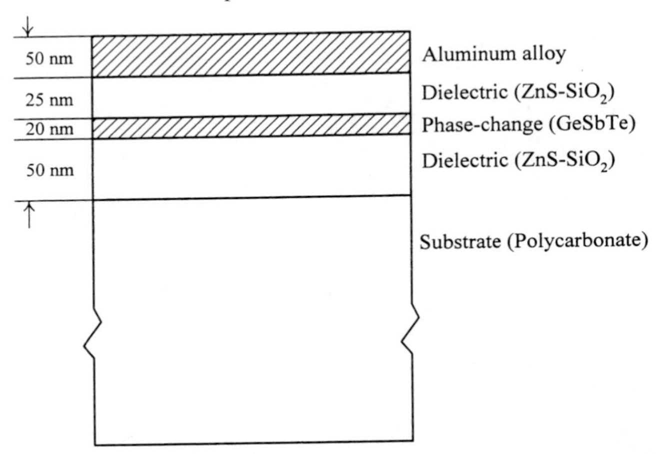

Figure 37.8 A quadrilayer stack designed for through-substrate phase-change recording at $\lambda = 400$ nm. The reflectance of the stack $R_a \approx 8\%$ when the $Ge_2Sb_2Te_5$ film is amorphous, and $R_c \approx 30\%$ when the film is crystalline. The absorbed optical power in the aluminum layer in either case is only about 1%. Thus, for all practical purposes, the fraction of the incident power that is not reflected is entirely absorbed in the PC layer.

at all times. Also to simplify the calculations further, we ignore the heats of melting and crystallization. These are reasonable approximations, but the final results may need slight corrections if more accuracy is desired.

The laser pulse sequence applied to this sample is $P_3(t)$, shown in Figure 37.4. Here the laser operates at three different power levels. At the highest level the pulse is strong enough to melt the PC film. In the low-power regime, occurring immediately after the melting pulse, the temperatures drop rapidly, causing the quenching of the molten material into an amorphous state. The intermediate power level is for annealing the pre-existing amorphous marks, which is required when overwriting a previously written track. (Such tracks contain both amorphous and crystalline regions, and it is necessary that all amorphous regions that are not being melted be annealed into the crystalline state.)

Figure 37.9(a) shows the computed isotherms in the PC layer at $T = T_{\text{melt}}$ for a disk speed $V = 25$ m/s and an ambient temperature of 25 °C. The solid and broken isotherms as before represent the heating and cooling cycles, respectively. The maximum temperature reached in this sample is $T_{\text{max}} = 1153\,°\text{C}$ at $t = 59$ ns. The isotherms are plotted at intervals $\Delta t = 0.5$ ns whenever $T \geq T_{\text{melt}}$ in the PC film. The first two molten regions are well separated from each other and from other molten pools; these will eventually quench to form two small amorphous marks. The cooling in these

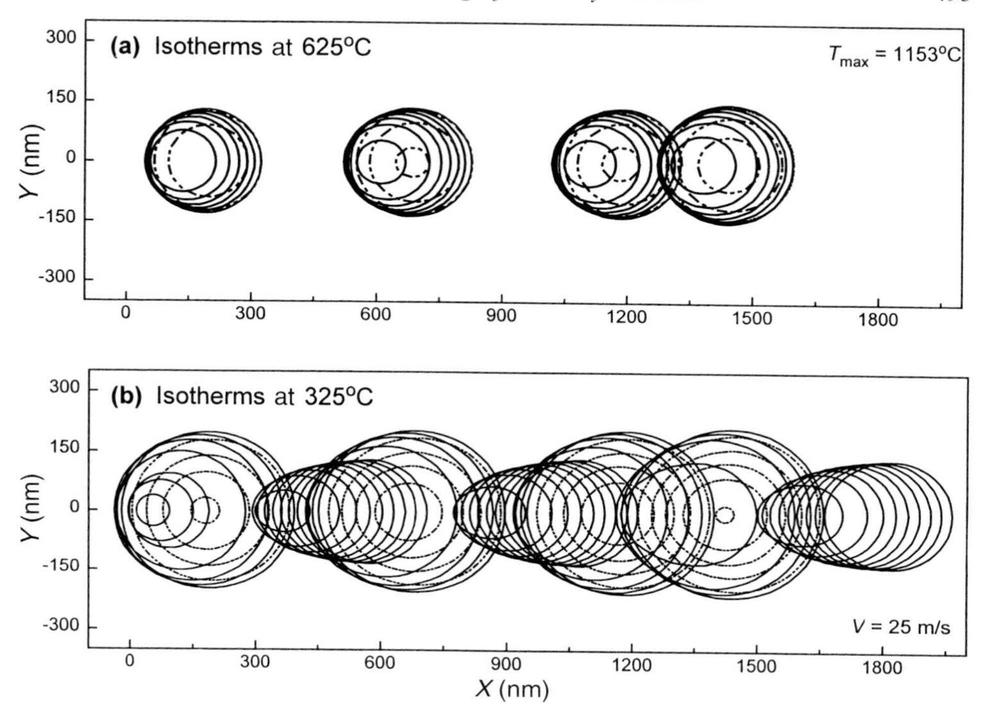

Figure 37.9 Computed isotherms in the GeSbTe film of Figure 37.8, subjected to the focused beam of Figure 37.2 and the pulse sequence $P_3(t)$ of Figure 37.4. The ambient temperature is 25 °C, and the quadrilayer disk moves at $V = 25$ m/s along the X-axis. The peak temperature $T_{max} = 1153$ °C in the film is reached at $t = 59$ ns. The solid (broken) isotherms correspond to the heating (cooling) phase of each pulse. (a) Isotherms at $T = 625$ °C, the melting point of the PC material. (b) Isotherms at $T = 325$ °C, the presumed (elevated) annealing temperature given the short annealing time

regions is rapid, and the temperatures return to the vicinity of T_{glass} in about 5 ns.

The third and fourth molten pools, however, have some degree of overlap. (In practice two or more short overlapping marks such as these are used to create a long mark.) The heat generated by the fourth pulse flows backward and affects the amorphous region being formed in the wake of the third pulse. In general, heat diffusion from the tail end of any long mark can anneal the leading edge as well as the mid-sections of the same mark, causing partial crystallization. This problem may be better appreciated by examining the $T = 325$ °C isotherms of the same system, shown in Figure 37.9(b). Here the annealing pulses (i.e., the medium-power levels of $P_3(t)$) appear behind the first two marks well after they have cooled. By then the annealed region is far enough from the previously molten pools that there is no danger of recrystallization. In contrast, the two large isotherms in Figure 37.9(b) corresponding to the third and fourth melting pulses partially overlap, causing the formation of a small, undesirable crystallite in the middle of the long amorphous mark. These are some of the issues with which the designers of optical disk drives must grapple, in order to create robust and reliable data storage systems.

References for chapter 37

1. T. W. McDaniel and R. H. Victora, eds., *Handbook of Magneto-optical Recording*, Noyes Publications, Westwood, New Jersey, 1997.
2. M. Mansuripur, *The Physical Principles of Magneto-optical Recording*, Cambridge University Press, UK, 1995.
3. H. S. Carslaw and J. C. Jaeger, *Conduction of Heat in Solids*, Oxford University Press, UK, 1954.
4. D. Chen, G. N. Otto, and F. M. Schmit, MnBi films for magneto-optic recording, *IEEE Trans. Magnet.* **MAG-9**, 66–83 (1973).
5. Y. Mimura, N. Imamura, and T. Kobayashi, Magnetic properties and Curie point writing in amorphous metallic films, *IEEE Trans. Magnet.* **MAG-12**, 779–781 (1976).
6. S. Yonezawa and M. Takahashi, Thermodynamic simulation of magnetic field modulation methods for pulsed laser irradiation in magneto-optical disks, *Appl. Opt.* **33**, 2333–2337 (1994).
7. S. R. Ovshinsky, Reversible electrical switching phenomena in disordered structures, *Phys. Rev. Lett.* **21**, 1450–1453 (1968).
8. J. Feinleib, J. deNeufvile, S. C. Moss, and S. R. Ovshinsky, Rapid reversible light-induced crystallization of amorphous semiconductors, *Appl. Phys. Lett.* **18**, 254–257 (1971).
9. T. Ohta, M. Takenaga, N. Akahira, and T. Yamashita, Thermal change of optical properties in some sub-oxide thin films, *J. Appl. Phys.* **53**, 8497–8500 (1982).

Index